The Energy
Sourcebook

The Energy Sourcebook

A Guide to Technology, Resources, and Policy

Edited by

Ruth Howes
and
Anthony Fainberg

American Institute of Physics
New York

This volume was prepared by a study group of the Forum on Physics and Society of The American Physical Society. The American Physical Society has neither reviewed nor approved this study.

American Institute of Physics
335 East 45th Street
New York, NY 10017-3483

Library of Congress Cataloging-in-Publication Data

The Energy sourcebook: a guide to technology, resources, and policy /
  edited by Ruth Howes and Anthony Fainberg.
    p. cm.
"Prepared by a study group of the Forum on Physics and Society of the
American Physical Society"—T.p. verso.
Includes bibliographical reference and index.
ISBN 0-88318-705-1 (case)   ISBN 0-88318-706-X (paper)

  1. Renewable energy sources. 2. Power resources. 3. Energy conservation. I. Howes, Ruth (Ruth Hega) II. Fainberg, Anthony. III. Forum on
Physics and Society.
TJ808.5.E53   1990
333.79—dc20   90-84610
                                              CIP

Contents

Technologies Related to End Use

Preface

In August 1990, President Bush ordered American soldiers to Saudi Arabia to deter the military takeover of the world's major oil exporter by Iraq. The price of oil on the international market surged upwards. Gasoline prices in the United States followed. As the troops shipped out, the popular press rediscovered U.S. dependence on foreign petroleum and our resulting economic vulnerability. Without a national energy policy in place, the U.S. faces the real possibility of an energy crisis in the 1990s.

During the 1970s, the American public and physicists, in particular, were bombarded by information on energy technologies, supplies, and use. The high price of oil brought about by the actions of the OPEC nations shocked the country into an awareness of coming shortages of fossil fuels. Federal agencies and state governments financed a number of crash programs to develop new energy sources and more efficiently use energy from old ones. The flood of research and development money spawned projects, some well-planned and technically sound, and some less carefully considered. In a time of plentiful funding, all thrived.

Prior to Iraq's invasion of Kuwait, the price of oil on the international market had dropped dramatically and the energy crisis seemed to have disappeared from public consciousness. Certainly the flood of information on new energy technologies and sources crossing the desk of non-specialists has decreased to a trickle. Funding for energy research has declined dramatically along with tax incentives for energy conservation measures and development of new energy sources. Well informed members of the public and the physics community, who are not specialists in energy issues, must search many sources to find up-to-date information on energy technologies, resources, and policies. Such studies as appear tend to treat a limited range of energy sources and uses and are all too often written by those with a vested interest in a particular technology or policy.

At the same time, the public has been growing increasingly aware of the social and economic costs of unrestrained energy use. The recent rash of well-publicized oil spills and their environmental effects have raised public consciousness of the dangers of petroleum production and use. Oil consumption is on the rise, and the United States now imports about half of the petroleum we use. The spiraling national trade deficit raises concerns over the expense of importing large amounts of oil. The military crisis in Saudi Arabia has underscored how politically vulnerable our foreign sources of petroleum are. Finally, a new recognition of the potential damage to the atmosphere from by-products of energy use and production has revived interest in renewable, non-polluting energy sources.

With this increase in public concern and interest, we feel that a broad review of U.S. energy production and utilization is needed. The Forum on Physics and Society of the American Physical Society undertook this study in an attempt to bring the interested community up to date on developments in energy technology, resources, and policy since 1973. Despite the drop in available funding, energy researchers have made important advances in several fields. Each of the chapters presents a review of the resources, the technologies, and the policy issues associated with an energy source or an area where energy is consumed. We have tried to present the material so that it is accessible to the interested non-specialist. Where it seems appropriate, we have made recommendations for energy policy.

The Energy Study Group of the Forum on Physics and Society of the American Physical Society is composed of independent researchers from a variety of institutions. We share a long-standing concern with energy issues. The Study has been funded only by the Forum. The opinions presented here and in the following chapters are those of the authors and do not represent those of their host institutions or any other organization, nor do they represent a view endorsed by the Forum or the American Physical Society. We disagree on several issues among ourselves. Certainly not all of us would endorse every chapter nor every policy recommendation.

Perhaps the greatest surprise in the course of the study has been the extent to which we have been able to reach consensus on what constitute the most pressing concerns in the energy field. Our conclusions are outlined in the Introduction to the study. The Introduction also indicates some of the places where members of the group disagree.

The chapters which follow have been prepared by individuals and, although there has been some opportunity for comment by other participants in the study, their contents are the responsibility of the authors. We believe that these chapters present a substantial body of information demonstrating the continued existence of an energy problem in the United States and suggesting an urgent need for research, development, implementation of policy decisions, and long-term planning aimed at reducing our dependence on fossil fuels.

About the study group

Samuel F. Baldwin received his Ph.D. in physics from the University of Maryland, College Park, in 1980. In 1981 he was the APS Congressional Scientist Fellow. In 1982–1983 he was the technical coordinator for an eight-country West African energy program. He was a member of the Senate staff and an independent consultant and analyst in 1984. He was a Hewlett Fellow, Center for Energy and Environmental Studies, at Princeton University from 1985 through 1988. From 1988 to the present he has worked in the Office of Technology Assessment, U.S. Congress.

H.H. Barschall is Emeritus Professor of Physics, Nuclear Engineering, and Medical Physics at the University of Wisconsin–Madison. He is a member of the National Academy of Sciences and a Fellow of the American Academy of Arts and Sciences. He currently serves as Secretary-Treasurer of the Forum on Physics and Society of the American Physical Society.

David Bodansky is Professor of Physics at the University of Washington, where he has been a faculty member since 1954 and Chairman of the Physics Department from 1976 to 1984. He received the Ph.D. in physics from Harvard University in 1950 and was at Columbia University from 1950–1954. His research interests have been in experimental nuclear physics, nuclear astrophysics, and energy and environmental policy. He is co-author (with F. H. Schmidt) of *The Energy Controversy: The Fight Over Nuclear Power* (1976) and co-editor and co-author of *Indoor Radon and Its Hazards* (1987). He is a Fellow of the American Physical Society and has served on the Panel on Public Affairs of the APS.

Alan Chachich has been involved with energy problems since the oil crisis of the 1970s. His interest in the interaction between technology and society originated with his contribution to the U.S. Senate Commerce Committee report on Radiation Health and Safety in 1978. His first contributions in energy were made during his graduate research at Rensselaer Polytechnic Institute. That work advanced ion beam diagnostics for nuclear fusion experiments. As a staff scientist at the BDM Corporation specializing in ultrasonic and microwave based sensor systems, he developed an ultrasonic locomotive fuel efficiency instrument for the American Association of Railroads. Currently, he is a systems engineer at the Massachusetts Institute of Technology, Lincoln Laboratory. He received his B.S. in physics at Oakland University (1979) and his M.S. in electrical engineering at Rensselaer Polytechnic Institute (1982).

Jamie Chapman is an independent consultant in the areas of power systems
embodying renewable and conventional generation sources, electronic power
conditioning and conversion systems, energy storage systems, and power
system control. As Vice President, Engineering for U.S. Windpower, Inc.,
from 1982 through early 1989, Dr. Chapman was responsible for the
development of large-scale, grid-connected electrical power stations driven
by the wind. Prior to his involvement with power systems, Dr. Chapman was
involved in electro-optical measurement and surveillance systems, radars,
and oceanographic instrumentation. Dr. Chapman serves on the following
committees: Solar Energy Research Institute (SERI) Wind Research Program Technical Review Committee, National Research Council Committee
on Assessment of Research Needs for Wind Turbine Rotor Materials Technology, and the EPRI/PG&E/USW Technical Review Committee for the
USW Variable Speed Wind Turbine Program. He obtained his undergraduate degree in physics from the University of California, Santa Barbara
and the Ph.D. degree in Earth and Planetary Sciences from the Massachusetts Institute of Technology.

Gary Cook has been writing on science and technology for more than 20
years. With a B.S. in physics and an M.A. in philosophy of science, he has
written articles, books, chapters of books, brochures, speeches, and Congressional testimony for many organizations and people, including universities and professors. For the last ten years he has been a senior writer for the
Solar Energy Research Institute (SERI), which boasts the world's premier
laboratory in photovoltaic research. While there, he has developed an expertise in photovoltaics, writing extensively on the field. He is currently writing
a book on basic photovoltaic technology.

Janet H. Cushman, also an ecologist at Oak Ridge National Laboratory, initiated the Herbaceous Energy Crops Program and has managed it since its
inception in 1981. She obtained an M.S. in Ecology and Evolution from
Yale University in 1972. She has over 20 publications on herbaceous and
wood energy crops, environmental impacts of energy facilities, and renewable energy assessments. She currently serves as U.S. representative to the
International Energy Agency on Forest Growth and Production, Herbaceous
Energy Crops Task.

John Dowling is Professor of Physics at Mansfield University, Mansfield, PA.
He has a B.S. from the University of Dayton, and an M.S. and Ph.D. from
Arizona State University. His primary research area is in applications of
physics to issues affecting society such as arms control and energy. His particular area of expertise is in ways to inform the public about these issues.
He was film review editor of the *American Journal of Physics* from 1976–
1982 and of the *Bulletin of the Atomic Scientists* from 1979–1984. He edited
the American Association of Physics Teachers book *The Cinescope of Physics,* coauthored the *1984 National Directory of AV Resources on Nuclear
War and the Arms Race,* co-authored the AAPT's *Resource Letter: Physics*

and the Arms Race, and co-edited the Forum on Physics and Society Study, *Civil Defense: A Choice of Disasters.* He served as editor of the Forum on Physics and Society Newsletter from 1980–1986.

Gerald L. Epstein directed the Congressional Office of Technology Assessment's study *Starpower: The U.S. and the International Quest for Fusion Energy* (1987). At OTA, he also worked on the study *Holding the Edge: Maintaining the Defense Technology Base* (1989), wrote major portions of the 1985 study *Ballistic Missile Defense Technologies,* and wrote the proceedings for OTA's 1984 workshop on *Arms Control in Space.* Since March 1989, Epstein has directed the Dual-Use Technologies Project in the Science, Technology, and Public Policy Program at Harvard University's Kennedy School of Government. Epstein received bachelor of science degrees in physics and electrical engineering from MIT and was awarded a Fannie and John Hertz Foundation fellowship for graduate study at the University of California, Berkeley, where he received M.A. and Ph.D. degrees in physics.

Anthony Fainberg received his Ph.D. in experimental particle physics from the University of California, Berkeley, in 1969. He spent 12 years in basic research at Berkeley, CERN, Syracuse University, and Brookhaven National Laboratory. In 1977 he switched to applied research at Brookhaven, working in nuclear safeguards. In 1983, as Congressional Science Fellow of the American Physical Society, he served on the staff of Senator Bingaman. Since 1985 he has worked for the Congressional Office of Technology Assessment where he has specialized in strategic defense questions and in the application of technology to counterterrorism.

David Hafemeister is Professor of Physics at California Polytechnic State University. He was a science advisor in the U.S. Senate during 1975–1977, working on the EPCA and ECPA energy laws. At Cal Poly, he teaches a course on energy and buildings for architectural students, and he runs the House Doctor Energy Laboratory. He co-edited *Energy Sources: Conservation and Renewables,* and has published in the *Scientific American* and elsewhere on energy in buildings. He spent part of 1985–86 in the Center for Building Sciences at the Lawrence Berkeley Laboratory. In addition he has been a Visiting Scientist in the State Department, MIT, Princeton, and Stanford, working on arms control issues. He was chair of the APS Forum on Physics and Society during 1985–1986.

Evans M. Harrell is Professor of Mathematics at Georgia Institute of Technology. He has degrees in physics from Stanford and Princeton and is a former A.P. Sloan Fellow. Before going to Georgia Tech, he taught at Haverford College, the University of Vienna, MIT, and Johns Hopkins. Harrell's research interests include mathematical quantum mechanics, operator theory, and chaotic dynamics. For several years he has been involved in energy and arms control issues with the APS Forum on Physics and Society.

He served on the Forum Executive Committee in 1986–88 and co-edited the 1986 Forum Study, *Civil Defense: A Choice of Disasters.*

George W. Hinman, D.Sc. is Professor of Physics and Director of the Office of Applied Energy Studies at Washington State University. He has worked on energy supply, demand, environmental, and policy issues for many different energy technologies. Most of his experience with unconventional petroleum resources occurred while he was in New Mexico serving as Director of the New Mexico Energy Research and Development Institute in Santa Fe and later as consultant to Los Alamos National Laboratory.

Ruth Howes is a Professor of Physics and Astronomy at Ball State University where she has been employed since 1976. She earned a Ph.D. in nuclear physics from Columbia University in 1971. Through teaching, she became interested in issues of physics related to society. In 1984–85, she served as a William C. Foster Fellow at the U.S. Arms Control and Disarmament Agency, and has since studied a variety of issues related to verification of arms control agreements. She served on the Executive Committee of the APS Forum on Physics and Society from 1986–88 and is currently vice-chair of that organization.

John G. Ingersoll obtained the Ph.D. degree in physics at the University of California, Berkeley in 1978, and joined the Applied Sciences Division of the Lawrence Berkeley Laboratory as a Scientist. As a project leader in the Indoor Air Quality Program from 1978 to 1980, he determined that the source of indoor radon in U.S. residential buildings is the soil rather than building materials as was initially believed. In 1980 he became project manager of the Residential Buildings Energy Efficiency Program, where he was responsible for the development of the large-scale application of computer models in the evaluation of energy efficiency techniques as a function of building type and climate. Since 1983 he has been a Senior Scientist in the Thermodynamics Department at Hughes Aircraft in Los Angeles where he has been in charge of numerous research and development projects in the areas of heat transfer, fluid mechanics, and energy systems. Of these projects the most notable ones include the conceptual design of a testing facility for the thermal-electric power systems of the NASA space station, the design and evaluation of the Georgetown University 300-kW photovoltaic system, and the thermal-structural analysis of failure of the solid electrolyte of sodium-sulfur batteries. Since the acquisition of Hughes by General Motors, he has been responsible for the development of state-of-the-art thermal and human comfort models of the passenger compartment of automobiles. More recently, he has been part of the GM Impact electric car team with responsibility for the thermal management of the vehicle. He is affiliated with the UCLA Department of Architecture, conducting research on energy systems. He has authored or co-authored over 60 papers, articles, and reports and has several patents (pending or awarded).

J. Warren Ranney, an ecologist at Oak Ridge National Laboratory, has managed the U.S. Department of Energy's terrestrial energy crop R&D efforts since 1978, the program's official beginning. He has managed projects on carbon dioxide balances, vegetation management, and energy in the U.S. and developing countries. He is U.S. technical representative to the International Energy Agency on Forest Growth and Production 1980–1985 and presently participates in the Wood Energy Qualities Task. He holds a Ph.D. in systems ecology from the University of Tennessee, Knoxville and has 60 publications on wood energy crops and natural forest dynamics in man-dominated landscapes.

Marc Ross is Professor of Physics at the University of Michigan and Senior Scientist in the Energy and Environmental Systems Division, Argonne National Laboratory. He obtained his Ph.D. in physics at the University of Wisconsin in 1952. He worked in theory of fundamental particles, primarily at Indiana University and Michigan, until 1972 when his interests shifted to energy and environmental problems. He co-directed the American Physical Society's study of efficient energy use in 1974. His research concerns the technology and economics of energy use. Major activities in the early 1980s were work on the Mellon Institute's project on Industrial Energy Productivity and on the Alliance to Save Energy's project on Industrial Investment in Energy Efficiency. A book on energy policy, *Our Energy: Regaining Control,* written with Robert H. Williams of Princeton University, was published by McGraw-Hill in 1981. More recently he has worked on issues of technology and energy demand for the Office of Technology Assessment of the U.S. Congress and the American Council for an Energy Efficient Economy.

Malcolm Sanders earned his Bachelor of Science degree in Engineering Physics from the University of Maine at Orono. Currently, he is finishing his doctoral research at the Yale University Applied Physics Department. He studies the manifestations of Hamiltonian chaos in atomic physics, but he has a strong interest in energy and environmental issues.

William Vogelsang (Ph.D., physics, University of Pittsburgh, 1956) is a Professor in the Nuclear Engineering and Engineering Physics Department of the University of Wisconsin–Madison. His interests have included reactor critical assemblies, neutron diffraction, and non-destructive reactor fuel assay. His research interests are in the radioactivity and safety problems of fusion reactors, the design, and the neutronics analysis of fusion and fission systems. He is a member of the American Nuclear Society.

Leonard Wall is a Professor of Physics at California Polytechnic State University, where he has been since 1969. He obtained his Ph.D. from Iowa State University in mathematical physics. He has been active in the study of energy efficient buildings since 1975, often collaborating with researchers at Lawrence Berkeley Laboratory, where he has spent two sabbatical leaves and numerous summers as a visiting scientist and consultant. His works include the computer modeling of residential space conditioning loads, the analysis of residential retrofit measures, and the energy performance of new commercial buildings. His publications have appeared in *Energy and Buildings, The ACEEE Summer Study on Energy Efficiency in Buildings,* and the *ASHRAE Journal.* He has presented testimony to the state energy commissions of California and Pennsylvania on the modeling and energy performance of buildings.

Summary findings

Summarized below are the findings of the Energy Study Group of the American Physical Society's Forum on Physics and Society. Specific points are discussed in more detail in the Introduction.

General findings

• The Energy Study Group feels strongly that the underlying energy problems that caused the crises of 1973 and 1979 have not disappeared.

• The Nation needs long-term planning and coordination among all sectors of our society to deal with the problem of future energy supplies and the problem of global climate and environmental change which our energy practices bring about.

• We cannot rely on any one or two energy sources or the improvement of end-use technologies alone to solve our future energy problems. Many sources and actions are needed.

• Improving the efficiency of end-use technologies for energy should continue to play an important, indeed an increasing, role in U.S. energy policy.

Sources of energy

• From an environmental point of view, coal and oil are undesirable energy sources and their use should be minimized. Taxes, incentives, regulations, and other policy initiatives should be established immediately to reduce the national dependence on fossil fuels.

• A shift away from combustion as an energy source is desirable.

• A tax on carbon dioxide production, a so-called carbon tax, should be given serious consideration.

• Particularly in view of the coming inevitable decline in domestic oil and natural gas production and of the United States' increasing dependence on foreign oil, it is time for a new look at priorities in energy R&D, both in absolute and in relative terms.

• Increased funding and attention should be given to renewable sources of energy which show promise of being capable of providing a significant fraction of our energy budget in the near future.

• It is probable that the need for nuclear power will increase over the next decades. We should take steps now to ensure that we will have the capability of achieving the expansion safely and at a rate commensurate with future U.S. energy needs.

• Risk assessment could be an even more important tool in comparing the desirability of different energy options than in the past, but it is a complicated technique and not well developed.

End-use technologies

• Research and development of alternatives to petroleum for transportation energy should be significantly increased.

• An unbiased study of the future economics of conservation is urgently needed to guide energy policy.

• We feel strongly that the government should take immediate actions, such as the implementation of higher CAFE standards and an increase in R&D support, to increase the efficiency of the automobile fleet.

• We should implement national standards for efficiency in buildings, such as building efficiency performance standards (BEPS) and requirements for energy efficient appliances.

Conclusion

The U.S. can avoid the economic and social trauma of sudden unpredictable and uncontrollable shortages in future petroleum supplies as well as severe environmental damage. To do so, we must undertake the coordinated, long-term development of a variety of energy sources and make a concerted effort to increase the efficiency of end-use technologies. We feel that it is extremely important to begin these efforts immediately.

Introduction: Yes, Virginia, there is an energy crisis

According to the American public, the energy crisis happened more than 15 years ago and we solved it. Low oil prices have relieved the sense of crisis we felt when energy prices skyrocketed in the early 1970s.

Like the grasshopper in the old children's story, Americans cheerfully drive inefficient cars, build homes that waste heat, generate electricity with fossil fuels, and import vast quantities of petroleum. Warnings of coming domestic shortages of petroleum bounce off public complacency. Provident ants, who speak of coming shortages or of the negative impact of fossil fuels on the environment, are largely ignored.

When OPEC suddenly increased oil prices in 1973, the United States responded by diving into a major recession. Crash energy policies funded research and development projects in nearly all conceivable areas. "WIN" (Whip Inflation Now!) buttons were decorative but not the hallmark of a carefully crafted policy to deal with the sudden increase in petroleum prices.

The sensitivity of market oil prices to minor disruptions in Western supplies of petroleum continues to be evident. The Iraqi invasion of Kuwait sent petroleum prices skyrocketing as the future of petroleum imports from the Middle East became cloudy. The oil spill from the Exxon Valdez in Prince William Sound cast doubt on the future Alaska supply, which, in any case, will begin to decline precipitously within two to three years. It also brought into serious question previous federal plans to explore the Arctic National Wildlife Refuge for possible large fields of recoverable oil. Together with recent slowdowns in North Sea production and corroding pipes in the supply lines, the Valdez spill scared producers and petroleum prices suddenly rose by around 25%. This rise was temporary, but we cannot ignore the possibility that political upheavals, accidents, or natural disasters will produce other such sudden increases.

In addition to other long-term environmental problems, many atmospheric scientists now believe that there is a real possibility that effective CO_2 doubling could cause an average global temperature rise of from 2–5 °C. This greenhouse effect will severely modify the Earth's climate. The predicted temperature rise may radically change the agricultural productivity of many vital areas (such as central North America), and cause a significant rise in sea level. Atmospheric CO_2 levels are now projected to reach twice their present concentration sometime in the next century. In addition, production of other greenhouse gases, such as methane, nitrous oxide, and chlorofluorocarbons, as well as the positive feedback of water vapor from the oceans will result in an earlier "effective doubling."

Atmospheric modelers cannot predict exactly when the global temperature rise will be detectable, how great the increase will be, or what will be its regional consequences. For example, the effect of increases in cloud cover, which could be produced by an atmospheric warming, might well act as a negative feedback by reflecting sunlight and reducing the magnitude of the warming. There is, however, consensus that at least some rise in global temperatures is likely. The hot dry summer of 1988 may have been a portent of things to come, or it could be simply a statistical fluctuation. However, the world (not just the U.S.) temperature trend for the last 20 years, and, in particular, for the last 10, presents us with the possibility that the early signs of the greenhouse effect may already be with us.

The tragedy of the greenhouse effect is that once the damage is done, many decades will be needed to reverse it. Indeed, a reversal may not be possible, and we may face adapting to a warmer Earth. Burning fossil fuels provides the major new source of CO_2 in the atmosphere. Because a greenhouse effect is well within the realm of possibility, and because the time required to reverse its effects would at best be decades, it seems prudent to start reducing CO_2 production now.

A brief history

Government, industry, and political pressure groups pursue "magic bullet" solutions to the twin problems of energy supply and damage to the environment. As a result, U.S. policy has been driven by a series of fads. The advocacy and adoption of each successive cure-all for energy problems and its subsequent failure form the recent history of energy policy.

In the 1950s nuclear fusion was touted as a possible clean energy source. The public and the government bought the dream and provided the substantial sums of money needed to start moving toward it. Today, we realize that, at a minimum, 30–50 more years are needed for fusion to begin to make an impact on our energy supply.

In the late 1960s and early 1970s there was a major, expensive, and possibly premature push to expand nuclear fission energy. Fission power promised cheap electricity to fuel economic expansion. The discovery that nuclear fission power costs enormously more than had been predicted and that it has potentially severe environmental side effects discredited it among large segments of the public. Faced with a dramatic public distrust and cost overruns, the electric utilities have stopped ordering new reactors. The U.S. cannot count on any significant increase in energy contribution from nuclear fission for at least another decade.

In the late 1970s and 1980s, optimistic views held by proponents and supported by popular press accounts held that renewable sources of energy, notably solar, would be available in the near future and would solve the then-current oil crisis. Thus far, 20 years after Earth Day I and in spite of research and development efforts and tax credits, not only have the impacts of new renewable resources been minimal but even a tried and true old renewable, hydropower, has declined in importance. Further, increased use of another old one, wood, has demonstrated that wood burning can disperse large amounts of noxious gases, particulates, and carcinogens per unit of energy produced, particularly when wood is burned in small residential stoves. The success of conservation techniques in flattening the rise in energy demand is the one bright spot in the recent story of U.S. energy production and use.

Most recently in the U.S., the long hot summer of 1988 struck many commentators as the beginning of the greenhouse heating of the world. The cool summer of 1989 has in some measure calmed popular fears in the U.S., but the scare of the summer of 1988 has led to a reassessment of energy policy, even to the extent that some environmentalists are reexamining the use of nuclear fission as an energy source.

Unfortunately, energy policy and funding for research and development have closely followed the fads recounted in this brief history. Despite public confidence induced by low oil prices, U.S. petroleum and natural gas supplies continue to decline. The fraction of petroleum we import has climbed to nearly 50%. Nuclear fission power cannot make up a shortfall quickly, and indeed the production of electricity will not solve all our energy needs since it is not clear that electricity is the most cost effective way to heat houses or propel automobiles. The renewable alternatives are not yet an economical means to fill the energy supply gap. Just as in 1970, the potential for an economically disastrous energy crisis exists today, and the effects of energy production on the environment concern a growing minority.

We feel strongly that the underlying energy problems that caused the crises of 1973 and 1979 have not disappeared.

In this volume, we review the progress and lack of progress that has been made in most of the areas of energy related research. Policy makers, teachers, and the general public must be informed so that the United States can develop a practical, long-term energy policy.

The energy crisis has not disappeared. It has just been forgotten for a while.

Dealing with energy problems

The Nation needs long-term planning and coordination among all sectors of our society to deal with the problem of future energy supplies and the problem of global climate and environmental change which our energy practices bring about.

In the past, we have maintained a consistent energy plan for no more than a few years at best. Business focuses on next year's profits, and the public sector worries about next year's election and this year's federal deficit. Planning within these time constraints does not provide workable long-range solutions to national energy problems.

We must plan now for future energy supplies. The bad news is that long development times, slow market penetration, and major capital costs delay the impact of most technical projects on energy supplies and many of them on end-use efficiency for at least ten years. Past complacency has cost us dearly. Fortunately there is also good news: improvements in the efficiency of some end-use technologies are easier to develop and less capital intensive than new energy supply technologies, and they can be implemented more quickly. For example, mandated efficiency standards for cars, buildings, and appliances could have earlier impacts. In spite of these more quickly effective measures, realistic long-range planning must allow for the time-lag between funding an energy research and development program and the contribution of the technology to the energy supply.

Energy problems, particularly environmental effects, are global in both impact and origin. These global problems will have to be dealt with on an international as well as a domestic level. International cooperation has become urgent and essential. We are happy to note that some efforts have already begun along these lines. More formalized cooperation is still needed, especially in the area of reducing CO_2 production. Finally, since the U.S. is the major contributor to global energy problems (our 5% of the population accounts for about 25% of the world's energy consumption), even measures taken only at home will greatly influence the world both by direct effect and by example.

We cannot rely on any one or two energy sources or the improvement of end-use technologies alone to solve our future energy problems. Many sources and actions are needed.

Economical energy resources are limited. Environmental problems attend every method of producing energy. We face uncertainties in projecting future economics of different energy sources as well as uncertainties in predicting the future at all. For example, our needs might be met and environmental problems avoided, at least for a few decades, by relying mostly on natural gas and conservation. It may also be that the world's needs cannot be met. We do not yet know. Therefore, it would be the height of folly to plan the economic survival of the Nation, and, by extension, the globe, based on these two measures alone.

Unforeseen problems may seriously reduce the contributions of any one energy source. To provide the flexibility we need to deal with an unpredictable future, we must develop a variety of energy sources and increase end-use efficiency.

Energy sources should not be ignored or discounted on the sole grounds that they cannot solve all the problems themselves. This limitation applies to all energy sources and is thus a poor criterion for ruling out potential energy technologies. Such arguments have been used to denigrate the ability of nuclear power to solve greenhouse problems. These arguments are disingenuous. Using more nuclear power would not eliminate the CO_2 problem, but it would reduce CO_2 production. Similar arguments have been used to belittle the importance of some renewable technologies like biomass. We cannot overlook potentially important options simply because the estimated impact on world emissions of actions taken in the U.S. alone appears small.

Improving the efficiency of end-use technologies for energy should continue to play an important, indeed an increasing, role in U.S. energy policy.

Improved efficiency of end-use technologies can save a large fraction of our diminishing energy supplies. During the 1970s and 1980s, energy use per unit of economic activity declined sharply. In particular, the energy/GNP ratio decreased by 28%. Had this ratio remained constant, the U.S. would require

40% more energy than we currently use. Technologies driving this change were more efficient new houses, more efficient new automobiles and airplanes, and improved efficiency of new and renovated factories. As old housing and appliances, vehicles, and factories continue to be retired, there will be a considerable opportunity for continuing the overall efficiency improvement.

New technology fosters efficiency improvement. Many, perhaps most, of the improvements now being made use technology not available in 1970. Thermodynamic analysis shows that only a few activities, such as reducing metal ores, in principle require substantial energy input. Other processes use energy but the energy is not required thermodynamically to make the process work. A particular furnace or motor may approach the theoretical limits of efficiency, but its job could be done quite differently, dramatically reducing the use of heat and power. More and more ideas are being explored for sharply improving the efficiency of activities now characterized by conspicuous inefficiency on the basis of physical principles. We expect technological opportunities for improving end-use efficiency to continue.

Moderate efficiency improvements are much cheaper than major new supply alternatives and tend to come in the form of smaller, more digestible chunks. For this reason alone, improvement of end-use efficiency has played a very prominent role in resolving the energy problems of the 1970s and early 1980s and may be expected to continue to do so.

In the long run, we must note that improving efficiency cannot replace new sources of energy, but it is an important complement to other energy measures. No single approach to the resolution of energy problems is sufficient.

Sources of energy

From an environmental point of view, coal and oil are undesirable energy sources and their use should be minimized.

Major reductions in coal use may not be practical for a considerable time. Coal is attractive because the United States has a large domestic supply and the cost per unit energy is low. Unfortunately burning coal produces chemical pollution and generates more CO_2 per unit energy than other fossil fuels. Pollution and acid rain from burning coal could be reduced if the U.S. were to emphasize production of synthetic gas from coal. However, the CO_2 production, local environmental damage, and costs per unit energy would be considerably higher for synthetic gas than for straight coal burning.

Oil generates more CO_2 and chemical pollutants than natural gas. In addition, the U.S. is now heavily dependent on foreign petroleum and is currently increasing its petroleum imports. Major petroleum supplies lie in politically unstable countries. The Oil Crisis of 1973 clearly demonstrated the drastic effect of political action on the petroleum market. The United States remains vulnerable to market manipulation today. Domestic oil shale resources have limited availability at acceptable costs and are unable to take the place of imported petroleum.

The use of natural gas as a major energy source is an energy policy option that has recently received considerable attention. Natural gas burns cleanly, producing few chemical pollutants. In addition, it produces just over half the CO_2 per unit energy that coal does. Technology for burning natural gas has been used for decades. Newer applications, such as natural gas for transportation, are feasible with current or near-future technologies.

The disadvantages of massive reliance on natural gas are twofold. First, it still produces CO_2 in large quantities, although only about half as much as coal for the same energy output. Replacing other fossil fuels with natural gas is thus a "semi-environmental" solution to reducing CO_2 production. Second, at current consumption rates, the domestic natural gas supply will last only about 15 years. It is true that known natural gas reserves increase each year almost to the point of compensating for that year's use. However, doubling or tripling natural gas use to replace oil and coal would almost certainly rapidly deplete the remaining domestic supply. Would the U.S. want to rely on foreign sources for natural gas? Probably not, since the world's largest producers are the Soviet Union and several states in the Middle East. Thus natural gas should probably be considered as an interim fuel whose rate of use cannot be greatly increased.

Taxes, incentives, regulations, and other policy initiatives should be established immediately to reduce the national dependence on fossil fuels.

Future adverse environmental consequences of the use of fossil fuels, such as acid rain and the greenhouse effect, are unavoidable, but prompt action on our part can reduce the damage. Again no single regulation or tax will solve all energy problems, but such incentives will be needed to control patterns of energy use and make environmentally desirable energy sources commercially attractive.

A shift away from combustion as an energy source is desirable.

Combustion, particularly of fossil fuels, can be replaced either by electricity generated without combustion (nuclear fission or solar) or by conservation measures. A shift to electricity has been underway since electricity was first introduced. Electricity offers advantages both in the mid-term and in the long-term. First, there is the obvious flexibility of electricity as an energy source for use on-site. Also, electric motors are usually substantially more efficient than combustion motors, even after taking energy losses incurred in generating electricity into account. Electric heating suffers from inefficiencies in electric generation, but heat pumps and other more efficient forms of electrical heating may compensate for this. Other means of electricity production, using cogeneration and future technologies, promise even greater efficiency. Unfortunately each of these technologies has limitations. For example, heat pumps are most effective in climates warmer than those of Kentucky and Tennessee, and cogeneration is not cost effective in all situations.

If electricity's role can be substantially increased for heating and if nonfossil fuel sources can be found for electric generation, CO_2 and acid rain emissions could be significantly reduced. In addition, we must strengthen the search for a substitute for fossil fuels to power automobiles and, if possible, trucks. At present there is no assurance that electricity will prove the most effective means to perform tasks such as heating buildings, and we need to consider other means to fulfill these needs. For example, insulating buildings can reduce the need for heating.

A tax on carbon dioxide production, a so-called carbon tax, should be given serious consideration.

Our society responds best to short-term economic rewards. Reducing CO_2 production, while essential for the long-range good, offers no immediate pay off. A well planned carbon tax could offer consumers a tangible, immediate reward for protecting the environment over the long run. As an example of this strategy, the sizable gasoline tax in Europe has been the main cause of their more energy efficient fleet of autos.

The details of a carbon tax would require careful thought. The cost to CO_2 emitters of production will increase, and these added costs will also be passed on to consumers. If only the U.S. implemented a carbon tax, the price increase might hurt our competitiveness. Should one place higher taxes on imported sources of carbon to promote a lower dependence on foreign sources? Or should one simply tax in proportion to CO_2 production? Should tax relief be provided to those disproportionately dependent on fossil fuels such as residents of New England or the poor? In setting tax rates, should one account for the specific levels of other pollution produced by a given energy source as well as CO_2 production?

A carbon tax would stimulate conservation, encourage the use of nonfossil fuel sources, mitigate the greenhouse effect, and reduce pollution. The revenues from the carbon tax could be devoted to energy R&D, or to the stimulation of energy conservation or mass transit. The economic impacts of a carbon tax must be carefully considered and great caution exercised in implementing such an approach. Nevertheless, the study group feels the idea of a carbon tax deserves serious exploration.

Research and development of alternatives to petroleum for transportation energy should be significantly increased.

Any discussion of reducing U.S. dependence on petroleum or fossil fuels must include the transportation industry. Unlike other sectors, transportation depends almost totally on petroleum. At present, the U.S. could not rapidly shift its transportation energy requirements from petroleum products. It may require a decade of work or more to provide alternative sources of transportation energy and even longer to convert substantial portions of the transportation infrastructure and vehicle fleet to make use of these new sources. Methanol cars, electrical cars, and even vehicles running on hydrogen have been designed, and some exist as prototypes, but large-scale use of these technologies is still in the future. Current or near-future technologies do offer useful alternatives. Certainly mass transportation should be strongly considered but construction requires time.

It is probable that the need for nuclear power will increase over the next decades. We should take steps now to ensure that we will have the capability of achieving the expansion safely and at a rate commensurate with future U.S. energy needs.

Without increased use of coal and natural gas to generate electric power in addition to the continued use of existing nuclear fission power plants, the United States cannot meet its energy needs for the next decade. Some regions of the country will need additional electric generating capacity during the coming decade. Natural gas is the likely energy source for this new capacity. Natural gas is particularly attractive since new turbine-based technologies such as combined cycles and steam-injected systems have increased efficiencies to nearly 50%. Nevertheless, as discussed above, natural gas is not a satisfactory long-term energy source.

Nuclear power plants currently produce about 20% of domestic electricity. In the United States, construction of new nuclear fission reactors has virtually stopped. Thus nuclear fission cannot significantly increase its contribution to our energy supply for at least a decade. Other nations produce a far higher fraction of electricity through nuclear power than does the U.S. (up to 75% for France), although none approaches the total fission power generated here.

Nuclear fission is the most controversial of energy sources. It suffers from the name "nuclear" and the association with nuclear weapons. In the past, advocates of nuclear fission power promised cheap, safe power. However, there have been two major accidents over the past decade. Although the first, at Three Mile Island, will probably have no detectable health effects, much of the public feels that nuclear fission is the most dangerous way to produce energy. Many people would like to ban nuclear fission power altogether, and many more are relatively skeptical about its possible future contribution to the Nation's energy supply. Recent enormous cost increases and uncertainties in waste disposal methods have further undermined public confidence.

If fission is to play a greater role in the future, public concerns must be addressed. We feel strongly that visceral opposition to any fission power is unwarranted. Health risks from fission power, even including catastrophic failures, do not appear to be more severe than those from acid and other poisonous emissions from fossil fuel sources. For example, the Office of Technology Assessment estimates that the number of deaths in one year which occur from sulfates produced by burning fossil fuels[1] is comparable to the number of deaths worldwide over the next 50 years anticipated from the Chernobyl disaster (both estimates have large uncertainties). Yearly deaths from "normal" indoor radon contamination may also be comparable with the total long-term number of deaths from Chernobyl.[2]

If nuclear fission power is not as deadly as many think, neither is it without potential hazards nor is it cheap. It does, however, have the advantage of being a relatively mature technology and of producing virtually no CO_2 or acid rain, a property shared only with hydroelectric power among today's large-scale energy sources.

There are several requirements for maintaining nuclear power as an option. First, a satisfactory means of disposal of high-level radioactive waste must be implemented. It appears to most members of the group that the technology for disposing of spent fuel rods is available or will be shortly, but there are some doubts about disposal of wastes from reprocessing. The political will and political cooperation to this end need to be established. Even if nuclear power were to cease operation today, we would still need to dispose of the spent fuel rods already generated.

Second, development of designs for the next generation of reactors, strongly relying on "inherent" or "passive" safety features, should be continued and intensified, with a view to putting such reactors into operation within 10–15 years. Detailed recommendations for streamlining licensing procedures and a selection from several designs suggested for adoption will have to come from expert consensus. However, a credible, standard, and tested design is needed to restore public confidence and make costs manageable.

Energy funding should not be a zero sum game, but in a country plagued by budget deficits, the question of how much R&D funding should be applied to nuclear fission power relative to other energy sources must be addressed. For instance, conservation R&D is a lot cheaper than fission R&D and would probably have a higher early payoff. Therefore conservation R&D should not be crowded out by very large fission R&D budgets. However, in the long run, fission power will also be needed, unless the development of solar power for the entire country proceeds at an unexpectedly rapid pace. Responsible deployment of future reactors requires a sustained program starting now rather than a crash program embarked upon in 10 or 20 years.

Risk assessment could be an important tool in comparing the desirability of different energy options, but it is a complicated technique and not well developed.

Since it is complicated and not transparent, risk assessment is easy to abuse, allowing analysts to cull data selectively in order to support a preconceived position. It is also hard to avoid some subjectivity in assessing and comparing the consequences of energy production. How, for example, does one compare the loss of farmland or wilderness area to a hydroelectric dam with the environmental stress from coal-produced acid rain, or high-level nuclear waste storage? How can one reasonably assign dollar values to damage to a beautiful view? Despite creditable attempts to develop techniques for these comparisons, risk assessment remains a young discipline.

Caution should be used in applying risk assessments to energy decision making. Good analyses will compare risks for alternative means of producing energy. For each alternative, the assessment should also be longitudinal, incorporating all risks involved in the entire energy cycle, including long-term environmental effects.

Particularly in view of the coming inevitable decline in domestic oil and natural gas production and of the United States' increasing dependence on foreign oil, it is time for a new look at priorities in energy R&D, both in absolute and in relative terms.

Since there has been no crisis for a decade, government and industry feel less pressure to fund research on energy sources and uses. Several energy demonstration projects, funded by the Department of Energy, have proven less than completely successful. Despite the problems, some of which result from the crash funding programs of the early 1970s and reduced funding since, technology in many fields, notably including solar thermal, solar photovoltaic, and conservation, has progressed.

We now discuss several sources that could or do contribute significantly to the Nation's energy supply.

Nuclear fusion energy was once seen as a magic bullet solution to all our energy problems. Despite decades of enthusiastic research, nuclear fusion energy remains an expensive research undertaking. An enormous supply of fuel, the deuterium in the oceans, is potentially available for fusion reactors. Fusion produces no CO_2. Although fusion produces and consumes tritium and although materials used in fusion plants will become radioactive, their half-lives are on the order of tens of years rather than tens of thousands of years characteristic of radioactive by-products of fission reactors. Researchers believe that nuclear fusion plants can be made intrinsically safer than nuclear fission plants since they will contain less inherently dangerous materials and will also have less stored energy that could potentially cause those materials to be released in an accident. For example, only enough fuel to run the reactor for a few seconds will be contained within the reactor vessel at any given time.

These striking advantages have prompted federal funding of programs to develop nuclear fusion reactors. Slow progress in the field has been and is being made, but the technology for constructing working power plants is not available and probably will not be ready in the next decade or two. Capital costs per kilowatt are very uncertain and will no doubt be high. Moreover, fusion's potential environmental and safety advantages will not be automatically realized. Even if the technology can produce a reactor that generates net energy, there may be safety or environmental problems that we do not foresee. We cannot look to nuclear fusion to solve our near-term energy problems, and even the longer term prospect is uncertain.

Geothermal energy has been exploited for decades in disparate places around the globe, including Iceland, New Zealand, Italy, and the United States. In some cases, the heat energy from underground water near the surface is converted to electricity, and in others, the hot geothermal fluid provides space heating or heat for agriculture or industry. The Geysers facility in California is the only significant geothermal electric plant in the U.S. Unfortunately there are some indications that the thermal field which supports the Geysers facility is being more rapidly depleted than had originally been calculated. The environmental problems associated with geothermal energy include contamination of ground water by geothermal brine, release of toxic gases, noise from the release of pressurized fluid, and depletion of steam fields and geysers that have been traditional tourist attractions. Most of these effects can be ameliorated by relatively inexpensive measures.

The real limitation on geothermal energy is geographical. Thermal fields suitable for direct exploitation at reasonable cost occur in a very limited number of places many of which are in sparsely inhabited regions. Researchers have attempted to create artificial steam fields by pumping water down deep wells onto hot dry rock. At best, the "hot dry rock" technique for producing geothermal energy has had limited success. Unless it can be much further developed, geothermal energy will continue to provide a very small fraction of the nation's energy supply.

Hydroelectric energy has been exploited for decades. It currently furnishes around 7% of the world's commercial energy. In the U.S., it furnishes about 4% of the total energy produced. However due to droughts and silting, the total energy generated by hydroelectricity has actually decreased since 1973. Hydroelectricity is clean and efficient, and the technology for using it is well-understood. In many facilities, it can be stored during slack hours and released to provide extra power during peak use periods. On the other hand, most hydro plants require construction of dams and artificial reservoirs which take up large areas of land. Dams and plants may also degrade the local environment and pose a threat to indigenous flora and fauna. Risk assessments indicate that the chance of accidents at some hydroelectric dams may significantly threaten human populations living downstream.

As in the case of geothermal energy, resources for producing hydroelectricity are geographically limited. Much of the U.S. resource that is suitable for large-scale production of electric power has already been exploited. Hydroelectricity is also somewhat dependent on climate, since a series of dry years can lower the water levels in the reservoirs used to produce hydroelectricity. It seems likely that most of the new hydroelectricity in the United States will be produced by relatively small operations. In some regions of the U.S., especially where dams have already been sited, hydroelectricity can be exploited to provide relatively modest additions to the Nation's energy supply with less damage to the environment than is found with many other energy sources. It cannot replace a large fraction of the electricity currently generated using fossil fuels.

Several additional renewable energy sources could provide significant energy to the U.S. in the near future. In each case, experts differ in the extent to which that energy source can be exploited.

Recent technical advances have led to widespread optimism about the future potential of **photovoltaics**. Some predict that photovoltaics could become economically competitive for baseline electric production by the turn of the century and could satisfy all needs for electric power within 40 or 50 years. Others claim this assessment is flawed in that it does not consider storage technologies and costs which would become vital when solar sources provide more than some 20% of the national electric power supply.

At this moment, photovoltaics cannot compete economically with cheap energy from fossil fuels. However, there are many reasons for thinking that, within a few years, large-scale photovoltaic electricity may well become economically feasible, at least for power peaking. More will be known by the mid-1990s when large demonstration projects are scheduled to be built and operating. Meanwhile R&D support should be funneled to this area as a high payoff could result within a short period of time. Not only is solar photovoltaic power renewable, but it also has few known environmental side effects, aside from the fact that large-scale photoelectric facilities will occupy large land areas and there will be some environmental damage from the manufacture of materials.

In recent months, **solar thermal power** has begun to show some significant entry into the California market. It has been applied on a building-by-building scale, and Luz International Ltd. recently opened an 80-MWe electric generating plant bringing its total solar thermal generating capacity to 274 MWe. There is little doubt that solar thermal energy can be used effectively in heating and cooling buildings where the building designs use passive solar energy as well as for active applications such as heating water. Much of the energy gain from exploiting solar thermal energy is likely to come from its use by the individual rather than its exploitation by large corporations.

Large-scale applications include generation of electricity, desalination of water, and promising techniques for the destruction of toxic chemical wastes. From a technical standpoint, large-scale exploitation of solar thermal energy will require development of less expensive and more durable coatings for optics. The solar thermal generation of electricity has been heavily researched, but the process is still in the developmental phase and its efficiency can probably be increased. Before solar thermal energy finds widespread acceptance, its economic practicality will have to be demonstrated to both individual consumers and the construction industry. As with conservation, solar thermal energy becomes more attractive as fossil fuel prices increase. Use of passive solar heating and cooling has led to aesthetic concerns, and some worry about indoor air quality in highly insulated buildings with little air infiltration. However, good architectural design and air-to-air heat exchangers can be used to mitigate these problems. As far as we know, the environmental effects of solar thermal electricity are confined to the need for large land areas to hold the collectors for commercial scale projects and the effects of mining the materials and manufacturing the technical equipment.

Like solar energy, the use of **energy produced from biomass** will depend on technological and manufacturing developments to lower its costs and/or on increasing fossil fuel prices. Energy crops offer the production of liquid fuel which may help reduce the dependence of transportation on petroleum fuels. Today biomass-fueled systems produce about 4% of the energy used in the U.S. and mostly burn wood or wood products. Researchers have made substantial progress in the 1980s in species selection, plantation management, erosion resistance, and especially in the processing of biomass into useful energy carriers. Processes for conversion into ethanol and into electricity have both been dramatically improved, reducing costs.

Fuels from biomass can be replaced by growing new crops that will remove at least as much CO_2 from the atmosphere as they release when burned except perhaps for the additional CO_2 released in processing biomass fuel. Some researchers believe that selected biomass fuels produce fewer chemical pollutants than fossil fuels. A potential danger from the wide-spread use of biomass to produce energy is that forests may be harvested without replacement. One of the factors currently facing major biomass-for-energy development is the economic need to replace the crops that used to be grown for export, particularly since exports have declined sharply in the 1980s. Another potential problem is damage to the soil due to over-intensive biomass production. In addition, claims that combustion of biomass-derived fuels causes much less local air pollution need careful evaluation. Biomass, in the form of ethanol or methanol, can certainly provide needed energy in the fairly short-term, particularly to replace petroleum as a liquid fuel for transportation where there are few other proven technologies.

Wind-driven power stations offer another renewable source of energy that has been developed over the last decades. Besides the fact that it is a proven, renewable resource, wind energy produces no chemical or radioactive by-products aside from some pollution released in manufacturing components. Wind generators are inherently modular and can provide incremental increases to the energy supply in short times. Unlike the solar energy technologies, wind energy is at least close to being economically competitive at present oil prices. Wind generators are easily integrated into existing electrical utility grids.

Like solar power production, wind generators require relatively large areas of land. Windmills can occupy the same area as other applications such as cattle grazing, so the land need not be completely lost for other uses. Wind energy production is limited geographically to areas where there is usually enough wind to drive generators. In addition, wind generation will require a backup energy supply during windless periods. Other known side effects of wind generation include noise from the generators and interference with local television. Despite these drawbacks, some feel that wind generators could make a substantial contribution to replacing fossil fuels in the production of electricity.

A final source of renewable energy is **energy from the oceans.** This energy source includes the thermal energy trapped in the warm surface layers of the oceans, the mechanical energy in waves, and the gravitational energy of tides. There have been several efforts to develop electric generating plants based on each of these energy sources. There are currently demonstration generating plants based on tidal and wave energy which have been shown to be economically viable. There are also development projects for Ocean Thermal Energy Conversion (OTEC) which convert the stored thermal energy of the oceans into electricity. Despite these early technical successes, the OTEC technology is not yet ready for commercial development nor has it been shown to be economically viable.

The great advantages of ocean energy are that it is non-polluting, except for the effects of mixing thermal layers of the ocean, and renewable. Construction of conversion plants necessarily affects the local environment, but the technologies have minor ecological consequences compared to fossil fuel or nuclear fission power plants. Like geothermal power, ocean power would be geographically limited first of all to the coasts and then to areas of strong tides and waves for the gravitational energy conversion processes or, for OTEC, to warm ocean areas.

End-use technologies

While new energy sources are essential, efficient end-use technology is probably the most immediately effective tool currently available for grappling with energy problems. It requires low capital investment and can be introduced in stages which are palatable to the public. Conservation is often referred to as an energy source. It is not, of course, but it can function as the equivalent of a source because reduced energy use implies that real energy sources that otherwise would be exploited can be freed. In particular, the capital costs of constructing new sources can be saved.

While we favor considerably more emphasis on R&D in end-use efficiency, we must also ask one general question about energy conservation. In the recent past conservation has been extremely successful in saving energy with low capital investments. But how much more energy can it save before its current cost advantages over energy generation disappear?

An unbiased study of the future economics of conservation is urgently needed to guide energy policy.

Such a study is beyond the scope of this volume, but should be undertaken as a matter of priority. This study should employ well-developed economic analysis, using life-cycle costing methods to compare the conservation and supply technologies.

On the road to end-use, energy must be transmitted and stored. **Energy storage technologies** cannot be classified as either energy sources or end uses. Nevertheless efficient storage can save substantial energy from existing sources. Storage technologies can distribute loads on power sources more evenly, make possible the exploitation of new sources far from application sites, and allow the use of energy sources, particularly renewables, which could not meet demand without storage. Efficient, economical energy storage will be needed to exploit solar electric technologies where the energy supply varies with time and weather conditions. It is already needed to meet peak power demands in utility networks. If we are to reduce our dependence on petroleum for transportation, we must develop an alternative fuel source or portable devices such as batteries or fuel cells.

Mechanical energy storage using flywheels, pumped water, or compressed air has been widely utilized since the industrial revolution. There are development efforts to increase the efficiency of these established technologies. Electrical storage technologies include devices such as large capacitor banks and superconducting coils as well as the more economically attractive electrochemical batteries. A major research and development effort is underway to produce more durable, efficient batteries. Thermal storage and thermochemical storage show promise in applications involving solar energy. Chemical energy storage includes such applications as the use of methanol and hydrogen as portable fuels for transportation. Finally the biological storage of energy can be used to produce fuels based on methanol and ethanol. Research and development are desirable at two levels: that of the individual consumer and that of the large industrial user.

We feel strongly that the government should take immediate actions, such as the implementation of higher CAFE standards and an increase in R&D support, to increase the efficiency of the automobile fleet.

Transportation uses 28% of the energy consumed in the United States and nearly two-thirds of the petroleum burned in the nation. The transportation sector is more vulnerable to a sudden disruption in petroleum supplies than other sectors. As emphasized above, we feel it is desirable to reduce U.S. dependence on petroleum. Clearly this means examining the energy efficiency of the transportation industry since there are no timely alternatives to petroleum for transportation use on the scale required by the U.S.

Because passenger cars consume most of the petroleum used, their energy use has been examined in detail. Energy savings are possible through changes in propulsion systems, engine design, and body design. Petroleum use could be reduced by employing alternate fuels in cars. The large increase in transportation efficiency (an average increase from 14 to 28 mpg for new automobiles since the embargo) has roughly compensated for increased use of transportation. More efficient automobiles are not currently in demand although designs for them exist. Low petroleum prices provide little incentive to the industry or the consumer to seek greater fuel economy. Federal incentives such as increased Corporate Average Fleet Economy (CAFE) Standards are needed to decrease petroleum use. In the long-term, petroleum use in transportation must be replaced by alternate fuels or electricity.

Government planners must, in addition, consider the various transportation systems as an integrated whole. Public transportation systems planned today will not be available for at least ten years. Should oil prices climb abruptly, they will be needed. We should also consider the use of new technologies, such as greatly increasing conferencing by phone or video which reduce the need for transportation.

American agriculture has been described as the art of turning petroleum into food. **Agricultural energy** is important to the well-being of the well-fed American consumer, who frequently forgets that our extremely efficient, mechanized agricultural system is heavily dependent on energy in general and petroleum in particular. Not only is energy used in farm equipment, but it is vital to preparing the fertilizers and agricultural chemicals that permit the high productivity of the U.S. farm. In considering replacements for petroleum, it will be necessary to consider its use in farm equipment and procedures. In particular, energy use often saves other scarce commodities such as arable land and fresh water. Agricultural technologies which save on fossil fuel inputs frequently require the use of chemicals which themselves pose a severe environmental threat and consume energy in their production.

Industrial energy use per unit of production has long been subject to efficiency improvement as industrial processes have been updated. Successful new industrial processes tend to reduce all costs including energy. Recent stricter environmental regulations provide important motivation for process changes that will also save energy such as waste minimization or closing gas and fluid cycles. In addition industry uses add-on conservation equipment, such as heat recovery devices, whose implementation depends sensitively on energy prices. Recycling avoids the initial energy-intensive processing of raw materials. Cogeneration, the generation of heat and shaft power together, is more efficient than separate generation if a steady nearby heat load is present.

Substantial reduction in the emission of CO_2 by industry could require strong policy incentives from the government and investment in the development of new technologies, particularly new process technologies. Although industry will continue to conserve energy, it will not change rapidly unless energy prices rise. The development of new industrial technology can speed up conservation endeavors.

We should implement national standards for efficiency in buildings, such as building efficiency performance standards (BEPS) and requirements for energy efficient appliances.

We use energy to heat and cool buildings as well as in a wide variety of appliances. Conservation measures, such as improved insulation and windows, have already reduced energy consumption in newer buildings. Such measures require an initial capital outlay which is paid back by energy savings over time. The cost of a building should be considered over its entire lifetime. Since many home owners occupy a house for only a few years, this argument is difficult to justify. Other incentives need to be devised.

While many technologies for constructing **energy efficient buildings** exist and their use should be encouraged, new technology can also have an important role. On new construction, the application of the DOE2 computer software is accurate to about 10%–15% on predicting energy consumption. On existing buildings, a "house doctor" can determine losses with blower doors and infrared thermography to reduce energy losses to about one-half. Policies are needed to encourage buyers to demand efficient buildings and to facilitate continued improvement of technology.

The **energy efficiency of appliances**, as well as buildings, can be greatly increased. As a case in point, the California appliance standards adopted after the energy crisis of 1973 have reduced the energy consumption of individual refrigerators to one-half of the pre-1973 level and are expected to reduce it to one-third before 1993, eventually saving the nation about 25 GW of electrical power. Such standards work relatively quickly to change energy use patterns and are a powerful tool in reducing our energy consumption. Policy considerations should include energy taxes, a method to compensate owners selling property for the investment they have made in energy conservation, and energy efficiency standards for buildings and appliances.

The **electric motor** is the largest single user of electricity. Improving the efficiency of the electric motor offers a significant opportunity to conserve energy. The revolution in materials technology makes a variety of improvements in the operation of motors possible, and adjustable speed electronic drives allow motors to operate at maximum efficiency under widely varying loads. More important, the control capabilities offered by these electronic drives enable system redesign and substantial reductions in the load on a motor in many applications. New, stronger permanent magnets offer opportunities to increase the efficiency of motors which, although much more efficient than combustion-powered motors, still have room for improvement.

The electric motor provides a case study in the impact of efficiency in one technical application on overall energy consumption. The kinds of improvements that are possible to make in the use of energy by electric motors can be extended to other applications. Electric motors also provide an excellent example of the impact of new developments in areas such as materials science on energy conservation.

Energy conservation in buildings and appliances and energy efficiency in many other end-uses may offer the promise of using only half the amount of energy utilized today to produce the same gross national product (GNP). In a sense, this number is quite believable since, for example, the Japanese use about half the energy per GNP that the U.S. does. However, we must be cautious in making the comparison with Japan because of differences in societal lifestyles and because the size, geography, and climate of Japan are different from the U.S.

More to the point, there are obvious off-the-shelf technologies, represented by efficient light bulbs and efficient freezers and refrigerators, which produce clear savings at a capital cost estimated to be five times less than the cost of installing the equivalent electric power supply to the grid. Similarly, increases in CAFE standards provide a proven and already successful way to reduce energy expenditures without reducing the quality of life.[3] In addition, improvements in industrial efficiency, beyond those that have occurred since the oil embargo of 1973, can certainly be realized. Many will be accomplished that have short pay-back times.

Conclusion

The U.S. can avoid the economic and social trauma of sudden unpredictable and uncontrollable shortages in future petroleum supplies as well as severe environmental damage. To do so, we must undertake the coordinated, long-term development of a variety of energy sources and make a concerted effort to increase the efficiency of end-use technologies. We feel that it is extremely important to begin these efforts immediately.

References and notes

1. "Acid Rain and Transported Air Pollutants: Implications for Public Policy" (Washington, DC: U.S. Congress, Office of Technology Assessment, OTA-O-204, June 1984).
2. Anthony Nero, *Physics Today*, April 1989.
3. If the situation demands, the quality of life can, of course, be reduced. However, unless there is a serious energy crisis, it is unlikely that energy conservation measures that impose discomfort or hardship would be sought or implemented.

Background Chapters

Overview of the energy problem: Changes since 1973

David Bodansky

1. Introduction

1a. Recognition of problems

The existence of an "energy problem" has been recognized for many decades. For example, Harrison Brown pointed out in 1954 that "in a length of time which is extremely short when compared with the span of human history . . . fossil fuels will have been discovered, utilized, and completely consumed."[1] This warning came at a time when the United States depended upon fossil fuels for about 95% of its energy.

Subsequent years brought widespread awareness of impending exhaustion of affordable fossil fuels, and by the early 1970s energy had become a matter of serious concern, albeit in limited circles. For example, the Ford Foundation at the end of 1971 commissioned a major energy study, which culminated in the 1974 report *A Time to Choose—America's Energy Future.*[2] In June 1973, President Nixon proposed a 5-year 10-billion dollar program of energy research and development.[3]

The public began to take a serious interest in energy only after the oil embargo imposed by the Organization of Petroleum Exporting Countries (OPEC) in Autumn 1973. Shortages of gasoline and the rapid rise in oil prices shocked the country into an awareness that we no longer were in control of our energy destiny. Not only were we heavily dependent on oil imports (for 35% of our petroleum products in 1973),[4] but OPEC determined the price and even the availability of these imports.

Two issues figured prominently in considerations of the energy problem: (a) our dependence on imported oil was dangerous and (b) sustaining an exponential growth in energy consumption would be difficult, and perhaps undesirable. (Energy consumption had doubled in the 20 years between 1953 and 1973.) Scientists and policy makers examined "hard" alternatives to oil, such as nuclear power or coal, and "soft" alternatives, such as conservation and solar power. Individual viewpoints varied on whether these alternative approaches were complementary or conflicting.

The "greenhouse effect," global heating caused by the buildup of CO_2 from the burning of fossil fuels, played little part in energy policy thinking of the early 1970s although the problem had been recognized at least since 1896.[5] During the 1970s and 1980s, the potential importance of the greenhouse effect became clearer. Professionals in atmospheric science and related fields gave increasing attention to the climatic consequences of carbon dioxide buildup and to means for avoiding or coping with it.

One of the predicted effects was drought in the U.S. midwest. When high temperatures and a drought occurred in 1988, some took this as a fulfillment of these predictions. Although natural climatic fluctuations are now generally believed to have been the main "cause" of the 1988 drought, the publicity attendant upon the drought transferred the pre-existing, longer term scientific concerns into the public domain. Now, in the minds of many analysts and some of the public, the greenhouse effect has become the most important of the energy-related problems and is considered more intractable than the problem of oil shortages. In fact, some would say that the greenhouse effect is *the* energy problem.

1b. Predictions and surprises

Almost all energy studies of the early and mid-1970s, both preceding and following the oil embargo, based projections on the assumption of a continued rise in energy demand. Demand might rise at a slightly lower rate than in the previous decade, but would still show substantial increases for the indefinite future. Table 1 displays a number of 1974 estimates of 1985 U.S. energy demand. The tabulated projections range from 88 to 123 quads. These all turned out to be gross overestimates. Actual energy consumption in 1985 was only 73.9 quads, slightly below the 1973 level of 74.3 quads, rather than well above it.

In the 1974 *Project Independence Report*,[6] the Nixon administration projected a 1985 energy demand of 94 to 110 quads, depending on the future price of oil, on measures taken to develop alternative supplies, and on the extent of conservation efforts. The 1974 Ford Foundation report *A Time to Choose*[2] considered three scenarios, termed the Historical Growth, the Technical Fix, and the Zero Energy Growth (ZEG) scenarios. The projections of the first two scenarios were not very different from those of other scenarios of the time (see Table 1), but the ZEG scenario marked a major break with prior thinking. It projected a demand of 88 quads for 1985. This was only moderately lower than some of the other 1985 demand projections, because it was accepted that even in "zero growth" demand would continue to increase temporarily. However in the ZEG scenario, growth stopped by the year 2000 at 100 quad per year, while in virtually all other standard scenarios substantial growth continued for the foreseeable future.

It may be that the main contribution of the Ford Foundation study was a conceptual one— imprinting the idea of *zero* growth on the public consciousness. While showing unusual foresight on the issue of energy demand, the study was not very prescient about the sources of energy supply. It substantially overestimated the production of domestic oil and gas; it underestimated coal use; and it projected a slight decrease in the fraction of primary energy used for

Table 1. 1974 projections for future U.S. energy consumption. [Consumption in base year (1973) = 74.3 quads.]

Source	Scenario	Energy in future years (quads)	
		1985	2000
Project Independence[a]	Oil at $7/barrel (1973 $)		
	Base case, WOC	109.1	
	Base case, WC	99.2	
	Accelerated supply, WOC	109.6	
	Accelerated supply, WC	99.7	
	Oil at $11/barrel (1973 $)		
	Base case, WOC	102.9	
	Base case, WC	94.2	
	Accelerated supply, WOC	104.2	
	Accelerated supply, WC	96.3	
Ford Foundation, EPP[b]	Historical growth	116.1	186.7
	Technical fix	91.3	124.0
	Zero energy growth	88.1	100.0
Nat. Academy Engineering[c]	Historical trends	123	
	With conservation	104	
Annual Energy Review	ACTUAL 1985	73.9	

a. Reference 6, p. 34. WC = with conservation; WOC = without conservation.
b. Reference 2, pp. 21, 46, 98.
c. "U.S. Energy Prospects: An Engineering Viewpoint," Nat. Acad. Engineering (Washington, DC, 1974), pp. 4, 130. (Converted from mbd of petroleum to quads at: 1 mbd = 2.12 quad.)
d. Reference 4. (See also Table 2, below.)

electricity rather than the major increase which took place (see below). Nor did it focus effectively on the greenhouse effect.

In short, all predictions were faulty to one degree or another and energy history since 1973 has been one of surprises: (a) the near-constancy in energy consumption until 1986; (b) the subsequent substantial rise in energy consumption from 1986 to 1989; (c) the continued rise in the use of electricity even during the period of static total energy demand; (d) an effective moratorium on the expansion of nuclear power; and (e) wide fluctuations in oil prices. These and other developments of the recent past will be discussed in more detail in the remainder of this chapter.

1c. The energy problem today

The following common assessments suggest the nature of our energy problems: (a) the availability of cheap oil is a transient phenomenon and future oil shortages are inevitable; (b) oil and gas imports contribute excessively to our balance of trade deficit; (c) even aside from CO_2, coal is undesirable because of the production of sulphur dioxide and other pollutants; (d) all use of fossil fuels produces CO_2, and with it a dangerous global warming; (e) alternatives to fossil fuels also have problems: nuclear power with safety, waste disposal, and weapons proliferation and solar power with cost and practicality; and (f) we have not succeeded in keeping energy consumption from rising, except when inhibited by high prices.

Different evaluations give different weight and credence to these assessments, and many of them are subject to debate. For example, it is variously argued that nuclear power can be made convincingly safe, that the burning of coal can be made much cleaner, that natural gas and unconventional sources of oil can forestall an oil crisis, that during the long time periods over which the greenhouse effect will take hold we will learn to adjust, and that conservation alone can solve most of our energy problems. The wide array of arguments and counter-arguments makes it difficult to find a consensus, even among "experts."

With the experts divided, with oil currently cheap and plentiful, and with seemingly more immediate problems such as AIDS, the destruction of the ozone layer, and drugs, it is perhaps natural that energy issues were on a back burner of the national agenda during the mid to late 1980s. It is the motivating belief of this study, and the thesis of this chapter, that continuing to neglect the energy problem is unwise and dangerous.

2. Historical record of energy consumption, 1950–1989

Before considering the future, it is pertinent to review the past and present. Table 2 presents data for energy consumption from 1950 to 1989. In contrast to the rapid rise up to 1973 (at an average annual rate of 4.4% for the preceding ten years), there was no overall increase in total energy consumption in the 13 years from 1973 to 1986. Despite the near-constancy in energy use, there was a 14% increase in population and a 36% increase in gross national product (measured in constant dollars). If the previous 4.4% annual increase had continued after 1973, energy consumption in 1986 would have been about 130 quads, instead of the actual 74 quads.

However, congratulations may be premature. Energy consumption was 8% higher in 1988 than in 1986, and slightly higher in 1989 than 1988.[7] It is not clear how much of this rise is due to increased economic activity and how much reflects lower prices. Between January 1986 and January 1988, crude oil prices to domestic refiners dropped 38% and coal prices to utilities dropped 8%.

In any case, true "zero energy growth," which in 1986 seemed perhaps to have been achieved, appears not yet in hand. 1987 and 1988 may have been anomalies. If so, we will return to a pattern where efficiency improvements compensate for the growth in population and GNP. However, the 1973 to 1986 period may have been the anomaly, caused by higher energy prices and the opportunity for easy reductions in a prodigal energy economy. Now we may be returning to a "natural" pattern of energy growth.

Comparative energy profiles for 1973 and 1989 are presented in Tables 3 and 4. The changes since 1973 have not been uniform across sectors. In particular, there was a 23% increase in residential and commercial energy use and a 19% increase in transportation, increases which are a little above the increase in population. On the other hand, industrial energy use dropped by 6%, despite the large increase in the GNP. As of 1989, energy was used in roughly equal amounts in the three major categories, a little more than a third each in the

Table 2. U.S. primary energy consumption, normalized to population and GNP:1950–1989.[a]

Year	Pop. (mil.)	GNP (billion 1982 $)	Total energy (quad)	Annual change (percent per year)		Norm. energy (1975 = 100)	
				Annual	10 yr av.	Per cap	Per $ GNP
1950	151.3	1204	33.06			66.8	105.0
1951	154.0	1328	35.47	7.2		70.4	102.0
1952	156.4	1380	35.30	-0.5		68.9	97.7
1953	159.0	1435	36.27	2.7		69.7	96.6
1954	161.9	1416	35.27	-2.8		66.5	95.1
1955	165.1	1495	38.82	10.1		71.8	99.2
1956	168.1	1526	40.38	4.0		73.4	101.1
1957	171.2	1551	40.48	0.2		72.2	99.7
1958	174.1	1539	40.35	-0.3		70.8	100.2
1959	177.1	1629	42.14	4.4		72.7	98.8
1960	179.3	1665	43.80	3.9	2.8	74.6	100.5
1961	183.0	1709	44.46	1.5	2.3	74.2	99.4
1962	185.8	1799	46.53	4.7	2.8	76.5	98.8
1963	188.5	1873	48.32	3.8	2.9	78.3	98.5
1964	191.1	1973	50.50	4.5	3.7	80.7	97.8
1965	193.5	2088	52.68	4.3	3.1	83.2	96.4
1966	195.6	2208	55.66	5.7	3.3	86.9	96.3
1967	197.5	2271	57.57	3.4	3.6	89.0	96.8
1968	199.4	2366	61.00	6.0	4.2	93.4	98.5
1969	201.4	2423	64.19	5.2	4.3	97.4	101.2
1970	203.2	2416	66.43	3.5	4.3	99.9	105.0
1971	206.8	2485	67.89	2.2	4.3	100.3	104.4
1972	209.3	2608	71.26	5.0	4.4	104.0	104.4
1973	211.4	2744	74.28	4.2	4.4	107.3	103.4
1974	213.3	2729	72.54	-2.3	3.7	103.9	101.5
1975	215.5	2695	70.55	-2.7	3.0	100.0	100.0
1976	217.6	2827	74.36	5.4	2.9	104.4	100.5
1977	219.8	2959	76.29	2.6	2.9	106.0	98.5
1978	222.1	3115	78.09	2.4	2.5	107.4	95.8
1979	224.6	3192	78.90	1.0	2.1	107.3	94.4
1980	226.5	3187	75.96	-3.7	1.3	102.4	91.0
1981	229.6	3249	73.99	-2.6	0.9	98.4	87.0
1982	232.0	3166	70.85	-4.2	-0.1	93.3	85.5
1983	234.3	3279	70.52	-0.5	-0.5	91.9	82.2
1984	236.5	3501	74.10	5.1	0.2	95.7	80.9
1985	238.7	3619	73.94	-0.2	0.5	94.6	78.0
1986	241.1	3718	74.24	0.4	0.0	94.1	76.3
1987	243.4	3854	76.85	3.5	0.1	96.4	76.2
1988	245.8	4024	80.20	4.4	0.3	99.7	76.1
1989	248.2[b]	4144	81.28	1.3	0.3	100.0	74.9

a. Data from Ref. 4 (pp. 23, 291) and Ref. 7 (pp. 15, 20); includes commercial energy only; excludes wood, waste, geothermal, wind, and other solar except if used by utilities for generation.
b. Estimated by extrapolation from 1988.

residential and commercial sector and in the industrial sector, and a little less than a third in the transportation sector.

In the residential and commercial sector, the decrease in energy use per capita may be attributed to the earlier impact of higher energy prices and the availability of technical options for improvements, such as weatherization, more

Table 3. U.S. energy consumption by sector, normalized to population and GNP: 1989 vs 1973.

	1973	1989	Change (Abs)	Change (%)
U.S. population (millions)[a]	211.4	248.2	36.8	17.4
GNP (1982 billion dollars)[b]	2744	4144	1400	51.0
Consumption by sector (quads)[b]				
Residential and commercial	24.14	29.61	5.47	22.7
Industrial	31.53	29.52	-2.01	-6.4
Transportation	18.61	22.15	3.54	19.1
Total	74.28	81.28	7.00	9.4
Consumption by sector (percent)				
Residential and commerical	32.5	36.4	3.9	
Industrial	42.5	36.3	-6.1	
Transportation	25.0	27.3	2.2	
Total	100.0	100.0		
Energy per capita (million BTU)				
Residential and commercial	114	119	5.1	4.5
Industrial	149	119	-30.2	-20.3
Transportation	88	89	1.2	1.4
Total	351	327	-23.9	-6.8
Energy/GNP (1000 BTU/$)				
Residential and commercial	8.8	7.1	-1.7	-18.8
Industrial	11.5	7.1	-4.4	-38.0
Transportation	6.8	5.3	-1.4	-21.2
Total	27.1	19.6	-7.5	-27.5

a. Data from Ref. 4, p. 23; 1989 population estimated by extrapolation from 1988.
b. Data from Ref. 7, pp. 20, 29.

efficient appliances, and more energy-efficient practices in building construction. Rosenfeld and Hafemeister, writing in the *Scientific American*, point out dramatic additional possibilities for conservation in this area.[8] Electrification has been a notable trend in the residential and commercial sector, as discussed in Section 3.

In the transportation sector, the fleet mileage for passenger cars has increased from 13.3 miles per gallon in 1973 to 20.0 miles per gallon in 1988,[7] and further improvements are to be expected as older cars are replaced. Lower fuel consumption per car more than compensated for the 38% growth in the number of registered passenger cars from 1973 to 1988.[4] However, the 84% increase in number of trucks (including small trucks and vans) outweighed the gains for passenger cars. In aggregate, motor fuel consumption rose 19%.

Transportation relies on petroleum for 97% of its energy (see Table 4). Its share of petroleum consumption rose from 51% in 1973 to 63% in 1989. Thus, moderation of petroleum use increasingly requires moderating its use in transportation. Present cars can be replaced with ones which are much more fuel-efficient. Another hope for the future lies in electrified mass transit and in electric vans and passenger cars, but little progress has been made in this direction since 1973. Local environmental policies put forth in California in 1989 could accelerate a move towards electrified transportation.

Table 4. Direct primary energy consumption, by sector and fuel: 1989 vs 1973.

Year and fuel	Direct energy use (in quads)[a]				Fraction from given fuel (in percent)			
	Res + comm	Indus[b]	Trans	Total direct	Res + comm	Indus	Trans	Total direct
1973								
Coal	0.25	4.05	0.00	4.31	1.1	12.8	0.0	5.8
Natural gas	7.63	10.39	0.74	18.76	31.6	32.9	4.0	25.3
Petroleum	4.39	9.10	17.83	31.33	18.2	28.9	95.8	42.2
Electricity	11.87	7.99	0.03	19.89	49.2	25.3	0.2	26.8
TOTAL	24.14	31.53	18.61	74.28	100.0	100.0	100.0	100.0
1989								
Coal	0.14	2.89	0.00	3.04	0.5	9.8	0.0	3.7
Natural gas	7.79	8.26	0.61	16.65	26.3	28.0	2.7	20.5
Petroleum	2.66	8.19	21.50	32.34	9.0	27.7	97.1	39.8
Electricity	19.02	10.18	0.05	29.25	64.2	34.5	0.2	36.0
TOTAL	29.61	29.52	22.15	81.28	100.0	100.0	100.0	100.0
Change (absolute)								
Coal	-0.11	-1.16	0.00	-1.27	-0.6	-3.0	0.0	-2.1
Natural gas	0.17	-2.13	-0.14	-2.10	-5.3	-5.0	-1.3	-4.8
Petroleum	-1.73	-0.92	3.67	1.02	-9.2	-1.1	1.2	-2.4
Electricity	7.15	2.19	0.02	9.36	15.1	9.2	0.1	9.2
TOTAL	5.47	-2.01	3.55	7.00	0.0	0.0	0.0	0.0
Change (percent)								
Coal	-44	-29		-30				
Natural gas	2	-21	-18	-11				
Petroleum	-39	-10	21	3				
Electricity	60	27	61	47				
TOTAL	23	-6	19	9				

a. Direct fossil fuel use does not include use for electricity generation; data from Ref. 7, pp. 31–35.
b. Small amounts of coal coke inputs and hydroelectric power (each 0.03 quads in 1989) included with coal and electricity, respectively.

The industrial sector, the one sector where total energy use has dropped since 1973, encompasses such disparate activities as manufacturing, mining, agriculture, and construction. Thus, without further disaggregation it is difficult to understand the main trends. Some of the complexity is illustrated and discussed in *Annual Review of Energy* articles by Williams, Larson, and Ross,[9] and by Kahane and Squitieri.[10] One difficulty is disentangling the relative importance of: (a) changes in the product mix and (b) improvements in the energy efficiency with which products are manufactured within a given industrial sector. Commenting on the 3.5% average annual decrease from 1973 to 1985 in the ratio of industrial energy consumption to GNP, Williams et al.[9] conclude " . . . that this decline was due almost equally to a shift of production from energy-intensive manufacture of materials to lighter industry (1.6% per year) and a reduction in sectoral energy intensity (1.9% per year)." In other words, the two factors were given roughly equal credit.

Overall, total energy consumption was essentially the same in 1986 as in 1973, but energy consumption rose substantially in 1987 and 1988 (see Table

2). This correlates with the trend in energy prices. Oil prices peaked in 1981, dropped gradually through 1985, and dropped sharply in 1986. Price appears to have been an important factor in fostering improved energy efficiency. Energy standards such as the mileage standards imposed on automobiles are an alternative to price in driving energy efficiency. If we are willing to impose state or federal energy efficiency standards, substantial further energy savings can be realized not only for cars but also for houses and appliances. In most cases, this would result in eventual savings for the consumer and would help decouple our energy economy from the vagaries of oil prices.

3. Electricity in the energy economy

Despite the near-constancy of total energy consumption, electricity generation has increased significantly, albeit at a slower rate than in earlier years. Table 5 shows the history of electricity consumption. In the ten years prior to 1973, the average rate of growth was 7.5% per year. Electric utilities expected that the historical rates of growth would continue after 1973. Instead, electricity sales increased only at an average rate of about 2.5% per year from 1973 to 1986. This mistaken prediction has partially caused the financial problems of the electric utilities during the past decade. Like energy consumption as a whole, electricity use increased more rapidly between 1986 and 1989, but this trend is too brief to foster a clear utility response.

The electricity growth, such as it was, has displaced the direct use of fossil fuels. Energy input to domestic utilities (including the 1% to 2% of electricity imported from Canada) increased by 47% from 1973 to 1989. Electricity's share of the national (primary) energy budget rose from 27% in 1973 to 36% in 1985, and has remained essentially constant at this fractional level through 1989. As can be seen in Table 6, increases have occurred in all sectors except transportation, where as yet electricity has made no appreciable inroads. At the same time, direct fossil fuel use decreased in the sectors other than transportation (see Table 4).

The rise in electricity use during a period when conservation has been so successful is surprising because from a conservation standpoint electricity is sometimes viewed as undesirable. In some critiques electricity from central power stations is considered to have negative social attributes.[11] More broadly, critics point out that electricity generation typically throws away about two-thirds of the available energy. Thus, in general, low energy-use scenarios have not presumed a major role for electricity.

A counter-argument, however, is that electricity is highly efficient at end-use. It can be delivered where one wants when one wants, and turned on and off quickly. During the 1970s, even without the widespread use of heat pumps, electrically heated houses required well under half the delivered energy for space heating as did comparable gas-heated houses.[12,13] Similarly, in some industrial applications, for example, in processing of steel and glass at high temperatures, electricity can add to the efficiency of the process.[14] It remains, therefore, a matter of debate whether the near-constancy in total energy use has occurred *despite* or *because of* the rise in the role of electricity.

Table 5. Energy input to utilities and sales by utilities, with normalization to population and GNP: 1950–1989.[a]

Year	GNP (billion 1982 $)	Energy total (quad)	Energy elec. (quad)	Fraction elec. (%)	Elec. sales TWh	Growth in elec. sales (% per year) Annual	Growth in elec. sales (% per year) 10 yr av.	Sales per cap MWh	Sales/GNP 1975 = 100
1950	1204	33.08	4.07	14.2	291			1.92	37.3
1951	1328	35.47	5.09	14.4	330	13.4		2.14	38.3
1952	1380	35.30	5.36	15.2	356	7.9		2.28	39.8
1953	1435	36.27	5.75	15.9	396	11.2		2.49	42.6
1954	1416	35.27	5.80	16.4	424	7.1		2.62	46.2
1955	1495	38.82	6.50	16.7	497	17.2		3.01	51.3
1956	1526	40.38	6.98	17.3	546	9.9		3.25	55.2
1957	1551	40.48	7.26	17.9	576	5.5		3.36	57.3
1958	1539	40.35	7.22	17.9	588	2.1		3.38	58.9
1959	1629	42.14	7.82	18.6	647	10.0		3.65	61.3
1960	1665	43.80	8.19	18.7	688	6.3	9.0	3.84	63.7
1961	1709	44.46	8.47	19.1	722	4.9	8.1	3.95	65.2
1962	1799	46.53	9.03	19.4	778	7.8	8.1	4.19	66.7
1963	1873	48.32	9.63	19.9	833	7.1	7.7	4.42	68.6
1964	1973	50.50	10.33	20.5	896	7.6	7.8	4.69	70.1
1965	2088	52.68	11.01	20.9	954	6.5	6.7	4.93	70.5
1966	2208	55.66	11.99	21.5	1035	8.5	6.6	5.29	72.3
1967	2271	57.57	12.70	22.1	1099	6.2	6.7	5.56	74.7
1968	2366	61.00	13.88	22.8	1203	9.5	7.4	6.03	78.4
1969	2423	64.19	15.18	23.6	1314	9.2	7.3	6.52	83.7
1970	2416	66.43	16.27	24.5	1392	5.9	7.3	6.85	88.9
1971	2485	67.89	17.15	25.3	1470	5.6	7.4	7.11	91.3
1972	2608	71.26	18.52	26.0	1595	8.5	7.4	7.62	94.3
1973	2744	74.28	19.85	26.7	1713	7.4	7.5	8.10	96.3
1974	2729	72.54	20.02	27.6	1706	-0.4	6.7	8.00	96.4
1975	2695	70.55	20.35	28.8	1747	2.4	6.2	8.11	100.0
1976	2827	74.36	21.57	29.0	1855	6.2	6.0	8.52	101.2
1977	2959	76.29	22.71	29.8	1948	5.0	5.9	8.86	101.6
1978	3115	78.09	23.72	30.4	2018	3.6	5.3	9.09	99.9
1979	3192	78.90	24.13	30.6	2071	2.6	4.7	9.22	100.1
1980	3187	75.96	24.50	32.3	2094	1.1	4.2	9.25	101.4
1981	3249	73.99	24.76	33.5	2147	2.5	3.9	9.35	101.9
1982	3166	70.85	24.27	34.3	2086	-2.8	2.7	8.99	101.6
1983	3279	70.52	24.96	35.4	2151	3.1	2.3	9.18	101.2
1984	3501	74.10	25.98	35.1	2278	5.9	2.9	9.63	100.4
1985	3619	73.94	26.48	35.8	2310	1.4	2.8	9.68	98.5
1986	3718	74.24	26.64	35.9	2351	1.8	2.4	9.75	97.5
1987	3854	76.85	27.55	35.8	2455	4.4	2.3	10.09	98.3
1988	4024	80.20	28.63	35.7	2568	4.6	2.4	10.45	98.4
1989	4144	81.28	29.22	35.9	2634	2.6	2.4	10.61	98.0

a. Data from Ref. 4 (pp. 23, 203, 207, 291) and Ref. 7 (pp. 20, 37, 87).

Quite apart from these general considerations, price affects electricity use. From 1973 to 1983, when electricity's share of the energy budget grew rapidly (see Table 5), average retail electricity prices rose by a factor of 3.2 (in current dollars) while natural gas prices rose by a factor of 6.6. From 1983 to 1989,

Table 6. U.S. electricity sales by sector and electricity generation by fuel: 1989 vs 1973.

	1973	1989	Change (Abs)	Change (%)
Primary electrical energy (quad)[a]	19.85	29.22	9.37	47
Net sales (TWh)[a]				
Residential	579	904	325	56
Commercial	388	724	336	86
Industrial	686	915	229	33
Other[b]	59	91	31	53
TOTAL	1713	2634	921	54
Fraction of sales (%)				
Residential	33.8	34.3	0.5	
Commercial	22.7	27.5	4.8	
Industrial	40.1	34.7	-5.3	
Other	3.5	3.4	0.0	
TOTAL	100.0	100.0		
Net generation (TWh)[a]				
Coal	848	1551	704	83
Petroleum	314	158	-156	-50
Natural gas	341	264	-76	-22
Nuclear	83	529	446	534
Hyrdoelectric	272	264	-8	-3
Other[c]	2	11	9	397
TOTAL	1861	2779	918	49
Fraction from sources (%)				
Coal	45.6	55.8	10.3	
Petroleum	16.9	5.7	-11.2	
Natural gas	18.3	9.5	-8.8	
Nuclear	4.5	19.1	14.6	
Hydroelectric	14.6	9.5	-5.1	
Other	0.1	0.4	0.3	
TOTAL	100.0	100.0		

a. Data from Ref. 7, pp. 37, 87, 88.
b. Includes sales to Government, railways, and street lighting authorities.
c. Electricity produced from geothermal and (nonhydro) solar sources connected to electric utility distribution systems.

electricity's share increased very little. In this period, electricity prices rose slightly and natural gas prices dropped.

The ratio of electricity sales to the gross national product (in constant dollars) has remained remarkably constant since 1975 (see Table 5). A similar relation between GNP and total energy use had existed from 1950 to the late 1970s (see Table 2). That coupling was broken after 1976. The present coupling between GNP and electricity use may have some deeper significance, or it may represent a transition between a period, before 1975, when electricity sales were rising more rapidly than the GNP and a future period when they will rise more slowly.

Changes in the institutional arrangements for electricity generation and distribution complicate the picture. The Public Utilities Regulatory Policy Act of

1978 (PURPA) has encouraged the development by non-utility organizations of electricity generation facilities which use cogeneration or renewable resources. In 1987, only 2% of the electricity supplied to the utility grid came from non-utility producers. However, this fraction is expected to rise in the 1990s, especially with the 1987 repeal of restrictions on the use of natural gas for electricity generation. This makes gas-fired plants an attractive option for the 1990s, both for utilities and for non-utility cogenerators. Such plants have high efficiency, short construction times, relatively low production of carbon dioxide, and virtually no production of sulphur dioxide. Unfortunately, they continue to contribute to CO_2 buildup and face prospective substantial increases in the cost of natural gas as demand increases.

4. Energy sources

4a. Historical trends

Until the 1880s, the main fuel in the United States was wood. With industrialization and urbanization, coal then became dominant and remained so until the late 1940s. By 1950, the increased importance of road travel and the convenience of oil in home heating promoted the use of petroleum which provided about 40% of our energy in 1950. Coal and natural gas, in rapidly decreasing order, supplied the remainder of our energy except for a small amount from hydroelectric power. Details of consumption of energy from various sources since 1950 are shown in Table 7. (This table, along with most similar tabulations, omits wood which supplied 2.6 quads in 1984;[15] see Section 4g.)

The dominance of oil has continued, rising to about 49% in the late 1970s and dropping to about 42% in 1989. Coal and natural gas now provide roughly equal amounts, each a little under one-quarter of our energy, and nuclear power and hydroelectric power together contribute the remaining 10%. Other sources, excluding wood, totalled well under 1%. For each energy source, Table 8 compares the consumption in the various sectors in 1973 and 1989.

4b. Oil consumption

The experience of sharply higher prices was in large measure responsible for a 2% drop in oil consumption from 1973 to 1989. In successive oil price shocks, the average refiner acquisition (constant dollar) cost of crude oil doubled from 1973 to 1974 and more than doubled again from 1978 to 1981.[16] From 1981 to 1988, there was an equally dramatic collapse in crude oil prices, which dropped at one point in 1986 to less than one-third of 1981 peak levels.[17] Overall, from 1973 to 1988 the average price per barrel of crude oil rose from $8.38 to $11.80 (in 1982 constant dollars). (In current dollars, the price rose from $4.15 to $14.71.) Oil consumption has followed the price trends, with some delay, dropping in every year from 1978 to 1983 and rising slightly between 1983 and 1989, but the changes in total consumption have been less extreme than those in price.

The lack of a closer correspondence between total oil use and price stems in part from difficulties in implementing conservation and fuel substitution. As

Table 7. U.S. primary energy consumption, in quads and percent by source (coal, gas, petroleum, hydro-electric, and nuclear): 1950–1989.[a]

Year	Total[b] quad	Coal quad	Gas quad	Petr quad	Hydr quad	Nucl quad	Coal %	Gas %	Petr %	Hydro %	Nucl %
1950	33.08	12.35	5.97	13.32	1.44	0.00	37	18	40	4	0
1951	35.47	12.55	7.05	14.43	1.45	0.00	35	20	41	4	0
1952	35.30	11.31	7.55	14.96	1.50	0.00	32	21	42	4	0
1953	36.27	11.37	7.91	15.56	1.44	0.00	31	22	43	4	0
1954	35.27	9.71	8.33	15.84	1.39	0.00	28	24	45	4	0
1955	38.82	11.17	9.00	17.25	1.41	0.00	29	23	44	4	0
1956	40.38	11.35	9.61	17.94	1.49	0.00	28	24	44	4	0
1957	40.48	10.82	10.19	17.93	1.56	0.00	27	25	44	4	0
1958	40.35	9.53	10.66	18.53	1.63	0.00	24	26	46	4	0
1959	42.14	9.52	11.72	19.32	1.59	0.00	23	28	46	4	0
1960	43.80	9.84	12.39	19.92	1.66	0.01	22	28	45	4	0
1961	44.46	9.62	12.93	20.22	1.68	0.02	22	29	45	4	0
1962	46.53	9.91	13.73	21.05	1.82	0.03	21	30	45	4	0
1963	48.32	10.41	14.40	21.70	1.77	0.04	22	30	45	4	0
1964	50.50	10.96	15.29	22.30	1.91	0.04	22	30	44	4	0
1965	52.68	11.58	15.77	23.25	2.06	0.04	22	30	44	4	0
1966	55.66	12.14	17.00	24.40	2.07	0.06	22	31	44	4	0
1967	57.57	11.91	17.94	25.28	2.34	0.09	21	31	44	4	0
1968	61.00	12.33	19.21	26.98	2.34	0.14	20	31	44	4	0
1969	64.19	12.38	20.68	28.34	2.66	0.15	19	32	44	4	0
1970	66.43	12.26	21.79	29.52	2.65	0.24	18	33	44	4	0
1971	67.89	11.60	22.47	30.56	2.86	0.41	17	33	45	4	1
1972	71.26	12.08	22.70	32.95	2.94	0.58	17	32	46	4	1
1973	74.28	12.97	22.51	34.84	3.01	0.91	17	30	47	4	1
1974	72.54	12.66	21.73	33.46	3.31	1.27	17	30	46	5	2
1975	70.55	12.66	19.95	32.73	3.22	1.90	18	28	46	5	3
1976	74.36	13.58	20.35	35.18	3.07	2.11	18	27	47	4	3
1977	76.29	13.92	19.93	37.12	2.51	2.70	18	26	49	3	4
1978	78.09	13.76	20.00	37.96	3.14	3.02	18	26	49	4	4
1979	78.90	15.04	20.67	37.12	3.14	2.78	19	26	47	4	4
1980	75.96	15.42	20.39	34.20	3.12	2.74	20	27	45	4	4
1981	73.99	15.91	19.93	31.93	3.10	3.01	22	27	43	4	4
1982	70.85	15.32	18.50	30.23	3.57	3.13	22	26	43	5	4
1983	70.52	15.89	17.36	30.05	3.90	3.20	23	25	43	6	5
1984	74.10	17.07	18.51	31.05	3.76	3.55	23	25	42	5	5
1985	73.94	17.48	17.84	30.92	3.36	4.15	24	24	42	5	6
1986	74.24	17.26	16.71	32.20	3.39	4.47	23	23	43	5	6
1987	76.85	18.01	17.75	32.87	3.07	4.91	23	23	43	4	6
1988	80.20	18.85	18.55	34.23	2.64	5.66	23	23	43	3	7
1989	81.28	18.95	19.50	34.03	2.86	5.69	23	24	42	4	7

a. Data from Ref. 4 (p. 11) and Ref. 7 (p. 15); includes commercial energy only.
b. Total includes "other" category (see Table 8); excludes most wood and waste (2.9 quad in 1984).

seen in Table 8, oil use has been cut drastically by electric utilities and in the residential and commercial sector, where substitution is relatively easy. It has dropped only slightly in the industrial sector and has increased in transportation. Even given the will, improving the average automotive mileage can at best be only slowly achieved, requiring at least the time for existing cars to be

Table 8. U.S. primary energy consumption, by source and sector: 1989 vs 1973.

| | Source[a] | | | | | | | |
	Coal	Gas	Petr	Elec	Hydro[b]	Nucl	Other[c]	Total
1973 (quads)								
Res + comm	0.25	7.63	4.39	11.87				24.14
Industrial	4.06	10.39	9.10	7.95	0.04		-0.01	31.53
Transportation	0.00	0.74	17.83	0.03				18.61
Electric utilities	8.66	3.75	3.52		2.98	0.91	0.05	19.85*
TOTAL	12.97	22.51	34.84	19.85*	3.01	0.91	0.04	74.28
1989 (quads)								
Res + comm	0.14	7.79	2.66	19.02				29.61
Industrial	2.86	8.26	8.19	10.15	0.03		0.03	29.52
Transportation	0.00	0.61	21.50	0.05				22.15
Electric utilities	15.95	2.85	1.68		2.83	5.69	0.22	29.22*
TOTAL	18.96	19.50	34.03	29.22*	2.86	5.69	0.25	81.28
Abs increase (quads)								
Res + comm	0.11	0.17	-1.73	7.15				5.47
Industrial	-1.19	-2.13	-0.92	2.20	0.00	0.00	0.04	-2.01
Transportation	0.00	-0.14	3.67	0.02				3.54
Electric utilities	7.30	-0.90	-1.83		-0.14	4.78	0.17	9.37*
TOTAL	5.99	-3.01	-0.82	9.37*	-0.15	4.78	0.21	7.00
Fract increase (%)								
Res + comm	-44	2	-39	60				23
Industrial	-29	-21	-10	28	-9			-6
Transportation		-18	21	61				19
Electric utilities	84	-24	-52		-5	525	376	47
TOTAL	46	-13	-2	47	-5	525	538	9
% to sector (1989)								
Res + comm	1	40	8	65				36
Industrial	15	42	24	35	1		12	36
Transportation	0	3	63	0				27
Electric utilities	84	15	5		99	100	88	36*
TOTAL	100	100	100	100	100	100	100	100

*Not included in totals.
a. Data from Ref. 7, pp. 15, 31–37.
b. Hydroelectric includes imports.
c. "Other" includes net coke imports and electricity from geothermal and (nonhydro) solar sources distributed by utilities.

replaced. Fuel substitution in transportation (e.g., electrified transit) would take even longer, and at present such substitution would not provide vehicles which are either convenient or economical for the typical user.

The price of oil depends on world demand for oil, the available sources of oil, and the existence of alternatives. World crude oil production, which was at a rate of 55.6 million barrels per day (mbd) in 1973, rose to 62.5 mbd in 1979, but by 1985 had dropped below 54 mbd and in 1988 still averaged only 58.5 mbd.[7] Worldwide use of other fuels, especially coal, natural gas, and nuclear power (see Table 9) rose about 66% between 1973 and 1988[18] while crude oil production rose only 0.7% in this period.[7] Different suppliers of oil have gained

Table 9. World commercial primary energy production: 1988 vs 1973.[*][a]

	1973 (quad)	1988 (quad)	Diff (quad)	Ratio 88/73	1973 (%)	1988 (%)
Crude oil	119.7	125.4	5.7	1.05	48.9	37.6
Natural gas liquids	4.2	7.2	2.9	1.69	1.7	2.2
Dry natural gas	43.2	67.1	23.9	1.55	17.7	20.2
Coal	61.8	93.0	31.2	1.50	25.3	27.9
Hydroelectric power	13.5	21.3	7.8	1.58	5.5	6.4
Nuclear power	2.2	19.1	16.9	8.77	0.9	5.7
TOTAL	244.7	333.1	88.4	1.36	100.0	100.0

*No adjustment made for 0.3% difference in number of days.
a. Data from Ref. 18; 1973 data adjusted for small differences in the two Ref. 18 series.

in importance. Mexico and the United Kingdom together added about 4.3 mbd between 1973 and 1988 and China and the USSR about 5 mbd.[7] Simultaneously OPEC's output fell by over 10 mbd between 1973 and 1988 (a decrease of 33%). Not surprisingly OPEC, at least temporarily, lost its ability to control oil prices.

Troublesome signs have developed, however. World crude oil production, which had hardly changed over the previous 14 years, rose 5.7% from 1987 to 1989 and OPEC accounted for the increase. Price rises in 1989 followed.

4c. Natural gas

Despite pressures to reduce oil consumption, there has been a drop in natural gas use in the United States, although there has been a worldwide rise in gas consumption. Natural gas prices have risen more than coal and oil prices since 1973,[7] and this has been accompanied by a drop in natural gas use in all sectors. It appears unlikely that it will be possible to significantly increase domestic gas production without further increases in price.[19] This is unfortunate, if true, because natural gas releases less carbon dioxide and other polluting gases per unit energy than do coal and oil (see Section 6, below) and has been viewed as a partial short-term solution to some of the environmental problems associated with coal.

4d. Coal

Coal has been the big winner as far as U.S. energy consumption is concerned. Its use increased by 46% from 1973 to 1989 due to the increased use of electricity and the dominance of coal in electricity generation (56% of all generation in 1989). In 1989, 84% of coal use was for electricity generation, up from 67% in 1973. The rise in total coal use occurred despite a large drop in its use in the industrial sector. Coal plays little role in residential and commercial use and virtually none in transportation. Some possible restrictions face coal as we nominally intensify our efforts to decrease the output of sulphur dioxide and other pollutants, but as yet this has not seriously inhibited its increased use.

4e. Nuclear power

U.S. nuclear power generation has increased dramatically since 1973. In 1989, nuclear power produced 19% of our electricity, second only to coal. However, the expansion in nuclear power use cannot continue for much longer, because very few additional reactors are in the procurement and production pipeline. In late 1989 there were nominally 110 operational reactors with a net capacity of 98 gigawatts (GW),[7] but some of these had been shut down for extended periods and the actual number of available reactors was closer to 100. Another 11 reactors, representing roughly an additional 13 GW of capacity, were in one stage or another of construction and licensing procedures. However, the status of a number of these reactors is highly questionable (for example, the WPPSS reactors in Washington). The total actual operating capacity in the early 1990s will probably be in the neighborhood of 100 GW and is unlikely to be as high as 110 GW. At this level, about 20% of our electricity production will be nuclear in the early 1990s.

Beyond that, we are assured of a substantial hiatus. This is in marked contrast to earlier predictions. Even relatively conservative 1974 projections would have placed our 1988 nuclear capacity above 200 GW. However, most of the 100 plants ordered in the 1971 to 1973 period have since been cancelled, as have all plants ordered after 1973.[20]

The worldwide nuclear power situation is very uneven. A comparison of nuclear generation in 1988 and 1977 is given in Table 10, along with the fraction of electricity derived from nuclear power.[21,22] Some countries, particularly France, have implemented very vigorous nuclear programs, and a substantial commitment to nuclear power continues as stated policy in Japan and the USSR. In most other countries the programs are now rather static, while in one, Italy, nuclear power has been at least temporarily abandoned.

Further nuclear expansion is controversial. Proponents of expansion point to the very good safety record of reactors in the United States, Western Europe, and Japan as well as to the cleanliness of nuclear power. There are no emissions of sulphur dioxide or carbon dioxide, and there is negligible emission of radioactive effluents in the absence of accidents. Opponents point to the potentially very severe consequences of an accident, such as at Chernobyl, and to the never zero possibility of an accident. Further, no repositories are yet in operation for the permanent disposal of the high-level radioactive wastes, in the United States or elsewhere.

There are also disagreements as to the economic practicality of nuclear power, with examples cited in support of either position. In well managed programs with predictable and short construction schedules, as for the standardized reactors being built in France and for some U.S. reactors, electricity from nuclear power is less expensive than electricity from its chief competitor, coal. However, other programs have been much less successful, for reasons including overestimates of future demand, poor management, and requirements for retrofits of existing plants or of plants under construction. In such cases, nuclear power can be prohibitively expensive.

Table 10. World nuclear power: capacity (net gigawatts), generation (net gigawatt-years), and fraction of all electricity.

| Country | Status (6/30/89)[a] | | Annual generation[b] | | Percent nuclear[c] 1988 |
	Number of units	Capacity (GWe)	(GWyr) 1976	(GWyr) 1988	
United States	110	97.6	21.8	60.1	19.5
France	55	52.6	1.7	29.7	70
USSR	56	33.8	2.7	23.3	13
Japan	38	28.3	4.0	18.7	23
Germany, West	24	22.7	2.7	15.7	34
Canada	18	12.2	2.0	8.9	16
Sweden	12	9.7	1.7	7.5	47
United Kingdom	40	12.4	4.0	6.3	19
Spain	10	7.5	0.8	5.5	36
Belgium	7	5.5	1.1	4.6	66
Korea, South	9	7.2	0.0	4.3	47
Taiwan	6	4.9	0.0	3.4	41
Switzerland	5	3.0	0.9	2.5	37
Czechoslovakia	8	3.3	0.0	2.4	27
Finland	4	2.3	0.0	2.1	36
Hungary	4	1.6	0.0	1.4	49
Bulgaria	5	2.6	0.5	1.3	36
South Africa	2	1.8	0.0	1.2	7
Germany, East	5	1.7	0.6	1.2	10
India	6	1.2	0.3	0.7	3
Argentina	2	0.9	0.3	0.6	11
Yugoslavia	1	0.6	0.0	0.4	5
Netherlands	2	0.5	0.4	0.4	5
Brazil	1	0.6	0.0	0.03	0.3
Pakistan	1	0.1	0.1	0.02	0.6
Mexico	1	0.7	0.0	0.00	0
Italy	2	1.1	0.4	0.00	0
World Total	434	316.4	46.0	202.4	17

a. Data from Ref. 21, p. 63.
b. Data from Ref. 18: IEA 1983, p. 26; IEA 1988, p. 24.
c. Data from Ref. 22, p. 1.

4f. Hydroelectric power

U.S. hydroelectric power use has fluctuated significantly since 1973 (see Table 7), largely because of varying precipitation patterns. Overall, power from hydroelectric sources was less in 1989 than in 1973 and its share of all electricity dropped from 15% to 9.5%. The more general trend, apart from changes in precipitation, is for constant hydroelectric generation in the U.S., because it is difficult to find acceptable new sites. The situation is somewhat better in Canada, where large hydroelectric projects are underway. For example, a major project planned for James Bay in Quebec is scheduled for completion in the mid-1990s, with electricity to be sold to New York at a cost of 12 cents per kWh.[23] This is well above the average cost of power from existing coal or nuclear plants.

4g. Biomass

The use of other energy sources is on a much smaller scale than that of the sources cited above. In terms of present energy use, the most important of these is biomass. For 1984, the *Energy Information Administration* reports U.S. use of 2.9 quad of biomass energy,[15] an amount not included in its main energy tabulations (sometimes being termed "non-commercial" energy). Of this, about 1.7 quad was wood consumed in the industrial sector, primarily accounted for by the use of wood by-products in wood-associated industries. Another 0.9 quad was consumed in home heating. The use of wastes was quite small, amounting to only 0.2 quad, and alcohol-type fuels (such as ethanol) accounted for only 0.04 quad.

4h. Solar heating

Direct use of solar power has been limited. Presumably there is a large and growing contribution of passive solar space heating as building designers make more effective use of this potential. However, it is hard to disentangle increases in passive solar input from decreases in heat loss, and there is no straightforward way to quantify passive solar heating.

Sales of solar collectors have been low. Only about 4 million ft^2 of solar collectors were shipped in 1987.[24] Of this, 76% was for swimming pools. Two states, Florida and California, accounted for 72% of the total. If one were to assume, in what would amount to a gross overestimate, that the utilization of the incident solar power was at an average rate of 250 W/m^2, or 0.7×10^6 BTU/ft^2 per year, then the 1986 solar collector shipments would add a potential of about 0.003 quad/yr. To make the picture even less promising, the trend in deployment of solar collectors is downward, from 20 million ft^2 in 1980[25] to one-fifth of this in 1987, perhaps because of a loss of tax incentives.

4i. Solar generation of electricity

In terms of actual sales alone, the picture is not much different for solar photovoltaic devices. U.S. production (excluding space satellite applications) in 1988 amounted to about 12 peak MW, less than 0.002% of our electric capacity.[26] However, photovoltaic technology is rapidly changing. The 1979 report of the American Physical Society Study Group on *Solar Photovoltaic Energy Conversion*[27] indicated a then current price for flat cell modules of $5 to $10 per peak watt, with a reduction to $0.10 to $0.40 necessary to be competitive with coal in the year 2000 (all in 1975 dollars). This implied a required cost reduction of something like a factor of 30. Since then, costs have been very substantially reduced,[28] but solar photovoltaic power remains too expensive to be competitive with other sources. As of early 1989, a summary for the Solar Energy Research Institute (SERI)[26] reported that "electricity produced by photovoltaics costs 25 to 35 cents per kilowatt-hour, depending upon the size of the system." Presumably these costs are for favorable regions, such as the U.S. southwest.

Especially in view of the progress over the past 15 years, SERI and others are optimistic about the prospects of achieving the required further reductions in cost. The goal of the DOE's National Photovoltaic Program has been " . . .

to gain the technology improvements needed to produce electricity at 12 cents per kilowatt-hour (in 1986 dollars) by the early 1990s."[29] Power at this price would be useful in specialized circumstances, where less expensive power from the grid was not available. A longer term goal is to reach 6 cents per kilowatt-hour, at which point solar photovoltaic power would be highly competitive.

There are also active programs of electricity generation through direct solar thermal technology, in which water or an alternative heat transfer fluid is heated to high temperatures and drives a heat engine. The attractiveness of these may depend strongly on both locale and on the price of competing fossil fuels. Again, cost is a key issue and the possibilities of major further cost reductions may be less than for the solar photovoltaic technologies.

4j. Wind power

There has been some progress in generation of electric power from wind turbines, but here there is less prospect for radical technological improvements. Most of the world's output of wind-generated electric power (presumably the contributions of independent individual turbines strictly for home use are not included in this accounting), comes from large wind turbine farms in California.[30,31] With a rated capacity of 1.44 GW, they produced 1700 GWh of electricity in 1987, for a capacity factor of 14%. Although still under 0.1% of total U.S. electricity generation, wind output has been increasing and exceeded 2100 GWh in 1989.[31]

4k. Geothermal power

The largest source of unconventional electric power in the United States has been geothermal power. In 1988, it represented a little over 0.2% of U.S. utility generating capacity (1.7 GWe) and provided 0.4% of electricity generation (about 10000 GWh).[4] Virtually all comes from the The Geysers development, in California. The Geysers utilizes relatively clean natural steam, an inexpensive but not widely available source of geothermal power. There are no announced plans in the U.S. for further geothermal projects of comparable size.

4l. Summary

Overall, our energy sources underwent no large basic changes between 1973 and 1989. The contribution from fossil fuels has dropped from 95% to about 90%, largely due to an increased use of nuclear power, but fossil fuels still dominate. Among the fossil fuels, the greatest changes have been a rise in the use of coal and a drop in the use of natural gas. The hopes for "alternative" sources of energy, advanced with varying degrees of enthusiasm in the 1970s, have not as yet been fulfilled. Leaving out wood, together they currently contribute well under 1% of our total energy budget. Assessments of future prospects vary widely, but it would appear that the chances for a major expansion are best for solar photovoltaic power, it being the area where scientific and technological progress are likely to make the largest differences. An expansion in the use of natural gas is also possible, especially because it can be readily used for electricity generation and produces less carbon dioxide per unit energy than coal or oil. Here, however, the prospects are for rising prices and possibly limited supply.

5. Fossil Fuel Resources

The fact that the largest share of our energy still comes from fossil fuels is at the root of our two key energy problems. In this section we discuss the problems of oil and natural gas shortages, which may impact us within the next few decades. In Section 6, the longer-term problem of carbon dioxide buildup will be considered.

5a. Resources and reserves

A complex terminology has evolved to describe the available "resource" of a given fuel — past, present, or future. The complexity arises from the need to make a variety of distinctions:

(1) A temporal distinction captured by the self-explanatory terms *original resource* and *remaining resource*. Their magnitudes differ by the cumulative production to date.

(2) An economic distinction based on the feasibility of extracting the resource at some presumed level of acceptable cost and technological difficulty. Here distinctions are made between "economic," "marginally economic," and "sub-economic" resources.

(3) Differences in degree of geological certainty as to the existence of the deposit. The remaining "identified" resources, i.e., those which have been demonstrated to exist and are economic under current conditions, are termed *reserves*. The remaining resources suggested by broad geological evidence but whose location has not been specifically identified are termed *undiscovered resources*.

No matter how carefully these terms are defined, uncertainties in geology, extraction costs, and competitive price levels blur many of the distinctions which are made. Usage varies as to the degree to which the terms "reserves" and "resources" incorporate considerations of economic or technical feasibility.

In the discussion below we will adopt a relatively standard usage,[32] in which the term *resources* is used to denote *original economic resources*, namely the total available amount, past and future, which can be recovered "under a specified set of costs and technologies" (usually those of the present and near future). Remaining resources are also restricted to economic resources. The term *resource base* will be used to denote the entire original resource, independent of feasibility of extraction. Reserves will not be further considered. They are of interest chiefly for making short-term economic plans because they describe what is relatively quickly accessible. For longer-term energy policy considerations, it is resources which are of primary interest. There are many ambiguities in specifying the magnitude of resources. There are options of going to less accessible regions (deeper in the ground or more inhospitable in locale), of drawing upon poorer grade deposits, and of extracting a higher fraction of the desired material from the deposits. The technologies required and the costs to be incurred are not clearly predictable. Resources similar in type and mode of extraction to those now being used are termed *conventional resources*. The term *unconventional resources* refers to forms, such as tar sands oil or gas in very deep locations, that require extraction techniques which differ substantially from current practice. Except where otherwise indicated, the discussion below will refer to conventional resources.

5b. The finiteness of the resources

It is generally believed that fossil fuels are a remnant of decaying organic matter. As such they have been deposited during the past 500 million years on the top of the then existing earth or seabed, and will not be found at great depths. An alternative view, advanced particularly by Thomas Gold,[33] is that the oil and gas on earth derive primarily from the processing of primordial hydrocarbon gases at high temperatures and pressures deep within the earth. In this view, oil and gas deposits are not restricted to the near surface of the earth, and the potential supplies, especially of gas, are much greater than inferred from the explorations at moderate depths which have taken place to date.

This, however, is a maverick opinion. Assuming the more standard view to be correct, the possible magnitude of the resource is limited. One of the first to emphasize the finiteness of fossil fuel resources was M. King Hubbert, a petroleum geologist.[34] His analysis utilizes a simple model, in which the production rate is plotted as a function of time. The area under the resulting curve represents the magnitude of all of the resource which is ultimately available. When use first begins, the rate of consumption is low. When it is almost depleted, the price will be high and the consumption rate will again be low. In between, the rate rises and falls. The resulting curve of production as a function of time is assumed to be bell-shaped. With a symmetric curve, knowledge of the early growth of the curve defines the remainder of its shape. Hubbert's triumph lay in his prediction, well in advance, that oil production in the continental U.S. would peak in 1971. In fact, U.S. crude oil production peaked in 1970.[4]

While Hubbert's picture is suggestive, it does not account for what happens in a world perturbed by deliberate policy actions, including price manipulation, which can accelerate or retard the use of a resource. Hubbert predicted an initial steady rise in world oil production continuing into the 1990s. In fact, world production has been nearly constant since 1973. Nonetheless, although such departures from the details of Hubbert's description may modify the time scale for depletion of oil and gas resources, they do not fundamentally change the long-term prospects.

5c. Conventional crude oil

The main petroleum resource has been crude oil, often including natural gas liquids. Most estimates made since 1965 have put worldwide crude oil resources in the neighborhood of 2000 billion barrels of oil (bbo), although there exist significantly higher and lower estimates. Comprehensive studies by Charles Masters of the U.S. Geological Survey and collaborators[35] and Joseph Riva of the Congressional Research Service[36] suggest a most likely value (mode) of about 1790 bbo for world crude oil resources, with a 5% chance that the resources are less than about 1600 bbo and a 5% chance that they exceed about 2300 bbo. A rounded estimate of about 2000 bbo falls within this range, and can serve as a working guide.

The U.S. share of this resource is relatively small, about 230 bbo. By far the largest share, estimated to be about 700 bbo, is in the Middle East, primarily in Saudi Arabia, Iraq, Kuwait, and Iran. Major resources also exist in the Soviet Union, Africa, and Latin America.

According to the estimates discussed above, the U.S. has already used up more than half of its resources. Average U.S. production in 1989 was 7.6 mbd or a little under 3 bbo per year.[7] If we take remaining resources to be about 90 bbo, this production could be sustained for roughly 30 years. Obviously, consumption would not continue at a flat rate until the oil suddenly runs out. Instead, it is expected that domestic production will gradually taper off and U.S. oil imports rise, with short-term fluctuations determined by the price at which imported oil is available. In fact, oil imports have been rising since 1985. Net crude oil and petroleum products imports were 66% higher in 1989 than in 1985.[7] Looking past the effects of short-term fluctuations in price, the DOE estimates (in its "base case" or central estimate) that net petroleum imports will increase by the year 2000 to 72% above the 1987 level and that we would then depend on imports for 60% of our oil.[37]

Remaining world oil resources are in the neighborhood of 1400 bbo. At the 1988 world rate of 21 bbo per year, this amounts to roughly a 70-year constant rate supply. Of this remainder, about 600 bbo are in the Middle East. With 1988 production at about 4.9 bbo per year,[7] these resources represent a 120-year supply. In view of the continuing availability in the region of large resources at relatively low extraction costs, Middle East oil will probably become progressively more important over the next half-century. An important component of our resource problem is the expected growing dependence of the world on this oil.

5d. Increased recovery of conventional oil

At present, only a relatively small fraction of the oil in a given region is actually recovered. Losses occur because of incomplete recovery of oil accessible to a given well and also because of oil in regions between wells. The former can be reclaimed by *enhanced recovery* and the latter by *infill drilling*. Extensive exploitation of these possibilities would substantially increase the resources considered above, although it would involve higher costs.

Primary recovery of oil depends upon natural pressure to drive out the oil. Riva[38] estimates that primary recovery extracts only about 20% of the oil in place worldwide, and about 28% in the U.S. (due to more permeable reservoirs). Secondary recovery uses flooding with water or gas to extract a greater fraction of the oil. It is now widely used to extract from 30% to 50% of the resource base. As determined by current practices and the nature of the rock formations, overall recovery in the U.S. now represents about 32% of the oil in place. "Extraordinary" techniques to bring out a higher fraction of the resource base than given by secondary recovery include injecting steam or other materials to improve oil mobility. This increases potential recovery to 40%–80%. If the U.S. could go from 32% to 50% recovery, this would add an amount which is greater than the total of the still remaining conventional oil. Enhanced recovery could make a sizable addition to world oil resources as well. However, it is not clear what fraction of this potential can be reached on an economically practical basis.

William Fisher[39] argues that the U.S. can also substantially increase its available resources by infill drilling of existing fields and by searching for new small fields. Together with enhanced recovery, this would make it possible to sustain current crude oil production levels in the lower 48 states for 50 years with only moderate increase in costs, and for longer times with larger increases. It should be noted, however, that not all analysts share Fisher's optimism.

5e. Unconventional oil resources

Apart from crude oil itself, there are so-called "unconventional" oil resources, such as shale oil, tar sands, and heavy oil. Oil shale is a sedimentary rock in which there is a significant concentration of kerogens (an amorphous hydrocarbon formed at relatively low temperatures). The oil can be extracted by heating the shale. Tar sands and heavy oils result when crude oil migrates to the surface and releases its volatile components. The two differ in their viscosity: heavy crude oils can flow while tar sands cannot.

The total resource base of shale oil is very large, equivalent to over 1 million bbo (i.e., 10^{15} barrels). The base becomes much smaller, but still is large, if restricted to resources of higher grade, i.e., those which yield over 25 gallons of oil per ton of shale (for crude oil at 7 barrels/ton, this corresponds to roughly 10% oil by weight). World resources of this quality shale oil are believed to be about 17,000 bbo, with about 3000 bbo in North America.[32]

However, even were the price of conventional oil to rise enough to make shale oil more economically attractive, the prospects for shale oil development are limited by environmental considerations, particularly the availability of water. The attainable rates of extracting shale oil are very uncertain, with Edmonds and Reilly[32] citing limits on U.S. production which range from 1 to 30 mbd. The report of the Committee on Nuclear and Alternative Energy Sources (CONAES)[40] puts recoverable resources at about 600 bbo but indicates an extraction limit by the year 2010 of about 1.5 mbd, even under conditions of a major national commitment (to have started by about 1980). It is to be noted that shale oil is also expensive in terms of the amount of carbon dioxide produced per unit energy (see Section 6).

Masters et al.[35] estimate recoverable oil from heavy oil and tar sands (bitumen) to be about 1000 bbo, with by far the largest shares being in Venezuela, Canada, and the Soviet Union. Again, costs will be higher than those of present conventional crude oil.

Proven techniques exist for recovering oil from the very large coal resource. In fact, during World War II, Germany was sustained by such oil. However, the costs and the production of carbon dioxide are both high. One process in current use is the SASOL Process employed in South Africa.[41] It is a follow-on to prior German projects and is motivated by the abundance of coal and the shortage of oil in South Africa.

Overall, the apparent large resources of unconventional oil are not being exploited because their cost is in excess of current crude oil prices. The buildup of a large stream of production from any of these sources requires long lead times. Although oil prices have a good chance of rising again to levels where these sources may be economically attractive, the volatility of oil prices makes it virtually impossible to plan the development of alternatives on the basis of costs alone. Even if prices become much higher for a period, the possibility will remain for many decades that Saudi Arabia and other major oil producers could again reduce prices and undermine the economic base of the alternative sources, as happened in the mid-1980s. Under these circumstances it seems unlikely that market forces alone will lead to the large-scale development of additional sources of oil soon enough to avoid vulnerability to another oil crisis. The federal government could resume efforts to foster development of these alternative sources, as a matter of national energy or security policy, but at present the federal government has no interest in such intervention, and private industry has no realistic incentive to take the initiative.

5f. Natural gas resources

To a rough approximation, conventional natural gas resources are equal in magnitude to crude oil resources, expressed in energy terms. Masters *et al.*[35] estimate remaining natural gas resources to be 8×10^{15} ft^3, equivalent to about 8000 quads or about 1400 billion barrels of oil. Of this, about 13% is in North America, corresponding to the equivalent of some 180 bbo. In addition, there may be very large resources of unconventional natural gas, but the magnitude, accessibility, and cost of such resources are not well established.

World natural gas production in 1988 was at a rate of 67 quad per year (see Table 9). This is about one-half the rate of petroleum production, and with roughly equal conventional resources the supply would last about twice as long at these use rates. The U.S. situation is qualitatively similar to the world situation and, unless there is large-scale substitution of gas for oil, the long-term prospects for domestic gas supplies are somewhat better than those for oil supplies.

5g. Coal resources

Conventional estimates of world coal resources are in the range of 10 to 12 trillion tons.[32,42] This coal is primarily in the USSR, the U.S., and China (in descending order). As discussed by Edmonds and Reilly,[32] the supply is sufficiently ample that in describing resources, no credit is given for resources yet to be discovered — in contrast to the situation for oil and gas where much of the resources are undiscovered but inferred.

At 22 quads per billion tons, the energy content of 10 trillion tons of coal is 220,000 quads, equivalent to about 40,000 bbo. This exceeds conventional oil resources by about a factor of 20. The U.S. has about 25% of the world's coal resources, amounting to the equivalent of 10,000 bbo. This dwarfs our remaining conventional oil resources of under 100 bbo.

Coal resources of this magnitude might be thought to provide a solution to the energy problem, especially as it is possible to burn coal in much cleaner ways than is current practice. Further, coal can be converted to liquid form, given a willingness to pay a greater price for transportation fuels. However, as discussed in Section 6, there remains the very important problem of carbon dioxide buildup.

5h. Perspective on fossil fuels

Were there no environmental constraints, in particular the greenhouse gas problem, aggregate world fossil fuel supplies would suffice for quite long into the future. While conventional crude oil is limited, there are large potential supplies from enhanced recovery and unconventional oil, from coal and synthetic fuels from coal, and possibly from unconventional natural gas.

The higher cost of future oil and other fuels does not in itself necessarily constitute a prohibitive barrier. As efficiency of use goes up, for example in better gas mileage per car, the total energy expenditures do not rise as fast as the price. Some countries have taxes which make gasoline prices triple those in the United States. The greater difficulty comes in exploiting these sources. As discussed above, there are large resources of inexpensive oil which can be used to undercut more costly alternatives, and it is therefore difficult to develop these alternatives in advance of future needs. In consequence, one can expect dependence on Persian Gulf oil to grow, with the attendant dangers to international stability.

Beyond this, there is the problem of carbon dioxide emission. From this standpoint, any use of fossil fuels is bad. The major fossil fuel alternatives to crude oil, such as shale oil and synthetic oil from coal, are even worse than crude oil. Thus, if we were to succeed in finding the institutional and technical means of tapping these large potential resources of fossil fuels, we would be solving the oil problem at the cost of creating a dangerous environmental problem.

6. Carbon dioxide and the greenhouse effect

6a. Greenhouse gases

CO_2 is largely transparent to solar radiation but absorbs an appreciable fraction of the outgoing infrared radiation. Therefore, the equilibrium temperature at the earth's surface rises as the atmospheric content of CO_2 increases. This is the greenhouse effect.

Quite apart from any activity of man, there is already a greenhouse effect operating in the atmosphere. The CO_2, water vapor, and ozone naturally in the atmosphere act to raise the average surface temperature of the earth to 288K. Without infrared absorption in these gases, equilibrium with incoming solar radiation would give a surface temperature of 255K. The "greenhouse effect" of current interest is the further rise expected with increased concentrations of CO_2 and other gases largely arising from the burning of fossil fuels.

An array of gases, collectively referred to as "greenhouse gases," act in a manner similar to CO_2 in absorbing infrared radiation. Greenhouse gases include ozone (O_3), methane (CH_4), nitrous oxide (N_2O), and several chlorofluorcarbons (e.g., $CFCl_3$ and CF_2Cl_2) in addition to CO_2. The expected contributions of these other gases to global warming are individually less than that of CO_2 but collectively of about the same magnitude.

A focus on CO_2 is nonetheless common for several reasons: (a) it is the largest single contributor; (b) it is the best understood of the greenhouse gases; (c) very large volumes are involved and curtailing the emission rate is more difficult for CO_2 than for some of the other gases (already steps are being taken to curtail CFC emissions to reduce ozone depletion), and (d) measures to reduce CO_2, i.e., less use of fossil fuels, will reduce the production of some of the other gases, particularly N_2O.

For these reasons, and because of its close relation to energy production, CO_2 will be emphasized in the remainder of this section. The other greenhouse gases serve to increase the global warming at a given date over that predicted on the basis of CO_2 alone. The magnitude of this further increase is difficult to gauge, particularly because of uncertainties in the rate of injection of these gases into the atmosphere and because of incomplete knowledge of their sinks. The mechanisms for their production are not fully understood, and we can predict neither their "natural" rate of buildup nor the success of efforts to restrict their emission.

Review publications, which provide an introduction to the greenhouse effect, have been prepared under the sponsorship of the Institute for Energy Analysis of the Oak Ridge Associated Universities,[44] the National Research Council,[45] the Oak Ridge National Laboratory,[46] and the International Council of Scientific Unions.[47] More recent overviews have been presented in *Science* articles by Ramanathan[43] and by Schneider.[48]

6b. Carbon dioxide reservoirs and fluxes

The concentration of CO_2 in the atmosphere, as of 1988, was about 350 parts per million by volume (350 ppm). From the end of the last ice age to the beginning of the 19th century, a period of about 10,000 years, the level is believed to have been between 260 and 290 ppm.[49] Starting in about 1800 and continuing to the present there has been a rise in the concentration, first probably from deforestation and more recently from fossil fuel burning.

An accurate record of CO_2 concentrations has been available since 1958 from systematic studies of infrared absorption in air carried out at Mauna Loa, Hawaii. From these and from other studies it is concluded that the atmospheric concentration rose from an average of 315.7 ppm in 1958 to 338.4 ppm in 1980[50] and has been rising during the past decade at a rate of about 1.5 ppm per year.

The ratio of the increase in atmospheric content of CO_2 to the amount injected is the effective airborne fraction (AF). (Note: 1 gigatonne of C in the

Table 11. Carbon dioxide emission rates for different fuels [in kg °C (in CO_2) per million BTU energy].

Fuel	kg °C per MBTU[a]	Adopted[b]
Natural gas	14–15	14.5
Liquid fuels from crude oil	19–22	20.3
Bituminous coal	25	25.1
Shale oil	30–110	
Liquids from coal	32–54	
High BTU gas from coal	34–43	

[a]From G. Marland, in Ref. 44.
[b]Used in IEA/ORAU model (32), p. 266.

form of CO_2 corresponds to 0.47 ppm of CO_2 in the atmosphere.) Allowing for the uncertain input to the atmosphere of CO_2 from the biosphere, the AF for fossil fuel emissions in recent decades is believed to have been roughly 0.5. The magnitude of the AF is dependent upon the rate of CO_2 production because the increase in CO_2 content is determined by the difference between the production rate and the rate at which CO_2 is taken up by the ocean. If fossil fuel burning were to be abruptly reduced, the effective AF would be reduced. Given the uncertainties both in the behavior of the natural sources and sinks and the rate of future fossil fuel burning, some authors treat the AF as an unknown parameter, lying between 0.4 and 0.7,[32] but it is common to carry out calculations on the basis of an AF = 0.5.

The rate of CO_2 production differs for different fuels, as seen in Table 11. In terms of CO_2 production, natural gas is the most desirable fossil fuel. Coal is much worse. The amounts of CO_2 produced are even higher for synthetic fuels and shale oil, but little use is being made of these at the present time. World energy consumption of fossil fuels in 1988 was about 290 quad[18] and, considering the breakdown by type of fuel, the carbon dioxide release corresponded to about 6 billion tonnes of carbon. If one assumes AF = 0.5, this corresponds to a 1.4 ppm increase in the atmospheric CO_2 concentration. Deforestation and other interchanges with the biosphere may raise the total increase to about 1.5 ppm per year or a little more.

Considerable attention has been given to the consequences of a nominal "doubling" of the CO_2 content, where "doubling" conventionally corresponds to the reaching of a CO_2 content of 600 ppm (550 ppm in some studies), twice a value intermediate between the preindustrial and present levels. At present rates of CO_2 increase from fossil fuel burning, it would take about 180 years to rise from 350 ppm to 600 ppm. With a 2% annual increase of CO_2 production (and a constant AF = 0.50) the 600 ppm level would be reached in 80 years, and at a 3% rate of rise it would be reached in about 60 years. Overall, unless there is a restraint on fossil fuel use, one can anticipate that "doubling" will be reached in the middle of the next century although it is conceivable that it could come even sooner. The effect of the other greenhouse gases can be viewed as advancing the time of "effective doubling."

6c. Climatic consequences of increased CO_2 levels

The climatic consequences of increased CO_2 concentrations are modeled with extensive calculations. Models take account of positive and negative feedback mechanisms, such as the lesser reflection from snow and ice, the complex role of the clouds, and the more rapid rate of photosynthesis which follow from higher temperatures and increased CO_2 levels. Earlier estimates suggested that an increase in the CO_2 concentration to 600 ppm will lead, at equilibrium, to an average global temperature rise of $(3\pm1.5)°C$ with smaller changes in the equatorial regions and higher changes in the polar regions.[45] More recent calculations based on the so-called general circulation models (GCMs) tend to favor the higher rather than the lower end of the earlier estimates. A current central estimate based on such models would correspond to a rise in the neighborhood of 4°C with an increase of 5°C, or even higher, not excluded.[43,48] Other recent calculations claim lower effects if clouds are correctly considered in the models.

The other greenhouse gases will accelerate the temperature rise over that from CO_2 alone, an effect expressible either as an advance in the time of effective doubling, as suggested above, or as an increase in the total temperature rise at the time of true CO_2 doubling. Overall, by the early or mid-21st century, we can expect higher average global temperatures and different rainfall patterns. Details of the longitudinal distribution of the temperature changes are not well established nor is the distribution of the expected changes in rainfall patterns. In addition, we expect a rise in sea level due to melting of snow and ice and the thermal expansion of the oceans.

Both temperature changes and rainfall changes will benefit some nations and damage others, but the existing atmospheric models do not suffice to identify the winners and losers with any certainty. Some models predicted a "drought belt" in the U.S. mid-west before the actual drought during the summer of 1988. However, there is no certainty that a correct model predicts a drought in this region nor is it established that the increases in CO_2 levels to date have been sufficient to be responsible for the 1988 U.S. drought or for the small rise in average global temperature which has been seen in recent years.

However it is clear that if we remain on the present path of large-scale and increasing fossil fuel use, we are initiating an unprecedented experiment in global change. There is no reason for confidence that the overall outcome will be desirable and definite reasons to believe that some of the outcomes will be undesirable, e.g., the rising sea level and the strains put upon plant and animal species by an unprecedentedly rapid change in their local climate.

Note that the thermal inertia of the global system, particularly the oceans, creates lags of several decades or more between the time of greenhouse gas injection and the time when the temperature reaches (or rather, gets close to) its new equilibrium value.[43] Thus, when there is an unambiguous temperature rise, there will have already been an inexorable commitment to a substantial further temperature rise. In addition, less inexorably but virtually unavoidably, the fossil fuel use which produces the CO_2 cannot be immediately terminated even given a decision to do so.

6d. Responses to the problem of CO₂ emissions

Until the 1988 regional droughts and heat waves triggered some public alarm, we were moving on a course of increasing worldwide fossil fuel consumption and impending greenhouse warming with an insouciance that was quite inconsistent with the approach taken towards other environmental hazards, even speculative ones. The experiences of the summer of 1988 in the U.S. have led to a substantial increase in calls for action, but a comprehensive national policy, much less a world policy, still is a long distance away.

Overall responses have fallen into several categories: (a) it is too soon to panic because we do not know enough about the problem; (b) it is too late to panic because we are irrevocably committed to higher CO_2 levels; (c) it is unnecessary to panic because nations learn to cope with new circumstances; and (d) it is timely to panic (although different words are usually chosen) because resolute action is needed to forestall possible environmental disaster.

It is generally agreed that it is not practical to remove and store CO_2 in the amounts produced through global fossil fuel use. Although some time could be gained by the slowing of deforestation and by the large-scale planting of trees, the only serious long-term prospect for addressing the problem is to restrain the use of fossil fuels.

It is not clear that matters here are within our control. For one, the problem is global. At present the U.S. is responsible for only about one-fourth of the CO_2 production, and this share is expected to drop (see Table 12). Reduction in fossil fuel use would require the cooperation of a significant number of the major users, present and future. The ability and willingness of countries to reduce their fossil fuel use will depend in large measure on the anticipated effects on them of the greenhouse changes and on the perceived practicality and desirability of the alternatives: conservation, solar power, and nuclear power (including fusion).

The United States, the Soviet Union, and China together hold over 80% of the world's estimated geological resources of coal, and coal is the fuel for which the known resources are sufficient to cause the largest CO_2 buildup. Agreement among these countries to control the global use of coal could have an important effect, both because of their own resources and their collective global influence. Further, these countries and the industrialized countries, which might follow a lead taken by them, are the main energy consumers and are responsible for over 80% of the current CO_2 release.

Of course, any policy will have to take into account the problems of the developing countries, including China, which have a very low per capita energy use and which see increased energy use as the key to economic progress. With about 75% of the world population, they are responsible for only about 25% of the world's energy consumption, excluding local biomass. Thus, a very substantial reduction in CO_2 production in the industrialized countries must be achieved, in order to make energy growth possible for the developing countries.

A variety of scenarios for future energy production and consumption have been developed to explore the practicality of reducing or restraining global fossil fuel consumption. To the extent that these scenarios have been normative,

Table 12. Geographical distribution of carbon dioxide emission from fossil fuel use, 1988 (in gigatonnes of carbon released to atmosphere).

Region	Carbon released	
	Gigatonnes[a]	Percent
U.S.	1.44	24
USSR	1.05	17
Western Europe	0.96	16
China	0.59	10
Eastern Europe	0.42	7
Japan	0.29	5
India	0.15	2
Canada	0.13	2
Other	0.98	16
World	6.0	100

a. Based on energy data from DOE/EIA (18) and on IEA/ORAU carbon emission rates (see Table 11).

that is they involve the fulfillment of a desired goal rather than the working out of relatively blind economic forces, they have tended to emphasize conservation and solar power (including biomass).[51] It is of course also possible to include nuclear power in order to better the overall prospects. Intermediate scenarios moderate the effects of fossil fuel use by substituting natural gas for coal and oil.[52]

Reducing CO_2 emissions can be considered concretely in the context of the 1988 U.S. energy budget, which is crudely representative of the situation for the industrialized world. Taking into account the differences among fossil fuels, the CO_2 came in roughly equal parts from three sources: (a) electricity generation (33%), mostly from coal; (b) the residential/commercial and industrial sectors (36%), mostly from oil and gas; and (c) the transportation sector (31%), almost exclusively from oil. Were CO_2 reduction a matter of high priority, over a period of several decades it would be possible to eliminate almost all fossil fuel use in electricity generation, and reduce substantially the consumption of oil and gas in the remaining sectors. The latter could be accomplished by more efficient utilization of energy (e.g., a factor of two or better improvement in automotive fleet mileage and much lower residential heating requirements) and by replacing fossil fuels used for heating with direct solar power or nuclear or solar electricity. Major changes of this sort in the U.S. and other industrialized nations could keep global CO_2 emissions at a constant or reduced level while permitting energy growth in the developing countries.

At this time it is not clear if any consensus will be reached, internationally or within the United States, on the importance of comprehensive energy programs to restrain CO_2 emissions, much less on a framework of effective incentives. However, without explicit action to control fossil fuel use, it almost inevitably will continue to grow. Given the uncertainties in the predictions of the temperature rise and of other climatic consequences, this may mean only a moderate change for most of the world or it may mean an unprecedented environmental catastrophe.

7. Establishing an energy policy

Planning for our energy future is frustrated by disagreements as to the nature of a desirable policy and, perhaps even more, by a belief that there is no need to formulate a coherent national energy policy. Certainly, there has been no coherence since 1973. To put matters in the simplest terms, the Nixon administration in the years through 1976 favored "energy independence" through the massive expansion of domestic nuclear power, coal, and coal-based synthetic fuels. The Carter administration, from 1977 through 1980, termed nuclear power the resource of last resort, and tried to shift the emphasis more towards "soft" energy alternatives, including conservation. The Reagan administration, from 1981 through 1988, was largely content to let matters take their course, in the apparent belief that the wisdom of the market place is greater than the wisdom of the planners. As of this writing, in early 1990, a characteristic Bush administration policy has not yet emerged.

It might appear that the hands-off attitude was successful. For reasons for which the United States may deserve little credit, oil prices dropped and the sense of crisis ended. The electric utilities have turned from being promoters of electricity use to promoters of conservation, recognizing that thereby they decrease their financial risks, avoid the need for expensive new generating capacity, and increase their image of social responsibility. Energy users, faced with higher prices and a greater awareness of the benefits of conservation, have restrained energy use, and in some areas significant advances have been made. It may be that an incremental approach, based largely on market forces, has been vindicated.

It may also be that it is sowing the seeds of disaster. There is no way to inject a concern over carbon dioxide buildup into such an approach. If coal is the cheapest fuel, then it becomes the fuel of choice, as has been the case in recent years for electricity generation. Attempts have been made to reduce some of the noxious emissions, such as sulphur dioxide, but attempts to grapple with the greenhouse effect are still in their infancy. Some constructive legislation is before Congress, but as of early 1990 no action has been taken. Even the success in restraining energy demand, which seemed so spectacular at the end of 1986, may be slipping away with U.S. energy consumption significantly greater in 1989 than in 1986.

In terms of energy security, an approach which relies on small-scale private initiatives cannot take into account the possibility of rapid changes in oil supply or price. Although we now have a strategic oil reserve good for several months, a new eruption in the Persian Gulf region or a new coherence on the part of OPEC could again limit supply or drive oil prices sharply upward. We could not quickly cut back on oil use or turn to major supplies of unconventional oil. The time needed for large-scale changes of this sort is a decade or more. Of course, on a much shorter time scale, we could deploy military forces in the hopes of securing oil supplies.

Even granted that formulation of a federal energy policy is a matter of major importance, there is no responsible consensus as to what the energy policy should be. There have been too many surprises in the past 15 years to make anyone sanguine about our ability to anticipate or prescribe for the future. To recapitulate, a major surprise has been the roller coaster in oil prices. There have been other important surprises as well, in part connected with the energy price changes, including: (a) the constancy in U.S. energy consumption from 1973 to 1986, followed by a resumption of growth in the next three years; (b) the increased electrification of the energy economy (excluding transportation); and (c) the virtual moratorium on further development of nuclear power.

A major obstacle to achieving consensus is the division which has developed between the advocates of "soft" and "hard" energy paths, epitomized in the debates over nuclear power. Each now can point to some success: conservation has worked, albeit not to its full potential, and electrification has proceeded (using "hard" sources: coal and nuclear power), albeit at a slower rate of growth than had been the case in earlier years.

The establishment of a reasonable national policy would be furthered by a consensus, or operational compromise, on a number of key questions:

• How much emphasis should be placed on reducing energy consumption through improved energy efficiency and a heightened ethic of energy conservation?

• For greater energy efficiency, should stress be put on mandatory standards or on the economic self interest of the energy consumer?

• How much emphasis should be placed on developing energy sources which minimize CO_2 production?

• Is natural gas a desirable compromise in terms of CO_2 emissions, or an unfortunate stopgap? How adequate are gas supplies?

• At what point, if ever, will large scale use of solar power be economically practical? Do the main hopes for solar power lie in the expansion of reasonably mature technologies (wind turbines, direct solar thermal) or from advances in emerging technologies (solar photovoltaics or ocean thermal)?

• Can nuclear power be made environmentally acceptable at reasonable cost? If so, should the next generation of nuclear power plants be based on advanced versions of conventional light water reactors or on radically different reactors with greater "inherent safety?"

• When, if ever, can fusion energy become a practical energy source?

There seems little chance of obtaining a broad consensus on these questions in the near future in view of technical uncertainties and "philosophical" differences. It may be possible, however, to reach an effective agreement on the existence of critical problems—too little domestic oil and too much global carbon dioxide—and to recognize that we cannot now decide how these problems can be best addressed over the next several decades. Rather than attempting to map out a full program now, substantial investments might be put into developing a broad variety of potentially promising options. As this development progresses, the environmental and economic implications of each of the options will become clearer, and it may then be possible to agree upon the vigor with which each should be pursued.

A program of this sort would require a large and consistently maintained financial commitment to a system of subsidies and pilot plant development. We are familiar with large national financial commitments to energy: in 1980, we spent over $100 billion dollars (in 1987 dollars) on oil imports. If we take that as a yardstick, then even inches of domestic investment would carry us far. As an illustration of a potential source of revenue, a U.S. CO_2 emission fee of $10 per tonne of carbon (corresponding to 0.26¢ per kWh for electricity from coal and 2.5¢ per gallon for gasoline) would yield $14 billion per year.

The eclectic approach suggested above is not aesthetically satisfying and almost by definition will prove to have been wasteful in some aspects. But a broad approach of this sort may represent a desirable compromise, given that agreement on the need to reduce oil consumption and CO_2 emissions is likely to be reached before there is agreement on the most effective reduction strategies.

Bibliography

Basic sources of data

The Energy Information Division of the U.S. Department of Energy publishes basic energy data for the U.S. and, less completely, for the world in: the *Annual Energy Review* (Ref. 4), the *Monthly Energy Review* (Ref. 7), and the *International Energy Annual* (Ref. 18).

Comprehensive energy tables for their member states are published annually by the Organization of Economic Co-operation and Development in *Energy Balances of the OECD Countries*. Broad, but sometimes incomplete, data are published on a three-year cycle by the World Energy Conference.

Energy demand and supply data, as of 1988, are summarized in two chapters of *A Physicist's Desk Reference*, 2nd Edition of *Physics Vade Mecum*, H. L. Anderson, ed. (American Institute of Physics, New York, 1989).

Reviews of energy issues

Important energy topics are reviewed on a continuing basis in the *Annual Review of Energy*. Broad perspectives on energy issues are presented in: *Global Energy: Assessing the Future* by J. Edmonds and J. M. Reilly (Ref. 32) and *Learning About Energy* by David J. Rose (Plenum Press, New York, 1986).

Major energy studies include: (a) IIASA study (global): *Energy In a Finite World, A Global Systems Analysis*, W. Häfele, Program Leader, International Institute for Applied Systems Analysis (Ballinger, Cambridge, 1981); and (b) CONAES study (U.S.): *Energy in Transition: 1985–2010* (Ref. 40).

References and notes

1. H. Brown, *The Challenge of Man's Future* (Viking Press, New York, 1954).
2. *A Time to Choose, America's Energy Future*, Energy Policy Project of the Ford Foundation (Ballinger, Cambridge, 1974).
3. For a history of these developments see: P. L. Auer, Annu. Rev. Energy 1, 685 (1976).

4. *Annual Energy Review 1988*, Report DOE/EIA-0384(88)(U.S. DOE, Washington, DC, 1989).
5. S. Arrhenius, 1896 (cited, for example, in Ref. 43, below).
6. *Project Independence Report*, Federal Energy Administration (U.S. FEA, Washington, DC, 1974).
7. *Monthly Energy Review*, December 1989, Report DOE/EIA-0035(89/12)(U.S. DOE, Washington, DC, 1990).
8. A. H. Rosenfeld and D. Hafemeister, Sci. Am. **258**, No.4, 78 (April 1988).
9. R. H. Williams, E. D. Larson, and M. H. Ross, Annu. Rev. Energy **12**, 99 (1987).
10. A. Kahane and R. Squitieri, Annu. Rev. Energy **12**, 223 (1987).
11. See, e.g., A.B. Lovins, *Soft Energy Paths: Toward a Durable Peace* (Friends of the Earth International, San Francisco, 1977).
12. C. C. Burwell, W. D. Devine, Jr., and D. L. Phung, Energy **7**, 961 (1983).
13. D. Bodansky, Energy **9**, 303 (1984).
14. C. C. Burwell, in *Electricity Use, Productive Efficiency and Economic Growth*, S. H. Schurr and S. Sonenblum, eds. (Electric Power Research Institute, Palo Alto, 1986).
15. Reference 4, Table 95.
16. *Ibid.*, Table 63.
17. Reference 7, Table 9.1.
18. *International Energy Annual 1983, International Energy Annual 1988*, Reports DOE/EIA-0219(83) and DOE/EIA-0219(88) (U.S. DOE, Washington, DC, 1984, 1989).
19. J. P. Riva, Jr., *The Enigma of Natural Gas*, Report No. 88-561 SPR (Library of Congress, Washington, DC, 1988).
20. "Historical Profile of U.S. Nuclear Power Development," *AIF Background Info* (Atomic Industrial Forum, Bethesda, 1986).
21. IAEA Bulletin **31**, No. 3, 63 (1989).
22. IAEA Newsbriefs, **4**, No. 4, 1 (May 1989).
23. "Big Quebec Power Plan in Phase 2," New York Times (March 9, 1988), p. 29.
24. *Solar Collector Manufacturing Activity 1987*, Report DOE/EIA-0174(87) (U.S. DOE, Washington, DC, 1988).
25. Reference 4, p. 231.
26. *Photovoltaic Energy Program Summary, Volume I: Overview, Fiscal Year 1988*, Report DOE/CH10093-40 (U.S. DOE, Washington, DC, 1989), prepared by SERI.
27. *Principal Conclusions of the American Physical Society Study Group on Solar Photovoltaic Energy Conversion*, H. Ehrenreich (chairman) (American Physical Society, New York, 1979); see also H. Ehrenreich and J. Martin, Phys. Today, September, 1979, p. 25.
28. See e.g., Y. Hamakawa, Sci. Am. **256**, No. 4, 87 (April 1987).
29. *Five Year Research Plan, 1987–1991, National Photovoltaics Program*, Report DOE/CH-10093-7 (U.S. DOE, Washington, DC, 1987).
30. D. R. Smith, Annu. Rev. Energy **12**, 145 (1987).
31. P. Gipe, "Wind Energy: No Longer an 'Alternative' Source of Energy" (to be published); Wind Energy Weekly **7**, No. 300 (April 24, 1988); and private communication.
32. J. Edmonds and J. M. Reilly, *Global Energy: Assessing the Future* (Oxford University Press, New York, 1985).
33. T. Gold, Annu. Rev. Energy **10**, 53 (1985).
34. For a discussion of the approach and citation of earlier references see: M. K. Hubbert, M. King, "Energy Resources" pp. 157–242 in *Resources and Man*, a Study and Recommendations by the Committee on Resources and Man, National Academy of Sciences/National Research Council (W.H. Freeman, San Francisco, 1969)
35. C. D. Masters, E. D. Attanasi, W. D. Dietzman, R. F. Meyer, R. W. Mitchell, and D. H. Root, Proceedings of the 12th World Petroleum Congress, Houston, Vol. **5**, 3 (1987).
36. J. P. Riva, Jr., *The World's Conventional Oil Production Capability Projected into the Future by Country*, Report 87-414 SPR (Library of Congress, Washington, DC, 1987).
37. *Annual Energy Outlook 1989*, Report DOE/EIA-0383(89) (U.S. DOE, Washington, DC, 1989).
38. J. P. Riva, Jr., *Enhanced Oil Recovery Methods*, Report 87-827 SPR (Library of Congress, Washington, DC, 1987).
39. W. L. Fisher, Science **236**, 1631 (1987).

40. *Energy in Transition, 1985–2010*, Final Report of the Committee on Nuclear and Alternative Energy Systems (CONAES), NRC/NAS (W. H. Freeman, San Francisco, 1980); H. Brooks and J. M. Hollander, Annu. Rev. Energy **4**, 1 (1979).
41. M. E. Dry and H. B. W. Erasmus, Annu. Rev. Energy **12**, 1 (1987).
42. H. H. Schobert, *Coal: The Energy Source of the Past and Future* (American Chemical Society, Washington, DC, 1987).
43. V. Ramanathan, Science **240**, 293 (1988).
44. *Carbon Dioxide Review: 1982*, W. C. Clark, ed. (Oxford University Press, New York, 1982).
45. *Changing Climate*, Report of the Carbon Dioxide Assessment Committee, National Research Council (National Academy Press, Washington, DC, 1983).
46. *Atmospheric Carbon Dioxide and the Global Carbon Cycle*, J. R. Trabalka, ed., Report DOE-ER/0239 (U.S. DOE, Washington, DC, 1985).
47. *SCOPE 29: The Greenhouse Effect, Climatic Change and Ecosystems*, B. Bolin, B. R. Doos, J. Jager, and R. A. Warrick, eds. (Wiley, Chichester, 1986).
48. S. S. Schneider, Science **243**, 771 (1989).
49. R. H. Gammon *et al.*, in Ref. 46, p. 27.
50. W. R. Emanuel *et al.*, in Ref. 46, p. 143.
51. See, e.g., J. Goldemberg, T. B. Johansson, A. K. N. Reddy, and R. H. Williams, Annu. Rev. Energy **10**, 613 (1985); I. M. Mintzer, *A Matter of Degrees* (World Resources Institute, Washington, DC, 1987).
52. J. H. Ausubel, A. Grubler, and N. Nakicenovic, Climatic Change **12**, 245 (1988).

Estimating the risks of producing energy

Evans Harrell

1. Why and how are risks estimated?

Risk assessment is part of a larger discipline known as systems engineering, which has sprung up over the last few decades. A professional organization, the Society for Risk Analysis, now exists, and several journals are devoted to the subject. Risk analysis originated in the aerospace industry when aircraft reached the level of complexity where no one person could understand all their mechanisms in detail. Earlier technologies were as likely to be dangerous as modern ones are, but intuition and experience ordinarily allowed people to discern hazards and to devise common-sense ways of coping with them. For example, fire hazards in the home are readily identifiable. They can be avoided by such measures as chimney sweeping and enactment of electrical codes or mitigated by construction of fire escapes and maintenance of fire departments. These measures evolved over time without resort to elaborate calculations or systems analysis. Since intuition worked sufficiently well and computational power was limited, modern risk calculations simply never crossed anyone's mind.

Contrast this state of affairs with that of complex man-made systems which cannot be understood completely in their design stage. It is often useful to treat such systems as natural objects for scientific investigation, devising theories and performing experiments on them. The success of this approach became apparent early on as telephone systems reached the level of complexity where hackers could figure out ways of making free calls or shutting down exchanges. Their inspired experiments revealed ways the system would respond that were never anticipated by its designers. When caught, these telephone hackers, like their somewhat less harmless descendants the computer hackers, were often given suspended sentences and hired at good salaries to help companies analyze complex systems. Planned experimentation has also become a commonplace way of learning about man-made systems. Somewhat irregular planned experimentation at Chernobyl, for example, produced interesting information about that complex system, although it does not appear likely to lead to professional advancement for those involved.

Risk assessment is an attempt to quantify the probabilities and degrees of harm arising from the operation of a complex system. In this discussion, the system will be a means of producing energy. The goals are to help make intelligent policy about energy supplies; to understand and improve existing power plants and designs; and to educate—some say persuade—the public and the politicians. The great majority of energy risk studies have been concerned with fission power, and many of the techniques of probabilistic risk assessment have been developed specifically for this purpose. In recent years risks from other sources of energy have begun to receive their share of attention, sometimes entailing different sorts of analyses.

The underlying philosophy of risk analysis is that of cost-benefit analysis and economic utility theory: the best decision among many choices is the one that maximizes a hypothetical utility function. A utility function satisfies certain axioms formalized by Ramsey, von Neumann, Morgenstern, and others. For economic decisions, it is made larger by increased benefits and smaller by increased costs, which can include risks. The utility function is typically equivalent to a profit and is denominated in dollars. The variables on which it depends are the quantities about which the decision is made — amounts of raw materials and labor to procure, level of production, etc.

Since economic decisions are inherently probabilistic, the utility function is ordinarily an expectation value of the return after a decision is taken. In a study of risk one may calculate a radiation dosage or a number of deaths instead of a monetary value, but the principles followed are analogous to those of an economic analysis. Disparate risks must be compared, and in order to do so they will be reduced to common terms, often explicitly or implicitly equivalent to a dollar measurement. Devising a utility function which correctly describes all the aspects of an energy technology is rarely a simple or unambiguous process. Risk analysts must reduce such diverse quantities as costs of fuel ten years in the future, damage to local wildlife, and the chance of cancer in the local population caused by plant emissions to a single unit such as dollars.

Several kinds of analysis go by the name of risk assessment, and it is important to distinguish them. One can take a global point of view and attempt to set the economic benefits of, say, a program of building a certain type of power plant against its costs including the risks that will be incurred. Alternatively one can analyze a particular aspect of the problem in detail. One might calculate the probability of a dam breakage or release of radioactivity into the environment from a nuclear plant, or assess the loss of life assuming that such a disaster has happened. Risk analysis is most straightforward when it addresses a single, carefully identified event in an engineering context. If a failure has happened frequently, good statistics are probably available, and mean casualty or loss figures can be stated with high confidence. For energy technologies this is often the case for hazards endured by workers either in the construction phase or in mining raw materials. The rather dangerous job of a coal miner is a case in point.

Risks from rare but possibly catastrophic accidents are more difficult to quantify. The analyst first guesses what accidents might occur. The next step is to write down the sequence of events leading to each accident. Then probabilities are assigned to the events. The analyst combines the event probabilities to calculate the probability that a particular accident will happen. Finally total risk is calculated from the probabilities that all known accidents will occur.

As a simplified example, let us imagine calculating the risk to a bedroom community in a canyon from destruction of an upriver dam. There are two major risks, failure of the dam for structural reasons and earthquake. The analysis assumes a vanishingly small chance of other accidents such as terrorist action against the dam. Suppose that major earthquakes occur locally with a frequency $p_1 = 10^{-3}$/year and that the particular type of dam is known to have a failure rate in the absence of earthquakes of $p_2 = 2 \times 10^{-4}$/year. Both figures might be uncertain by a factor of 2 either way: earthquake frequencies are only roughly estimated from historical and geological data. Failure rates depend not only on the type of dam but also on the quality of the construction and maintenance at a particular site which is not known.

The two events are roughly statistically independent, although of course shoddy construction would allow smaller earthquakes to destroy the dam. The joint probability is very small, certainly much smaller than the uncertainties in our estimates, so our best estimate of the failure rate would be roughly $p_1 + p_2 = 1.2 \times 10^{-3}$/year. The high estimate would be 2.4×10^{-3} and the low estimate would be 6×10^{-4}/year.

Containment and mitigation add additional complications to the risk calculation. Suppose local authorities design an evacuation plan which requires warning and transportation, and which may also depend on whether the accident occurs during working hours or not. These probabilities would not be independent of the mode of the accident. Dam deterioration is easier to notice and does not interfere with the possibility of evacuation, while an earthquake could block evacuation routes and would probably strike without warning.

Let us assume a population of 6000 during working hours (9 a.m.–5 p.m., one-third of a day) and 10,000 otherwise. Suppose that, with sufficient warning, 75% of the people can be safely evacuated. In the event of an earthquake, perhaps only 10% can be evacuated. Thus in the worst case 9000 lives are lost. The probability of this worst case occurring is $2p_1/3$, or between about 3.3×10^{-4}/year and 1.3×10^{-3}. The overall expected death rate for the worst case is about 1.3/year. This expectation value may be useful for actuarial purposes or as a standard of comparison, but it does not describe a real situation. In this example, deaths would be either 0 or 1500 or more.

The energy hazard for which there have been the most studies is the release of fission products to the environment in a nuclear power plant accident. Although similar to calculations for the dam model, these calculations are much more involved. The sequence of events leading from some initial event to an identified failure such as a core melt now has several stages. For example, water must be lost and core degradation must occur before the reactor vessel will fail.

A complicated calculation involving sequences of events is typically aided by a *tree analysis*. This begins with a map of the accident with boxes representing particular events connected by lines or arrows leading to other boxes representing events happening later in the sequence. Both the timing and the physical or logical progression of events must be worked out in detail. When a specific chain of events must occur for a system failure, the diagram of that sequence of events is referred to as an "event tree." When many sequences of events can lead to a particular failure, with some sequences occurring independently of one another while others are correlated, the resulting "fault tree" diagram is often equipped with Boolean elements in the style of digital circuit design. Such a tree can be as multiply connected as the many-trunked banyan.

In a study of reactor accidents, the problem of containment of a malfunction is analyzed as a semiautonomous second stage of the tree diagram. Depending on the nature of the event and the availability of information, probabilities are assigned to each event based on experience, laboratory results, or theoretical calculations. It is important to determine both the probability estimates and the confidence which can be placed in those estimates.

A tree diagram for the simplified model of a dam failure would consist of two branches, each one having a box for the type of accident, labeled "structural failure" or "earthquake" and a box labeled "evacuation." A diagram for a reactor source-term study is shown in Figure 1.

The procedure has many variants depending on the purpose of the risk calculation and who carries it out. Knowing the probability that a disaster will occur leads immediately to a different question about risk: how much risk can the public tolerate? One thinks primarily of health risks, but economic, environmental, and social risks are also significant. In any case, medicine, economics, and social science are quite different disciplines from physics, and these contingent risk assessments can be nowhere nearly as quantitative as at least some parts of the strictly engineering and physical analysis. Secondly, the methodology for assessing a whole energy technology is quite different from that of an assessment for a particular power plant. Even in the same context, the procedures vary considerably from study to study, since risk analysis is still very much a developing discipline and has not been standardized. This unfortunately makes it easier for expert opinion to be sharply divided and polarized.

Even if the absolute magnitudes of a risk cannot be reliably estimated, systematic analysis can be quite valuable simply for identifying where improvements can be made in design and operation. Nuclear engineers generally feel that probabilistic risk assessment has led to substantial improvements in the design of fission plants, and that this can be counted as a great success of the discipline.

2. Historical overview of energy risk analysis

Until several years ago, almost all serious energy-related risk assessments focused on nuclear fission, most often on evaluating *source terms*, i.e., the inventory of radionuclides released to the environment. The first fission study, WASH-740, was produced by the Atomic Energy Commission (AEC) in 1957 when the first nuclear power plants were being planned. This study was part of an effort to begin the program of commercial nuclear fission under the Atomic Energy Act of 1954, which required consideration of the liability that the government and industry might face in the event of an accident. A related effort led to the Price-Anderson Act of 1957 which amended the Atomic Energy Act to limit all liability for such accidents to $560 million.

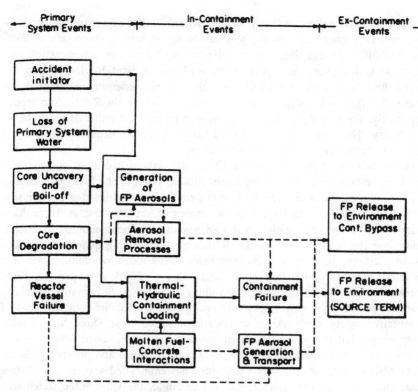

Primary System Events — **In-Containment Events** — **Ex-Containment Events**

FIG. III.A.9. Core-melt accident sequence. Thermal-hydraulic processes are indicated by solid lines, fission-product-aerosol processes by dotted lines.

Figure 1. Typical diagram used in a risk assessment for a course-term study, from R. Wilson, Report to the American Physical Society for the study group on radionuclide release from severe accidents at nuclear power plants. Rev. Mod. Phys. **57** No. 3. Part II (July, 1985), p. S29.

WASH-740 was not very detailed. A full-scale attempt at probabilistic risk assessment awaited the Rasmussen Report, officially entitled the *Reactor Safety Study* and designated WASH-1400 by the AEC. The study began in 1972 and was released in 1975, both years in which there was major Federal legislation on liability of nuclear power plants. The Rasmussen report broke new ground in reactor-safety analysis both methodologically and in its attempt at comprehensiveness. It uncovered many previously unsuspected hazards, notably breaks in small pipes and system stresses due to transients. In the aggregate, the unsuspected risks posed larger risks of core melt than all of the previously known hazards, which were "design basis accidents."

Critics immediately attacked the Rasmussen report for not going far enough in this direction and for understating errors. For example, the Union of Concerned Scientists published a critique of the report's methods and analysis that appeared within two years, and the Nuclear Regulatory Commission established a Risk Assessment Review Group shortly after that. The report of this commission called the procedures of the Rasmussen report into question, and eventually the NRC quit regarding the estimates of the Rasmussen report as reliable. Importantly, however, the report of the Risk Assessment Review Group endorsed the concepts and methods of probabilistic risk assessments although the authors felt that they had not been fully implemented in the Rasmussen report. Apparently, the techniques had not been well understood either at the NRC or in industry. These groups opposed their use until probabilistic risk assessment had been blessed by Federally funded studies.

On March 28, 1979, a reactor at Three Mile Island had a serious accident which has established itself as the major turning point in the history of energy risk assessment. After the accident, President Carter appointed a commission headed by J. Kemeny to look into the causes of the accident at Three Mile Island. The commission collected a vast amount of documentation and testimony over the next several months. Rather than attempting quantitative analysis for itself, the Kemeny Commission made broad recommendations as to the organization of the NRC and other larger issues of risk management. Soon afterwards, the NRC's Advisory Committee on Reactor Safeguards officially endorsed probabilistic risk assessment and recommended the establishment of quantitative safety goals. After analyzing the accident at Three Mile Island, many people felt that if techniques of risk assessment had been systematically applied, the accident sequence might have been foreseen and prevented. Since Three Mile Island, the Advisory Committee has enunciated several quantitative guidelines based on risk assessment. For example, the Committee stated that "the chance of a very large release of radioactive materials to the environment should be less than 10^{-6} per reactor-year,"[1] just before the Chernobyl disaster in 1986. (Of course the NRC is only concerned with American reactors.)

Probabilistic risk assessment has spread throughout other industries and other branches of the government, most notably to the Environmental Protection Agency (EPA), which now regards it as a central responsibility, and to the National Aeronautics and Space Administration (NASA). The EPA and the chemical and drug industries routinely sponsor or carry out risk assessments, primarily of health risks associated with exposure to particular substances. In the energy industry there have been 30 or so probabilistic risk assessment studies of core-melt frequencies for particular nuclear power plants, giving estimates of the probability of a core-melt accident ranging from 2.4×10^{-5} to 1.5×10^{-3} per year.[2] (Containment substantially reduces the risk to the public from core melt. These figures can be compared to a probability of about 10^{-4} per year for hydroelectric dam failure.) On one occasion, a full-scale risk assessment for particular plants has been central to a licensing board hearing, *viz.*, Indian Point reactors 2 and 3 in New York. Unfortunately, preparing full-scale risk assessments for the energy industry is vastly more complex than preparing the relatively routine analyses in some other fields, which are increasingly becoming the foundation for developments in risk assessment.

More recent major studies, especially those by the American Nuclear Society in 1984 and by the American Physical Society in 1985, continue to react to the Three Mile Island accident and the controversy over the Rasmussen report. The ANS study was the report of an international committee, commissioned "to examine the state of knowledge relative to the source term; to assess the methods and assumptions used to describe fission product behavior and retention associated with various phenomena, plant systems, and structures in a severe reactor accident; to provide a summary of source term results obtained by various investigators; and to compare these methods, assumptions, and results to those in WASH-1400." The APS source-term study was the result of a two-year effort to "review the adequacy of the technical base upon which the phenomenological models for radionuclide release from postulated severe reactor accidents are constructed, the adequacy of the models themselves, and the correct use of the computer codes that incorporate these models in the analyses of accident sequences." The study specifically attempted to incorporate the experiences of Three Mile Island. It did not calculate probabilities of particular accidents, although it did endeavor to state which accident sequences were most significant. While explicitly claiming not to be the final word, both the ANS and APS studies revised source-term estimates dramatically downward in some cases.

In this spirit, a 1986 article[3] observed that after 1500 reactor years since Three Mile Island without a core melt, we can "say with about 50 percent confidence that Rasmussen's upper limit (1 in 2000 reactor years) is not too optimistic." In April of that year there was a another partial core melt at Chernobyl, so this upper limit is actually fairly close to our experience to date. In fact, the risk assessment studies of U.S. reactors do not apply to the Soviet design in use at Chernobyl, which lacked many of the safety features that are standard in American reactor designs. The accident at Chernobyl, like the one at Three Mile Island, has been attributed in large measure to "human error," and recent Federal analyses of reactor risk have made efforts to take this into account more fully.

3. Comparative, longitudinal risk assessment

Since the early risk analyses by the AEC, the scientific community has become much more sophisticated in its thinking. It has been recognized that in order to make intelligent energy policy we need studies that are both comparative and "longitudinal." Comparative studies use quantitative risk assessment to compare the costs and benefits of various energy technologies such as building coal-powered, hydroelectric, or fission plants to meet anticipated energy needs. Longitudinal studies examine risks from all phases of energy production and examine the whole fuel cycle from raw materials to end use. Even inaccurate risk assessment studies are useful for improving plant design, but for comparative purposes, accuracy at every step is essential. Each possible power plant poses a host of diverse hazards to life, limb, and pocketbook, from each of the following stages:

1. Obtaining raw materials for plant construction
2. Building plants
3. Obtaining raw materials for fuel

4. Refining or processing fuel
5. Transporting materials
6. Maintaining and operating plants
7. Pollution control and waste management
8. Defense against accidents (or arsonists, terrorists, etc.)
9. Delivery of power to industry and the home
10. Ensuring consistency of the power supply
11. Decommissioning old plants

In principle one should compare opportunity costs and secondary hazards such as those arising from economic, social, or political effects of major decisions about energy. One can imagine, in caricature, a country adopting an extremely safe and reliable source of energy which was so expensive that it caused a serious decline in the standard of living or used up essential supplies needed for housing or food, thus causing a rise in mortality. The effects of these shortages could easily be larger than risks posed by more direct hazards, but it is not easy to see how they could be forecast with much confidence. It should be noted here that risk analysts have usually neglected the decommissioning of old plants. It is now realized that decommissioning might be enormously expensive for fission plants, especially for defense plants, which have been operating behind a veil of secrecy that has kept safety standards far below those now current for commercial power reactors.

Inhaber was among the first to push the important idea of longitudinal, comparative studies. In particular, he startled the scientific community and the public with the claim that many "soft" technologies, such as solar power, were significantly more hazardous than fission.[4] As one would imagine, this claim aroused indignation, especially since much of the calculation centered on ensuring the consistency of the power supply. By many measures, coal power is both the dirtiest and most hazardous of the choices available to us. Inhaber reasoned that since solar energy is inconsistent, a large reliance on solar energy would require the use of other technologies, especially coal, as a back-up. This led to a high risk for solar energy. Inhaber's critics pointed out that the marginal risk of solar energy would be very much less than Inhaber's calculations since back-up power would not be important until solar energy became dominant. The details of Inhaber's calculations and his assumptions have also been criticized. Although Inhaber evidently has an axe to grind, his book is a remarkable and important paradigm of comparative risk assessment.

Comparative, longitudinal risk assessments for the various forms of energy production cannot be said to be definitive yet, but attempts have been made. As a practical matter, the most important comparisons are among nuclear fission, coal, and conservation, since they are the only proven energy technologies which can economically be exploited in the next several decades for additional energy needs. Figures 2 and 3 enumerate the principal risks and reproduce some quantitative comparisons.

Because it frees energy for new needs, conservation should be considered as an alternative, and potentially very large, source of energy, and it is not without

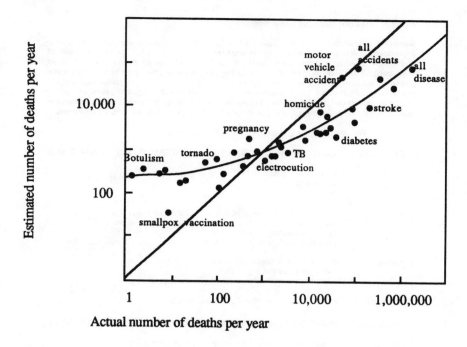

Figure 2. Principal risks of various technologies. Some risks of various energy technologies are listed (the order does not systematically indicate the importance of the risks). Quantative assessments are partially summarized in Figure 3. In addition to these, some less clear-cut but possibly equally important risks have not been listed. These have to do with such things as the economic consequences of having finite resources; the need for back-up power, especially for some of the currently minor technologies, and terrorism or acts of war. As a general rule, all the important technologies involving centralized storage or production could be tempting targets for hostile activity. It should also be remembered that additional risks will emerge as some of the newer technologies, such as fusion, are developed.

hazards. Better home insulation has probably led to increases in indoor air pollution and radon exposure, for example. Conservation, however, consists of such varied things as insulation; better efficiency in appliances (particularly refrigerators), in automobiles and other machinery, and in electrical storage and transmission; as well as changes in our personal habits. Hence it is the concern of several industries, many of which are small and can respond rapidly to economic and technological change. Longitudinal studies are feasible for conservation, but will inevitably be rather different from those for energy sources requiring power plants or similar large directed public projects. Although primary hazards from manufacturing may resemble those of other energy technologies, conservation has many unique and important secondary effects. For instance, if we were to drive less or more slowly or were to generate less waste because of conservation, some of these secondary effects would entail *negative* risks, i.e., greater safety. Obviously this could make conservation an extremely attractive choice. Unfortunately, the risks of conservation have not been documented as carefully as those of even some marginal production technologies.

Estimating risks of producing energy

Figure 3.   Principal risks of various technologies.

Some risks of various energy technologies are listed (the order does not systematically indicate the importance of the risks).   Quantitative assessments are   partially summarized in Figure 4. In addition to these, some less clear-cut but possibly equally important risks have not been listed. These have to do with such things as the economic consequences of having finite resources; the need for back-up power, especially for some of the currently minor technologies, and terrorism or acts of war. As a general rule, all the important technologies involving centralized storage or production could be tempting targets for hostile activity. It should also be remembered that additional risks will emerge as some of the newer technologies, such as fusion, are developed.

ECONOMICALLY IMPORTANT TECHNOLOGIES

TECHNOLOGY	PRINCIPAL CHRONIC RISKS	PRINCIPAL CATASTROPHIC RISKS
Biomass	air and water pollution toxic wastes fires	greenhouse effect
Coal	air and water pollution, acid rain toxic wastes miner's respiratory disease construction or operational accidents	greenhouse effect major mine accidents
Conservation	more severe automotive accidents indoor air pollution, e.g., radon (Conservation may enhance safety in some respects.)	
Hydropower	construction or operational accidents	dam failure
TECHNOLOGY	PRINCIPAL CHRONIC RISKS	PRINCIPAL CATASTROPHIC RISKS
Natural gas	construction or operational accidents	greenhouse effect explosion of drilling, storage and transport facilities
Nuclear fission	construction or operational accidents radiation exposure in mines, mills, and reprocessing plants radioactive wastes	core meltdown or other catastrophic release of radiation transportation accidents nuclear proliferation
Oil (conventional)	air and water pollution toxic wastes economic stress acid rain construction or operational accidents	greenhouse effect tanker spills political stress leading to war
Passive Solar	construction or operational accidents air and water pollution (from materials production)	
TECHNOLOGY	PRINCIPAL CHRONIC RISKS	PRINCIPAL CATASTROPHIC RISKS
Geothermal	air pollution (H_2S) construction or operational accidents occupational noise hazard	
Nuclear fusion	radiation exposure in processing plants, waste management	catastrophic release of radiation nuclear proliferation
Ocean Thermal	environmental degradation construction or operational accidents	climate change (if on large scale)
Oil from shale, tar sands, etc.	(Risks vary, but are. similar to conventional oil and coal, except for international economic and political risks)	
Photovoltaic	air and water pollution (from materials production) toxic wastes	
Wind	construction or operational accidents	

Figure 3. Some quantitative risks of different energy technologies as estimated by Inhaber and Holdren, from L.D. Hamilton, Comparative Risks from Different Energy Systems: Evolution of the Methods of Studies. International Atomic Energy Agency Bulletin 22 (Oct. 1980) 35–71.

4. Difficulties of the comparative program

The comparative program's big problem is that the technologies being compared vary immensely. It may simply be easier to evaluate risks from one energy technology than from another, either for practical reasons or perhaps just because of differences in the amount of past scrutiny of one technology. In particular, the tremendous attention that has historically been given to fission power in comparison with all other sources of power may give a misleading impression of the safety of those other sources. There are many other knotty problems having to do with different types of risks and differences in the confidence we can have in our estimates:

The differences in confidence can be huge. Suppose a society must choose between two alternatives for which the risk assessments display very different confidence levels. Would it be better to make decision A, predictably costing 100 lives a year, or to make decision B, ordinarily costing 10 lives a year, but for which a disaster occasionally takes 1000 lives? Suppose that the best expert estimates of the probability of the disaster of choice B range from one disaster in 5 years to one disaster in 50 years—an order of magnitude difference is typical of estimates of rare risks. Depending on the actual probability, choice B would be expected to cost between 30 and 210 lives per year. In this case we simply do not know which choice will lead to fewer lives lost in the long run.

Curiously, some traditional and reasonably common risks are poorly documented because they were not hot topics at the time when risk analysis was developing. Consider hydroelectric power, certainly one of the most familiar energy technologies. While there have been studies of dam failures, and there is sufficient experience to establish a failure probability of about 10^{-4}/year with some confidence, it turns out that the materials used in dams are poorly documented. Hence the results for one dam are difficult to extrapolate to another, and longitudinal studies are as yet impossible.

The nature and timing of the harm that occurs may vary from one accident to another. Suppose that decision A is compared with decision C. For decision C there will be no immediate deaths, but there will be many mining injuries and health problems from pollution. Even if, as is somewhat customary, one focuses on deaths alone, it is not clear that all sorts of deaths are equivalent. Deaths many years from now are generally preferable to deaths that occur immediately. Voluntary risks, faced as a condition of employment and presumably in return for better pay, may be more acceptable than accidental and unpredicted deaths of innocent bystanders. Even if the average numbers are the same, 100 predictable deaths a year may not be equivalent to a one in 10,000 chance of 1,000,000 deaths. The death of the last remaining people with a unique culture or skills may not be equivalent to the same number of deaths of average suburbanites.

The problem of making a decision among choices that are not comparable, even when accurately projected, is not so much a matter for science or engineering analysis as for philosophy and politics. Such comparisons rarely lead to unanimity. Therefore scientific analysis should carefully separate the calculations of different sorts of risks along several dimensions, rather than make an overall rank ordering. In addition to distinguishing among deaths, nonfatal injuries and other health problems, and social and economic disruption, a *risk profile* could portray differences according to whether they are

1. voluntary or involuntary;
2. catastrophic or chronic;
3. spread evenly or primarily afflicting certain groups of people.

The distribution of damage raises the question of whether the victims can be identified individually or only statistically, as is the case for a low-level radioactive exposure. Victims are likely to be different from beneficiaries, which raises issues of fairness.

5. The controversy over risk assessment: Theory

Professional risk assessors quickly admit some limitations of their trade. However, they make the case that imperfect estimates are preferable to no estimates at all and go on to describe the techniques and assumptions they use to make their estimates as reliable as possible. Reactors have certainly been made safer as a result of risk assessment, even though it remains difficult to obtain consensus on how safe they are in absolute terms. Critics generally agree that risk assessment is important and necessary but point both to methodological shortcomings, which can in principle be rectified, and to intrinsic limitations, which cannot. It is of great importance to consider the intrinsic limitations, especially if the results of risk assessment are to be incorporated into energy policy.

The most obvious intrinsic limitations concern rare or unforeseen catastrophes. There are necessarily large relative uncertainties about their likelihoods, and, paradoxically, efforts to make catastrophes rare can make understanding them more difficult. In addition to unusual engineering failures and natural accidents, it is also difficult to quantify dangers of human origin, such as war or terrorism. A recent example of an inadvertant threat is the finding that injections of industrial waste into the ground may have caused an unusual earthquake that shook a nuclear power plant in Ohio in 1986.

It can be difficult to determine even what the significant risks are. If a risk assessment had been proposed for coal power in 1950, many dangers, such as mining accidents and air pollution, would have been readily identified. In the last decade it has been widely realized that burning of fossil fuels will affect the chemical balance of the atmosphere (notably increasing CO_2, but recent suspicion also points to decreasing stratospheric ozone[5]), yet the danger was little noticed then. Burning coal is certainly one of the more familiar energy technologies. How many other such dangers have we failed to notice? In the opinion of Bodansky[6]

> By far the greatest risks in the energy domain are the ones that we do not understand well enough to evaluate in quantitative terms: those of war, of carbon dioxide buildup, and of the social and economic disruption that could accompany severe energy shortages.

The fault tree analysis mentioned above is designed to organize the identification of a presumably rare mishap by working backwards from an identifiable end result as well as to organize the analysis once the mishap is known. If there are many branches in the tree, as is typical for a complex system, the number of possibilities to consider can be immense. Each one may have a nearly negligible probability, but to make a useful estimate of the total risk, we need to be rather precise about these very small probabilities. Unfortunately we simply do not know how great some risks are, other than that they are small. In practice some less than completely quantitative judgment is made that certain conceivable scenarios are *de minimis*, that is, so unlikely that they can safely be left out. In determining which risks are *de minimis*, we compare them to the background level of risks that are run independently of energy production. If we never resorted to a *de minimis* standard, we would have an infinite number of calculations of scenarios involving such things as meteors striking power plants while an accident is occurring. Nonetheless, since careful calculations are not actually made, the use of *de minimis* assumptions is somewhat controversial.

Sometimes a risk may be well studied but in circumstances which differ from those of energy production, necessitating extrapolation. This is especially the case for health risks from pollution and radiation. In high doses, it is fairly clear that radiation, for example, can cause cancer. There is a range of exposures for which the probability of developing cancer is approximately linear in the dosage. There are also extensive studies of the effects of radiation on animals. If one is interested in the danger to the public from a low-level exposure, there are serious differences of opinion about how to extrapolate from the hard data at high doses to very low doses or from animals to humans. The extrapolation must account for threshold phenomena, the presence of a nonzero background rate, and the like. For nonlinear systems, even very simple ones like the periodically driven pendulum, linear techniques like extrapolation and superposition can be wildly misleading, and can especially give a wrong impression about stability.

A common alternative to the use of the *de minimis* standard is to aggregate rare events together and then make some conservative assumptions about the risk posed by the aggregate. This procedure uses the background level of risk as a reference, attempting to ensure that the additional risk of, say, building a power plant, is much less than the background. The reliance on the background is officially sanctioned as the measure of acceptable risk. The 1986 NRC Policy Statement advocates that the public should "bear no significant additional risk to life and health" and stipulates that risks "from nuclear power should be comparable to or less than the risks [of] competing technologies and should not be a significant addition to other societal risks."[7] The NRC policy does not set quantitative standard of acceptability, but a typical figure used by the EPA is 20% of background level. Paradoxically, expressing standards in terms of background risks means that standards could become stricter or looser as our understanding of other risks changes. Since the known background radiation exposure has jumped considerably due to measurements of radon in the home, should we tolerate more radiation from mining, power plants, and other sources? Also, if aggregate risks are assessed unevenly in different systems that are being compared, a greater risk may be substituted for a lesser one.

The statistics called for in risk assessment can be subtle. For rare or previously unsuspected events, the uncertainties of the estimates are typically much greater than the estimates themselves. Yet the uncertainty of an estimate is generally much more dependent on assumptions than the estimate itself. Hence a fundamental problem, and a common debating point, is how to deal with uncertainty about uncertainties. There appears to be no easy answer. In addition, although much effort goes into calculation of expectation values, the existence of the insurance industry is proof that expectation values are not what ultimately matters to people who confront risks. No matter how favorable an expected return, it is hard to imagine a probability threshold below which the total destruction of mankind would be a rationally acceptable risk (the arms race notwithstanding). Economic utility theory, however, encourages reliance on expectation values.

The philosophical underpinning of cost-benefit, or risk-benefit, analysis is problematic in other ways besides its emphasis on expectation values. Serious and puzzling paradoxes have been recognized in utility theory from its beginning at the time of Bernoulli, and are certainly not resolved today. People simply fail to act in accordance with what utility theory regards as rational. Studies show that people will choose differently from among the identical choices of risks, depending on whether they are stated in terms of lives saved or lives lost. Secondly, cost-benefit analysis offers no way to cope with matters that cannot be measured or quantified, such as the quality of life, and in effect it encourages setting them aside. It also has trouble dealing with risks for which estimates are impossible for practical reasons. An example of the latter is the risk of hostile human actions, such as attacks by terrorists or arsonists, or ancillary political effects such as nuclear proliferation. No one is ready to attempt to calculate such risks at present, but they certainly occur, and are recognized as very significant. Indeed, von Hippel has termed nuclear proliferation the greatest danger from fission power.[8]

In cost-benefit analysis, the utility function to be calculated is heavily dependent on philosophical and political assumptions. For instance, should one weight everyone's opinions equally, or give more weight to experts' opinions? Some mathematical consequences of various ethical systems for risk-benefit analysis have been presented by Schultze *et al.*[9] These authors also make the point that the usual practices of cost-benefit analysis have the effect of emphasizing short-term thinking and are affected in some unsettling ways by economic conditions of the moment. For example, they report on a calculation using a standard financial risk-benefit formula which demonstrated that a complete loss of the world's entire gross economic product 100 years hence would have a present value of only about -$1,000,000. (At that time the economy was growing slowly and interest rates were high.)

Being quantitative, risk assessment looks scientific, but it differs from hard sciences like physics in many ways. As time goes on, our best estimate of the risk of a given event fluctuates either because new information and experience become available or else because external conditions change, as when the population near a power plant changes. The calculated uncertainty of the estimates does not necessarily decrease. This is very different from the usual situation in physics where further study of a physical quantity such as the speed of light or the gyromagnetic ratio of the electron only increases the accuracy.

Clark[10] and Weinberg[11] have pointed out other ways in which the methodology of risk assessment differs from what is usual in science. For instance, once a risk is conceived, experimentation does not stop until some risk is found. This will almost inevitably happen as measurements are refined. This argument applies especially to the biomedical part of a risk assessment, as in a study of excess cancers caused by radiation or pollution. Researchers use massive doses to produce statistically significant effects in animals, and increase the dose until side effects are seen. In physics, if a hypothesis is not verified, it is not pursued so relentlessly over alternative hypotheses. Clark expresses the fear that the drive for a riskless society may become a mass mania like that of witch hunting, which killed half a million people before playing itself out. He also makes the interesting suggestion that risk assessment itself should be scrutinized from a cost-benefit perspective, and advocates a major retrospective study.

6. The controversy over risk assessment: Practice

As remarked above, historically, the conclusions of essentially every significant risk assessment have been vehemently opposed. Citizens' and environmental groups have been quick to criticize government and industry studies, and have produced their own analyses, which have as quickly been assailed by the other side. While some of the criticism has dealt with the theory of risk assessment as discussed above, the more heated arguments have tended to be over shortcomings of specific studies. There are two reasons for this. As opposed to criticism of risk assessment in principle, practical criticism raises the possibility of doing better and therefore inspires reaction and further studies. Secondly, practical criticism is bound to seem more *ad hominem* than philosophical criticism and is often explicitly coupled with innuendo that the study being criticized is intellectually shoddy.

The most pervasive charges and countercharges are incompleteness and bias or hidden ethical assumptions. Indeed, as we have seen, it is hard to imagine a risk analysis which is truly complete or which makes no assumptions about how to compare disparate sorts of risks. The best we can hope for in this regard is clarity and honesty about the assumptions on which a study is structured. The marvel of the adversarial process is that questionable assumptions will be pointed out by critics if not by authors.

It is not possible here to evaluate specific charges against specific studies, but a number of groups concerned in the debate over risk assessment have listed what they see as generic problems with the methods of existing studies. Some of these charges have already been discussed. One recurring charge is that analysts have not sufficiently considered the human element in building or operating a plant. Human error includes actions that are both intentional and unintentional, and can be either internal to a power plant or external. It is clearly very difficult to project or quantify these human factors. In view of findings that human error was one of the principal problems at Three Mile Island, Chernobyl, the Teton Dam, and many other disasters, human factors must be carefully considered. The NRC has taken this very seriously, and its latest guidelines on risk assessment attempt to be more systematic in accounting for human error.[12]

Another recurring charge is the uneven scrutiny of competing systems, the central point in the controversy raised by Inhaber's studies. It is quite true that risk assessments for such familiar technologies as hydroelectric and coal power are in a surprisingly primitive state. Scrutiny can also be uneven within the same technology when some systems are protected for reasons of national security. An early and egregious example of this was the British Government's suppression of the facts of the Windscale accident in 1957. Defense nuclear reactors continue to have their own sets of standards and receive vastly less scrutiny than civilian reactors. A National Research Council report on safety of defense reactors stated, "Many problems have been obscured for years by an 'ingrown . . . loose-knit . . . self-regulated' system whose errors have been hidden from view."[13] The problem is particularly well illustrated by the continuing revelations at the Savannah River Plant. It has been reported that Du Pont researchers at the Savannah River Plant "knew by 1981 that the assumptions underlying the design of the emergency system were faulty . . . . For the next 5 years they worked slowly and quietly . . . to establish the validity of the 1979 data. They failed. Meanwhile the reactors ran at full power."[14] Because of particularly bad practices allowing contamination, there is now the question of who should pay the bill for cleanup around defense reactors.

Other charges commonly made against risk assessments include:

- Failure to make the study completely longitudinal
- Ignoring important specific risks
- Failure to distinguish risks that differ in kind or occur at different times
- Treating failures with possible common causes as statistically independent
- Overlooking variations in design
- Overlooking the aging of structures
- Overlooking corrections of problems and other benefits of experience
- Overlooking the possible evolution of new risks or decline of old ones
- Poor use of statistical technique and extrapolation
- Overlooking partial failures
- Poor technical understanding of engineering problems

7. The perception of risk

Scientists, engineers, and other technically trained people tend to view risk differently from laymen. Scientists are trained to think in quantitative terms, and are comfortable with and even fond of notions such as the "expected value of the mean rem dosage" or "median worker-days lost per megawatt-year." To understand and compare risks, one must calculate the values of many such quantities. For many citizens such words and numbers are incomprehensible and more terrifying than any power plant. The problem is exacerbated because risk assessment should involve sophisticated statistics and not just calculations of averages.

In scientific thinking details are important. Politicians find details annoying. Those making the decisions about energy production and many other matters often find scientific analyses painful and frustrating. On one memorable occasion, Senator Muskie called for more "one-armed scientists" whose Congressional reports were less in the style, "on the one hand this, and on the other hand that."[15] Undeniably, the public has a poor understanding of numbers: In a recent examination at an Atlanta law school, most students felt that a 1% probability of death was a "negligible risk." These students are certainly not more innumerate than average, and many of them may become politicians. Yet how many of them would dare to fly if, given a comparable risk, an airplane crashed with no survivors on average once an hour at Atlanta's airport?

In fact, people, including scientists, ordinarily make decisions according to rules of thumb or gut feelings rather than by quantitative formulas. Psychologists refer to these feelings as *heuristics*, and many interesting studies have examined how heuristics for facing risk lead to very different impressions than scientific investigation. For example, according to the *availability heuristic*, a risk which is easily imagined is felt to be probable. Dramatic events are also perceived as more probable than they are in reality. The best-known overestimated risk is that of airplane accidents, although homicide, tornadoes, and pregnancy are likewise perceived to cause many more deaths than is actually the case. In contrast, asthma and stomach cancer are greatly underestimated as causes of death. Radon pollution is now recognized as a major source of public exposure to radiation, but as yet people do not feel much urgency about it. (These very words were composed in a basement with a somewhat elevated radon level.)

Figure 4 shows the result of a study comparing the perceived and actual death rates from various causes. This, of course, has great effects on public policy—the public is willing to spend $1,000,000 per life saved on measures connected with air safety but indifferent as to auto safety measures costing $100,000 per life saved. The situation with respect to radioactivity from power plants versus that from radon is analogous and indeed orders of magnitude more discrepant.

The public perception of risk appears unsystematic and inconsistent owing to irrational fear of, or faith in, technology, quirks in press coverage, confusion of vaguely similar technologies (e.g., nuclear power and nuclear bombs), etc. There is no realistic hope that this state of affairs will change in the near future. The educational system would require a full generation to produce a public with a good understanding of science even if, by some miracle, education were to become an immediate national priority. Past errors by experts have certainly encouraged public mistrust and inconsistency. In the light of this, public skepticism about expert testimony is not much more irrational than that of professional managers who must choose whether or not to listen to experts when making decisions. In a democracy, it can be argued, persistent public opinions should be the basis for decisions, even if experts disagree. In a litigious democracy, public opinion will be the basis for some decisions, regardless of what experts think.

For risk management, it is important to understand and respond to public attitudes towards risk. The effort to understand the public injects another layer

	Inhaber	Holdren et al.
Nuclear		
WDL	1.7–8.7	3.1–12
PDL	0.3–1.5	0.3–70
Ocean Thermal		
WDL	23–30	2.2–4.6
PDL	0.8–1.4	0.40–0.90
Solar Heating		
WDL	91–100	11–17
PDL	4.6–9.5	2.1–5.5
Methanol		
WDL	220–350	6.3–6.6
PDL	0.05–0.14	0.005–0.015
Solar-Thermal-Electric		
WDL	62–100	7.4–15
PLD	9.4–520	1.0–2.7
Photovoltaic		
WDL	140–190	5.0–14
PDL	10–510	0.9–2.2
Wind		
WDL	220–290	9.7–10
PDL	22–540	0.21–0.58
Oil		
WDL	2–18	3–19
PDL	9–1900	9–1000
Coal		
WDL	18–73	19–43
PDL	20–2000	20–1500

*   Figures in worker-days lost (WDL), or public person-days lost (PDL) per megawatt-year of output, rounded to two significant figures.

Figure 4. Comparison between the perceived and actual number of deaths per year for various risks, from P. Slovic, B. Fischhoff, and S. Lichtenstein, Rating the Risks, *Environment* **21** (3) pp. 14–20, 36–39, 1979.

of uncertainty into risk management, however. Public attitudes are not always easy to determine. In addition, changeable polls, letters from the public to politicians, and votes are notoriously subject to biases. Press coverage, which plays a great role in forming public opinion, can be both biased and inaccurate. This was the case in particular in the coverage of the Kemeny Report after Three Mile Island. Finally, even if public values could be understood with confidence and built into some sort of utility function representing the benefit to the nation, there remains the paradox that people notoriously fail to act in accordance with utility theory.

Experts belong to the same species as laymen. While the heuristics used by experts tend to be more quantitative and more honed by experience, they are subject to many of the same shortcomings. Fischhoff *et al.*[16] cite some interesting experiments in this regard, including one by Hines and Vanmarke. Hines and Vanmarke asked several prominent engineers to estimate the height of an embankment that would cause a clay foundation to fail, and to specify upper and lower bounds so that there would be a 50% probability of enclosing the right answer. The estimates were both high and low of the mark but not one contained the true figure within its confidence bounds. In light of findings that the 1976 collapse of the Teton dam was due to "unwarranted confidence of engineers," this result is of more than academic interest. In addition, experts on occasion react defensively to opposition, and become arrogant towards public opposition they regard as foolish.

Both the public's apparent inconsistency and the tendency of scientists and engineers to have too many hands for politicians' liking are partially due to risk assessment's multidimensionality. A major catastrophe is not the same as many individual tragedies. Immediate deaths and long-term health problems are not really comparable. Even though actuaries can and do reduce all risks to common monetary values depending on present value of lifetime loss of earning and the like, it does not follow that these comparisons have any validity beyond enabling insurance companies to operate. In recognition of this, the results of public-opinion surveys are increasingly portrayed as risk profiles with categories similar to the ones that have been advocated for risk analyses.

8. Risk management

Complete avoidance of risk is neither practical nor desirable, since we gladly face risks every day in order to achieve benefits. Hence the goal of risk assessment, as recognized by the NRC, is to control and live with risks, not to eliminate them absolutely. In theory there are four appropriate techniques for managing risk:

1. reducing risk;
2. mitigating damage;
3. accepting and sharing risk; and
4. redressing damage.

Reduction of risk can take the form of designing safe power plants, siting plants away from population centers, etc. Society is less disrupted by tragedies that are widely distributed. For example, compare the problems of hospitals in a hurricane or earthquake with those coping with an equivalent number of ordinary traffic accidents around the country. We should certainly prepare for the possibility of accidents. Licensing procedures for power plants require crisis plans to be drawn up. Education of the public about the risks of energy and other technologies has often been advocated in order to mitigate damage if an accident occurs, but it is not clear that education has had much effect.

Sharing risk raises issues of fairness since ideally risks should be borne by the beneficiaries. Because a power plant or waste site is necessarily only in some people's backyards and not endangering everyone equally, it is sometimes argued that people should be compensated monetarily for risks in advance of injury. Risks can also be shared through insurance or taxation and are in effect shared through the economic system by rises in prices reflecting industrial expenditures for safety. Redressing damages raises similar issues except that it refers to actions taken after an accident rather than in anticipation of one. Part of the mission of the NRC has been to promulgate specific standards and recommendations for risk management in these four categories.

In practice, two other techniques for coping with risk, in addition to the four listed above, are especially popular:

5. litigating, and
6. sticking one's head in the sand.

Litigation, of course, can be a vehicle for managing risk under any of the first four items mentioned, but it is terribly costly and time-consuming and is rightly notorious for being inconsistent. While some plaintiffs have received little compensation for their injuries, others have been compensated merely for the fear of cancer in a case that could have significant implications for the nuclear-power industry. Fear of litigation may be far more significant than litigation itself, and may lead firms and agencies to spend time protecting themselves which might be better spent pursuing safety.

Those of us who are not at the moment waiting in gas lines or suffering power shortages may prefer to think that it is best not to worry too much about difficult questions having to do with energy production. We can find support from eminent authorities who claim that resources are not finite, the environment is not disrupted, and things are generally rosy.[17,18] It is only natural that our politicians might take similar comfort, especially since stands on controversial issues may lose more votes than they gain. Perhaps this helps explain why laws on such topics as dam safety are often too vague to use, or why the California legislature has become so fond of avoiding hard questions of risk assessment and similar technical matters by leaving them to voter initiatives like California's Proposition 65, which requires businesses to label products for a host of possible carcinogenic and chemical hazards.

Actually, a serious argument that can be raised in favor of inaction is that the record of risk management is mixed. Efforts at controlling risk not infrequently make things worse. Coates[19] estimates that major technological blunders occur about once a week, and a large fraction of these are attributed at least in part to attempts at risk management. The examples cited in the references are not primarily related to energy (Love Canal, flood damage being exacerbated by efforts at flood control, and other environmental disasters), but there is no reason to think that the potential for mismanaging complex power systems is less than that in these other situations. Indeed, the accident at Chernobyl was touched off by unnecessary testing of safety in the event of a hypothetical power failure. In addition to the difficulties in understanding complex systems and the ever-present possibility of simple error, the failures of risk management are attributed by most authors to bureaucratization. It does not seem, however, that there is a useful alternative to having a bureaucracy of some kind oversee the nation's efforts at risk management.

9. Conclusions

Risk assessment, with all of its limitations, is here to stay. It has become an integral part of engineering whenever there is a question of safety. The public has benefited from it through improved designs of power plants and many other manufactured systems. Researchers in the field constantly learn from experience, whether of the engineering or the legal and political sort, and so risk assessment will no doubt become more and more professional, sophisticated, and useful although there will always be occasional errors.

Difficult questions remain in

(1) forecasting and quantifying rare risks in absolute terms,

(2) comparing the quite disparate risks of competing technologies as well as the risks of inaction, and

(3) using risk analysis properly in management and the public arena, in particular balancing risks intelligently against benefits.

There has been significant progress, for instance in the realization that comparisons have to be longitudinal, from raw materials to end use of energy. It is not likely that the limitations of risk assessment in the area of policy will disappear. Yet it is difficult to conceive of a good alternative to making assessments and attempting to manage the risks we know. Determining the value of risk assessment is no easier than assessing risks themselves, and it will always be controversial.

Bibliography

Science **236**, 267–300 (17 April 1987). Many good articles on various aspects of risk assessment from some of the leaders of the field.

Health Risks of Energy Technologies, C. C. Travis and E. L. Etnier, eds. (Westview Press, Boulder, Colorado, 1983). *AAAS Selected Symposium*, Vol. 82. A collection of useful articles on various subjects, and containing some calculations.

R. Wilson, Chairman, "Report to the American Physical Society of the study group on radionuclide release from severe accidents at nuclear power plants," Rev. Mod. Phys. **57**, No. 3, Part II (July, 1985). This contains not only analysis but a good review of previous studies.

References and notes

1. As quoted by D. Okrent, "The Safety Goals of the U.S. Nuclear Regulatory Commission," Science **236**, 299 (1987).
2. Tabulated in S.C. Sholly and G. Thompson, *The Source Term Debate*, A Report by the Union of Concerned Scientists (Union of Concerned Scientists, Cambridge, Massachusetts, 1986).
3. A. Weinberg, "Science and its Limits: The Regulator's Dilemma," in *Hazards: Technology and Fairness* (National Academy Press, Washington, DC, 1986), p. 12.
4. H. Inhaber, "Risk with Energy from Conventional and Nonconventional Sources," Science **203**, 718 (1979); *Energy Risk Assessment* (Gordon and Breach, New York, 1982).
5. R. Monastersky, "New Chemical Model, New Ozone Fear," Science News **134** (September 3, 1988), p. 148.
6. D. Bodansky, "Risk Assessment and Nuclear Power," J. Contemp. Stud. **26** (Winter 1982).
7. D. Okrent, *loc. cit.*
8. F. von Hippel, "Global Risks from Energy Consumption," Chap. 5 of *Health Risks of Energy Technologies*, C. C. Travis and E. L. Etnier, eds., AAAS Selected Symposia, Vol. 82 (Westview Press, Boulder, Colorado, 1983).
9. W. D. Schultze, D. S. Brookshire, and R. C. d'Arge, "Economic Valuation of the Risks and Impacts of Energy Development," Chap. 6 of C. C. Travis and E. L. Etnier, eds., *op. cit.*
10. William C. Clark, "Witches, Floods, and Wonder Drugs: Historical Perspective on Risk Man-

agement," pp. 287–313 in *Societal Risk Assessment: How Safe is Safe Enough?*, R. C. Schwing and W. A. Albers, eds. (Plenum, New York and London, 1980).

11. A. Weinberg, *op. cit.*
12. *Reactor Risk Document*, NUREG-1150 (U.S. Nuclear Regulatory Commission, Washington, DC, 1987).
13. As quoted by Eliot Marshall, "Safety of DOE Reactors Questioned," Science **238**, 714 (1987).
14. *Ibid.*
15. B. Fischhoff, P. Slovic, and S. Lichtenstein, "Judging the Health Risks of Energy Systems," Chap. 1 of C. C. Travis and E. L. Etnier, eds., *op. cit.*
16. *Ibid.*
17. A. Hite, "Chicken Little Was Wrong About Oil, Too," Wall Street Journal, p. 22 (3 February 1988).
18. J. L. Simon, "Resources, Population, Environment: An Oversupply of False Bad News," Science **208**, 1431 (1980).
19. J. F. Coates, "Why government Must Make a Mess of Technological Risk Management," Chap. 3 of *Risk in the Technological Society*, C. Hohenemser and J. X. Kasperson, eds., AAAS Selected Symposia, Vol. 65 (Westview Press, Boulder, Colorado, 1983).

Source Technologies

Fossil fuels: Coal, petroleum, and natural gas

Anthony Fainberg

1. Introduction

Since the 19th century, fossil fuels[1] have provided the U.S. with the lion's share of its energy. Coal has been mined for centuries. Petroleum has been extracted for energy use in the U.S. since 1859 when the first well was established in Titusville, Pennsylvania. Since the end of the 19th century, natural gas, often found in conjunction with liquid petroleum, has provided useful energy. In 1925, it contributed only 5% of the fuel supply, but this fraction rose to 30% by 1970. The U.S. now relies upon fossil fuels for some 89% of its energy consumption (see Table 7, Chapter 1). Petroleum and petroleum products provide about half. The two important nonfossil energy sources in the U.S. are nuclear power and hydroelectric power. Renewable resources (other than hydro) supply some energy, but after nearly two decades of great interest and much research, their market penetration is still insignificant.

There are two fundamental problems with U.S. reliance on fossil fuels. United States production of petroleum and natural gas appears to have peaked, and the petroleum production rate is now in decline. As a result, the U.S. now imports nearly 50% of the petroleum that it uses. Foreign dependence has increased markedly since the mid-1980s. It is questionable whether it is wise for America to rely so heavily on foreign suppliers for what amounts to its lifeblood.

Energy from coal could, in principle, substitute for a major portion of this foreign petroleum. However, burning coal has undesirable effects on the environment. Burning any fossil fuel releases unwanted, even deadly, chemicals into the atmosphere. Some of these products, such as sulfates, may cause emphysema and other upper respiratory diseases; some are carcinogens; and one, carbon dioxide, is accepted as the principal cause of the greenhouse effect, that is, atmospheric heating due to changes in the infrared absorption properties of the atmosphere. Another combustion product, nitrous oxide, also contributes to greenhouse heating as do methane, ozone, and chlorofluorocarbons. The greenhouse effect could cause severe local climate changes that may result in famine, flood, death, and destruction. Among the fossil fuels, coal produces the most pollutants.[2]

Heavy reliance on coal would exacerbate most of the fossil fuel environmental problems. Natural gas, on the other hand, may be offensive mainly in its production of carbon dioxide, and even in this respect it produces less, per unit energy, than oil or coal (about 58% of coal's CO_2 production and 69% of oil's).[3] Most of the negative environmental effects of fossil fuel production (except for carbon dioxide production) could be greatly reduced by use of filters, scrubbers, or advanced technologies, such as fluidized bed combustion or synthetic gas.

This chapter will attempt to put these issues into perspective and will attach numbers and uncertainties to specific supply and environmental problems, so that the reader may better understand the advantages and disadvantages of our heavy reliance on fossil fuels. This understanding, together with the understandings gained from chapters dealing with alternative energy sources, should aid the reader in making her/his own decisions regarding energy supply policies.

2. Brief review of U.S. and world fossil energy supplies

Since much of this information is already contained in Chapter 1, the material in this section will be condensed, particularly regarding domestic supplies.

2a. Petroleum

In terms of conventional recoverable resources of crude oil, the U.S. has already produced 142 billion barrels out of an estimated original supply of 226 billion. Thus, about one-third remains.[4] The U.S. produces approximately 8.1 million barrels per day (mbd) and 1.6 million barrels of natural gas liquids (NGL—liquids found with natural gas, such as ethane, pentane, propane, etc.) per day. About 43% of the total energy budget comes from oil and petroleum products. In 1988, the U.S. consumed about 17.2 mbd and imported about 7 mbd of crude oil and refined products.[5] Principal suppliers include the United Kingdom, Norway, Mexico, Venezuela, Saudi Arabia, and Nigeria. Of these, Saudi Arabia has by far the most remaining resources.

Of the foreign suppliers, the U.K. and Norway might be considered the politically most secure sources followed by Venezuela and Mexico. Saudi Arabia and Nigeria will probably be relatively sure sources barring major shifts in political leadership and in attitudes towards the U.S. Unfortunately, these possibilities cannot be excluded. If oil shortages were to develop for political or other reasons, even the sure suppliers might not be able to supply the U.S. with sufficient oil. Further, the U.K. has already produced about one-fourth of its resources (7 billion of 27 billion barrels). Its North Sea field will probably peak in production within 5–10 years and then decline fairly rapidly. Venezuela (+ Trinidad) and Mexico have produced about 40% and 20% of their resources, respectively—about 55 billion barrels remain to each.[6]

The source of about one-fourth the domestic U.S. supply, the North Slope of Alaska producing 2.2 mbd in 1988, began to decline in 1989. The decline will steepen by 1993–1995. A recent set of projections of crude oil production in the U.S. for 1995 ranged from 5.2 to 6.9 mbd; for 2000, the range was 4.5–6.1 mbd.[7]

Moreover, estimates of undiscovered U.S. resources have dropped from 82

Table 1. United States oil resources (in billions of barrels).

Orignal proven conventional recoverable resources	226
Already produced	142
Remaining	84
Estimated undiscovered resources	46
Domestic production (per year)	3
Domestic consumption, including imports (per year)	5.5
Years left. under current production conditions, and no increase in imports	28–43

Note 1: If imports decrease or use increases, the number of years left will be smaller.
Note 2: As supplies shrink, increasing costs will decrease use, so reserves will increase number of years left.

billion barrels, in 1981, to 46 billion barrels in 1987. When the estimated 84 billion remaining barrels run out (at present rates, in 28 years, although production levels will decline drastically much earlier), those 46 billion barrels, if economically recoverable, would provide 15 years supply at current production rates. Table 1 summarizes this information.

A limited set of choices will face the U.S. in as little as 5–10 years when the Alaskan supply begins to drop significantly: greater reliance on foreign sources, development of unconventional and more expensive oil resources, substitution for petroleum by other energy sources, or conservation. The largest foreign petroleum reserves are held by Saudi Arabia and the Soviet Union. Political and financial constraints may render unconventional resources unattractive, and most advanced extraction techniques are unlikely to become significant within the next 5–10 years.

Conservation can already claim clear accomplishments. Petroleum consumption peaked in 1978 at 18.8 mbd. From 1981 to 1987, consumption was significantly lower (15.2–16.7 mbd) while the Gross National Product has continued to grow. A shallow minimum in petroleum consumption coincided with the recession year of 1983, and a slow upward trend followed. In 1988 and 1989, the upward drift accelerated. Although consumption is still (at this writing) below the peak of 1978, the recent upward trend fueled by relatively low oil prices shows that the lessons of the oil shortages of the 1970s might have been forgotten until the Persian Gulf Crisis of 1990.

Nuclear fission power now provides the primary energy equivalent of about 2.5 mbd of oil. Although it is often difficult to prove a "one-for-one" substitution of nuclear power plants for oil-fired electric generating plants, it is likely that nuclear power has reduced oil consumption by at least 1 mbd. Political opposition and economic considerations have halted the expansion of nuclear power in the U.S. Contrary to a widespread impression in this country, many other nations including Japan, France, China, and Korea continue to expand their use of nuclear power. Since nuclear plants now appear to require 8–10 years from concept to kilowatts, a major increase in energy from nuclear power is unlikely for at least 10–15 years.[8] In any case, direct electricity generation provides few remaining opportunities to save petroleum, although increased use of electricity in home heating could, in principle, substitute nuclear or other non-oil generated energy for oil.

Conservation can reduce petroleum use for transportation. Substitution of other transportation fuels, such as natural gas and coal, is feasible although it would require the manufacture of synthetic liquids or retooling the automobile and truck manufacturing industries. Substitution of natural gas or even coal for heating oil would be relatively easy to accomplish. Efforts are underway to advance the use of methanol instead of petroleum, but on environmental, not conservation grounds. Renewable resources might replace some petroleum, but the outlook for significant market penetration within ten years looks dim barring major unexpected price reductions. Fossil fuel substitution and conservation appear to be the best options for addressing oil shortages in the near term.

2b. Natural gas

The U.S. currently produces about 17 trillion cubic feet (tcf) of natural gas — mainly methane CH_4 — per year, and some 23% of U.S. energy comes from natural gas. Estimates of total conventional resources vary widely. In the lower 48 states, there are around 267 tcf of known reserves recoverable at less than \$3 per 1000 cubic feet. Alaska might add about 12 tcf. In addition, some 105 tcf are available for extraction at less than \$5 per 1000 cubic feet (commonly denoted mcf), approximately a 6 year supply at present consumption rates. The current price is about \$1.70 per mcf. These figures imply that the U.S. will run out of natural gas in less than 25 years at current use rates. However, each year, estimates of known recoverable reserves grow, recently by some 14–15 tcf per year.

In addition to the known conventional resources, there may be 353 tcf (229 tcf at less than \$5) of undiscovered conventional resources and a possible 146 tcf of unconventional resources at \$5 per mcf.[9] However, a preliminary 1987 U.S. Coast and Geodetic Survey Report that reassessed undiscovered conventional resource estimates reduced them from 427 tcf to 254 tcf. Other estimates of conventional (known plus undiscovered) resources in the lower 48 states vary by a factor of nearly four from 244 tcf to 916 tcf.[10] Different bodies of experts continue to differ significantly on the question of remaining natural gas resources in the U.S.

The feasibility of developing unconventional sources within the next decade or two remains questionable. The Congressional Office of Technology Assessment has estimated that "tight gas," that is, gas trapped in rock of low permeability, could contribute 1–4 tcf per year in the year 2000. Coal seam methane is found in conjunction with coal deposits. There may be hundreds of tcf of methane in this form, although only a small fraction may be recoverable economically in the near future. Recent advances in methane recovery lead to a real possibility that some of it may be recoverable at low prices.

Natural gas has been adsorbed onto Devonian shale structures in the Appalachian, Illinois, and Michigan basins. Releasing it would entail fracturing the rock and overcoming the low permeability of the shale to extract the gas. Esti-

Table 2. United States natural gas resources (in trillions of cubic feet).

Proven conventional recoverable resources (including Alaska, and at less than $5 per thousand cubic feet)	384
Production rate (per year)	17
Recent yearly addition to proven recoverable resources	14–15
Estimated total remaining conventionally recoverable resources (lower 48)	400–900
Estimated unconventional recoverable resources (price of recovery not determined, but probably high)	140–700
Total estimated resources	540–1600
Years left at current rate	35–95
Years left at double current rate	17–47

Note: Current cost of natural gas is about $1.70 per thousand cubic feet.

mates for production from this source range from 0 to 1.5 tcf per year by the year 2000.[11] Other more exotic sources of natural gas are highly speculative.

After examining many estimates, the Office of Technology Assessment has assigned a range of 400–900 or more tcf for recoverable conventional resources remaining in the U.S. with exotic resources at 140–700 tcf or more. At 17 tcf per year, this implies 34–94 years of supply at current levels of use. If we replaced one-half of petroleum energy with domestic natural gas, keeping the same energy use pattern including the assumption of zero energy growth, the U.S. would need to double its natural gas production. In that case with the above assumptions, the total domestic supply would last from 17–47 years.

The largest foreign reserves of natural gas are in the Soviet Union which produces some 26 tcf annually and has about 2600 tcf left. The Middle East has another large potential supply, about 2100 tcf.[12] All estimates have significant uncertainties associated with them. In the Middle East, a considerable amount of natural gas is wasted by flaring, burning natural gas associated with liquid petroleum. Thus, production could be increased fairly quickly by curtailing this practice. Flaring is not common practice in the U.S. Table 2 summarizes these data.

2c. Global view of petroleum and natural gas supplies

As of early 1987, an estimated 1227 billion barrels of oil remained in the world, about two-thirds the original supply.[13] Some 57 million barrels are produced per day.[14] At this rate, 60 years' supply is left. Undiscovered and unconventional resources might double the ultimate available resources. However, third world use of petroleum is expected to increase considerably within the next decades with a demand arising from increased population and a strong desire for a far better standard of living. In that case, barring widespread substitution and significant conservation among major users, severe pressures on petroleum supplies are likely to occur within at most 30–50 years.

The same estimates[15] indicate that about 88% of the world's natural gas supplies remain: some 8100 tcf. About 68 tcf are produced per year, leaving 119 years' supply. Note, though, that substitution of natural gas for petroleum

Table 3. Global resources.

Oil (in billions of barrels)	
Original resources	1900
Produced	673
Remaining	1227
Production (per year)	21
Years left at current rate	60
Note: Undiscovered and unconventional resources could approximately double the total supplies, but at undetermined cost.	
Natural gas (in trillion of cubic feet)	
Estimated remaining resources	8100
Production rate (per year)	
Years left at current rate	120

would reduce this number considerably, as would a greatly increased world energy consumption.[16] Table 3 displays the data for oil and natural gas.

2d. Coal — U.S. and world

In terms of proven reserves, the U.S. has a disproportionately large amount of coal — nearly one-quarter of the world's supply. At current production levels of 0.7 gigatonne equivalent per year, domestic coal would last for 300 years. If all current U.S. energy were derived from coal combustion, the supply would still last 75 years. In addition, geological considerations indicate the possibility of an additional supply of over ten times more.

The world's supply of proven coal resources would last about 250 years at current rates of consumption; an estimated 14 times as much coal could exist in potential geological resources. Therefore, coal could, in principle, supply the world's energy needs for centuries, even allowing for greatly increased use and for replacement of other fossil fuels. But could the biosphere reasonably stand the consequences of the massive release of carbon dioxide and sulfates into the atmosphere that such heavy coal burning would entail?

3. Current use of fossil fuel

3a. Petroleum

Petroleum products are used for transportation, for producing heat in industry and residences and commercial establishments, and, to a limited degree, for producing electricity. In 1986, 63% of petroleum use went into the transportation sector, 24% into industrial uses, 8% into the commercial and residential sectors, and less than 5% into electricity production.[17,18]

Since the 1950s, gasoline for transportation (automobiles and light trucks) has accounted for about 40% of petroleum products. Distillate fuel oil, including diesel fuel, is used for heat and transportation and has consumed a steady 18% over the past 30 years. Residual fuel oil left over after the refining process constitutes about 11% of petroleum products. This highly viscous oil must usually be preheated so that it can be pumped. This limits its utility to large boilers, including those used on ships, and its use has declined by roughly a factor of two over the last three decades. Another major petroleum use is for jet fuel, but this accounts only for about 7% of petroleum production.[19]

Looking at the breakdown of consumption another way, nearly all gasoline is used for transportation; distillate fuel oil is roughly evenly split between transportation and space or industrial process heating; and residual fuel oil is used to generate electricity, on ships, and for industrial heat and steam. Direct use of heat is far more efficient than using heat produced by petroleum burning for other purposes. Electric generation is about 30% efficient, whereas boilers have (theoretical if not always practicable) efficiencies in the 80%–90% range.

In 1984, the Office of Technology Assessment estimated petroleum use could be cut on a crash basis by about 5 million barrels per day through substitution over a five year period by natural gas, coal, wood, and electricity.[20] Undertaken to discover how much petroleum product use could be reduced in case of a major disruption of oil imports, the study also found that increased efficiency in building insulation and automobile transportation could reduce oil use considerably. After five years of non-emergency effort, the savings in heating use could be about 0.2 mbd. Transportation savings from more efficient automobiles could be up to 17% or about 1 mbd. After the same switching period, industrial savings could be about 0.5 mbd.

3b. Natural gas

Natural gas is used primarily for residential and industrial heating, but also for electric generation.[21] Its use for electricity production increased until the late 1970s, when supply concerns led to a cutback. In 1988, only about 17% of natural gas production was used to produce electricity. Gas-fired generators usually provide power only when there is demand for an additional increment of generation. In the same year, only 10% of the Nation's electricity came from natural gas, down from 19% in 1973.

Gas turbines with capacities of tens of megawatts convert the energy content of natural gas to electricity. These turbines are similar to jet engines which use gas combustion to turn the turbine blades. The turbines are directly connected to electric generators that produce electricity by placing rotating conductors in a strong magnetic field. When the turbine is operated in a combined-cycle system, the gas turbine is combined with a steam generator for heat recovery and to drive a steam turbine. Combined cycle systems may obtain efficiencies as high as 50%.[22] Steam-injected gas turbines, in which some produced steam is injected directly into the combustion chamber, may reach efficiencies that are somewhat higher.[23]

Remaining natural gas consumption is about evenly divided between residential/commercial heating and industrial processing. Natural gas and natural gas liquids, such as propane, are commonly used in homes for cooking and hot water heating. Natural gas is used as an energy source in the field by the petroleum industry and for petroleum refining. Natural gas liquids, along with petroleum, are also an important feedstock for the petrochemical industry, producing drugs, plastics, and textiles. Of the 22.5 quads of natural gas produced in 1973, 3.75 quads were used in electricity generation. In 1986, only 16.7 quads were produced, of which 2.7 were used for electricity.[24]

3c. Coal

Coal use peaked in the early part of the 20th century at around 600 million tonnes per year. As other sources took its place in transportation (railroads), industrial processes (particularly steel), and home heating, its use dropped irregularly until the 1960s when increasing demand for electricity drove coal production levels up again. Coal electric production was less than 9 quads in 1973 and grew to over 14 quads by 1986. Today over 80% of coal use is for electric production; and the total coal component of U.S. energy production has risen from 13 quads in 1973 to 17 quads in 1986.

Although coal used to be widely utilized for transportation and home heating, cleaner fuels have replaced it in both these sectors. Apart from electricity, the only other significant use of coal is in industrial processes, primarily for coke production in steel mills.[25] Like NGL, coal is also used as a feedstock for drugs and plastics.

4. Opportunities for improvements

4a. Residential/commercial heating

Improving the efficiency of existing systems, such as boilers and furnaces, can reduce the demand for fossil fuels. These systems are already above 80% efficient if they are properly adjusted, which is often not the case. Insulation and sealing of buildings have the potential for much larger energy savings.[26]

Another technology currently making inroads in those parts of the country with moderate winters is the heat pump. Heat pumps work much like refrigerators by moving heat from one location to another with the alternate condensation and vaporization of a circulating gas, such as freon. Whereas a furnace *creates* heat by chemical combustion, a heat pump uses far less energy because it just *moves* the heat from one place to another. In the summer, heat is removed from indoors and released outdoors to provide air conditioning. In the winter, heat is removed from the colder outside air and deposited indoors. Commercial heat pumps become far less efficient at outside temperatures of 0 °C and below so they are usually supplemented with electric heating. Current heat pumps use electricity as an energy source, and the efficiency of about 30%–35% in converting the fossil fuel to electricity at the power plant must be considered when calculating relative efficiencies of heat pumps and in-house furnaces (except, of course in those cases, such as hydro-powered electricity, which do not have the thermal energy conversion problem). Most current heat pumps move 1.7 units of heat per unit of electricity, but the most efficient unit on the market in 1982 could move 2.6 units. These devices are efficient enough to compensate for the inefficiency of fossil fuel electricity production, and make the heat pump about as efficient as excellently tuned in-house furnaces or boilers.[27]

4b. Industrial

Cogeneration. One can combine electricity-producing turbines with steam production for process heat, using steam turbines or steam-injected gas turbines. Cogeneration as a strategy for saving on energy bills is becoming increasingly popular in U.S. industries. Costs for installed systems can be as low as $1000 per kilowatt for systems on the order of 10 MW capacity, and considerably less for larger systems. The overall fuel use efficiencies of cogenerating systems can be on the order of 60%–80% or higher.[28] If electricity alone is produced, fuel efficiency is usually only on the order of 30% for classic technologies, and 50% for advanced turbines.

Other improvements in industrial processes are also quite feasible. Computer control systems that are able to monitor and regulate energy-intensive industrial processes can realize added efficiencies. Improved housekeeping techniques (maintenance and repair) can also reduce energy use, as can recycling waste heat. The Office of Technology Assessment estimated that:[29]

"Industry has the potential to replace about 25 percent of its 1985 level of fuel oil consumption through increased efficiency and process changes . . . while maintaining production levels."

4c. Transportation

New U.S. passenger cars travel an average of 24–28 miles per gallon (mpg). Twenty years ago, they averaged 12 mpg. Significant improvements are still quite feasible. Many Japanese models now exceed 40 mpg, and research indicates that mileages of 60–90 mpg may be possible. Doubling the average mileage could reduce transportation energy costs by half, although increased driving might reduce this gain.[30]

Mass transportation has a lower energy cost per passenger mile than private vehicles. Since the energy crises of the 1970s, use of mass transportation has increased only slightly. Bus routes have reappeared in many suburban and semi-rural areas from which they had disappeared a decade or two earlier. In addition, many cities have shown renewed interest in urban transportation, and several subway systems have been constructed in recent years. Light rail systems have been built or are being considered in such moderate-sized U.S. cities as San Diego and Sacramento. Increased use of mass transportation could have a small but significant impact on fossil fuel consumption for transportation. Note that such systems often use electricity rather than fossil fuels. To the degree that the local electricity is not fossil fuel generated, a reduction in fossil fuel use will result.

Biomass can potentially replace some fossil fuels for transportation. Notably, biomass-generated ethanol or methanol could replace gasoline. Ethanol has been employed in Brazil, but the positive energy budget from this process relies on labor-intensive (rather than energy-intensive) methods of agriculture and the widespread cultivation of sugar cane, neither of which may be feasible in the U.S. Ethanol is still used in the U.S. as an additive to gasoline, but because of its high cost and the drop in the price of oil, its utilization has dropped considerably over the past few years.

Methanol, produced from coal, wood, or natural gas, is being investigated as a possible future energy option. Some air pollution problems, in particular ozone production, could be reduced by widespread use of this fuel. Emissions of NO_x from cars might be reduced significantly. On the other hand, methanol produces formaldehyde, which is converted by sunlight into smog, and more hydrocarbons and carbon monoxide than diesel fuel.[31] Recent attention has been focused on methanol substitution by local authorities in California and by President Bush. The Bank of America has a fleet of 500 methanol-fueled cars, and the General Services Administration is considering the purchase of 5000.[32] Methanol-fueled buses are being tested by Triboro Coach in New York.

The use of electric cars or hydrogen cars with the hydrogen produced by electrolysis of water would shift energy use from fossil fuels to electricity. However, if the electricity were produced by fossil fuels, there might be a loss in efficiency rather than a gain because of the low efficiency of electric generation.[33] If the electricity were produced by other means (nuclear, solar, hydroelectric, wind), fossil fuel use would be reduced. The difficulty with electric cars appears to be that they are still a long way from commercially practical development except for such restricted uses as limited urban driving. The large battery masses required and the time needed to recharge have imposed practical limits. Cars have been produced which can cruise at 50 miles per hour, but acceleration is often slow and ranges are still on the order of 50–100 miles. Recharging requires several hours at best. Recently, however, both General Motors and Volkswagen have expressed new interest in this technology. General Motors intends to market an electric car, the Impact, within a few years.[34]

Hydrogen-based cars are in an early stage of development. Although some projections of the technology are quite optimistic, hydrogen is difficult to store and handle safely.

4d. Electricity

Aside from higher-temperature boilers, which increase efficiencies marginally, cogeneration presents the biggest potential for increasing the efficiency of electric power plants. Instead of using waste heat to produce electricity for local use as is done in industry, the power plant would use the waste heat from electrical production to provide moderate temperature heat for use by communities around the plant. This has been done in many countries for many years. Sweden and the Soviet Union are considering power plant designs which accomplish cogeneration economically. For years, Consolidated Edison has used waste heat to produce steam for heating purposes in New York City. The cost of retrofitting existing plants for cogeneration might present economic difficulties. However, any new electric plant, either fossil or nuclear, should be seriously considered for this option.

Efficiencies up to twice those of boiler-fired plants could result from the use of gas turbines to produce electricity while using the excess heat for industrial processes. This would be an attractive alternative if one were sure of a long-term domestic supply of natural gas.

On the negative side, the overall cost of energy from a coal or oil-fired plant would be increased by about 10% with the addition of scrubbers to reduce acid rain-producing sulfates by 90%. NO_x emission reductions would require other technologies that might raise costs more.

5. Risks

5a. Greenhouse effect

Carbon dioxide and other gases in the atmosphere are more transparent to shorter wavelengths of electromagnetic radiation than to some of the longer infrared wavelengths. The shorter wavelengths from solar radiation penetrate to the surface of the Earth. The reflected radiation from the Earth, which is substantial in the infrared region (around 10–20 um), is trapped near the Earth's surface by those gases in the atmosphere. As a result of this "greenhouse effect," the Earth's equilibrium temperature is higher than it would be without those gases. Water and carbon dioxide are the most important of these greenhouse gases. Others are methane, nitrous oxide, ozone, and chlorofluorocarbons. About half the greenhouse temperature increase would be due to man-made CO_2, which raises atmospheric concentration above the preindustrial level, and half due to the increases in concentrations of the other gases. Among the others, $CFCl_3$ and CF_2Cl_2 have the largest effects per mole. Methane, whose atmospheric concentration is increasing by about 1% per year for reasons largely not understood, also has a much larger effect than CO_2.

If one calculates the mass of the atmosphere and assumes that one-half of the injected CO_2 is retained by the atmosphere, one finds that the release of 1 gigatonne of carbon is equivalent to the addition of 0.47 parts per million CO_2 by volume. Each year, about 5.5 gigatonnes of carbon are burned, and an increase of about 1.3 ppm is observed. This implies that about one-half of the CO_2 added to the atmosphere is retained, neglecting the effect of deforestation which is not well known.[35] The preindustrial level of CO_2 was about 275 ppm, and the current (1987) value is 350 ppm, an increase of 25%.[36]

An empirical estimate for future concentrations[37] finds the formula

$$C = 330 \ e^{0.0056(t-1975)}$$

where t is the calendar year and C the concentration of CO_2. This formula is only based on past trends in consumption and is probably quite wrong in forecasting the future, but provides a useful basis for discussion of what would happen if we continue as we are. It predicts a value of 450 ppm atmospheric CO_2 by 2030, or an increase of 28% over the current level. Together with the increase in other trace gases, an average global temperature rise on the order of 3 °C is predicted.[38]

If there were a switch from petroleum to coal, the increase would be greater since coal produces 25% more CO_2 than petroleum. Given current U.S. apportionment of fossil fuel use, and extending this assumption to the global scale, a switch from oil to coal would increase by about 12% the amount of CO_2 put into the atmosphere, and increase the temperature proportionally (by 6% — since only one-half of the increase in temperature will be due to CO_2 increases).

Analyses of the greenhouse effect differ in detail, but most estimate that a doubling of CO_2 concentrations (**not** considering trace gases) over current levels would increase average global temperatures by 2–5 °C. If trace gases are considered and if they roughly double the effect, the empirical formula would predict an effective CO_2 increase of about 41% over current levels by the year 2047.

The poles are expected to warm up by larger amounts — 5–8 °C, melting part of the Antarctic and Greenland ice caps. Recently large blocks of the Ross Ice Shelf have broken off into the Antarctic Ocean (one recent chunk was 150 km long), and there are other indications of a reduced year-round ice cover in the polar regions.[39] Whether these events actually indicate a trend resulting from man-induced global warming is not yet established. More recent observations indicate the Greenland ice pack has actually increased slightly over the last decade, but this is consistent with increased temperatures, since higher temperatures would lead to increased precipitation over polar regions.

Higher concentrations of CO_2 could trigger a "runaway" to even more catastrophic temperature levels. Positive feedback would occur when snow or ice were present at higher latitudes for shorter periods of time. More solar energy would then be absorbed by the darker ground or ocean than would have been absorbed by white snow or ice.

Another potential positive feedback mechanism may be the release of methane trapped in the permafrost. This is one suggested source for the observed increase in atmospheric methane. Methane from organic material that was alive as recently as a small fraction of the half-life of ^{14}C (about 5700 years) should have roughly the same ratio of isotopes as found in atmospheric carbon dioxide. Data showing a small decrease in the ratio of the unstable ^{14}C isotope in atmospheric methane to the stable ^{12}C and ^{13}C isotopes appear to indicate that some 20% of the methane added to the atmosphere today has been isolated from the biosphere for thousands of years and contains no ^{14}C.[40] This quantity of methane cannot be easily accounted for by coal mines or gas wells. If the source is the methane trapped in the permafrost, more methane will be released as the Earth increases in temperature. The methane, in turn, increases the temperature further through the greenhouse effect.

Yet another positive feedback mechanism (although probably of marginal importance) would occur when hotter summers increase energy demand for cooling. In the absence of mitigating effects like conservation or use of fuels that do not produce CO_2, the extra energy demand would put more CO_2 in the atmosphere and thus increase the greenhouse effect further, making future summers hotter still.[41] Moreover, in some areas, such as the central part of the North American continent, significantly less rainfall is predicted in a hotter world. Thus a prime energy alternative to fossil fuels, hydropower, would be reduced.

There is also the possibility of a negative feedback effect with the increased production of clouds. Hotter temperatures would evaporate more water from the oceans to form clouds that would reflect incident sunlight and reduce the amount of solar radiation reaching the Earth. On the other hand, clouds would increase the trapping of infrared radiation by the atmosphere. Increased cloud cover might tend to mitigate the greenhouse effect globally. However, the net effect is not well understood and is the subject of current research.[42]

Relative to coal, petroleum produces about 20% less CO_2 per unit energy produced. Use of oil instead of coal would increase the CO_2 doubling time, assuming no change in total energy production. Domestic and world supplies would likely cause shortages in the U.S. long before this period elapsed. At best, one could reasonably plan on 30–50 years of continued heavy reliance of oil followed by a switch to other energy sources.

Natural gas is the cleanest fossil fuel. Per unit energy, it creates about 55% of the CO_2 produced by coal. Replacing the current energy mix with natural gas alone would increase the doubling time at constant energy production. It is not clear that domestic supplies would be sufficient to satisfy energy demand for the 60–70 year period required.

5b. Acid rain

Acid emissions from fossil fuel plants lower the *p*H of natural rain causing damage to the biosphere. Coal is the worst offender, oil is next, while natural gas produces the lowest rate of acid emissions. Coal plants could add scrubbers and other innovations to reduce sulfate and NO_x production by up to 90%, greatly ameliorating but not eliminating the problem. Costs of electricity would increase somewhat to pay for the scrubbers. Oil-fired electric plants can be treated in the same way. However, that part of the acid rain problem due to petroleum combustion may be ascribed principally to the transportation sector — mainly to cars and trucks.

In spite of widespread emissions control programs, 25%–30% of the acid rain-producing emissions are still produced by the Nation's automotive fleet. Significantly improved emission control may not be practical either in terms of engineering or cost. Alternative fuels for locomotion would have to be found. Use of methanol might decrease acid emissions by factors of 2 to 3. Ethanol would also reduce sulfate emissions significantly.

In the longer term, other solutions might include hydrogen-based or electric or fuel cell-driven cars. The first two have been mentioned above and have problems to resolve before they would be feasible. Research into fuel cells as automotive energy sources is expanding and should be watched for progress. Investigation is at an early stage now.

Natural gas produces virtually no sulfates and less NO_x. Thus substitution, when compared with coal-fired electric plants, of natural gas for other fossil fuels would help the acid rain problem immeasurably. From the point of view of technology, natural gas can already substitute for oil- and coal-fired electric plants. Natural gas plants produce some 11% of the Nation's electricity and used to produce a larger share before questions about long-term supplies forced a cutback. The problem of substitution by natural gas is only a problem of supply, not technology.

Likewise, practical automobiles today can be powered by natural gas. They have been tested and marketed on a small scale (e.g., taxis in Las Vegas). Current limitations include costs of fleet conversion and operation.

5c. Hydrocarbon emissions — carcigenicity

Combustion of fossil fuels inevitably results in hydrocarbon production. Some of the hydrocarbons will be carcinogens or mutagens or teratogens (or any combination thereof) and dangerous to humans even in small quantities. Many hydrocarbons are known to be harmful in doses as small as milligrams per kilogram of body weight. The compounds in question are mostly polycyclic hydrocarbons or aromatics. Only some of these have been adequately investigated for health effects. Fewer still have been studied for concentrations in emissions from fossil fuel plants, furnaces, or automobiles. Benzo(a)pyrene is one of the best studied and understood of this class of chemicals. It is a prime suspect as a cause of cancer in cigarette smoke, and known to be present, although at very low levels (tens of nanograms per cubic meter) in emissions from coal electric plants.

Molecules of this size and complexity are easier to produce if one starts with a fuel composed of larger molecules. More of these pollutants tend to be produced by combustion of biomass, such as wood, than combustion of fossil fuels. Wood smoke is far more carcinogenic, per unit energy delivered, than smoke from petroleum or coal combustion. Petroleum is somewhat better in this respect than wood. Coal, whose combustible compounds are generally composed of smaller molecules or simple carbon atoms, is better still from the hydrocarbon point of view. A small part of coal is coal tar which contains larger molecules whose combustion produces most of the hydrocarbons from coal use. Petroleum produces more benzo(a)pyrene than coal per unit energy in an electric plant. Natural gas, which is mainly methane (CH_4), produces far fewer carcinogenic compounds than coal or oil, and is relatively innocuous in this regard.

Presumably, scrubbing or other technological fixes can be developed that would significantly reduce emissions of these carcinogens. However, little research of this sort has apparently been carried out.

5d. Transportation of liquid natural gas

One serious difficulty with natural gas lies in the potential for catastrophic accidents during transportation, particularly in the liquid phase. The energy content of a ship laden with natural gas is equivalent to that of a good-sized hydrogen bomb. The most serious scenario would have a slow unignited leak from such a ship disperse over a city under circumstances that permit an explosive mixture of gas and oxygen to cover a large area. Eventual ignition could kill hundreds of thousands of people. The probability of such an accident is, of course, extremely low, since just the right weather conditions must prevail and many layers of protection on the ships would have to fail. Nevertheless, the consequences are extremely serious, as are the consequences of serious accidents at nuclear power plants. Nuclear plants, however, are generally located tens of kilometers from large cities, while in Boston at least, liquid natural gas ships dock and unload within hundreds of meters of a densely populated downtown area. Some studies of the probabilities of serious accidents associated with the transportation of liquified natural gas have been carried out and indicate that the chances of such an event, while very small, may not be negligible.[43] Further analyses are probably in order.

5e. Supply and substitution

Given current use patterns, the U.S. supply of petroleum appears most tenuous of the three fossil fuels. Limited supplies threaten to force massive changes in use within a decade. Conservation, substitution, or increased financial hemorrhage abroad will occur.

Natural gas, the highly desirable substitute for the other two fuels, is currently available; technologies for substitution are practical. However, large uncertainties also hang over the domestic supply of this fuel source. Significant political uncertainties would underlie a heavy reliance on foreign sources (e.g., the Soviet Union and the Middle East) for gas and petroleum.

The remaining fossil fuel, coal, will be domestically available to the U.S. for centuries. However, some of its environmental disadvantages have been outlined above. Additional environmental problems resulting from greatly increased reliance on coal include land use for strip mining, cost of mine rehabilitation, and disposal of toxic fly ash, which contains many poisonous and carcinogenic heavy metals and elements as well as radioactive thorium and uranium. Substitution of coal for oil or gas would certainly have negative effects on atmospheric concentrations of CO_2 and would probably worsen acid rain, or at least preclude its elimination, even with the introduction of scrubber technologies.

6. Conclusions

Fossil fuels provide almost 90% of U.S. energy. Domestic supplies of oil are limited and can be expected to drop significantly, beginning within the next five to ten years. Coal could provide a domestically available substitute for the other fossil fuels for centuries. However, its massive use would have extremely negative environmental implications, not only for the U.S., but for the Earth, since the U.S. consumes about one-fourth of the world's energy. In particular, the greenhouse effect and acid rain problems would be greatly aggravated, although acid rain could be reduced by use of scrubber technologies. Use of natural gas instead of coal and oil would, in principle, greatly reduce stimulation of the greenhouse effect and could greatly reduce acid rain.[44] However, there is probably not enough domestic natural gas to permit a total switch. If a total conversion were made, and optimistic scenarios for developing domestic supplies did not pan out, the Nation would suffer a severe economic dislocation when domestic supplies become exhausted. Therefore, switching entirely to natural gas now would be unwise.

In ameliorating large-scale or global environmental problems such as the greenhouse effect, acid rain, or the depletion of the ozone layer, the time for changes to occur is long—probably years in the case of acid rain and tens of years in the case of the greenhouse effect. This means that from the time emissions are reduced or even eliminated to the time that positive impacts on the environment will occur, many political and social cycles may take place. The earlier remedial action can be implemented, the less severe the effect will be. The best time to act is before the problem becomes extremely serious. Otherwise one will have to live with a major, dislocating effect for a long time.

Prudence dictates reliance on multiple sources of energy to guard against the possible catastrophe that could result from excessive reliance on just one or two sources. First, conservation should become a national policy priority both for environmental and national security reasons. However, conservation is not, ultimately, an energy source. In the end, some energy must be produced, not just conserved. A high priority must be placed on developing energy sources that produce little or no increase in atmospheric CO_2 or other forms of pollution. Waiting for a decade or so to begin making major decisions to develop multiple energy sources and implement conservation measures may have extremely serious consequences.

Until renewable sources become available to supply a significant fraction of today's energy demand, or unless massive substitution by nuclear fission power becomes an economical and politically acceptable option, fossil fuels will be used by the U.S. on a very large scale. During this period, great care will need to be taken in deciding upon the optimal allocation among the three fuels. Market forces will not be a sufficient guiding hand because they do not work well when time scales are long, and this is the case with both fossil fuel resources and environmental effects.

Bibliography

The following provide useful readings:

H. A. Bethe and D. Bodansky, "Energy Supply," in *A Physicist's Desk Reference*, 2nd Edition of *Physics Vade Mecum*, H. L. Anderson, ed. (American Institute of Physics, New York, 1989).

Energy Resources: Conservation and Renewables, D. Hafemeister, H. Kelly, and B. Levi, eds., American Institute of Physics Conference Proceedings No. 135 (American Institute of Physics, New York, 1985).

J. Goldemberg, T.B. Johansson, A.K.N. Reddy, and R.H. Williams, *Energy for a Sustainable World* (Wiley Eastern Limited, New Delhi, 1988).

U.S. Congress, Office of Technology Assessment, *U.S. Oil Production: The Effect of Low Oil Prices*, OTA-E-348 (U.S. Government Printing Office, Washington, DC, 1987).

U.S. Congress, Office of Technology Assessment, *U.S. Natural Gas Availability: Gas Supply Through the Year 2000*, OTA-E-245 (U.S. Government Printing Office, Washington, DC, 1985).

U.S. Congress, Office of Technology Assessment, *U.S. Vulnerability to an Oil Import Curtailment*, OTA-E-243 (U.S. Government Printing Office, Washington, DC, 1984).

Useful periodicals include The Oil and Gas Journal, The International Energy Statistical Review (published monthly by the CIA) and the Annual Review of Energy (published by the Department of Energy).

References and notes

1. Fossil fuels are those combustible fuels, coal, petroleum, and natural gas, which are generally thought to have been produced from the remains of organisms that inhabited the Earth millions of years ago and more.
2. This is true with the possible exception of chemical carcinogens, which are produced in higher abundances per unit energy in oil combustion. Burning coal also releases more radioactivity per unit energy than is permitted from normally operating nuclear plants. However, the level of radioactivity, from trace quantities of thorium and uranium in the coal, is too small to have a significant health effect on the public, and is dwarfed by public exposure to radon in homes.
3. See Table 11 of Chapter 1.
4. See H. Bethe and D. Bodansky, *Physics Vade Mecum,* H. Anderson, ed. (American Institute of Physics, New York, 1989), Chapter 11.
5. Central Intelligence Agency, *International Statistical Energy Review,* April 1989. The numbers do not add up precisely, presumably because of some uncertainty in imports and in definitions of refined products.
6. Figures from H. Bethe and D. Bodansky, *op. cit.*
7. U.S. Congress, Office of Technology Assessment, *U.S. Oil Production: The Effect of Low Oil Prices,* OTA-E-348 (U.S. Government Printing Office, Washington, DC, September 1987), p. 34.
8. There are some partially or totally completed nuclear plants around the country that could, in principle, be turned on to produce electricity much more quickly were a political decision taken to do so. But they only total an equivalent of about 0.4 mbd capacity, and would not affect the overall picture much.
9. Office of Policy Planning and Analysis Report, *An Assessment of Natural Gas Resource Base of the United States,* DOE/W/31009-H1 (U.S. Department of Energy, Washington, DC, 1988).
10. U.S. Congress, Office of Technology Assessment, *U.S. Natural Gas Availability: Gas Supply Through the Year 2000,* OTA-E-245 (U.S. Government Printing Office, Washington, DC, February 1985), p. 19.
11. Both of the exotic source estimates are from U.S. Congress, Office of Technology Assessment, *op. cit.,* p. 7.
12. H. Bethe and D. Bodansky, *op. cit.*
13. H. Bethe and D. Bodansky, *op. cit.*
14. Central Intelligence Agency, *op. cit.*
15. H. Bethe and D. Bodansky, *op. cit.*
16. Using the U. S. as an example, if natural gas provided the 43% of our energy supply now produced by burning petroleum as well as the 23% that gas now gives us, the use of natural gas would nearly triple. If this increase were reflected worldwide, only 40 years' worth would be left. Or, one could take the current world petroleum consumption of 56 million barrels per day, and assume that this were substituted by natural gas. This quantity translates to 0.33 tcf of gas per day, or about 119 tcf per year. This would leave a 46 year supply. Of course, conservation could, in principle, stretch this number out greatly.
17. U.S. Department of Energy, *Assessment of Costs and Benefits of Flexible and Alternative Fuel Use in the U.S. Transportation Sector* (Department of Energy, Washington, DC, 1988), DOE/PE-0080, p. 26.
18. Corresponding numbers are given in Table 8, Chapter 1 for 1988.
19. Much of the information in this paragraph from U.S. Congress, Office of Technology Assessment, *U.S. Vulnerability to an Oil Import Curtailment* (U.S. Government Printing Office, Washington, DC, 1984), pp. 45–48.
20. U.S. Congress, Office of Technology Assessment, *op. cit.,* p. 93.
21. Most of the numbers in this section come from tables in Chapter 1.
22. W.M. Burnett and S.D. Ban, "Changing Prospects for Natural Gas in the United States," Science **21**, April 1989, pp. 305–310.
23. E.D. Larson and R.H. Williams, "Technical and Economic Analysis of Steam-Injected Gas Turbine Cogeneration," in *Energy Resources: Conservation and Renewables,* D. Hafemeister,

H. Kelly, and B. Levi, eds. American Institute of Physics Conference Proceedings No. 135 (American Institute of Physics, New York, 1985).

24. Again, figures for 1988 may be found in Table 8, Chapter 1.
25. Coke is produced when coal is heated to 1000–1300 °C in the absence of air. More volatile products, such as coal gas, are driven off and coke remains. Coke is an excellent fuel for use in blast furnaces that produces iron; the carbon monoxide produced on combustion reduces iron oxides in ores to iron.
26. See Chapter 18 and A.H. Rosenfeld and D. Hafemeister, "Energy Efficient Buildings," Sci. Am. April 1988, p. 78.
27. J. Goldemberg, T.B. Johansson, A.K.N. Reddy, and R. H. Williams, *Energy for a Sustainable World* (Wiley Eastern Limited, New Delhi, 1982), p. 113.
28. W.M. Burnett and S.D. Ban, *op. cit.* and E.D. Larson and R.H. Williams, *op. cit.*
29. See U.S. Congress, Office of Technology Assessment, *U.S. Vulnerability to an Oil Import Curtailment, op. cit.* p. 113. Additional details on fuel savings possibilities in industry are presented on pp. 111–115.
30. On the other hand, increased used of light trucks and vans reduces the overall efficiency of the fleet. For more on this, see the chapter on Transportation.
31. M.L. Wald, "When Methanol is in the Tank," New York Times, May 17, 1989, p. D8.
32. *Carnegie-Mellon REPPort,* Summer, 1988.
33. This might be balanced by increased efficiency of the electric engine, but a net loss or gain would have to be calculated for a particular car design. For example, heavy battery weight will lower the effective efficiency of the car.
34. See Wilson, MacCready, and Kyle, "Lessons of *Sunraycer*," in Sci. Am., March 1989 for a discussion of prospects for electric automobiles.
35. The gradual elimination of the world's rain forest is thought to contribute to the increase in global carbon dioxide because there is progressively less capacity on the part of the world's green plants to convert CO_2 to oxygen by photosynthesis.
36. These numbers and the projections for future CO_2 levels come from V. Ramanathan *et al.,* J. Geophys. Res. **90**, 5547–5566 (1985).
37. D.J. Wuebbles, *Scenarios for Future Anthropogenic Emissions of Trace Gases in the Atmosphere,* UCID-18997, Lawrence Livermore Laboratory, Berkeley, CA, 1981.
38. V. Ramanthan, B.R. Barkstrom, and E.F. Harrison, "Climate and the Earth's Radiation Budget," Phys. Today, May 1989.
39. J. Gleick, "A Huge Chunk of Antarctica is Heading Out to Sea, at a Glacial Pace," New York Times, February 9, 1988, p. C4, discusses the Ross Ice Shelf event; the extent of polar ice is reported in a drawing in "Science for the Citizen," Sci. Am., January 1989. Richard Houghton and George Woodwell in "Global Climatic Change," Sci. Am., April 1989, report on these and other evidences that warming is occurring and argue that the warming is due, at least in part, to human activity.
40. F. Pearce, New Scientist, 6 May 1989, pp. 37–41.
41. There would also be less of a demand for space heating in the winter, counteracting this effect. However, in most of the country, electricity demand peaks in the summer, so the net effect would appear to be a positive feedback.
42. V. Ramanathan *et al., op. cit.*
43. E. Drake and R.C. Reid, "The Importation of Liquified Natural Gas," Sci. Am., April 1977.
44. A skeptic, however, might refer to the tactic as "semienvironmentalism," since use of natural gas, even if universal, could only reduce CO_2 emissions by half.

Unconventional petroleum resources

George W. Hinman

1. Secondary and enhanced oil recovery

1a. Background information

Approximately 25 billion barrels[1] of proven oil reserves and 80–100 billion barrels[2] of undiscovered oil resources remain in the United States. These amounts refer to oil that can be pumped out of the ground by conventional means. In addition about 300 billion barrels will remain in the ground after recovery by conventional methods is complete. Enhanced oil recovery (EOR) may yield an additional 50–100 billion barrels[3] of this otherwise unavailable oil.

In the past, descriptions of oil production have included three successive stages — primary production, secondary recovery, and tertiary recovery. Recently, the terms "conventional" for primary and secondary and "enhanced" for tertiary are frequently used. In primary production the flow of oil into the bore hole of a well occurs through natural forces, usually pressure exerted either by the oil itself or by adjacent gas or water. Producers plan well spacing to maximize profits or total yield. Sometimes new circumstances dictate changes in the original plan. For example, the producer may decide to drill more wells in an already developed reservoir, a procedure called "in-fill" drilling. However, beyond a certain point natural forces will no longer move the remaining oil efficiently, and the well owner will introduce an external driving force.

Secondary recovery restores the driving force by injecting water (water flooding) or natural gas (pressure maintenance) into the oil reservoir through a series of injection wells. The pressure restored in this way pushes the oil out through a separate series of production wells. Water flooding and pressure maintenance can suffer from uneven progress of the pressurizing fluid from the injection to the production well, leading to a bypass of most of the reservoir oil. Producers use various techniques to avoid this problem. In steeply dipping formations gravity often can be used. In other cases chemical additives will help. Oil production in the United States makes extensive use of secondary recovery techniques. Table 1 shows the effect on the recovery of oil from typical reservoirs.[4]

In cases where primary and secondary recovery are no longer effective, the well operator may resort to more elaborate procedures collectively called enhanced or tertiary recovery.

Table 1. Oil recovery by primary, secondary, and enhanced processes (percent of oil originally in place).

Oil type	Primary recovery	Secondary recovery	Enhanced recovery chemical & miscible	Enhanced recovery thermal	Average total
Extra heavy	1–5	· · ·	· · ·	0–40	23
Heavy	1–10	5–10	· · ·	0–40	33
Medium	5–30	5–15	0–15	· · ·	35
Light	10–40	10–25	0–15	· · ·	50

Source: Royal Dutch Shell Group, *The Petroleum Handbook* (Elsevier Press, New York, 1983), pp. 96, 109.

Thermal processes. Cyclic steam, or "huff and puff," processes have two stages. In the first (huff) stage, the operator injects steam or a mixture of steam and hot water into a well, after which it is shut in and allowed to "soak." During this time, the hot injected fluid heats the oil and reduces its viscosity. In the second (puff) stage, the operator reopens the well to produce the oil released by the heat treatment.

Steam drive and hot water flood are similar to the water flood used in secondary recovery except that the fluid injected is steam, hot water, or a mixture of the two. The hot fluid is injected through one or more injection wells, and the oil and fluid are removed through production wells.

Combustion or fire flooding processes also use injection and production wells. Hot air, injected into the reservoir, ignites some of the oil. Hot gases moving ahead of the combustion front distill the reservoir oil and move it toward the production wells. In the wet combustion version, air injection alternates with water injection. The water picks up heat that otherwise would remain unproductive in the reservoir matrix and sends it forward to help heat the oil.

Miscible processes. Miscible fluids mix with each other in all proportions. A fluid miscible with reservoir oil, when it passes through the reservoir, can eliminate the capillary trapping of oil droplets, releasing the oil from the matrix. Alcohol, hydrocarbons, and carbon dioxide have been used with varying degrees of success as injection fluids. Miscible recovery requires injection and production wells. As an economy measure, the operator may not inject the miscible fluid continously but rather in slugs alternating with a less expensive driver fluid. These processes, like water flooding, can suffer from uneven progress of the injected fluid.

Some fluids, for example, carbon dioxide, are not miscible with oil at low pressure but become so at high pressure. Even if such a fluid is injected below the miscible pressure range, the result can be beneficial. For example, below the miscible range, carbon dioxide will dissolve in the oil causing it to swell

Table 2. Costs of crude oil production.

Production method	Range of costs (1982 $/bbl)
Conventional oil	1–20
Thermal	15–30
Miscible	15–32
Polymer	6–16
Surfactant	25–48

Source: Royal Dutch Shell Group, *The Petroleum Handbook* (Elsevier Press, New York, 1983), p. 109.

and reducing its viscosity. Both of these changes contribute to increased oil mobility.

Chemical processes. Injected fluids sometimes contain chemical additives to decrease capillary forces on the oil or to increase the viscosity of the fluid. The front between the driving fluid and the oil is more stable and uniform when the driving fluid has higher viscosity. Polyacrylamides and polysaccharides are polymers used as thickeners in polymer flooding. Surfactants, which are soap-like chemicals, reduce capillary trapping of oil droplets by forming microemulsions with the oil. The surfactants can be introduced directly into the formation or can be created *in situ* by the interaction of injected alkaline chemicals with natural acids in the oils.

Other processes. Nitrogen acts as a miscible fluid at high pressure and as a driving gas at lower pressure. It has the advantage that it is available at low cost from the air. Some well operators have used the waste gas from combustion sources or carbon dioxide derived from it as an injection fluid, but this "flue gas" usually is contaminated by sulfur oxides which are corrosive. In addition producers have investigated foams as fluid thickeners and microbes as agents to produce *in situ* surfactants.

Summary. Table 1 shows the amounts of oil recovery possible as a result of primary, secondary, and EOR techniques. Polymer, surfactant, and miscible processes are usually applied to the final stages of production of light or medium oil reservoirs where primary and secondary methods are relatively effective. Thermal methods are normally used in heavy oil reservoirs where primary and secondary methods are not very effective. The costs of enhanced processes compared to conventional production in 1982 dollars appear in Table 2.

1b. Current status of enhanced oil recovery

Current production practices already include enhanced oil recovery on a commercial scale. Table 3 and Figure 1 show production in barrels per day since 1980. The 1988 rate of 637,000 barrels per day[5] is about 7.5% of the United States' crude oil total.[6] Activity in enhanced crude oil recovery responds to

Table 3. United States enhanced oil recovery production (thousands of barrels per day).

	1980	1982	1984	1986	1988
Thermal					
Steam	243	288	358	469	455
Combustion *in situ*	12	10	6	10	7
Hot water	...	...	...	1	3
Total thermal	256	299	365	480	465
Chemical					
Micellar-polymer	1	1	3	1	2
Polymer	1	3	10	15	21
Caustic/alkaline	1	1			
Total chemical	2	4	13	17	23
Gases					
Hydrocarbon miscible/ immiscible	...	...	14	34	26
Carbon dioxide miscible	22	22	31	28	64
Carbon dioxide immiscible	...	...	1	1	
Nitrogen	...	...	7	19	19
Flue gas (miscible and immiscible)	...	...	29	26	40
Total gases	75	72	83	108	150
Other					
Carbonated water flood	...	...	...	...	...
Grand total	333	375	461	605	637

Source: Oil and Gas Journal, April 18, 1988, p.46.

price as Figure 1 shows. The number of active EOR projects decreased from 512 in 1986 to 366 in 1988[5] as world oil prices declined from 28 $/bbl to about 17 $/bbl.[7] EOR production increased slightly in 1988 because several new carbon dioxide projects came to fruition. Furthermore, even with the relatively low oil prices prevailing in early 1988, U.S. and foreign producers plan to spend $966 million for EOR projects during 1988 and 1989.[5] In the United States, enhanced oil recovery using thermal processes is employed almost entirely in the San Joaquin Valley of California where this technique is especially applicable to the heavy, viscous deposits. The principal use of carbon dioxide recovery is in the Permian Basin of Texas and New Mexico where the oil is relatively light and large supplies of carbon dioxide are available. Chemical methods are used with light and medium oils in various parts of the country.

1c. Environmental effects

Secondary and enhanced oil recovery take place in oil fields where crude oil production already has produced environmental effects, but enhanced recovery methods will bring additional impacts. Air emissions from combustion of the fuels used as heat sources accompany thermal methods. These processes contribute an appreciable fraction of the nitrogen oxide emissions in the San

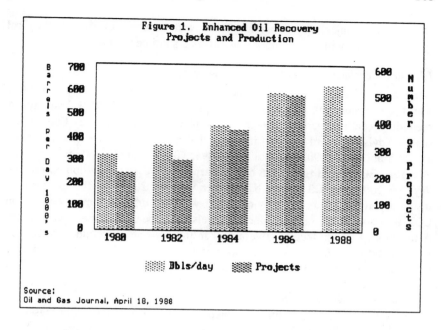

Figure 1.

Joaquin Valley. Sulfur emissions have declined as the operators have switched to natural gas. The only EOR method, other than thermal processes, that results in significant air emissions is CO_2 injection. The amount of CO_2 injected to produce a barrel of oil is similar to the amount produced by the subsequent combustion of the oil. Not all of this CO_2 is released after one pass however. The larger projects recapture, repressurize, and reinject the CO_2. CO_2 releases during these EOR processes effectively add to the total contribution that the oil produced makes to the greenhouse effect.

Enhanced oil recovery techniques consume one to six barrels of water per barrel of oil produced, equivalent to 50,000 to 300,000 acre feet per year for an EOR industry producing a million barrels of oil per day. At this rate water is not an overwhelming consideration in these processes. Most of the areas with EOR operations, e.g., the San Joaquin Valley in California and the Permian Basin in Texas and New Mexico, are regions where water is scarce. However, the great disparity between the high value of water to EOR operations compared to its value in agriculture means that it should be possible for the oil industry to buy water from farm users if it is unavailable from other sources. A common source of water for secondary or enhanced recovery is saline ground water, in some cases formation water that comes to the surface along with the oil. The use of brines, where it is possible, reduces the load on fresh water sources.

Secondary and EOR operations can contaminate ground and surface waters, but past experience with water flooding suggests that these operations will not create a major problem. Regulations in producing states generally control water cleanup and reinjection operations. There will be some effects of chemical spills associated with transportation of chemicals to injection sites.

1d. Future prospects for enhanced oil recovery

The use of enhanced oil recovery techniques will be very dependent on future world oil prices, tax benefits, and improvements in EOR technology. A scenario recently described by the Department of Energy (DOE) Energy Research Advisory Board (ERAB) indicates that if future oil prices (1987 dollars) are 20 $/bbl until 1990, 30 $/bbl until 1995, 40 $/bbl until 2000, and 50 $/bbl after that, EOR could rise to 0.6 billion barrels per year (1.5 million barrels per day) by 2000 and 1.0 billion barrels per year by 2010.[8] If the government institutes favorable tax changes or industry makes dramatic improvements in EOR technology, the same production levels could be achieved at lower prices. A recent step by the federal government in the opposite direction was the Tax Reform Act of 1986, which removed investment tax credits important to both R&D and deployment of EOR projects.

1e. Regulations and policies affecting enhanced oil recovery

Unitization refers to the planned production of an oil field owned or leased by many parties. Policies relating to unitization can affect implementation of enhanced oil recovery projects because oil can flow across the boundaries of any individual property. Planned production can increase the overall yield of oil in all phases of recovery including enhanced oil recovery, but producers are of course interested in the effects on their own profits. Disagreements about sharing can and do arise. Unitization agreements can be set up either voluntarily or by compulsion, but legal entanglements that arise can preclude or delay the initiation of EOR projects. On the other hand, the absence of a unitization agreement also can delay approval of an application for a new EOR project because of objections either by adjacent parties whose production may be affected or by a government agency that has environmental concerns. Injection of liquids into sources of drinking water is an example of a special environmental concern of EOR. While these legal issues can in principle inhibit EOR, they usually have not seriously impeded use of these techniques where other factors are favorable.[8]

Policies relating to stripper wells are also important because the cost of a new EOR project is considerably less if wells are already available for injection and production. Thus maintaining stripper well operation until EOR begins, rather than plugging wells and having to reestablish access to the oil, is desirable. Government agencies have recognized this problem in recent years, especially since the decline in oil prices in 1986. For example, the state of Texas has extended the grace period after a stripper well ceases operation until the well must be plugged (or a bond posted or operation resumed) from 30 days to a year with an optional extension for another year as a way of providing for possible changes in oil prices or plans for EOR.[9]

1f. Conclusions

Secondary recovery, i.e., water flooding or pressure maintenance, is a standard production process, and oil producers will continue to use it extensively in the future. Enhanced oil recovery techniques are currently sufficiently well advanced and cost effective to be used commercially at the present time. Oil prices, government tax policies and other incentives, and cost improvements made through R&D will determine future use. Future oil production in the U.S. by EOR probably will be 0.5 to 2.0 million barrels per day with total recovery of 50–100 billion barrels over a period of many decades.

2. Oil shale

2a. Background

Technology already exists to produce fuels and other petroleum products from oil shale. In fact the history of utilization extends back into the 14th century when the Austrians and Swiss pyrolyzed oil shales to produce "petro oleum," i.e., "rock oil."[10] Commercial applications of the material appeared during the 1800s and 1900s in many countries of the world. Uses have ranged from burning the crushed rock directly as a boiler fuel to production of fuel oil, gasoline, fuel gas, and chemicals. However, since the modern oil industry developed, shale oil has not been able to compete costwise with conventional crude oil pumped directly out of the ground, and the companies that originally produced it profitably finally have succumbed to competition, in some cases as late as the 1960s.

The oil shale resources of the United States are enormous when measured in terms of energy content as Table 4 shows. The ore is widely distributed in low concentration over broad areas of the country, but the most concentrated deposits are located in the Green River formation, which underlies parts of western Colorado, northeastern Utah, and southwestern Wyoming. This report discusses the recovery of shale oil from this formation. See Figure 2. If it were available in utilizable form and reasonable cost, oil shale could supply the nation's liquid and gaseous fuel requirements for hundreds of years at current rates of consumption. However, so far in the United States only one technology for producing a high-quality refinery feedstock has been demonstrated on a commercial scale. Even in this case the cost is so high that large-scale production is not likely in the next 10 to 20 years.

Government and industry have extensively explored the Green River formation.[11] They have drilled hundreds of holes and analyzed thousands of cores to determine the amount of hydrocarbon in place. Table 5 shows the distribution of the resource by quality. Although highest grade ores, on which most development projects are based, represent a substantial addition to the nation's energy resources, they are not by themselves a long-term solution to the United States' requirements for liquid fuels.

The layered appearance of the Green River oil shale arises from the layering of organic matter and not from any layering of the inorganic matrix during

Table 4. Potential shale oil in place in the oil shale deposits of the United States (billions of barrels).

Location	Range of shale oil yields (gallons per ton[a])		
	5–10	10–25	25–100
Colorado, Utah, and Wyoming (the Green River formation)	4,000	2,800	1,200
Central and eastern states (includes Antrim, Chattanooga, Devonian, and other shales)	2,000	1,000	(?)
Alaska	Large	200	250
Other deposits	134,000	22,000	(?)
Total	140,000+	26,000	2,000(?)

[a]Order of magnitude estimate. Includes known deposits, extrapolation and interpolation of known deposits, and anticipated deposits.
Source: Reference 1 as reported in Reference 2.

deposition. The "oil" is primarily a solid organic material called kerogen, which consists of hydrocarbon molecules cross-linked by sulfur and oxygen to form macromolecules with molecular weights of several thousand. The organic phase gives the ore part of its mechanical strength. When it is removed from the richest ores, the remaining inorganic matrix is quite weak and easily crushed to sand.

Table 6 shows the approximate composition of a typical Colorado oil shale. In a Fischer assay, such as the one reported in the table, the analyst heats a sample of material in the absence of oxygen, a process called pyrolysis, and measures the amounts of decomposition products of the organic phase. An ultimate analysis, on the other hand, measures the amounts of the different chemical elements that are present.

The results in Table 6 show that pyrolysis releases a large fraction of the organic constituent of the ore from the matrix as a liquid. This liquid will contain nitrogen, sulfur, and oxygen which must be removed to produce a good refinery feedstock. Three methods for commercial scale pyrolysis have been proposed and tested to some extent. The most common method is to mine the ore and heat it in an above ground retort.[10,11] Developers have tried various retort designs and processes. Some supply heat from outside the retort; others burn part of the ore itself. Furthermore, there are variants in the ways by which the processes convey heat to the pyrolysis region, feed the raw ore, and remove the final products.

A second method for pyrolyzing the ore is "*in situ*" processing. This method supplies the heat and removes the products while the ore remains in place underground. Successful application of this method requires that the ore deposit

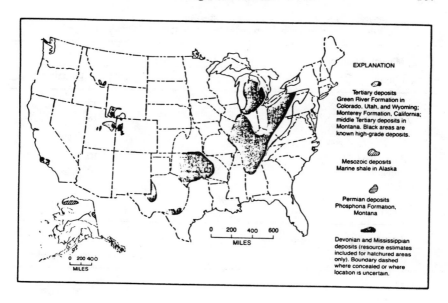

EXPLANATION

Tertiary deposits
Green River Formation in
Colorado, Utah, and Wyoming;
Monterey Formation, California;
middle Tertiary deposits in
Montana. Black areas are
known high-grade deposits.

Mesozoic deposits
Marine shale in Alaska

Permian deposits
Phosphona Formation,
Montana

Devonian and Mississippian
deposits (resource estimates
included for hatchured areas
only). Boundary dashed
where concealed or where
location is uncertain.

Figure 2.

Table 5. Potential shale oil resources of the Green River formation (billions of barrels).

Location	Resource class[a]				
	1	2	3	4	Total
Piceance basin (Colorado)	34	83	167	916	1200
Uinta basin (Colorado & Utah)	...	12	15	294	321
Wyoming basins	...	...	4	256	260
Total	34	95	186	1466	1781

[a]1 Deposits at least 30 ft thick and average 35 gal/ton.
2 Deposits at least 30 ft thick and average 30 gal/ton.
3 Similar to 1 & 2 but less well defined and not as favorably located.
4 Poorly defined, ranging down to 15 gal/ton.
Source: "An Initial Appraisal by the Oil Sale Task Group 1971–1985." U.S. Energy Outlook — An Interim Report, The National Petroleum Council, Washington, DC, 1972. Reported in Reference 2.

Table 6. Composition and pyrolysis products of typical Colorado oil shale.[a]

Mineral constituents

Mineral	Weight percent of minerals
Dolomite	32
Calcite	16
Quartz	15
Illite	19
Low-albite	10
Adularia	6
Pyrite	1
Acalcime	1
Total	100

Ultimate analysis of organic constituent

Element	Weight percent of organics
Carbon	76.5
Hydrogen	10.3
Nitrogen	2.5
Sulfur	1.2
Oxygen	9.5
Total	100.0

Yields from Fischer assay pyrolysis

Decomposition product	Weight percent of organic constituent in raw shale	Weight percent of total raw shale
Oil	63	10.4
Noncondensable gas	15	2.5
Fixed-carbon residue	13	2.2
Water vapor	9	1.4
Total	100	16.5

[a]Pyrolyzed by the standard Fischer assay at 932 F: oil yield = 26.7 gal/ton.
Source: T.A. Sladek, "Recent Trends in Oil Shale — Part 1," *Mineral Industries Bulletin,* Vol. 17, No. 6, November, 1974, pp. 4–5. As reported in Reference 10, Table 16.

have high permeability so that gases and liquids can flow through it. Preparation for *in situ* processing generally involves fracturing the deposit by explosives or other means. The heat is supplied by injecting a hot liquid or igniting a portion of the oil shale itself. The heated fluid or combustion air is introduced through an injection well, and the shale oil is withdrawn through surrounding production wells. So far this method is not well developed.

The third method for pyrolysis is "modified *in situ*" processing. In this case the developer increases the permeability of the deposit by first mining out a part of the ore and then rubbling surrounding material to create an underground chamber filled with ore rubble. The ore in this "retort" is then pyrolyzed and the resulting liquid drawn off. Occidental petroleum with other partners has

Table 7. Major U.S. oil shale projects in 1980.

Project	Location	Technology
Rio Blanco oil shale	Colorado	MIS and Lurgi-Ruhrgas above ground retorts
Cathedral Bluffs oil shale	Colorado	Occidental MIS
White River shale project	Utah	Above ground retorts
Colony development	Colorado	Above ground retorts
Long Ridge	Colorado	Above ground retorts
Superior oil	Colorado	Above ground retorts
Sand Wash	Utah	Above ground retorts
Paraho development	Colorado	Above ground retorts
Logan Wash	Colorado	Occidental MIS
Geokinetics, Inc.	Utah	True *in situ*
BX oil shale	Colorado	True *in situ*

MIS = Modified *in situ* process.
Source: Reference 10, Table 17.

tested this concept at Logan Wash in Colorado. Rio Blanco Oil Shale Co. and Multi Mineral Corp. also have experimented with this technique in the Green River formation.

Pyrolysis converts the kerogen into a combination of low BTU gas, raw shale oil, and a residual carbonaceous solid. The proportions of the three products depend on the raw material and the processing conditions. The raw shale oil must be upgraded by some combination of heating and hydrogenation before a refinery can use it as a feedstock. The objectives are to reduce the viscosity, convert unsaturated to saturated hydrocarbons, and remove such impurities as sulfur, oxygen, nitrogen, arsenic, and trace metals. In addition it may be necessary to add agents to the oil in order to make the product sufficiently fluid to ship by pipeline.

Most of the development work on western oil shale has concentrated on mining and retorting processes. Table 7 shows a list of major projects in 1980. While all of these projects have been exploratory, all but Logan Wash, Geokinetics, and BX were intended to proceed to commercial status if they were successful technically. However, the weakening of OPEC and subsequent decline in oil prices, combined with the termination of large-scale federal subsidies, has led to the closure of almost all of the commercially motivated oil shale projects. Only the Unocal Corporation, which has a ten year federal contract providing price supports, has continued to produce raw shale oil at its Parachute Creek site, upgrade it at a nearby location, and ship it by truck and pipeline to the company refinery at Lemont, Illinois near Chicago.[12] There it is refined to produce a conventional slate of petroleum products. The government pays Unocal the difference between a formula based price, currently about $46 per bbl, and the price of Wyoming sweet crude, a feedstock of comparable quality. Figure 3 shows the Unocal facility on Parachute Creek.

Unocal has experienced a number of technical problems with its retort, but the company has been producing up to 5000 barrels of raw shale oil per day

Figure 3. Photograph courtesy of Unocal.

and by late summer of 1988 had accumulated total production of a million barrels. Efforts are continuing to reach the design capacity of 10,000 barrels per day.

In the summer of 1988 there was one other active oil shale project in the Green River formation. The New Paraho Project was producing asphalt which was being tested as a road surfacing material. All of the other previously active commercially oriented projects are now defunct, and in some cases the facilities have been dismantled. Thus a revival of demonstration scale oil shale activities in the Green River area will require new startups for most of the technologies that were being tested in the 1970s and early 1980s.

On the other hand, R&D efforts are continuing on oil shale supported partly by the Department of Energy and partly by other parties.[13] The 1988 DOE budget for oil shale work is approximately $15 million and includes eastern as well as western resources. For example, the Institute of Gas Technology (IGT) has a $2.2 million DOE sponsored project to study shale oil production from eastern shale using an advanced fluidized bed concept. A fluidized bed levitates, or fluidizes, granules of the material to be processed in an upward moving stream of a processing fluid.

2b. Potential emergence of a shale oil industry

Eventually shale oil will become competitive with conventional oil as the latter becomes scarce and expensive and as technological development reduces the cost of the former. There have been periodic episodes of enthusiasm for oil shale development in this country, but so far all have eventually faded. The latest occurred during the "energy crisis" of the 1970s when oil prices rose sharply. During this period most of the large oil companies undertook projects in the Green River formation. Commercial development, and even large-scale demonstration, generally depended partially on federal support because industry feared the price break that eventually did occur and was unwilling to bear the high risk of strictly private ventures.

The federal government supplied research and development funds through the Department of Energy and offered some support for commercial ventures through the Synthetic Fuels Corporation, although actually SFC supported very few projects before it was discontinued. These ventures proved to be costly failures for those who undertook them. OPEC could not maintain the high price of conventional world crude on which commercialization of shale oil depended, and the prospects for the industry again disappeared. Based on this experience, industry will be very wary of any new push for development in the future, especially if the new stimulus is again a rise in oil prices brought about by OPEC restrictions on output rather than a rise in the actual cost of marginal new supply.

Many government policies and regulations affect the prospects for a commercial oil shale industry. Certainly the most important are those relating to taxes and subsidies. Oil shale development is extremely capital intensive. The Unocal facility represents an investment of more than $650,000,000 or $650 for each barrel of oil produced by the summer of 1988. Exxon's project, discontinued in 1983, would have cost more than a billion dollars. However, federal policies that effectively share the risk are less favorable now than they were in the early 1980s. The Synthetic Fuels Corporation is gone as are the favorable investment tax benefits available earlier. Recent government policy has been to place the risk of commercial development squarely on the developers and not to assume any risk on the basis of national security. Under these conditions industry probably will continue some research to reduce process costs but wait until oil prices stabilize at much higher levels before it again invests in shale oil production on a commercial scale.

The last round of activity to make federal leasing policy more favorable to lessees stimulated some proposals that could have affected oil shale development. Current laws governing leasing are the Mineral Leasing Act of 1920 and the Federal Land Policy and Management Act (FLPMA) of 1976. The Mineral Leasing Act effectively restricts a firm to leasing an area no larger than 5120 acres. The FLPMA requires that all overburden and waste disposal, as well as siting of surface facilities, be located on the lease from which the ore is being taken. In a large scale, long-term commercial development these provisions reduce flexibility of operations. The Department of Interior planned to request changes in the laws to remove these restrictions, but the changes never took place although a special act of Congress did permit off lease storage of material at the C-a lease in the Piceance Basin. The Department also intended to replace its Prototype Leasing Program with a permanent program, but subsequent events overtook these plans as well. By the summer of 1988 there was no activity on federal leases in the western oil shale region.

Actually most of the major commercially oriented oil shale projects so far have not been on federal land but on private holdings developed from pre-1920 mining claims on the southern edge of the region. Here it was easy to support a mining claim because the oil shale was readily visible on the cliffsides. Furthermore mining at these sites does not require sinking a vertical shaft to reach the ore, which is at considerable depth below the horizontal surface. Instead the mine entrance can be a horizontal tunnel (adit) bored in from the cliff face. Nevertheless the majority of the best ore is on federal land to the north, and thus leasing policy will be important in any large-scale future activity.

2c. Environmental effects

Environmental and socioeconomic effects and the availability of water are very important aspects of shale oil development. They too are subjects of government policies and regulations. Mining and above ground retorting produce huge amounts of solid wastes. Producing a barrel of shale oil from mined ore generates more than a ton of spent shale, which requires disposal in an environmentally acceptable way. Ten years of operation by a commercial oil shale industry producing two million barrels of shale oil per day would generate enough spent shale to cover 100 square miles of ground to a depth of approximately 400 ft. In addition to the large land requirements for storage of this material, there are problems of erosion, fugitive dust, and ground and surface water contamination by leachates from the piles. Rio Blanco Oil Shale Company is undertaking a project on physical and chemical stabilization of spent shale piles.

Processed oil shale wastes generally have a number of undesirable characteristics. The particle size is small so that erosion and siltation are possibilities. The alkalinity is high enough to discourage plant growth, and the wastes may contain leachable organic residues hazardous to people or soluble salts toxic to plants. The usual dark color leads to high soil temperatures, rapid soil drying, and poor seed germination. Modified *in situ* or true *in situ* processing can eliminate part, but not all, of these environmental impacts, but in their place underground processing introduces the possibility of aquifer disruption and contamination. Cleanup of a large aquifer after it has been contaminated is usually impractical.

There are point sources of waste water in addition to the distributed liquid waste from spent shale leachates. Combustion of hydrogenous fuels, chemical changes in the ores, and blowdown of cooling water systems produce waste water. Disposal of these liquids after necessary cleanup can be done by recycling, evaporation, injection into underground aquifers, or release to surface waters. A variety of water quality regulations control discharge of waste waters from point sources.

Air pollution from processing would be a major problem of a large-scale western shale oil industry. There are serious questions about how much development would be consistent with the Prevention of Significant Deterioration limits on increases in ambient pollutant concentrations imposed by the Clean Air Act. In the case of the Piceance Basin, the limitation appears to be the effects of development on the Flat Tops Wilderness, a national wilderness area less than 40 miles from the edge of the Basin. The Office of Technology

Assessment[11] estimated that a 400,000 barrel per day overall production program in the Green River formation might satisfy existing air quality regulations but that a 1,000,000 barrel per day industry could not.

An important disadvantage of oil shale as a major energy source is its contribution to the greenhouse effect. The carbon dioxide generated per unit of energy delivered is considerably higher for shale oil than for coal or conventional oil because of the combustion involved in processing the ore.

Water requirements for shale oil production range from two to five barrels of water per barrel of shale oil. A 50,000 barrel per day facility would need 5000 to 12,500 acre feet of water per year, and a 500,000 barrel per day regional output would use roughly 85,000 acre feet per year. This is a relatively small amount of water. For example, it is about 0.5% of the virgin flow of the Colorado River at Lee's Ferry, the boundary between the Upper and Lower Colorado River basins. Considering the availability of both surface and ground water, the Office of Technology Assessment concluded that water would not be a serious impediment to the development of a 500,000 barrel per day or even a 2,000,000 barrel per day shale oil industry except that such development would require new reservoirs and pipelines to make the water available on a year round basis.

2d. Socioeconomic effects

Very few people live in the western oil shale region. The total population of Rio Blanco and Garfield counties, which have most of the resource, is about 25,000. The only medium size town in the area is Grand Junction, Colorado. Under these conditions the growth of a large shale oil industry could produce severe effects on people. The Colorado West Area Council of Governments estimated in 1980 that the development of a 500,000 barrel per day industry over a ten year period would mean a growth in the two county population to about 108,000. Examples from western coal areas have demonstrated that rapid and extensive growth can lead to such stressful and uncomfortable living conditions that workers lose efficiency and citizen morale declines. Community planning activities and financial support for new schools, more community services, etc., can partially mitigate these effects. Both government and industry in the area made extensive provisions for rapid growth during the 1980s, but the adequacy of the effort cannot be judged because the growth did not take place.

2e. Permits and licenses

Shale projects require a myriad of permits and licenses before they can begin. During the 1970s and 1980s, when preparations were being made for new projects, industry spokesmen contended that licensing procedures were burdensome, unnecessary, and obstructionist. They made a number of recommendations for change such as require regulatory agencies to make decisions in a specified period of time, "grandfather" projects under development to make new laws and regulations inapplicable to them, create an energy board or authority to overrule federal regulatory decisions, and limit litigation. During that period some regulatory agencies did in fact move to simplify, or at least coordinate, applications for permits and licenses, but many proposals died with the decline in interest and activity.

2f. Conclusions

A number of shale oil production technologies are partially developed at the present time, and Unocal has demonstrated technical feasibility and production on a commercial scale for underground mining together with one version of above ground retorting and subsequent upgrading to a high grade refinery feedstock. At some future time a combination of technological development of lower cost oil shale processing and higher cost conventional oil extraction will create favorable conditions for commercial development of the oil shale resource. Government sharing of the risk of commercial development will be necessary unless it is clear that conventional oil prices are stable enough to justify purely private investment. When and if production of shale oil becomes economically attractive in the Piceance and Uinta basins, there will almost certainly be problems of maintaining the environment and socioeconomic status of these regions as they now exist because the capacity for physically processing the ore exceeds the ability of the region to absorb the effects without major alteration. People will have to limit production to preserve the current environment or else accept transformation of the area into a center of mining and ore processing.

3. Tar sands

3a. Background

Tar sands are sedimentary rocks containing bitumen (solid or semisolid hydrocarbon) or other heavy petroleum that conventional petroleum production methods cannot recover. The largest deposits of tar sands are in Canada, where the total resource is estimated to be more than one trillion barrels,[14] much of which is close to the surface. Venezuela has a resource base estimated to be 0.7 trillion barrels of heavy oil, a slightly less viscous hydrocarbon than tar sand. Resources in the United States, mainly in Utah at depths of several hundred feet, total about 53 billion barrels (21.6 billion measured and 31.1 billion speculative),[15] but economic and physical constraints will probably limit recovery to 5% to 10% of this amount.[16]

Developers have proposed both *in situ* processing and mining plus above ground processing for tar sand projects. *In situ* processing technologies have included cyclic steam (huff and puff), steam drive, *in situ* combustion, and wet combustion, all of which have been previously described in the section on enhanced oil recovery. Surface extraction techniques include retorting, hot water extraction, and solvent extraction.

Hot water extraction has been used on a large scale with the Athabasca tar sands of Canada. The success of this process depends on the fact that in these particular deposits the sand grains are wet. Thus the bitumen is not sticking directly to the matrix grains but is separated from them by a film of water, making the extraction using water possible. The hot water process begins with a "conditioning" of the sand in a rotating drum containing water, sodium hydroxide, and steam. The effluent from this drum is screened to remove large particles, mixed with more hot water, and sent to another chamber where the sand settles out and the bitumen floats to the surface. The final step skims off the bitumen and separates it from the remaining water by heating or centrifugation. Tailings ponds receive the waste water.

Solvent extraction is possible with tar sands, in contrast to oil shale, because bitumen is soluble in various common organic liquids while kerogen is not. This process begins with preparation of the tar sands by grinding the ore and adding the solvent to dissolve the hydrocarbon component. The sand is then separated off and the bitumen recovered by evaporating and steam-stripping away the solvent. The solvent steam mixture is condensed; the two components are separated; and the solvent is recycled. Very high solvent recovery is necessary in order to keep the processing cost within reasonable limits.

Lurgi-Ruhrgas dry distillation could be used to retort surface or underground-mined tar sand. This pyrolysis process yields liquid and gaseous hydrocarbon products. A retort distills the bitumen from hot crushed ore. The liquid and gaseous products of the pyrolysis leave the system in one direction for purification and recovery while the remaining char, sticking to the sand matrix, drops through the bottom opening in a hopper and into a combustion chamber where it heats the sand. The hot sand feeds back into the retort, along with the fresh ore, where it provides the heat for the distillation.

The bitumen produced by hot water or solvent extraction is a very heavy and viscous material, which is not transportable by pipeline and which has a very limited number of uses. Environmental restrictions on sulfur and nitrogen content and competition with coal limit its marketability as an industrial and utility fuel. Industry can use it to make asphalt or coke for some applications, but for most petroleum markets it requires upgrading. Bitumen has a hydrogen to carbon atom ratio of approximately 1.4. This ratio must be increased to 1.6 to produce a heavy fuel oil and to 1.8 to 2.0 to produce diesel fuel or gasoline. In addition processing must remove sulfur and nitrogen.

Carbon removal (coking) or hydrogenation can ugrade bitumen to an acceptable product for pipeline transportation and refinery feedstock. Carbon removal is accomplished either by heating the bitumen to release higher hydrogen fractions, leaving the carbon behind as a coke residue, or by using a catalyst to produce the separation. In either case coke is a by-product of the process. Hydrogenation, on the other hand, adds hydrogen to the bitumen to raise the H/C ratio. It utilizes more of the carbon content but requires a source of hydrogen, which is expensive. Several of the operations in Canada use a coking technique to upgrade the bitumen. In these cases excess coke produced, beyond the amount sold or used in the process itself, has been stored for later final disposition by burial or other means. Operations in the United States have not had to face large-scale upgrading problems of bitumen, but the techniques involved would be similar to those used for treating heavy oils.

3b. Current status

By the summer of 1988 there was no commercial scale tar sands activity in the United States. *In situ* projects conducted earlier in California, Texas, Wyoming, and Utah using EOR production techniques have ended. A surface mining and retorting project of Getty Oil in Kern County, CA also has stopped. There is, however, some bench scale R&D underway. The total Department of Energy FY88 budget for heavy oil and tar sands R&D was $5.6 million.[17]

3c. Environmental effects

The environmental effects of *in situ* tar sand extraction and processing in the United States will be qualitatively similar to the effects of thermal methods for enhanced oil recovery. The difference between the two is a matter of degree. The bitumen in tar sands is at least as dense and viscous as heavy oil so that the need for heat to mobilize the material is as great or greater. Air pollution is therefore potentially a factor in a large-scale *in situ* tar sands industry. *In situ* processing is favored in Utah and parts of California where most of the resources are under several hundred feet of overburden. Retorting and solvent extraction also use combustion of fuels to produce elevated temperatures at which the processes proceed. Thus they will contribute to air emissions to some extent.

Processing tar sands above ground will generate large amounts of solid waste, a situation similar to that faced with oil shale. Future work will have to establish the physical and chemical stability of the tailings piles.

The hot water process has a special environmental problem. The waste water from the process, at least in Canada, contains suspended fine sand and clays that have very poor settling properties. The disposal of this water therefore requires construction of very large ponds for long-term storage. This concern is probably not important in the United States where the deposits are less suitable for this process.

Thermal processing of tar sands to produce usable liquid and gaseous products will raise the total carbon dioxide release per unit of usable energy available from these fuels. Thus these unconventional fuels would make a special extra contribution to the greenhouse effect.

3d. Prospects for future U.S. commercial development

Because most U.S. tar sands are located at depths of several hundred feet, the prospects for commercial tar sand production appear to depend on the economics of *in situ* thermal processing although some work has been done to evaluate underground mining. Therefore, except for local deposits near the surface, tar sand development will occur only after appropriate enhanced recovery operations have been performed at current and former producing oil fields. Even during the late 1970s and the early 1980s when the "energy crisis" was stimulating commercial scale projects, tar sands activity lagged behind EOR and oil shale. It seems unlikely that tar sands will develop commercially for many decades.

3e. Conclusions

The United States possesses tar sands resources containing about 53 billion barrels of bitumen, most of which is located at depths of several hundred feet. *In situ* thermal processes probably will be used for extraction at some future time when oil prices have risen to several times their present value. Recovery of 2.5 to 5 billion barrels of bitumen may be possible at that time. The role of government incentives in stimulating commercial activity will be similar to that previously described in the section of this report on oil shale.

4. Synthetic fuels from coal

4a. Background

Coal is the largest fossil fuel resource in the United States already undergoing extraction and use. Total resources are about 2.5 trillion tons with an energy content of 55,000 quadrillion BTU. At the present rate of consumption the nation's coal would last for hundreds of years.

Although most coal now is used to generate electricity, in the future it could become the principal source of liquid and gaseous fuels. However, if the nation chooses this course, it must reconcile using synthetic fuels with the atmospheric warming that will follow from continued carbon dioxide releases to the atmosphere.

Technology to liquefy and gasify coal has been known for decades. In some cases the basic processes date from the last century. Liquefaction and gasification have been used on a limited scale at various times and places when economic conditions have been favorable or natural fluid fuels unavailable, but on the whole synthetic fuels from coal have not been economically competitive with conventional oil and gas up to the present time. As natural fuel supplies dwindle, comparative economics will shift toward synthetic fuels.

4b. Technology description

Increasing the hydrogen carbon ratio produces liquid and gaseous fuels from coal. This ratio, approximately 0.8 for coal, must be raised to 1.4 to 1.8 to produce a liquid or to a value close to 4 to produce synthetic natural gas. Pyrolysis is one way to raise the ratio. The temperatures and pressures used determine the proportions of liquids, gases, and remaining solid char produced.

Adding hydrogen directly or indirectly to the coal is the second way to raise the ratio. In the direct method, coal is mixed with a hydrocarbon liquid under pressure; gaseous hydrogen is usually added; and the coal absorbs hydrogen from the combination. The process produces various hydrocarbon gases, liquids, and solids together with sulfur and ash.

In the indirect method steam and oxygen, or air, react with the coal to break it down chemically into synthesis gas, a mixture of carbon monoxide, carbon dioxide, hydrogen, residual steam, and nitrogen if air is used. The proportions of these constituents depend on conditions in the reactor used for the process. Synthesis gas can serve directly as a fuel or as the reactant in subsequent

Figure 4.   Lurgi Gasifier

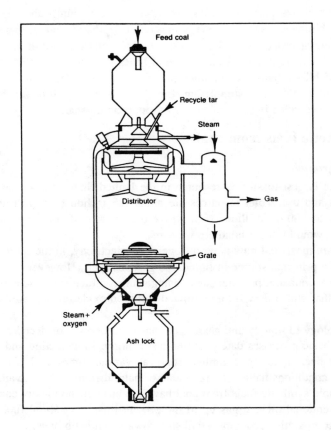

Source: Clark, D., "South Africa's Coal Gasification Prospects
          for the Future," <u>Hydrocarbon Processing</u>, vol.58, June
          1979, 56C-56J.

Figure 4.

chemical reactions to produce various hydrocarbon gases and liquids through
interaction with catalysts. The sulfur and ash in the original coal separate as
waste or by-products during the processing.

Industry and government organizations have developed many different versions of pyrolysis and direct and indirect synthesis of hydrocarbons from coal
to a pilot plant stage or beyond. There are a few examples described below, but
this field is too extensive to cover in detail in this summary chapter. Several
books are available on this subject.[18,19]

Lurgi gasifier. The SASOL plants in South Africa have used this type of unit
for years to produce synthesis gas. Figure 4 shows a typical Lurgi gasifier. The
process takes place at elevated temperature and pressure (350–450 psi, 1150–

1400 F) in the central chamber shown in the figure. Crushed coal enters the upper chamber through a pressure lock system. Steam and oxygen enter at the bottom. The gas produced in the center region leaves the main chamber and is scrubbed and cooled in a spray of water before leaving the Lurgi unit. The synthesis gas at this point has a heating value of 300 BTU/standard cubic foot. A typical composition is H_2 39%, CO 22%, CH_4 9%, CO_2 28%, N_2 1%, and H_2S 1%. The SASOL plants use the output of the Lurgi gasifiers to produce liquid products via the Fischer-Tropsch process, which is discussed below.

The Great Plains Coal Gasification Plant in Beulah, North Dakota, on the other hand, uses Lurgi gasifiers to produce synthesis gas that is further upgraded to pipeline quality substitute natural gas (SNG). This plant started operation in 1984 and has a nominal output of 125 million scf/day. It uses North Dakota coal as fuel, gasifies it in Lurgi units, conditions and cleans the raw gas, and methanates it to pipeline quality. The Great Plains Coal Gasification Plant was recently sold to Basin Electric Power Cooperative by the federal government under an agreement whereby the government and Basin Electric will share profits. Basin Electric will operate the plant until 2009 so long as revenues cover operating costs, and the government provides certain other guarantees.

To upgrade the synthesis gas from a Lurgi unit to final liquid or gas form, the processor typically first adjusts the relative amounts of H_2 and CO and perhaps other compounds to the best values for subsequent catalytic chemical reaction. This process often involves further reaction of the gas with steam, during which CO is converted to CO_2 with release of H_2, the so-called shift reaction. "Scrubbing" the gas using an appropriate liquid in a counterflow absorption column removes the acid gases (CO_2, H_2S, COS) from the main stream. Catalytic chemical reactions produce the final product. For example, in the case of methanation, the gas passes over a nickel catalyst to convert the H_2 and CO to methane and water. The methanated product, when purified, is a substitute for natural gas.

Other operating conditions and catalysts can produce methanol rather than methane from the synthesis gas. Copper- and zinc-based catalysts have worked well in both high- and low-pressure reactions. Another catalytic reaction using a process developed by Mobil can convert the methanol to gasoline.

Fischer-Tropsch synthesis. Catalytic conversion of synthesis gas into a variety of hydrocarbon liquids is called Fischer-Tropsch synthesis. The main reactions are of the type:

$$2n\text{CO} + n\text{H}_2 \rightarrow (\text{CH}_2)_n + n\text{CO}_2.$$

Possible catalysts include iron, cobalt, molybdenum, and nickel. In the Synthol version of a Fischer-Tropsch process synthesis gas containing a catalyst flows through a fluid bed reactor where it reacts to produce mainly liquid hydrocarbon products. SASOL II and SASOL III use Synthol fluid bed reactors to produce liquified petroleum gas (LPG), gasoline, and diesel oil. At these two largest SASOL plants, close to 100,000 tons of coal per day are processed to produce considerably more than 100,000 bbl/day of motor fuels plus a variety of other petroleum products.

Integrated coal gasification combined cycle concept (IGCC). At Daggett, CA the Cool Water Project, a 100 MWe IGCC, operated from May 1984 until recently. This demonstration plant used coal converted to gas in a Texaco gasifier as fuel. This type of gasifier operates at much higher temperature than the Lurgi, introduces the reactants in a different way, and melts the ash in the coal to produce a slag form of solid waste. The coal, slurried with water, flows under pressure into the gasifier along with oxygen produced in an adjacent oxygen plant. Gasification (partial combustion) occurs at about 550–600 psi and 2500 F, producing a medium BTU gas (300 BTU/scf) mainly consisting of H_2 and CO. In the Cool Water plant this gas was cleaned of particulates and condensates, desulfurized, and used as fuel. Waste heat from the gasifier generated additional steam for the steam turbine component of the plant.

This particular application of medium BTU synthetic fuel gas to electricity generation worked well. The plant operated successfully using both low and high sulfur coals and also a high fusion temperature coal. Air emissions of particulates and sulfur and nitrogen oxides have been far below new source performance standards. The plant had achieved a plant capacity factor of 60% by 1986. Heat rates have been about 11,000 BTU/kWh, but through the use of a new combustion turbine design operating at higher temperature it should be possible to reduce this figure to 9000 BTU/kWh.

The modular features of the IGCC should be quite attractive to utilities. A utility can construct a unit to run on natural gas initially and add the coal gasification component when the economics justifies the added investment. Furthermore a utility can make capacity additions in relatively small steps which eases the problem of matching supply to demand. Three utilities, Florida Power and Light Company, Virginia Electric Power Company, and Potomac Electric Power Company have announced plans to develop IGCC units in stages.

Direct liquefaction. Bergius in Germany discovered one of the earliest direct liquefaction processes in the 1920s, and Germany used it to provide liquid fuels during World War II.[21] A large plant also was built in England, but the product could not compete with inexpensive oil discovered in the Middle East in the 1950s. The latest round of development, which began after the Arab oil embargo, centered around three technologies, solvent refined coal (SRC-I and -II), Exxon donor solvent (EDS), and H-Coal. Developers carried each of these processes to the pilot plant stage. The H-Coal Process appears to have some advantages over the other two and is still the subject of active development in this country.

In the original H-Coal process coal is crushed, dried, and slurried with oil recycled from the process itself. The slurry is pressurized, hydrogen is added, and the mixture is heated and charged into the bottom of an "ebullient-bed" reactor, which is a liquid version of a fluidized bed reactor. A solid catalyst, suspended in the upward moving liquid stream, brings about the hydrocracking. The product separates at the top of the reactor into gaseous and liquid/solid components. Hydrogen is separated from waste gases, recompressed, and returned as part of the hydrogen supply. Waste gases are processed or refined as necessary. Mineral matter and undissolved coal separate in hydrocyclones, and the liquid product feeds into a distillation unit. Some of the liquid product is recycled back to be the slurry medium.

Researchers at a pilot plant in Wilsonville, Alabama have modified the original H-Coal process to improve distillable liquid yields and reduce hydrogen consumption. They have tried new catalysts and introduced two stages of liquefaction at the plant, which a number of organizations sponsor jointly. The workers at Wilsonville hope to reduce the cost of product from their process to less than $30 per barrel, but even if they do so, it is uncertain when the economics of oil will justify major investment in this technology.

Underground coal gasification. Much of the coal resource in the United States is located at such great depth or in such steeply sloping seams that conventional mining cannot economically extract it. Research and development groups have studied underground coal gasification as a method for recovering the energy from this fuel.[18] The chemical processes are basically the same as those used in above ground gasification, and the methods involve injection and production wells. Initially the mine operators must establish contact between the injection and production wells in some way, for example, by explosive fracturing, horizontal drilling, or an initial burn. In the latter case they ignite the coal at one of the wells and introduce air or oxygen at the second well. Oxygen must make its way, inefficiently at first, to the burning front to extend it and thus create a pathway for gas production. After they have established a satisfactory connection, the mine operators pass air or oxygen into the injection well and withdraw coal gas from the production well. If necessary they inject water also, but in some cases the seam itself may contain the water required.

Experiments on underground coal gasification of western coals began in 1973. With air as oxidizer, the gas produced has an energy content of 125–170 BTU/scf. When oxygen is used, the energy value doubles. The work done so far suggests that this technique can extract a large fraction of the energy in a coal seam. There is still some research and development being done on UCG in this country, but the current level of support is low.

This method of coal gasification has some special environmental problems. It creates a potential for overburden subsidence and leaves ash and various hydrocarbons behind where they can contaminate ground water. Both of these problems should be carefully evaluated in any commercial scale development of this process.

4c. Environmental effects

Effects on the environment from synthetic fuels include the consequences of coal extraction, subsequent impacts of processing the coal, and effects of using

Table 8. Potential pollutants from coal gasification.

Air	Water	Land
Sulfur oxides	ammonia	slags
Nitrogen oxides	hydroxides	ash slurries
Particulates	cyanides	fines
Carbon monoxide	thiocyanates	dry residues
Hydrocarbons	biological oxygen demand	waste treatment
Hydrogen sulfide	chemical oxygen demand	sludges
Carbonyl sulfide	suspended solids	spent catalysts
Ammonia	dissolved solids	
Hydrogen chloride	organic compounds	
Hydrogen cyanide	pH (activity)	
Metals	hydrogen sulfide	
Organics	sulfur oxides	
	chlorides	
	fluorides	
	alkalinity	
	oils/grease	

Source: References 19 and 22.

the fuels produced. Because coal extraction has been discussed in an earlier chapter, this section will cover only the latter two effects.

One of the major concerns with coal-derived liquids is the toxic nature of some of the compounds that may be included in them. Direct liquefaction processes produce polyaromatic organics, whose heavy, high boiling fractions are carcinogenic. Researchers on direct liquefaction are trying to eliminate these heavy fractions by recycling them through the liquefaction process. Synthesis of liquids through the indirect route using synthesis gas is less troublesome because it tends to produce chain hydrocarbons, which are less toxic.

Processing coal to produce synthetic products will lead to releases of residuals (wastes) in solid, liquid, and gaseous form. Part of these residuals will come from boilers, heaters, incinerators, etc. and the rest from the process stream. These residuals in general are similar to those produced during direct combustion of coal and processing of oil shale and tar sands.

Potential pollutants from coal gasification are shown in Table 8. Important potential contributors to the waste streams in coal gasification are:

Pretreatment of the coal (crushing, possible heating) releasing particulates and sulfur dioxide to air.

Water quenching of the raw synthesis gas to remove particulates and condensables producing waste water streams.

Disposition of residuals from the facilities used to control primary waste streams including:
- waste water treatment system
- sulfur recovery system
- cooling tower
- flares and incinerators
- slag or ash

4d. Regulations

Gasification and liquefaction facilities will have to satisfy regulations imposed by federal, state, and local governments. Some of the important federal regulations restricting releases of wastes are:

Air: Clean air act with amendments. Both ambient concentrations and emissions are covered. Although new source performance standards do not at present apply directly to coal gasification facilities, they may well do so in the future and even now some facilities are affected. For instance Cool Water is covered because it is also an electrical generating plant.

Water: Federal Water Pollution Control Act Amendments of 1972. U.S. Public Health Service Drinking Water Standards
Solids: 1976 Toxic Substances Control Act (TSCA) and Resource Conservation and Recovery Act (RCRA)
A critical issue is which wastes are designated hazardous. So far electric power plant wastes are exempt from such designation, but coal gasification plants are not. The distinction is important because hazardous wastes must be specially handled and disposed of at special disposal sites.

There have been very few commercial scale synthetic fuel plants built so far in the United States. Therefore not much direct evidence about environmental effects of these plants has been collected, but the Great Plains Coal Gasification Plant and the Cool Water Project do provide some data.

The Great Plains Coal Gasification Plant has had trouble in the past meeting SO_2 regulations.[22] The cause has been a failure of the Stretford unit that processes the sulfur gases, probably a result of trace compounds in the gas stream. There are no liquid discharges from the plant. All water is recycled. Dewatered cooling tower and wastewater treatment sludges are incinerated and disposed of with solid residues by deep well injection. Gasifier ash is classified as nonhazardous and is disposed of in a conventional landfill.

Operators of the Cool Water (IGCC) plant[21] treated condensables from the gasifier in a sour water stripper, whose effluent was sent to an evaporation pond along with clarified water from the slag and ash handling system. The gasifier slag was nonhazardous according to standard leaching tests for trace elements and organics. Most of the slag was put into a landfill, but the plant operators did investigate some potential applications of this material. For example, the slag, which is a sand, is suitable as aggregate in lightweight roofing tiles, a high valued product. The principal air emission sources in the plant were the heat recovery steam generator and the plant incinerator, but, as has been previously mentioned, the emissions were far below new source performance standards.

The pollutant that is common to all synfuels is carbon dioxide. A measure of the carbon dioxide pollution potential of a synthetic fuel is the amount of carbon dioxide released per million BTU of energy delivered when the fuel is burned (lb/MMBTU). The carbon dioxide released during manufacture of the synfuel must be included in the total. The value for direct burning of coal ranges from 200 to 210 lb/MMBTU. Because very little heat is wasted in the IGCC, the factor is only slightly larger, about 215 lb/MMBTU. The factor is 50% to 100% larger, i.e., 300 to 400 lb/MMBTU, for a synthetic gas or liquid produced as a separate fuel for later use because much of the carbon in the coal is converted to carbon dioxide in the synthesis process and is not part of the synthetic fuel. Thus total carbon dioxide releases are much larger for synthetic fuels than for either natural gas (116 lb/MMBTU) or residual fuel oil (170 lb/MMBTU), their natural fuel counterparts.

4e. Research, development, and demonstration

Research organizations are still undertaking research and development projects on coal liquefaction and gasification techniques under sponsorship by the federal government, oil companies, the Electric Power Research Institute, and the Gas Research Institute. The FY89 U.S. Department of Energy budget includes $32 million for coal liquefaction, $21 million for surface coal gasification, and $1.3 million for underground gasification, but large-scale investment in demonstration plants by industry and the federal government is no longer taking place. In the early 1980s, the U.S. government planned to spend $85 billion on demonstration plants through the Synthetic Fuels Corporation. However the Corporation sponsored very few projects. In 1986 it was discontinued partly because of problems in the corporate management, partly because the program was rejected by the administration that came to power in 1981, and partly because of falling world oil prices.

Coal gasification and liquefaction can benefit from a myriad of R&D activities. Recent reviews[22,23] cover these in considerable detail, but generally they fall into the categories of process improvement and cost reduction.

4f. Prospects for future use of synthetic fuels from coal

Synfuels, which represent a huge potential addition to resources of liquid and gaseous fuels, eventually will compete economically with naturally occurring oil and gas. However, although forecasts of energy prices suggest that the break even point for liquid synfuels may occur by the year 2000, the OPEC cartel and not marginal production costs presumably would still be setting the oil price. Thus investors in synfuel plants will still be vulnerable to undercutting by OPEC. It is questionable whether industry will risk an investment in liquid synfuel production under these circumstances without protection of some kind from the federal government. The situation probably will be similar to the 1970s and early 1980s when an important issue was whether the federal government should foster a synfuels industry on the basis of national security. This time the matter will be further complicated by the contribution that synfuels will make to the greenhouse effect.

Coal gasification is more attractive than liquefaction. The IGCC is an example of an application that is already interesting to utilities. Furthermore clean gaseous fuels are better protected than liquids from OPEC oil price cutting. Clean gas is less polluting than oil and is not a direct substitute. The problem of the greenhouse effect remains of course.

Bibliography

Office of Technology Assessment, *Enhanced Oil Recovery Potential in the United States* (U.S. Congress, Office of Technology Assessment, Washington, DC, January 1978). This study covers technical, environmental, and policy aspects of oil recovery techniques. It is written for a nontechnical audience.

Royal Dutch Shell Group of Companies, *The Petroleum Handbook* (Elsevier Science Publishers, Amsterdam and New York, 1983). This book provides comprehensive coverage of conventional and enhanced oil recovery techniques currently in use.

Office of Technology Assessment, *An Assessment of Oil Shale Technologies* (U.S. Congress, Office of Technology Assessment, Washington, DC, June 1980). This study covers technical, environmental, and policy aspects of shale oil recovery primarily from the Green River formation in Colorado, Utah, and Wyoming. It is written for the nonspecialist.

Heavy Oil and Tar Sands Recovery and Upgrading, M.M. Schumacher, ed. (Noyes Data Corporation, Park Ridge, New Jersey, 1982). This volume provides extensive coverage of tar sands extraction and upgrading technologies. See especially the chapter "An Assessment of International Tar Sands Recovery and Upgrading Processes," by Booz-Allen and Hamilton, Inc.

Sami Matar, *Synfuels, Hydrocarbons of the Future* (PennWell Publishing Company, Tulsa, Oklahoma, 1982). This book describes the various processes available to convert primarily coal into liquid and gaseous fuels.

Hunt, Daniel V., *Synfuels Handbook* (Industrial Press, New York, 1983). This book provides comprehensive coverage of the means for producing synthetic liquid and gaseous fuels primarily from coal.

References and notes

1. Oil and Gas Journal, December 28, 1987.
2. Energy Research Advisory Board, *R&D Initiatives for Energy Competitiveness*, U.S. Department of Energy Report DOE/S-0061, March, 1988, p. 26. ERAB refers to "undiscovered reserves," but "undiscovered resources" is more consistent with the nomenclature adopted in this report. See the discussion by Bodansky.
3. Reference 2, p. 27. Also Office of Technology Assessment, *Enhanced Oil Recovery Potential in the United States*, Congress of the United States, Washington, DC, January 1978. Extensive use was made of this OTA study in the preparation of this report.
4. Royal Dutch Shell Group of Companies, *The Petroleum Handbook* (Elsevier Science Publishers, Amsterdam and New York, 1983), p. 96.
5. Oil and Gas Journal, April 18, 1988, p. 33.
6. Energy Information Administration, *Annual Energy Outlook 1987* (Energy Information Administration, U.S. DOE, Washington, DC, March 1988), p. 2.
7. Energy Information Administration, *Annual Energy Review 1987* (Energy Information Administration, U.S. DOE, Washington, DC, March 1988), p. 2.
8. Reference 2, Exhibit 32.
9. Telephone conversation with Brian Schaible, Texas Railroad Commission, September 6, 1988.
10. Office of Technology Assessment, *An Assessment of Oil Shale Technologies* (Office of Technology Assessment, Congress of the United States, Washington, DC, June 1980). Extensive use was made of this study in the preparation of this report.
11. *Oil Shale Technical Data Handbook*, Perry Nowacki, ed. (Noyes Data Corporation, Park Ridge, New Jersey, 1981).

12. Oil and Gas Journal, July 6, 1987, p. 24; telephone conversation with Susan Alvillar, Unocal Corporation, Parachute, Co., August 1988.
13. Telephone conversation with J. Ramsey, U.S. Department of Energy, September 1988.
14. Booz-Allen and Hamilton, Inc., "An Assessment of International Tar Sands Recovery and Upgrading Processes," prepared for the International Technology Assessment Program, Lawrence-Livermore National Laboratory, December 1980. Reported in *Heavy Oil and Tar Sands Recovery and Upgrading*, M.M. Schumacher, ed. (Noyes Data Corporation, Park Ridge, New Jersey, 1982), pp. 268 & 271. Extensive use was made of this study in the preparation of this report.
15. Lewin Associates, *Major Tar Sand and Heavy Oil Deposits of the United States*, prepared for the Interstate Oil Compact Commission in cooperation with the U.S. Department of Energy, July 1983.
16. Telephone conversation with George Stosur, U.S. Department of Energy, Washington, DC, September 1988.
17. Reference 1, p. 385.
18. Sami Matar, *Synfuels, Hydrocarbons of the Future* (PennWell Publishing Company, Tulsa, Oklahoma, 1982).
19. Daniel V. Hunt, *Synfuels Handbook* (Industrial Press, New York, 1983).
20. D.F. Spencer, S.B. Alpert, and H.H. Gilman, "Cool Water: Demonstration of a Clean and Efficient New Coal Technology," Science **232**, 609–612 (1986).
21. Robert E. Lumpkin, "Recent Progress in the Direct Liquefaction of Coal," Science **239**, 873–877 (1988).
22. S.S. Penner, S.B. Alpert *et al.*, *Coal Gasification: Direct Applications and Synthesis of Chemicals and Fuels: A Research Needs Assessment*, Report DOE/ER-0326 (U.S. DOE, Washington, DC, June 1987).
23. U.S. Dept. of Energy, *Coal Liquefaction: A Research Needs Assessment*, Report DOE/ER-0400 UC-108 (U.S. DOE, Washington, DC, February 1989).

Nuclear power

W. F. Vogelsang and H. H. Barschall

1. Situation 1973 to date

1a. Reactor types

Power plants based on nuclear reactors convert the energy released during the fission of heavy nuclei into electric power. Actual fission takes place in the core of the reactor which contains the nuclear fuel. Heat is carried away from the core by the coolant which transfers the heat to steam which then drives a turbine much like that found in a fossil fuel generating plant. Reactors require a control mechanism for regulating the amount of heat produced in the core. Each fission is triggered by a neutron. Reactors are designed to operate with neutrons of a particular energy. The material called the moderator in the reactor adjusts the energy of the neutrons for optimum operation of the reactor. Reactor types are identified by the coolant, moderator, and fuel they use.

By 1973, utilities in the U.S. had decided that reactors for civilian electric power production would be based on light-water (H_2O) types using uranium oxide (UO_2) enriched to 2%–3.5% in ^{235}U as fuel. Other reactor types had been tried in small demonstration plants, found wanting for one reason or another, and discarded. Only the gas-cooled reactor appeared to be a possible competitor to the light-water types for the near future. In the more distant future when a shortage of uranium would limit the further deployment of light-water reactors, the fast breeder reactor would be needed.

The development of the light-water reactor had proceeded along two different lines, the pressurized water reactor (PWR) and the boiling-water reactor (BWR). The designs established by 1973 were much as they are today.

Pressurized water reactors. Pressurized water reactors use light water both as a moderator to slow neutrons down to thermal energies (thermalize) and as coolant to remove the energy produced by the reactor (Fig. 1). The water exits the reactor vessel, passes through heat exchangers which serve as steam generators, and then is returned to the reactor by coolant pumps. The temperature of the coolant flowing through the reactor rises approximately 35 °C. To obtain the temperature necessary for an acceptable thermal efficiency, the reactor vessel is pressurized to about 150 atm. Saturated or slightly superheated steam at about 70 atm from the steam generators functions as the working fluid in a Rankine cycle with a thermodynamic efficiency of approximately 32%.

127

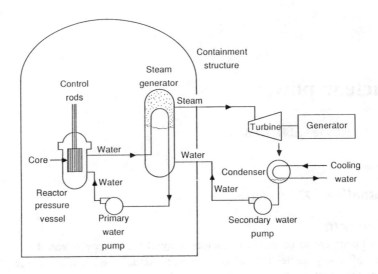

Figure 1. Schematic diagram of a pressurized water reactor power system.

Most of the reactors operating in 1973 were in the 500–600 MWe range. However, in response to the perceived desires of the utilities and the expected economies of scale, the industry had built units in the 700–900 MWe range and was constructing 1000 MWe units. Almost all orders for later plants specified higher capacities up to 1275 MWe. The energy crisis of 1973 combined with the expected growth of electrical power usage led to an upsurge in orders. Meanwhile plants already ordered were being completed. But the usage of electricity was not growing at the anticipated rate, reactor costs were greater than predicted, and opposition to reactors was growing. While new orders were being processed, other orders were being cancelled. The development and current status of PWR electric generating plants is summarized in Table 1.

The basic design of the PWR has changed very little since 1973. However, changes have been made to upgrade safety systems, to reduce the release of radioactive material, to comply with stricter or new regulations, or to take advantage of experience with operating plants.

Boiling-water reactors. The BWR, like the PWR, uses light water both as a moderator and coolant. However, in the BWR the water boils in the reactor vessel, and the steam is sent directly to a steam turbine (Fig. 2). Since boiling takes place in the reactor, the pressure in the reactor is lower than in the PWR. Water returning from the condenser after passing through the feed-water heaters travels to the reactor where it is mixed with that part of the coolant which had not boiled and directed back into the core. Pumps maintain internal circulation.

Unit capacity of the BWR has followed a pattern similar to that of the PWR. The initial plants were in the 500–600 MWe range and were followed by increasingly larger units so that by 1973 most of the plants under construction

Table 1. Summary of the deployment and current status of light-water reactors in the U.S. (November 31, 1988).

| | | PWR | | BWR | | Total | |
		number	GWe	number	GWe	Number	GWe
Operating	1973	18	11	12	7	30	18
Operating	1978	41	31	23	15	61	46
Operating	1988	72	64	36	31	108	95
In Progress	1988	6	7	2	2	8	9
Planned		2	2			2	2
Cancelled or							
Deferred		62	68	32	36	94	104

were in the 800–1000 MWe range. The same factors that affected the growth of the number of PWRs also affected the BWR. While new orders were being placed, other were cancelled. Table 1 also shows the growth and current status of BWR capacity.

The design of the BWR has undergone a series of modifications while still retaining its basic features. These changes have been influenced by the same factors that have led to modifications in the PWR. The most obvious changes have occurred in the design of the containment structure. The designs of systems to protect against loss of coolant accidents have also been modified.

Other reactor types. The high-temperature gas-cooled reactor (HTGR) is a possible competitor to the light-water reactors. In this type of reactor the moderator is graphite and the coolant is helium gas. The HTGR offers the possibility of better fuel utilization, higher steam temperatures and therefore higher plant efficiencies, and better response to certain types of accidents. One plant of 330 MWe (Fort St. Vrain in Colorado) was built and operated until 1989 in an effort to develop a practical HTGR. Numerous difficulties, principally with the gas circulating system, have plagued this plant. There are no plans in the U.S. to build additional commercial power reactors of this type.

The liquid-metal fast breeder reactor development has also been slower than anticipated. In the fast breeder, neutrons which have not been thermalized induce fissions. In contrast, thermal neutrons induce most of the fissions in the previously described reactors. Fission initiated by fast neutrons produces more neutrons than fission initiated by thermal neutrons and fewer fast neutrons are captured by the materials in the reactor core. The surplus fast neutrons can be captured by a blanket of ^{238}U to form ^{239}Pu, a fissionable fuel for the fast reactor. Thus the reactor can produce more fuel than it burns up. A liquid metal, such as sodium, may be used as a coolant since it will contribute little to slowing down the fast neutrons.

A considerable development effort has focused on the fast breeder. Test reactors have been built and operated and their safety problems have been studied in great detail both through analysis and experiment. Although the results have generally been favorable, questions remain. For example, all experience and analysis indicate that the large fast reactor will be more expensive than conven-

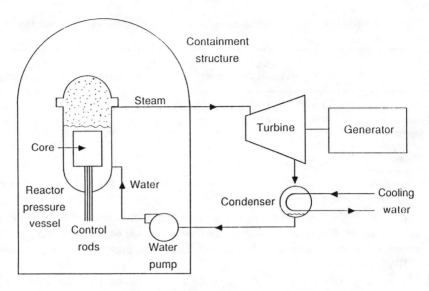

Figure 2. Schematic diagram of a boiling-water reactor power system.

tional types. The early experience with heat exchangers, particularly steam generators, was not satisfactory (molten sodium is corrosive), but considerable improvements have been demonstrated.

Light-water reactors do not require the reprocessing of used fuel although it may be desirable. Fast reactors, however, need reprocessing to take advantage of their ability to produce fuel. There are no plans to reprocess fuel from operating commercial reactors in the U.S. at present. All these factors combined with the slow growth in electricity demand and the anticipated difficulty of licensing the fast breeder reactor have delayed its introduction in the U.S. indefinitely except for small pilot plants, although fast breeder plants operate in Europe.

The CANDU reactor has been developed and deployed principally in Canada. No plans to introduce this type of reactor in the U.S. exist, but since several of them are located close to the U.S. and a portion of the power from them is supplied to the U.S., a brief description follows. The CANDU reactor uses heavy water (D_2O) as both moderator and coolant. Heavy water contains deuterium which is a good moderator and has a very low absorption cross section for thermal neutrons. Light water (H_2O) is an even better moderator than heavy water but its neutron absorption cross section is much higher. The heavy-water moderator makes it possible to use natural uranium as fuel and avoid the need for isotopic enrichment.

The reactor design differs greatly from the light-water reactor. The ratio of volume of moderator to volume of fuel must be greater to attain criticality. To avoid an extremely large and expensive pressure vessel, fuel in the CANDU is placed in pressurized tubes which contain only enough moderator to act as coolant. A large stainless-steel cylinder containing horizontal tubes houses the bulk of the moderator. The pressure tubes containing the fuel pass concentrically through these horizontal tubes. Thus most of the moderator is cool and near atmospheric pressure. Only that fraction needed to remove the energy from the fuel is at high temperature and pressure.

The pressure tubes are connected through manifolds, and the coolant from the pressurized tubes travels to steam generators similar in design to those used in the PWR. The natural uranium fuel is in the form of UO_2. Lower operating temperatures and pressures than in the PWR reduce the efficiency of the CANDU to 29%. Whereas light-water reactors must be partially dismantled to be refueled, the pressure tubes of the CANDU allow on-line refueling.

In 1973 there were four plants in operation with a total capacity of 2060 MWe, all at one site on Lake Ontario. Since that time another four plants with equivalent capacity have been added at the same site. Eight somewhat larger units with a total capacity of 6412 MWe have been put into service at one site on Lake Huron with two units of 1278 MWe capacity in service in the provinces of Quebec and New Brunswick. Four additional plants with a capacity of 3524 MWe are under construction.

1b. Fuel cycle

Mining and resources. For light-water reactors to be economically viable, there must be an adequate supply of uranium. The early proponents of reactors were not so concerned about this question for two reasons. First, they felt that additional uranium could be found if the effort were made. Second, they regarded the light-water reactor as a stop-gap which would shortly be replaced by the fast breeder. By 1973 additional uranium had been found but it was clear that the deployment of the fast breeder would be delayed. Thus there would be a large number of light-water reactors. The problem of uranium resources became more acute, and efforts to quantify the resource base were intensified, not only in the U.S., but worldwide. As of 1986 the Department of Energy (DOE) estimated there were 133,000 t of uranium in the U.S. reasonably assured plus 518,000 t estimated additional at a cost of less than \$80/kg of uranium. In addition there are 280,000 t reasonably assured and 394,000 t estimated additional at a cost between \$80 and \$130/kg uranium. This most expensive uranium would contribute about 4 mils/kWh to electricity cost.

The lifetime fuel requirements of a light-water reactor are about 4100 t of uranium per GWe. Thus the present installed and planned capacity of 113 GWe would require 463,000 t of uranium. If this requirement is compared with the resource estimates, the domestic supply for the present reactors seems to be adequate. Clearly uranium supplies will finally limit the deployment of light-water reactors so that advanced types with lower uranium requirements must eventually be developed.

The U.S. has been a major supplier of uranium for reactors. Between 1938 and 1984, the U.S. produced 39% of the uranium mined worldwide. Between 1965 and 1975 production was relatively constant at about 10,000 t/yr. In anticipation of an increasing demand, production was increased in 1980 to 17,000 t. The anticipated demand did not materialize and there would be a surplus of uranium if this level of production were sustained. In addition, foreign suppliers were offering uranium at lower prices. The result has been a sharp decline in production in the U.S. as the producers closed the more expensive mines. By 1985 production had dropped to 4352 t. The requirements in that year were 14,800 t with the difference between production and requirements made up by the use of inventories and purchases from foreign sources.

Enrichment. All enrichment in the U.S. uses the gaseous diffusion process. The current enrichment capacity is about 50% greater than the expected requirement in the year 2000, i.e., it is more than adequate for present reactors. Competition from foreign sources has challenged the use of this capacity as utilities strive to obtain fuel at the lowest cost. The relative cost of enriched fuel from domestic and foreign sources depends on DOE pricing policies and dollar exchange rates at the time of purchase.

The gaseous diffusion process has also experienced competition from other less energy-intensive enrichment processes such as the centrifuge and the laser. After a long period of development, DOE decided in 1985 to drop the gas centrifuge program and to continue the development of the Atomic Vapor Laser Isotope Separation process. In the 1988 budget, DOE proposed the termination of this program. Congress voted to continue funding for it, and this support has been extended in the 1989 appropriation.

Reprocessing. In the U.S., there is at present no reprocessing of commercial light-water reactor fuel. In the late 1970s the government decided that reprocessing exacerbated proliferation concerns. Studies showed that, at the then current uranium prices, there was no strong economic advantage to reprocessing fuel.[2,8] Some European nations and Japan reprocess spent fuel to provide plutonium for fast breeder projects, to reduce the volume of high-level wastes, and to provide fuel for the light-water reactors.

Waste treatment and disposal. At the present time all commercial spent nuclear fuel remains in storage racks, primarily at the reactor sites. Low-level wastes from a plant are reduced in volume and sent to a low-level waste site for burial.

Two processes in a nuclear power plant generate radioactivity. The fission process itself produces radioactive products. The fuel rods contain all but a very small amount of this activity and are classified as high-level waste. Fission neutrons activate nonfuel materials such as the reactor structure, corrosion or wear products carried by the moderator, and the moderator itself. This activity plus any activity which may leak from the fuel rods is classified as low- or intermediate-level waste.

The treatment of these two types of waste is quite different. The low- or intermediate-level wastes, which may range from demineralizer resins to discarded parts or tools, are compacted and sent to a near surface burial site. Present plans for the high-level waste call for burial in deep geologic structures. Fuel would not be separated from the fuel rods nor would it be reprocessed to recover the reusable portion which forms the bulk of the used fuel.

The Nuclear Waste Policy Act of 1982 and the Nuclear Waste Policy Act

Amendments of 1987 govern the establishment of a geologic storage facility. At present Congress has directed DOE to "characterize," i.e., make a detailed study, of a site on federal land at Yucca Mountain in Nevada. This site is one of three candidate sites identified by DOE under the original act. If the characterization shows that this site is not satisfactory, the Secretary of Energy must recommend an alternative to Congress.

The Acts have a number of important provisions. They require that DOE cooperate and consult with the states and affected Indian tribes. States have veto rights although Congress can override a state veto. A state agreeing to host such a facility will receive financial incentives. An assessment of 1 mil/kWh on all nuclear generated electricity will finance the program. Other provisions of the Acts deal with oversight, licensing, transportation, alternatives, and other aspects of state relations.

The Acts also authorize an interim monitored retrievable storage (MRS) facility. An MRS would alleviate the fuel storage problem at the plants and put the spent fuel into a form suitable for disposal. However, the Acts impose a number of conditions. They establish a commission to determine the need for and impacts of an MRS. When this commission has reported, DOE may begin a new site survey. The size of an MRS is limited, and no construction may be done until the NRC has issued a permit. The states and other affected parties have the same rights as for the repository itself. Financial incentives are available.

The program for the disposition of spent fuel has suffered from technical controversies and political problems. The Waste Policy Act with its later amendments gives hope that in the next decade significant progress can be made to resolving this issue.

The problem of the disposal of waste fuel is not unique to the United States. Similar political and technical problems exist everywhere. Consequently, many countries have launched efforts to find acceptable solutions. Almost all of the work is based on storage in geologic structures similar to those considered in the U.S. The form of the material to be stored depends on whether the fuel is reprocessed. For example, the French are anticipating reprocessing so that the waste form will be fission products in a glass matrix. There are also differences in approach in the containers for the fuel. The Swedish proposal relies more heavily on the engineering of the container to retain the fission products in place. Numerous studies have evaluated proposed sites. These studies indicate that suitable sites can be found but no site has been developed and put into use.

Low-level wastes generated by reactor operations also require disposal sites. Although reactors produce the largest amount of these wastes, other users of radionuclides generate a significant amount. Thus this problem is not specific to reactor operations. The Low-Level Waste policy Act of 1980 as amended in 1985 assigns the responsibility for establishing low-level waste sites to the states. The states were directed to form regional compacts and to share a site located in one of the states in the compact. Any state not joining a compact must develop its own site. The Act specified certain deadlines which had to be met for the state to be eligible for federal subsidies.

Establishing the compacts and finding site locations in the compacts have been fraught with difficulty and controversy. States failed to meet many of the original deadlines, and new deadlines have been set up. The system is making progress, and it appears that additional low-level waste sites will be available.

1c. Construction experience

Costs. The construction cost of power plants is usually expressed in terms of dollars/kW electric power. When the construction time is ten years, the value of the dollar may have shrunk by a factor of 2, which makes cost estimates in constant dollars difficult. Cost estimates also depend on the effect of interest on borrowed capital, which in turn depends on inflation, interest rates, and construction time. Often interest charges during construction are not included in quoted costs. In spite of these difficulties in determining the cost per kW, there is no doubt that the average capital cost (excluding interest charges) has increased enormously in constant (1982) dollars. Construction costs have risen from about $500/kW for nuclear plants completed in the early 1970s to $1000/kW for plants completed in the late 1970s to $2000/kW for plants completed in the mid-1980s. Another startling fact is the variation by a factor of about 5 in the cost of plants completed in recent years. The costs range from about $1000/kW to $5000/kW. The contribution to the cost of electricity due to capital cost and financing, amortized over the life of the plant runs from $0.01/kWh for older reactors to $0.1/kWh or even more for the more expensive recently built plants.

The reasons for the increase and variability of costs have been studied extensively. Increases in cost (in constant dollars excluding interest charges) are caused by the more rapid increase in the price of nuclear components than general inflation, an increase in the amount of construction material per kW, and increase in the number of work hours per kW. The latter two increases arise in large part from regulatory requirements, including quality control procedures and paperwork. For example, the average length of electrical cable has increased from 0.6 m/kW in 1971 to 1.6 m/kW in 1985, and concrete use has risen from about 0.06 m^3/kW in 1971 to 0.16 m^3/kW in 1985, while work hours/kW have grown from 3.5 to 22 in the same period (Ref. 3, Chapter 3). However, these numbers for plants built at about the same time vary by a factor of 2.

Variations in cost among plants built at about the same time can be traced to several factors: different labor costs in different regions; previous experience of a utility with building a nuclear plant; presence of a previously built plant at the same site; construction time; and competence of the builder of the plant. The rules imposed by many public service commissions which do not allow utilities to incorporate into their rate structure the construction costs of uncompleted plants also contribute to this variation.

Average construction times of nuclear plants have increased from about 5 years for the early plants to 10 years for plants which received construction permits in the early 1970s. This increase was caused in part by the 1971 decision to require environmental impact statements, the new regulatory requirements following the 1976 Browns Ferry fire and the 1979 TMI accident, and in some cases deliberate delays by the utilities because of slow growth in power demand. More recently the average construction time has decreased to about 8 years. The long construction times increase the construction cost in constant dollars because cost of components increases more rapidly than general inflation and because interest charges on capital are higher than general inflation.

Electricity costs. An unidentified scientist is often quoted as having predicted that nuclear power would make electricity "too cheap to meter." This prediction, if it was ever made, is obviously absurd. Electrical energy produced by the earliest nuclear plants was, however, much cheaper than coal-generated electricity, its chief competitor. The very different contributions of capital costs and fuel in generating electricity from coal and nuclear plants complicates a comparison between the two cases. The contribution of fuel costs proves easier to establish than that of capital costs which depend on accounting rules and practices. These rules result in very high capital charges when the plant first operates, followed by a decreasing rate. On the other hand, the cost of coal varies by a factor of about 4 depending on the distance from the coal mine.

Presently the cost of electricity from nuclear plants completed in the early 1970s is much lower than that from coal plants. For the recently completed plants the relative cost depends on the construction cost of the nuclear plant, the distance from a coal mine, and the length of time since the plant came on line. Electricity from some plants completed during the last decade is much cheaper than electricity from a coal plant at the same location. In other cases, especially for plants with the highest construction costs, nuclear power is more expensive. Furthermore, the operation and maintenance costs of nuclear plants have recently been increasing and in certain cases have become comparable with, or even greater than, those of fossil fuel plants. However, fuel costs remain much lower for nuclear plants.

Performance. Two measures of the performance of a power plant are widely used: Availability is the fraction of time during which the plant was available for operation independent of whether or not it was actually operating and whether or not it was capable of operating at full power. The capacity factor is the ratio of electrical energy actually produced to the design output at full power during the same period. Since nuclear plants usually provide the base load, i.e., operate at constant power, the capacity factor is a better measure of performance than the availability. Typical values for large power plants, either coal or nuclear, are availabilities of 70% and capacity factors of 60%. For nuclear plants of the LWR type, neither availability nor capacity factor can be close to 100% because of the need to turn off the plant for refueling. In addition, equipment failures can result in lengthy periods of downtime, especially problems with steam generators, and regulatory changes that require plant modifications while the plant is not operating. Following the TMI accident as many as 180 changes were ordered on each existing plant.

Since the cost of the generated electricity is sensitive to the capacity factor, the utilities have made a strong effort to increase it. The lifetime capacity factors of nuclear plants have varied by more than a factor of 2, from 80% to 34%. These are averages covering many years. Clearly, a well-operated nuclear plant can be a reliable source of electricity.

1d. Decommissioning

Nuclear plants operate for a finite useful life. Rebuilding plants may allow them to be operated after they exceed their original design lifetimes. When it is closed, a plant must be put into a condition such that it presents no additional hazard to the public from the radioactivity produced while it operated. This radioactivity comes from components that have been activated by neutrons escaping from the core and items that have become contaminated by radioactive corrosion or wear products or by fission products which have escaped from the fuel rods. The bulk of the radioactivity which is in the fuel rods will have been removed for permanent disposal. The NRC has specifically defined decommissioning as placing the site, in a timely and safe way, in a condition where it can be released for unrestricted use.

Several of the older small plants have been or are being decomissioned, most recently Shippingport, near Pittsburgh, and the LaCrosse BWR. The experience with these plants has served to define the tasks required and to develop techniques that may be applied to larger reactors in the future. The costs incurred in these previous decommissioning efforts should not be scaled directly to the larger reactors but should be used as a part of the input along with the definition of the tasks to prepare more realistic cost estimates.

Different schemes proposed for decommissioning range from immediate dismantling of the plant and returning the site to unrestricted use to using the reactor containment as the permanent disposal facility. Each of these has a number of arguments in its favor. Immediate dismantling (DECON) means that all radioactive material is removed from the site shortly after operation has ceased. Thus there is less chance of exposure to the general public. The site is available for other uses, and the costs may be more readily charged to those who benefited from the plant. Since the radioactivity has had little time to decay, the radiation dose to those who do the work is expected to be greater than if the work were done at a later time. The costs are likely to be greater because of the need for more shielding and more remote operations.

A second option called SAFSTOR would have the facility placed and maintained in a condition that would allow the site to be decontaminated at some future time, perhaps 30–50 years later. In this time some of the radioactivity would decay thereby reducing the total exposure to the workers as well as reducing the volume of material to be sent to a disposal site. The site is not available as soon as with DECON. If the site is to be used for some other generating facility, this may not be of consequence. During the waiting period seismic or other events might disturb the site which might result in the release of radioactivity.

A third option (ENTOMB) is to encase the radioactive part of the facility in a structurally long-lived material and carry out appropriate maintenance and surveillance until the radioactivity has decayed to the point where the site can be released for unrestricted use. Unfortunately it may be many years before the radioactivity has decayed sufficiently. In another option which is similar to the ENTOMB option the containment of the facility would be used as the encasement. All penetrations would be sealed and no maintenance or surveillance would be necessary. The proponents of this approach argue that there would be less personnel radiation exposure involved, and costs would be reduced. Containments are very highly engineered structures that could well be expected to last for the required time with no further attention, and even if it were breached at some future distant time, the radiation levels would be low enough to present little if any problem. This last option does not fit the definition of decommissioning used by the NRC and has not been considered by them.

The NRC has evaluated these options and the costs associated with them and has issued regulations for their implementation. They have endorsed the DECON and SAFSTOR options as preferred but have set a time limit of 60 years for the completion of decommissioning under the SAFSTOR option unless it is necessary for public safety to extend it. The ENTOMB option is not ruled out if it can be shown to be preferred in a specific case. The NRC has also asked the plant operators to show that sufficient funds will be available for decommissioning. Several methods may be used to provide the amount that the NRC feels to be reasonable. The operators may use a prepayment method, an external sinking fund, or a surety or guarantee method. The amount must be at least $105 million for PWRs and $135 million for BWRs in 1986 dollars. The amount must be adjusted annually to account for inflation according to a prescribed schedule. Five years prior to decommissioning, a plan for the process must be submitted including an estimate of the cost. If the estimated costs exceed the amount set aside, the operator must indicate how the difference is to be made up.

1e. Safety

Government regulations. Nuclear power plant construction and operation is the most intensely regulated activity in the U.S. Federal, State, or local governments regulate every aspect of the activity. On the Federal level the Nuclear Regulatory Commission (NRC) must issue a construction permit and an operating license. Later the NRC inspectors keep track of the operation of the plant and assure compliance with NRC regulations.[4] The NRC imposes fines for any violations of its regulations and may shut down a plant if the violations are serious. The purpose of the regulations is to assure that people living near nuclear plants as well as the general population are not exposed to unacceptable risks. The effectiveness of the NRC regulations has been the subject of much debate. For example, there is a question whether the many changes ordered in existing plants ("backfitting") really improve the overall plant safety.

Power plants. An operating power plant, whether nuclear or fossil fueled, must be provided with systems to avoid, or at least minimize, damage in the event that some part of the plant fails. For fossil-fueled plants, the realization that this damage can lead to injury to those involved in running the plant or in the vicinity is expressed in part by various regulatory actions, such as boiler codes, licensing, and inspection, that have existed for many years. For nuclear plants the need for these measures increases because of the presence of large amounts of radioactivity in the fuel. Even if there were no radioactivity concerns, safety analysis would have to consider the large amount of stored energy in the fuel and coolant.

Safety features of power plants. If the radioactivity is contained, the hazards peculiar to a nuclear plant are minimized. The design of the plant must accomplish this both for normal operation and for accidents. To prevent any release of radioactivity, the designs incorporate the use of multiple barriers. The fuel itself forms the first barrier. Uranium dioxide is a ceramic with a high melting temperature so that most of the fission products are retained within the fuel at normal operating temperatures. This barrier suffices for normal operation. The fuel cladding which prevents fission products which diffuse out of the fuel from entering the coolant acts as the second barrier. The next barrier is the pressure boundary surrounding the coolant, e.g., the pressure vessel, the piping, and the steam generator tubes in a PWR. This system as well as the systems used to purify and replenish the coolant are closed circuits isolated from the environment. The final barrier, the containment, usually is reinforced concrete with a steel liner.

The establishment of these barriers does not sufficiently guarantee public safety. Further steps assure that the barriers remain intact or, if one fails, the next will not be affected. To evaluate the adequacy of a barrier or the system designed to protect it, the possible threats to its integrity must be considered. To this end the NRC has established some basic principles: Design for maximum safety in normal operation and maximum tolerance for system malfunction. Assume that accidents may occur and provide safety systems to minimize their effects by maintaining core cooling and thereby isolation of radioactivity. Provide additional safety systems to control or mitigate the effects of hypothetical accidents assuming that some of the protective systems fail as a result of the accident. Applying these general rules, the NRC has developed "General Design Criteria" for reactor designers (Ref. 4, part 50).

As a result of applying these principles and criteria, safety considerations influence almost all phases of the designs. The physical laws provide some safety features. For example, the reactors must be designed to have a negative temperature coefficient, i.e., an increase in temperature of the reactor results in a reduction in the reactivity of the reactor. The ability to monitor the behavior of the system provides another important safety feature. In addition, all plants have systems which have no function other than to avoid, control, or mitigate a potentially dangerous situation.

While it is necessary to plan for a wide spectrum of events which may never be encountered, designers must assure safety in normal operations. Radioactivity releases to the environment and radiation levels inside the plant have to be controlled. NRC regulations require that the radiation exposure from the plant to a person living on the plant boundary be less than 5 mrem/yr (approximately 2% of natural background). Radiation releases have been kept within license limits almost without exception and modifications to operating plants have further reduced releases in many cases. Although the radiation received by plant operating personnel has been kept within the prescribed limits, unanticipated maintenance requirements have led to more workers being exposed than was anticipated in the plant design, i.e., the person-dose from plant operation is higher than expected. The introduction of equipment permitting remote or semi-remote maintenance operations is expected to alleviate this problem.

Plant accidents. An operating power reactor is a heat generator with a high power density. The fuel rods produce this power. If for any reason the ability to remove this energy as fast as it is being generated is lost, the fuel temperature will rise. If this heat-up is not terminated, it will result in failure of the cladding and eventual melting of the fuel. Thus at least one of the barriers would be breached, and radioactivity would be released into the reactor coolant.

Shutting the reactor down terminates the chain reaction. Decaying fission products in the fuel are a source of energy that is present even after the reactor is shut down. One hundred seconds after a reactor is shut down, the energy from fission products produces about 3% of the initial power. Some coolant flow must be maintained to accept this decay energy and transfer it to a heat sink and ultimately to the environment. The safety systems must ensure that the reactor shuts down under these circumstances and that coolant stays in contact with the fuel. For example, in a PWR, there are systems which inject a strong neutron absorber into the coolant if the pressure in the reactor is too low or if additional cooling water is needed. These systems back up the normal shut down mechanism of inserting control rods. Other sources of cooling water are activated to replenish the water lost from any break. Heat is rejected either through the steam generators or by coolers in the containment. The proper actions of these emergency systems depend on the reliable operation of pumps and valves.

If, for some reason, activity gets from the fuel rods into the containment, the containment forms a final barrier. A release of radioactivity to the containment through a break in the reactor cooling system means that the hot, pressurized coolant has also escaped into the containment raising both the temperature and the pressure there. To avoid conditions which would exceed the design limits of the containment, active or passive systems must be present to remove sufficient energy from the escaping fluid to limit temperature and pressure excursions.

Accident analysis. Since the response of these safety systems cannot always be tested in actual occurrences, designers must rely on analysis and extrapolation from experiment. In one approach, analysts postulate a number of design basis accidents. The response of the reactor to each of these conditions may be studied. If the design allows control of each accident, the design is adequate.

One may then make the assumption that all or some part of the safety systems did not work properly and determine the effect on the remainder of the plant, especially containment and the consequences to the public. If the design basis accidents are chosen properly, one may assume that any other accident is of lesser magnitude and may also be contained.

The initial design of reactor safety systems used this technique. While various steps were taken to control other events, the loss of coolant from a primary coolant line break was considered to be the most severe accident in a PWR, and most of the work performed dealt with the consequences of this accident. This approach, while very useful, has some deficiencies. Reactors are very complicated. Hence analysts must make assumptions and approximations which lead to questions as to the validity of the analysis. Experiments can answer some of the questions, but then concerns regarding scaling from the experiments arise. Another problem with the design basis accident approach is that some other event which is equally important may be missed.

Probabilistic risk assessment (PRA) techniques for reactors (see Chapter 2) can also be applied to reactor safety studies. In 1972 the AEC initiated the first comprehensive analysis of reactor safety in which a realistic modeling of an actual reactor was attempted.[5] This study (RSS), also known as the Rasmussen report, has been reviewed by several groups including the APS.[7] These reviews served to bring out both the strengths and weaknesses of the analysis. Most reviewers felt that PRA was an appropriate technique and could provide useful results. However, the method depends on knowing the probabilities of failure of systems or components and on the modeling of the response of the reactor to the actions of the safety systems. Some felt that the values of the probabilities used could not always be well justified, that the uncertainties in these probabilities were not properly treated, that the models used for the reactor might not always lead to conservative results, and that some accident sequences might have been overlooked. These studies also pointed out areas where additional research was needed.

Since the preparation of this first PRA study, work has continued to refine, improve, and extend the application of PRA to reactor systems. The results provide quantitative confirmation of previous opinions regarding the low probability of core damage and have been used to upgrade certain parts of the safety systems. These techniques are also being used to evaluate the advanced reactor concepts now under development. With PRA it becomes possible to estimate in a quantitative way the degree to which advanced concepts are superior to present designs.

In 1987 the Nuclear Regulatory Commission published the most recent comprehensive study of accident risks of commercial nuclear power plants.[10] This report reevaluates the safety analysis of five operating power plants. It takes into account changes made following the TMI accident, as well as the knowledge acquired about the consequences of the core damage in this accident.

TMI accident. In the course of operating the present-day U.S. reactors, the safety systems have seldom been seriously challenged. There have been numerous occasions where a part of the system has operated, but this has usually been a conservative response to a minor problem. The most serious exception to this statement was the accident at Three Mile Island. Here a combination of the failure of a component to operate properly, mistakes on the part of the operating personnel which prevented the safety systems from performing as designed, and design features peculiar to that type of plant led to a loss of coolant accident which resulted in almost complete destruction of the core and the release of activity into the containment. Personnel mistakes were in part due to inadequate monitoring and display of reactor conditions. Activity was also released from the auxiliary building into the atmosphere.

Although the amount of activity released resulted in no observable health effects on residents in the plant's vicinity, the psychological effects, both locally and nationally, were severe. Considering the amount of core damage, a relatively small fraction of the radioactivity was released to the containment, much less than had been estimated in earlier reactor safety studies. Much work has gone into trying to understand the reasons for this low release and whether such low releases may be expected in all light-water reactor accidents.[9]

Reviews such as the Kemeny Report present a detailed examination of the course of the accident and the deficiencies in design and operating procedures that contributed to it.[6] As a result of this process the NRC required changes to be made in equipment and procedures in all plants. There is now more emphasis on training operators with the use of plant specific simulators, on the proper display of information most needed to cope with unusual conditions, on instruments that provide this information, and on the analysis of a wide variety of potential accidents rather than concentrating on one design basis event. The TMI accident also showed that even though the safety systems functioned to protect the public, the financial loss to the owner can be intolerable if there is significant internal damage to the reactor.

As a result of the TMI accident the utilities sponsored groups to aid in the operation of the plants. One such group, the Institute of Nuclear Power Operations, aids the utilities in setting and maintaining good operating and training practices. The Nuclear Safety Analysis Center performs or sponsors studies on questions which have industry-wide application and are beyond the capability of any one member. The work of these groups as well as that of the NRC and its contractors has resulted in a much clearer understanding of the response to and consequences of potential accident initiating events.

Chernobyl accident. A discussion of the accident at Chernobyl falls beyond the scope of a review of the situation with respect to U.S. reactors. The radiological impact of the accident has been summarized by Anspaugh, Catlin, and Goldman[11] as follows:

"Global impacts to health from the Chernobyl accident may be characterized as acute (nonstochastic) effects or as delayed, stochastic effects predicted on a probabilistic basis. No acute effects have occurred outside of the Soviet Union where 237 cases of acute radiation sickness, including 31 deaths, were reported.

Outside of the immediate Chernobyl region, the magnitude of radiation doses to individuals is quite small, leading to extremely low incremental probabilities of any person developing a fatal radiogenic cancer over a lifetime. — Probably no adverse health effects will be manifest by epidemiological analysis in the remainder of the Soviet population or the rest of the world. The major global impacts from Chernobyl appear to be economic and social."

The reactor at Chernobyl was graphite moderated, used boiling water in pressure tubes for cooling, and had no containment. Thus its design is very different from the U.S. light-water reactors which are water moderated and cooled, use a pressure vessel, and have containment. Because the accident was so severe, it is worthwhile to consider what effect it might have on the design and operation of reactors in the U.S. Since the accident, there have been several analyses of the cause and the likelihood of a similar accident to the U.S.-style light-water reactors. Because of the differences in the type of reactor, the design philosophy used for the safety systems, and the method of operation, the general opinion is that no modifications to U.S. plants are needed.

This is not to say that the Chernobyl accident has had no effect. It emphasized the need for a well-trained operating staff working with well-developed procedures for operation. If it had not been justified before, it reinforced the concept of barriers and the decision to provide reactors with complete containment. Elsewhere it has led some of the European reactor licensing bodies to consider adding provisions to allow controlled venting of the containment through filter beds to avoid excessive pressure buildup and possible containment failure.

Mining. Mining of uranium gives cause for concern in three areas: the hazard associated with any mining operation, the additional hazard from the presence of radioactivity in the ore, and the hazard from ore left behind after the uranium has been extracted. The first two are occupational hazards and the last is a hazard to the general public.

The occupational hazards associated with the basic mining operations for uranium are the same as those of other types of mining operations, e.g., poor air quality with pollutants such as dust and diesel fumes, and the threat of cave-ins. However, since the energy content per unit mass of uranium ore excedes that of coal by a factor of about 30, uranium miners endure less exposure to mine conditions and consequently a reduced hazard per unit of energy produced.

The principal radioactive hazard in mining comes from ^{222}Rn, a naturally occurring product in the ^{238}U decay chain. Radon is a noble gas and diffuses into the atmosphere of the mine as the uranium is extracted. ^{222}Rn has a half-life of 3.6 d and, since it is not accumulated in the body to any great extent, itself poses little threat. Subsequent decays in the chain lead to the emission of three more alpha particles. The daughter nuclei are not gases but may attach themselves to dust particles. The particles may be inhaled and remain in the lung exposing the surface of the lung to radioactive decay products. Risk to the surface of the lung is confirmed by an observed increase in lung cancer in miners with high levels of exposure. Radon levels may be reduced by proper ventilation. Current standards set the maximum exposure per year to miners at a level which corresponds to residential radon concentrations that are not unusual. Radon is also released in the mine exhaust. Conservative estimates show that even at a very high level of uranium mining, the increase in radon levels in the atmosphere would be only 1% of the natural radon background.

After the uranium has been extracted from the ore, the bulk of the material remains as mill tailings at the site of the mine. All of the long-lived daughters of the uranium originally present in the ore now reside in the depleted ore. The material generally leaves the mill in the form of a liquid sludge and is allowed to dry in settling ponds. The tailings associated with the production of 1 GWe year of electricity would occupy a volume of 3×10^5 m^3. This material is in a form that is susceptible to erosion, to leaching, as well as to the emission of radon. All of these effects have been observed in tailings that have been left untreated. Because of the potential hazard of this material, which in some ways approaches that of spent fuel after the fission products have decayed, the NRC has set up regulations for the design of storage sites and their management to meet EPA requirements.

A site selected for the tailings must ensure long-term isolation. The preferred option is to place them either in mines or in pits dug for the purpose. If that is not possible, regulations allow above ground disposal providing appropriate precautions are taken. The tailings must have an earthen cover so that the release of radon is less than 20 pCi m^{-2} s^{-1}, and the site must be designed to provide reasonable assurance of control of radiological hazards for 1000 years. Because of the finite size of the tailings from a mine, this rate of release results in a radon concentration at large distances from the tailings several orders of magnitude less than natural background.

Transportation. There are three areas where transportation safety is of potential concern: the movement of uranium from the mill to the power plant, the removal of low-level wastes from the reactor to a low-level waste site, and the transportation of spent fuel for disposal. Since most of the radioactivity resides in the fuel, the movement of spent fuel has the highest potential hazard and has been considered in most detail. A small fraction of the fission products are gaseous and diffuse into the gap between the fuel and the cladding. A rupture of fuel rods combined with a failure of the shipping container would result in the release of this activity. A subsequent rise in temperature might release some of the remaining fission products.

To minimize the possibility of such a release, the NRC has imposed rather severe requirements on the design of shipping containers. In specifying these requirements, the NRC examined the conditions which might be found in a wide spectrum of accidents. Various agencies and groups both here and abroad have repeated the analysis of these accidents since the regulations were established.

A recent study by the NRC focused on the adequacy of the present requirements. Among the topics considered were the probability of a severe accident, the number of such accidents per year, and the probable response of a container to severe accidents that have occurred involving non-nuclear materials. The calculational models for the casks were based on existing designs rather than the proposed more modern designs. The results of this analysis confirm the previous analysis used to set the standards and indicate that it may have been conservative. The recent study estimated that in the 65 million km of truck travel needed to transport spent fuel presently at reactors, there would be one accident involving cask damage leading to a release of radioactivity within that allowed by the standard. In 130 million km of travel, there would be an accident resulting in a release slightly above the standard. The study also considered what would have happened if a cask had been involved in some of the severe transportation accidents that have recently occurred in the U.S., e.g., the September 1982 railroad accident in Louisiana in which a fire burned for several days and two tank cars exploded. In most cases there would have been no radioactive release, and in the worst case the release would have been only slightly above that permitted.

The containers are designed to withstand some very severe prescribed tests, e.g., a 10 m free fall drop onto a flat unyielding surface with the cask oriented so that maximum damage will occur. The design techniques have been verified by a series of engineering tests. In light of the care that has gone into the physical aspects of transportation and the numerous requirements for route choice, notification of national and local authorities, and the training of the drivers involved, the transportation of fuel should be one of the safest aspects of nuclear power.

Reprocessing. The question of the safety of a reprocessing plant is in a sense moot since there are no plans to reprocess commercial reactor fuel in the U.S. This topic has been considered, e.g., in an APS study on fuel cycles and waste management.[8] As an example of the hazard from such a plant, this report assumed that all of the ^{3}H, ^{85}Kr, ^{14}C, and a fraction of the ^{129}I escaped from a plant of 1500 t/y capacity (a plant of this size would service approximately fifty 1000 MWe reactors). According to this report, the global average per capita whole body dose increment from this release after 50 years of operation would be 0.003 mrem/y. This may be compared with the average exposure of the U.S. population to a dose of about 360 mrem/y. The technology exists to retain much of this material in the plant although provision must be made for the eventual disposal of the very long-lived isotopes, ^{14}C and ^{129}I.

Waste. High-level wastes such as spent fuel rods could be placed into an underground repository. Because these wastes contain radioactive species with very long half-lives, one must worry that at some time in the future radioactive material might be released and enter the environment. Thus accidental or intentional access must be highly improbable, and the waste must be sufficiently isolated from any circulating ground water. The standard for performance is not more than one death per ten years over a 10,000 year period under conditions of no maintenance or surveillance. The Nuclear Waste Policy Act Amendment of 1987 directs the Department of Energy to investigate the characteristics of the Nevada site to determine if these objectives can be met.

The response of a particular site to the storage of spent fuel depends on the nature of the material being stored and the nature of the host body.[12] The proposed form of the fuel is spent fuel rods, unprocessed, in a storage canister. Initially the decay of the fission products dominates the activity of the fuel. After a period of about 500 years, most of the fission products have decayed leaving the heavy elements as the major sources of activity. The average concentration of this activity is not large and may be comparable to that of some natural uranium ores.[8] The retention of these heavy nuclides over a very long time span is the more difficult problem. The concept used in the design and evaluation of a site is one of multiple barriers. The fuel rod material itself is the first barrier. Activity escaping from the fuel is held up by the cladding and then by the canister.

Since radioactive material from the fuel and fuel canisters may be released into the host over a long period of time, the transport properties of the radionuclides through the host into the environment must be determined. These transport properties depend not only on the host, i.e., chemical composition, porosity, physical form, water content, etc., but also on the chemical state of the radionuclides. Other factors such as the distance to circulating ground water, the velocity of the ground water, and the presence of intervening layers of different geologic formations have to be considered. The likelihood of significant tectonic or volcanic activity which might alter the properties of the site must be taken into account as must the probability of accidental intrusion.

The emplacement of the fuel may change the environment. The mining operation may fracture the host rock and alter its transport properties. The heat from the fuel has an effect. The host rock may affect the integrity of the canisters. Suitable methods of backfilling and closing the site must be developed. Preliminary work indicates that these and other related concerns may be satisfied at the selected site in Nevada. If they cannot, then the characterization of other sites must be made.

1f. Public concerns and attitudes

Although the present study is concerned primarily with the technology of energy sources, the use of nuclear power depends not only on technical and economic considerations, but also on public attitudes. Of all energy sources, its deployment faces the most widespread opposition. Public opposition is strong enough to have prevented the operation of completed plants even where there is a strong need for their output. Hence the discussion of public concerns forms an integral part of a discussion of nuclear power.

Surveys of public attitudes towards nuclear power and of the reasons for the attitudes have produced ambiguous results which vary with time in response to recent media coverage of the energy situation and of nuclear plant incidents. Reactor accidents, such as TMI and Chernobyl, result in opposition to nuclear power while news of power shortages and brownouts, of the greenhouse effect and acid rain, tend to make nuclear power more acceptable.

Opponents of nuclear power frequently cite concerns with radioactive waste disposal, thermal pollution, and diversion of nuclear materials for the manufacture of nuclear explosives, but the basic public concerns involve the release of radiation (or radioactive nuclides) in the event of a reactor accident, often associated with the fear of a nuclear explosion of a reactor. This public wariness is probably due to the fact that nuclear radiation cannot be detected by one's senses and that it can induce cancer and genetic abnormalities. In spite of all assurances that a power reactor cannot explode like an atomic bomb, several surveys have found that about 40% of those surveyed believe that power reactors can do precisely that.

Attitudes toward nuclear power vary widely among various groups. Those who consider themselves environmentalists often oppose nuclear power, while several surveys of scientists and engineers report that a majority strongly support rapid development of nuclear power with only 10% opposing all nuclear power development. Of course there are some vocal critics among the opponents, including some who have worked in the nuclear industry. Some surveys show strong support for nuclear power among those residing near nuclear plants. The overall public support for building new plants has steadily declined over time. This decline coincides with the decline of interest on the part of utilities in building new plants, based largely on economic considerations. On the other hand, some surveys indicate a wide-spread view that nuclear power will become more important in the future.

There have been numerous state-wide referenda aimed at closing operating nuclear plants. So far all have been defeated. However, in 1989, 53% of the voters in Sacramento, California, voted to close the municipally owned nuclear plant which had had an abysmal operating record with a lifetime capacity factor of less than 38%. This vote came a year after 51% of the voters had favored keeping the plant open for another year. In the meantime the plant had operated at a capacity factor of only 26%. The closing of Maine Yankee was voted on in 1980, 1982, and 1987. The vote for closing the plant varied from 40% to 45%; the percentage for closing was higher before Chernobyl than afterwards. The 1988 vote on closing the nuclear plants in Massachussetts resulted in a two-to-one majority against closing the plants. On the other hand, referenda restricting the construction of new nuclear plants have been approved in several states, and several states have passed legislation restricting the construction of new plants, most frequently requiring prior demonstration of safe radioactive waste disposal and/or requiring voter approval of new construction in a referendum.

2. Future direction in reactor design

All but one of the nuclear power plants in the U.S. are light-water reactors (LWR). Although the LWRs are of two different designs (PWR, BWR), their size, performance, cost, and difficulties are quite similar. If and when new power reactors are ordered, the utilities would like improved versions of the current designs because they have had several decades of experience with them. Efforts to accomplish such improvements are underway both in the U.S. and other countries. The primary aims of these efforts are greater safety and lower cost. Other related aims are more efficient use of uranium, and higher capacity factors.

2a. Improvements in current reactors

To improve safety, new designs aim to assure that the coolant will remove decay heat even if the circulation system fails and to reduce the chances that the core might become uncovered as happened in the TMI accident. Although the most important safety concern before the TMI accident was the possibility of a pipe break in the cooling system, the TMI accident showed that a valve failure combined with operator errors could have equally serious consequences. Design improvements seek to reduce the probability of an accident by at least a factor of 10. At the same time, design changes aim to increase the time between refueling as well as reducing the time it takes to refuel, thereby increasing the availability of the plant. Other proposed changes would reduce the radiation exposure of the workers and increase the reliability of the plant. The hope is that simplification of some parts of the system will compensate for the increased costs of other features so that the overall cost of the plant would not be appreciably greater than current costs.

The Electric Power Research Institute with the support of the DOE and the NRC has undertaken a somewhat more innovative approach to the design of advanced light-water reactors. J. J. Taylor recently summarized these efforts.[13] The design seeks to achieve significant improvements in safety, reliability, operation, and power costs. The reactor designs of about 600 MWe incorporate recent developments in control and display of reactor parameters. The safety systems operate in a passive manner with minimum reliance on valves, pumps, and electric power to operate them.

These designs, as exemplified by the AP600 PWR of Westinghouse and the small BWR of General Electric, show promise of meeting the goals that have been established. For example, the AP600 design requires no safety pumps compared to 25 in a conventional plant of the same size, 130 remote valves (5 cm diameter) compared with 350 in a conventional plant, and 3400 m of piping (5 cm diameter) compared with 9500 m. The number of heat exchangers and tanks is also reduced. These changes not only reduce capital costs and construction time, but would also lead to reduced operation and maintenance costs. A PRA analysis of the AP600 design predicts the probability of core damage to be at least a factor of 10 lower than the best current plants and the probability of a significant release of radioactivity from containment failure to be approximately 100 times less.

In addition to the LWRs, a high-temperature gas cooled reactor (HTGR) has operated in the U.S., and there is development work on this type of reactor underway in the U.S. and in other countries.[14] HTGRs have many desirable features which make them attractive for future use. They operate at higher temperature than LWRs, hence can achieve higher thermal efficiency. HTGRs are intrinsically safer than LWRs. The core has a lower power density and larger heat capacity so that the temperature of the core would rise more slowly in the event of a failure of the cooling system. Experience at Fort St. Vrain, the only HTGR that has been operated in the U.S., has shown exceptionally low radiation exposures of workers. Interruption of the forced helium cooling has caused no damage to any component of the reactor. Although the plant has experienced a number of problems related to its many new design features, none of the problems were serious. However they have resulted in the plant having very poor performance. On the other hand, HTGRs which have operated in Germany have performed well.

The design of a smaller HTGR, often referred to as modular HTGR (MHTGR), is under development. There is a proposal to build two such a reactors in Idaho for tritium production. They could be a demonstration project for later use of a MHTGR for power generation.

The MHTGR exemplifies what is often called an "inherently safe" reactor. This term describes reactors which would not be damaged and would not release radioactivity in the event of failure of the cooling system. In conventional LWRs the decay heat generated, even after the chain reaction is stopped, can melt the fuel elements unless the fuel elements are covered with water. In the helium-cooled MHTGR, the fuel elements would not be damaged if the helium flow were stopped. Some advocates hope that an MHTGR would not require a containment dome, nor would there be any need for emergency evacuation plans. The fuel particles would have a ceramic coating which would be expected to prevent the escape of radioactivity even at a temperature as high as 2000 °C.

The term "inherently safe" applies only to the behavior of the fuel if there is no cooling. The "melt-down" of the core remains the principal concern in reactor safety, but there may be other possible failures that could have serious consequences for the plant.

The MHTGR is an example of a small reactor. The term modular describes reactors which are intended to serve as modules for larger plants. Most present nuclear power plants are designed to produce more than 700 MWe. The cost of building one large plant was expected to be much less than building several smaller plants. Modular designs offer several advantages. There is a smaller financial commitment. The plant size can be increased by adding more modules to match the demand. If there is more than one module, an outage of one module does not shut down the entire plant. There is some evidence that the capacity factor of smaller plants is greater than that of larger ones. The principal advantage of the modular design is, however, that small reactors could be built in a factory rather than on-site, and that factory construction is cheaper and probably of higher quality. The question of whether small reactors are economically competitive with large reactors remains controversial.

2b. Heavy-water reactors

Nuclear power plants in Canada use heavy-water reactors (HWR), known as CANDU reactors.[15] These reactors have an excellent performance record; their capacity factor is on the average much higher than that of LWRs. This is partly due to the fact that these reactors can be refueled without being turned off. The question often arises why they are not used in the U.S. Should they be considered by utilities, if and when the utilities order new nuclear power plants? Several obstacles block the use of HWRs in the U.S. There are no domestic vendors, and there is no domestic experience with them. No safety study clearly shows the superiority of HWRs over LWRs. Perhaps the most serious problem is that the U.S. licensing procedures do not fit HWRs, hence it is not clear that an HWR could be licensed in the U.S. without major design changes.

2c. Inherently safe reactors

The term "inherently safe reactor," used in the discussion of the MHTGR, denotes a reactor that would not be damaged if the cooling system used in normal operation failed. Design efforts are underway to develop inherently safe LWRs. The design that has progressed farthest is known as PIUS (Process Inherent Ultimately Safe) and was developed by a Swedish firm. Its basic concept is similar to a PWR, but the PIUS reactor uses a much larger pool of water. The reactor is placed at the bottom of a very large water-filled concrete vessel which also contains all components of the primary cooling system. The only connections are at the top of the vessel. In the event of a failure of the primary cooling system, the large pool will cool the core even if there is no electric power available, and cooling failure requires no action by an operator. For automatic shut down of the reactor, neutron absorbing boron stops the chain reaction. Most of the technology of present LWRs applies to the PIUS reactor, especially the fuel handling. Whether or not the cost of a PIUS reactor is comparable with that of a conventional LWR remains an open question. The very large vessel would increase the cost, while some of the safety features of conventional LWRs could be eliminated, thereby reducing cost.

A quite different type of inherently safe reactor, which has been designed at Argonne National Laboratory, is known as the Integral Fast Reactor (IFR).[16] In contrast to LWRs and HTGRs, the IFR uses fast neutrons to maintain the chain reaction, i.e., it does not have a moderator. The coolant is liquid sodium at atmospheric pressure and at a temperature between 350 and 510 °C. The design of the IFR rests on the experience with the Experimental Breeder Reactor II (EBR II). In tests performed with the EBR II reactor, the inherently safe features of this reactor were demonstrated in one case by stopping the sodium coolant flow without inserting the control rods and in another case by not removing the heat from the sodium while the reactor was operating. In both tests the chain reaction stopped itself without damage to the reactor because of the negative temperature coefficient (see above).

The fuel of the IFR and EBR II is in metallic form rather than an oxide like the fuel in LWRs. The high thermal conductivity of the metal compared with the oxide results in much lower operating temperature for the metal. The lower temperature produces a smaller increase in reactivity of the fuel when the operating power level is reduced, a feature which is an important contributor to the improved safety of the IFR.

The IFR, like all fast reactors, will require reprocessing of the fuel. However, the use of metallic fuel results in a much simpler and less expensive process which may even be performed at the plant site. The reprocessing process proposed would leave the long-lived transuranic elements with the uranium and plutonium to be returned to the reactor. Thus the waste to be sent to a repository would contain only fission products. These isotopes decay much more rapidly than the transuranics so that the time for which storage must be assured is reduced to several hundred years. The simplification and safety features of the IFR quite possibly will lead to costs that are comparable to those of the LWR and make licensing at least no more difficult than for present reactors.

Implementations of the IFR concept have been developed by General Electric (Power Reactor Inherently Safe Module) and by Rockwell International (Sodium Advanced Fast Reactor). These designs have similar aims and approaches.

3. Outlook for nuclear power

One can ask two separate questions about the future of nuclear power in the U.S.: (1) How long will the currently operating plants remain in service? (2) When will new plants be ordered? The answers to these questions depend primarily on political and economic decisions rather than on technical considerations. Hence all predictions are likely to be wrong. At least that has been the experience of the past. Wrong predictions of demand for electric power, of costs, and of the environmental hazards of various methods of power generation, and of public attitudes led to wrong predictions of the construction of new plants.

Termination of the operation of existing plants may be the result of legislation, of the economics of operating the plant, or the expiration of the operating license. Only the third of these possibilities involves technical questions. It is, of course, possible to replace components of a plant that are no longer reliable. In view of the very large investment in the plant, large expenditures (of the order of $1 billion) for refurbishing an existing plant could be justified, since constructing a new plant, whether coal or nuclear, would cost more.[17] A substantial effort is underway to study methods for extending the life of nuclear plants. In some cases an existing plant might be relicensed to operate at reduced power after the current license expires.

Before addressing the second question, one needs to review the reasons why so many plants which had construction permits, and in some cases were partially built, were cancelled beginning around 1975. As shown in Table 5, page 31, at that time the growth rate of electricity demand suddenly decreased so that there was no need for many of the plants under construction. At that time, too, most of the plants being built were nuclear, and each involved a very large investment. Hence their completion was deferred or they were abandoned.

The TMI accident, a few years later, resulted in new regulations that increased the construction costs and required costly changes in operating plants. Other regulations increased operating costs by requiring a larger and more highly trained staff and much more paperwork. In addition some utilities lacked adequate management competence to handle the problems encountered in operating nuclear plants. With increasing costs nuclear power lost its economic advantages over coal. In addition, utilities with nuclear plants faced financial disaster if there were an accident as at TMI or if a completed plant could not be operated as at Seabrook and Shoreham. As a consequence utilities will not order new nuclear plants in the near future.

David Lilienthal coined the term a "second nuclear era" and Alvin Weinberg has written a book with this title.[18] The second nuclear era describes a future period when nuclear plants will again be built in the U.S. The second nuclear era will occur only if existing plants maintain a good safety record and new plants are expected to produce power at a competitive cost. In addition, public attitudes must be supportive of new nuclear plants. Otherwise regulatory and legal obstacles will interfere with bringing new plants into operation. The perceived safety and economy of new plants and risks of alternative technologies will influence public attitudes which will also be shaped by fears if energy shortages develop again. Dangers from the greenhouse effect and acid rain have to be compared with dangers of reactor accidents, of radioactive waste, and of nuclear weapons proliferation. The attitudes can shift rapidly so that predictions of public attitudes are likely to be wrong.

Both the economic and safety aspects of new reactor designs are under study, as are the economics of smaller "modular" plants. Many of the currently operating plants were designed before the widespread use of computers and automation and relied on human intervention for routine or emergency operations. Computerized systems being installed at many plants should increase safety and reduce costs, since operator errors caused the major nuclear plant accidents in the past. In a second nuclear era one would expect extensive automation and robotics to reduce the operating staff and their radiation exposures.

When new nuclear plants may be ordered in the U.S. is quite uncertain. Even a prototype of an improved reactor design is unlikely to be operating much before the end of the century so that substantial numbers of new plants could not be ordered until the next century.

References and notes

Most primary references to information discussed in this chapter are reports that are not readily accessible. We therefore give references to books and readily available journal articles which in turn contain references to original work.

Textbooks on nuclear energy

R. L. Murray, *Nuclear Energy: An Introduction to the Concepts, Systems and Applications of Nuclear Processes*, 3rd ed. (Pergamon Press, New York, 1988) (Elementary).

J. R. Lamarsh, *Introduction to Nuclear Engineering*, 2nd ed. (Addison Wesley, Reading, MA, 1983) (Intermediate).

S. Glasstone and A. Sesonske, *Nuclear Reactor Engineering*, 3rd ed. (Van Nostrand, New York, 1981) (Advanced).

In the late 1970s two critical studies of Nuclear Power were carried out under the auspices of the National Academy of Sciences and of the Ford Foundation:

1. *Energy in Transition 1985–2000* (National Academy of Sciences, Washington, DC, 1979) ("CONAES report").

2. Ford Foundation, *Nuclear Power Issues and Choices* (Ballinger, Cambridge, MA, 1977).

A more recent study of costs, regulations, and public acceptance was issued by the Office of Technology Assessment:

3. *Nuclear Power in an Age of Uncertainty*, OTA-E-216 (Washington, DC, 1984).

4. Federal regulations pertaining to nuclear power are contained in: *Code of Federal Regulations* (CFR), Title 10, especially Part 40, Domestic Licensing of Source Material; Part 50, Domestic Licensing of Production and Utilization Facilities; Part 100, Reactor Site Criteria.

The first comprehensive reactor safety study was prepared for the AEC by a group at the Massachusetts Institute of Technology:

5. *Reactor Safety Study* (RSS), Study Director N. C. Rasmussen, WASH-1400 (U.S. Nuclear Regulatory Commission, Washington, DC, 1975).

The investigation of the TMI accident was published in:

6. J. G. Kemeny, *The Accident at Three Mile Island; the need for change; the legacy of TMI. Report of the President's Commission on the Accident at Three Mile Island* (Washington, DC, 1979).

The American Physical Society has carried out three studies related to nuclear power:

7. "Report to the American Physical Society by the study group on light-water reactor safety," Rev. Mod. Phys. **47**, S1 (1975).
8. "Report to the American Physical Society by the study group on nuclear fuel cycles and waste management," Rev. Mod. Phys. **50**, S1 (1978).
9. "Report to the American Physical Society of the study group on radionuclide release from severe accidents at nuclear power plants," Rev. Mod. Phys. **57**, S1 (1985).

References for specific topics:

10. *Reactor Risk Reference Document*, NUREG-1150 (U.S. Nuclear Regulatory Commission, Washington, DC, 1987).
11. L. R. Anspaugh, R. J. Catlin, and M. Goldman, Science **242**, 1513 (1988).
12. K. B. Krauskopf, *Radioactive Waste Disposal and Geology* (Chapman and Hall, London, New York, 1988).
13. J. J. Taylor, Science **244**, 318 (1989).
14. H. M. Agnew, Sci. Am. **244** (6), 55 (1981).
15. J. A. L. Robertson, Science **199**, 667 (1978).
16. C. E. Till and Y. I. Chang, Adv. Nucl. Sci. Technol. **20**, 127 (1988).
17. B. I. Spinrad, Science **239**, 707 (1988).
18. D. Lilienthal, *Atomic Energy, A New Start* (Harper & Row, New York, 1980).
19. A. M. Weinberg *et al.*, *The Second Nuclear Era: A New Start for Nuclear Power* (Praeger, New York, 1985).

Fusion technology for energy

Gerald L. Epstein[1]

Fusion energy's appeal was amply illustrated early in 1989 as newspapers and magazines around the world carried reports that scientists had produced net fusion energy using simple equipment at room temperature. Although fusion's new-found attention faded as the claims for "cold fusion" were shown to be premature, its long-standing potential to provide the world with an attractive source of electricity continues to motivate a considerable world-wide research effort.

Great progress has been made in almost 40 years of fusion research. However, it is still too early to tell whether the advantages of fusion energy, which could be significant, can be economically realized. It appears that at least three more decades of additional research and development will be required before a prototype commercial fusion reactor can be demonstrated.

1. The allure

Fusion power could be an attractive source of energy because it might offer a combination of benefits unmatched by other fuels. Compared to other energy technologies, fusion could be:

• *Unlimited.* The deuterium found naturally in a gallon of water contains the fusion energy equivalent of 300 gallons of gasoline.

• *Clean.* Fusion reactors may be significantly cleaner than those using either fossil fuels or nuclear fission. Fusion, unlike coal combustion, does not produce carbon dioxide, and it will not contribute to acid rain or other potential environmental damage associated with fossil fuels. Fusion power plants will have to make and use radioactive material for fuel, and their reactor structures will become radioactive as a result of continued operation. However, fusion does not inherently produce long-lived radioactive wastes, and structural materials may be developed that would minimize the long-term radioactivity generated.

• *Safe.* It may be possible to design fusion reactors whose safe operation is assured by their physical properties, rather than by active safety systems that might fail due to malfunction, operator error, or natural disaster.

To be weighed against these advantages, however, is the extraordinary difficulty of developing a fusion reactor and the uncertainty as to whether it can ever be done economically without compromising fusion's potential environmental and safety advantages.

2. The fusion reaction

The nuclei — or central cores — of atoms combine or fuse together under appropriate physical conditions to produce the fusion reaction. The total mass of the final products is slightly less than that of the original nuclei, and the difference (less than 1% of the original mass) is released as energy. In the easiest fusion reaction to initiate, two hydrogen isotopes, deuterium and tritium, combine to form helium and a free neutron (Figure 1). This D–T reaction releases 17.6 million eV of energy[2], 14.1 million of which is carried off as kinetic energy by the neutron while the remainder is imparted to the helium nucleus, or alpha particle. By way of comparison, burning a single atom of carbon (the combustible material in coal) releases only about 4 eV, a factor 4 million times smaller.

Out of every 6700 hydrogen atoms on Earth, one is the deuterium isotope. Tritium is practically nonexistent in nature, but it can be manufactured from the element lithium in nuclear reactors — including fusion reactors. It is radioactive, with a half-life of 12.3 years.

The most straightforward application of fusion power will be to convert the kinetic energy of the neutron formed in the fusion reaction into heat, which can be used to generate electricity. In the process, the neutron can be captured by a lithium nucleus in the structure surrounding the reaction chamber, creating an atom of tritium to replace the one consumed in the reaction.

However, not all the neutrons will be captured by lithium. Some will become captured in, or otherwise react with, other materials in the reactor structure, making them radioactive. Most of the environmental and safety issues associated with fusion power relate to the use of radioactive tritium as fuel and to the radioactivity induced in the reactor structure.

Fusion reactions using fuels other than deuterium and tritium have the advantages of producing few or no neutrons and not requiring radioactive fuels.[3] However, these reactions are significantly more difficult to sustain, and are not likely to be utilized in the first generation of fusion reactors.

Fusion reactions can only occur when the requirements of temperature, confinement time, and density are simultaneously satisfied. The minimum temperatures, confinement times, and densities needed to produce fusion power have been known for decades. Achieving these conditions, however, has proven extremely difficult.

Temperature. Because atomic nuclei all have positive electric charges, they repel each other. To fuse, they must be heated to temperatures on the order of 100 million degrees Celsius (C) to become energetic enough to overcome this repulsion. Matter cannot exist in solid, liquid, or even gas form at fusion temperatures; individual atoms are broken down — ionized — into their constituent electrons and nuclei. Matter in this state is called a *plasma.*

Confinement time. If a plasma cools too rapidly, exorbitant amounts of heat must be added to maintain fusion temperatures. The cooling rate of a plasma is measured by its confinement time — the length of time in which the plasma would cool by a certain fraction if no additional heat were added. A plasma with an insufficient confinement time cannot generate net fusion power.

Figure 1 —The D-T Fusion Reaction and a Fission Reaction

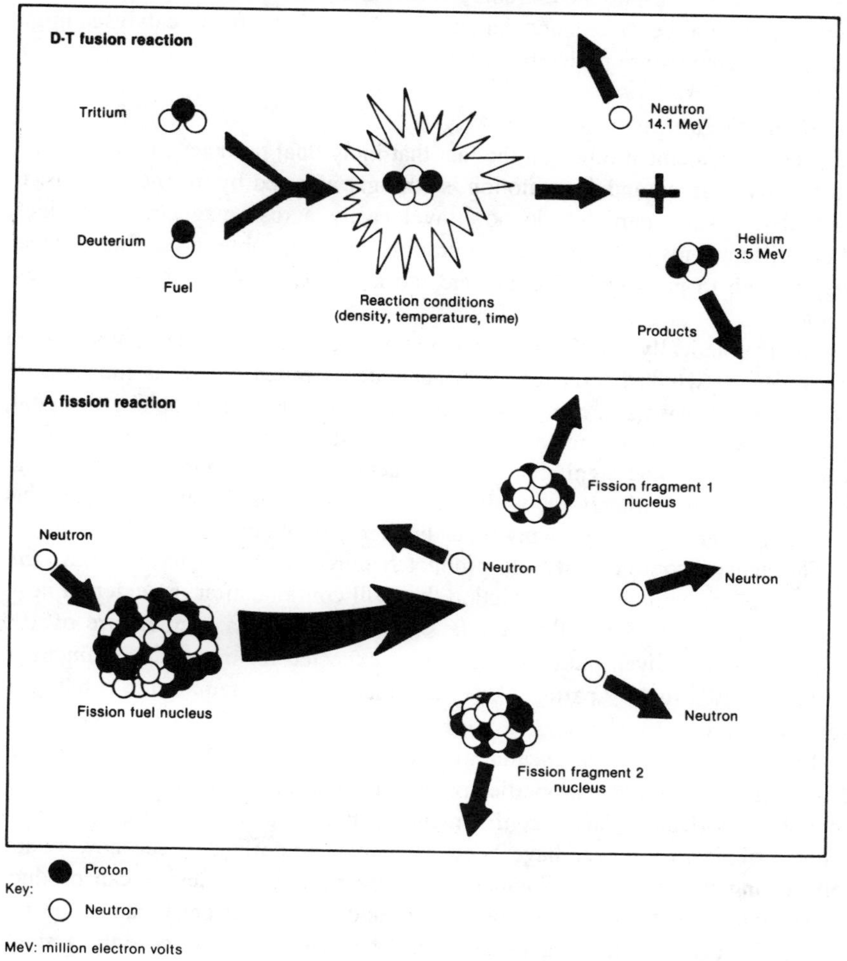

Figure 1. Source: U.S. Congress, Office of Technology Assessment, *Starpower: The U.S. and the International Quest for Fusion Energy,* OTA-E-338 (U.S. Government Printing Office, Washington, DC, October 1987).

Density. The exact confinement time required depends on plasma density. The fusion reaction rate, and therefore the amount of fusion power produced, goes up rapidly as density increases. If the product of density times the confinement time exceeds a certain value (which depends on temperature), the plasma can produce more power than is needed to keep it hot.

3. Confining fusion plasmas

A fusion plasma cannot be contained in an ordinary vessel. No matter how high a temperature the vessel can withstand, the plasma will be instantly cooled far below the minimum temperature required for fusion whenever it comes into contact with the vessel. There are two primary methods for confining fusion plasmas on earth despite the extreme temperatures involved — magnetic confinement and inertial confinement.[4] An additional technique, muon catalysis, might finesse the temperature problem entirely.

3a. Magnetic confinement

Magnetic confinement relies on the fact that individual particles in a plasma are electrically charged and their motion is strongly affected by magnetic fields. In particular, plasma particles do not travel easily across magnetic field lines. Many different magnetic field configurations might be able to confine plasmas well enough to produce fusion power, and a variety of them are under investigation.

No magnetically confined plasma has yet reached the point called *breakeven*, at which fusion reactions generate as much power as must be injected into the plasma to compensate for energy losses. Even reaching this goal, however, will not prove fusion to be a useful source of energy; a commercial reactor must generate significantly more energy than it consumes. A more significant milestone, therefore, will be creation of a "high-gain" reaction that generates an energy output many times greater than the energy input.

The density-confinement time product required for a plasma to reach *ignition* — a self-sustaining fusion reaction that will continue after the external heating is shut off — is 3×10^{14} particle s/cm^3 for a plasma temperature of 100 million degrees. Given the typical densities expected for magnetic confinement reactors of 10^{14} to 10^{15} particles/cm^3, confinement times required to reach ignition are on the order of 1 s.

The principal scientific uncertainties remaining to be resolved with magnetic confinement involve the properties of plasmas undergoing high-gain fusion reactions. Historically, plasma confinement has been found to degrade when new sources of heating power have been used; it is important to see how fusion self-heating will affect confinement. Because no existing device can produce significant amounts of fusion power, this uncertainty currently cannot be explored. Simply reaching breakeven would be insufficient, since only a small fraction of the total heating power in a breakeven plasma comes from fusion self-heating.[5]

The magnetic confinement concepts listed in Table 1 are briefly described below. They differ from one another in aspects such as the configuration of the confining magnetic fields, the means by which those fields are generated (external magnets as opposed to electrical currents induced within the plasma or the surrounding container), plasma shape, and eventual reactor geometry. Figure 2 shows the results (in terms of plasma temperature and density-confinement time product) that various experiments embodying these concepts have achieved. Curves in Figure 2 also show the combined values of these parameters that are necessary to reach breakeven and ignition.

Note that the "breakeven" and "ignition" curves shown on Figure 2 are for plasmas containing tritium, whereas the experimental results have been obtained in experiments that are fueled solely with ordinary hydrogen or deuterium, without tritium. Experimental procedures are greatly simplified by

Table 1. Classification of magnetic confinement concepts.

Well-developed knowledge base	Moderately developed knowledge base	Developing knowledge base
Conventional tokomak	Advanced tokamak	Spheromak
	Tandem mirror	Field-reversed configuration
	Stellarator	
		Dense Z pinch
	Reversed-field pinch	

Source: Adapted from Argonne National Laboratory, Fusion Power Program, Technical Planning Activity: Field Report, commissioned DOE, Office of Fusion Energy (OFE), ANL/FPP-87-1, 1987, p. 15.

avoiding the radioactive fuel and the radioactivity that actual fusion neutrons would generate in the experiment structures. The extrapolation to plasmas containing tritium is believed to be well understood for plasmas below the "breakeven" line but will not hold much above it.

Confinement concepts either have *closed* geometries, in which the plasma is contained by magnetic lines of force that are entirely contained within the device, or *open* ones, where the confining fields extend well outside the device. Since plasmas move freely along magnetic field lines, some additional mechanism is required to keep plasmas from escaping out the ends of open confinement concepts.

3a.1. Closed concepts. All closed configurations have the basic shape of a torus. A magnetic field can encircle a torus in either of two directions: "toroidal" fields run the long way around the torus in the direction that the tread runs around a tire, while "poloidal" fields run the short way around perpendicular to the toroidal field.

The most widely studied confinement concept is the *tokamak*, a closed configuration in which the principal confining field is a toroidal field generated by large external magnets encircling the plasma. However, a weaker poloidal field generated by electrical currents within the plasma itself is also required. The *advanced tokamak* incorporates features that can improve performance such as shaping the plasma cross section in the form of a "D" or a bean. In the *stellarator*, both toroidal and poloidal fields are generated by external magnets. The *reversed-field pinch* has a toroidal field generated primarily by external magnets and a poloidal field comparable in strength generated primarily by plasma currents. The toroidal field reverses direction near the outside of the plasma.

The *spheromak*, in essence, is a reversed-field pinch in which the external magnet coils that run through the central "doughnut hole" and encircle the plasma are removed and the central hole is shrunk to nothing. The plasma chamber, essentially spherical in shape, lies entirely within the external magnets. The *field-reversed configuration*, like the spheromak, has no magnets penetrating a central hole. Unlike the spheromak (and all other closed concepts), however, its internal magnetic field has no toroidal component. The

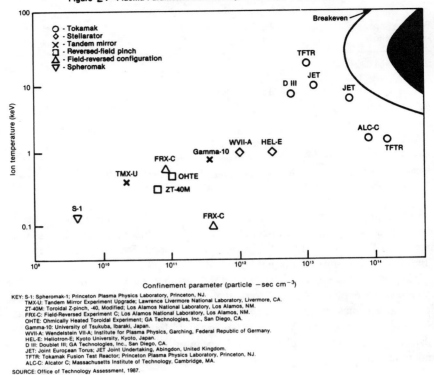

Figure 2.—Plasma Parameters Achieved by Various Confinement Concepts

KEY: S-1: Spheromak-1; Princeton Plasma Physics Laboratory, Princeton, NJ.
TMX-U: Tandem Mirror Experiment Upgrade; Lawrence Livermore National Laboratory, Livermore, CA.
ZT-40M: Toroidal Z-pinch, -40, Modified; Los Alamos National Laboratory, Los Alamos, NM.
FRX-C: Field-Reversed Experiment C; Los Alamos National Laboratory, Los Alamos, NM.
OHTE: Ohmically Heated Toroidal Experiment; GA Technologies, Inc., San Diego, CA.
Gamma-10: University of Tsukuba, Ibaraki, Japan.
WVII-A: Wendelstein VII-A; Institute for Plasma Physics, Garching, Federal Republic of Germany.
HEL-E: Heliotron-E; Kyoto University, Kyoto, Japan.
D III: Doublet III; GA Technologies, Inc., San Diego, CA.
JET: Joint European Torus; JET Joint Undertaking, Abingdon, United Kingdom.
TFTR: Tokamak Fusion Test Reactor; Princeton Plasma Physics Laboratory, Princeton, NJ.
ALC-C: Alcator C; Massachusetts Institute of Technology, Cambridge, MA.
SOURCE: Office of Technology Assessment, 1987.

Figure 2. Source: U.S. Congress, Office of Technology Assessment, *Starpower: The U.S. and the International Quest for Fusion Energy*, OTA-E-338 (U.S. Government Printing Office, Washington, DC, October 1987).

plasma is greatly elongated along the poloidal direction and from the outside looks like a cylinder.

3a.2. Open concepts. A *magnetic mirror* is an open-ended tube of plasma surrounded by magnets that generate a field running down the length of the tube. Plasma particles cannot easily cross the field to escape out the sides; strengthening the field near each end tends to "reflect" some of the particles running out the ends back into the center. However, other particles will escape out the ends unless some other mechanism prevents their escape. In a *tandem mirror*, each end is plugged by an end cell (actually another magnetic mirror) holding a plasma at a different electrostatic potential than the main plasma. The potential difference generates electric fields that prevent the main cell plasma from escaping. Losses in the end cells are high but can be compensated by injecting new particles.

In the *dense z pinch*, a fiber of solid deuterium–tritium is suddenly vaporized and turned to plasma by passing a strong electrical current through it. This current generates a strong magnetic field encircling the plasma, "pinching" it long enough for fusion reactions to occur. Although this technique was tried and discarded in the early days of fusion research, the combination of more advanced and controllable power supplies, along with the use of solid rather than gaseous D–T mixtures, has focused new attention on this approach.

3a.3. Confinement concept evaluation.

At this stage of the research program, it is not known which confinement concepts can form the basis of an attractive fusion reactor. The tokamak, which since the 1970s has been the mainstay of fusion programs around the world, has attained plasma conditions closest to those required in a fusion reactor. For the near future, studies of reactor-like plasmas will have to be done in tokamaks since no other concepts have yet demonstrated the potential to reach reactor conditions. Although important aspects of tokamak behavior — including the primary energy loss mechanism — have not yet been fully explained theoretically, it may well be possible to design reactor-scale tokamaks on the basis of experimental performance in smaller tokamaks.

Research on alternatives to the tokamak listed in Table 1 continues because it is not clear that the tokamak will result in the most attractive fusion reactor. However, none of these other concepts is being supported at anywhere near the level of effort devoted to tokamak studies, nor do any alternate concept devices approach the size of the largest tokamaks. Therefore, it is difficult or impossible to compare them directly to the tokamak.

Funding pressures have created a "Catch-22" situation: the expense of successively larger experiments will prohibit more than a very few concepts from being tested at large scale. (Indeed, given the present state of knowledge and existing levels of effort, the entire world fusion budget cannot support more than one true reactor-scale device.) However, at their present level of support it will be very difficult for alternate concepts to demonstrate results impressive enough to command substantial funding increases. Understandably, allocating resources among the competing concepts, and striking a balance between advancing understanding of the front-runner tokamak as opposed to supporting a diversity of alternate approaches, remains quite controversial.

3a.4. Reactor development.

Proponents of various alternate concepts argue that reactors based on those concepts could have substantial advantages over tokamak-based reactors. At the least, these other concepts offer important insights into the fusion process. It remains to be seen which alternate concepts will be able to reach the level of performance already attained by the tokamak, whether the relative advantages of these concepts will be preserved as the concepts are extrapolated to reactor scale, and what the costs of that extrapolation will be. Nor is it known what the ultimate capability of the tokamak concept itself will be.

Just as an automobile is much more than spark plugs and cylinders, a fusion reactor will contain many systems besides those that heat and confine the plasma. Fusion's overall feasibility will depend on all of the support systems that maintain the reaction, convert the power released in the reaction into usable energy, and ensure safe, environmentally acceptable operation. These

Figure 3 —Systems in the Fusion Power Core

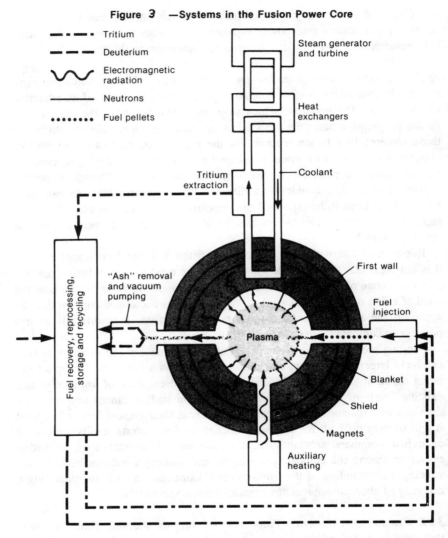

Figure 3. Source: U.S. Congress, Office of Technology Assessment, *Starpower: The U.S. and the International Quest for Fusion Energy,* OTA-E-338 (U.S. Government Printing Office, Washington, DC, October 1987).

systems include the ones that heat the plasma, fuel it, drive electrical currents within it if those currents are necessary for confinement, transfer energy to turbines and generators, breed tritium fuel, remove wastes and impurities from the plasma, provide shielding, cool the core if necessary, and control the reaction (Figure 3). Developing and building these associated systems and integrating them into a reactor will require a technological development effort at least as impressive as the scientific challenge of understanding and confining fusion plasmas. In particular, one of the key requirements for many of these systems is the development of suitable materials that are resistant to the intense neutron radiation generated by the plasma. The environmental and safety aspects of fusion reactors depend significantly on materials choice.

To convert heat from the fusion reaction into electricity, conceptual reactor designs have systems similar to the steam turbines found in existing electricity generating stations. However, fusion technology may permit generation of electricity by systems that are qualitatively different from the ones in use today. For example, plasmas operating at temperatures several times higher than needed to ignite deuterium and tritium could generate substantial amounts of synchrotron radiation at microwave frequencies, which could be rectified directly into electricity. Alternatively, the microwaves could be used to heat a working fluid to higher temperatures that would be reached by the coolant circulating in the blanket of a "first-generation" deuterium–tritium fusion reactor such as the one illustrated in Figure 3.

3a.5. Future plans and facilities. Additional facilities will be required to investigate specific confinement systems as well as to resolve general issues not associated with a single confinement concept. The latest major facility proposed for the United States, the short-pulse Compact Ignition Tokamak (CIT), will explore fusion's principal remaining scientific uncertainties, those associated with igniting a self-sustaining fusion reaction.[6] Successfully reaching ignition in CIT will establish fusion's scientific feasibility. However, CIT will not resolve many of fusion's engineering issues, nor will it address physics aspects — such as the evolution of a plasma under steady-state burn or the buildup and removal of reaction products — that would require an ignited plasma lasting hundreds of seconds.

A device to explore these long-term burn and nuclear technology issues is now being designed by a multinational effort under the auspices of the International Atomic Energy Agency. The major fusion powers (the United States, Western Europe, Japan, and the Soviet Union), along with countries such as Canada with smaller fusion programs, are together developing a conceptual design of an engineering test reactor called the International Thermonuclear Experimental Reactor (ITER). No commitment has been reached on building the device, estimated to cost several billion dollars. Given the magnitude of the project, construction and operation of ITER would require international collaboration on a scale that is unprecedented for the United States.

Even if an engineering test reactor such as ITER is built, materials testing will require a separate device capable of irradiating candidate reactor materials at radiation levels several times higher than those to be found in a reactor. Such a facility will be needed to complete accelerated lifetime irradiation tests in a reasonable amount of time.

3a.6. Schedules and budgets. Research to date has concentrated much more heavily on the physics of plasma confinement and heating than it has on the technological aspects of a future fusion reactor. A major fusion communitywide study has identified the technical tasks and facilities required to establish fusion's feasibility — both from a scientific and a technological perspective — that can enable a decision to be made early in the next century to start the commercialization process.[7] The study estimated that the worldwide cost of this research effort would be about $20 billion. Since the U.S. budget for magnetic fusion is less than $350 million a year, adhering to this schedule will require either substantially increased U.S. funding or wide-scale collaboration among the world fusion programs.

The requirements and schedule for establishing fusion's subsequent commercial feasibility are more difficult to project, and they depend on factors other than fusion research funding. Conceivably, if the research program provides the information necessary to design and build a reactor prototype, such a device could be started early in the next century. After several years of construction and several more years of qualification and operation, a base of operating experience could be acquired that would be sufficient for the design and construction of commercial devices. If the regulatory and licensing process proceeded concurrently, vendors and users could begin to consider manufacture and sale of commercial fusion reactors sometime during the middle of the first half of the next century. Even then, it will take decades for fusion to penetrate energy markets since power plant lifetimes are so long.

Under the most favorable circumstances, it does not appear likely that fusion will be able to satisfy a significant fraction of the nation's electricity demand before the middle of the 21st century. And this schedule contains a number of assumptions:

• Sufficient financial support or international coordination must be attained so that the research needed to establish technological feasibility can be completed early in the next century.

• Research must proceed without major difficulty and must lead to a decision to build a reactor prototype.

• The prototype must operate as expected and prove convincingly that fusion is both feasible and preferable to its alternatives.

3a.7. Status of the world programs. The United States, Western Europe, Japan, and the Soviet Union all have major programs in fusion research that are at similar stages of development. Each program has built or is building a major tokamak experiment. The U.S. Tokamak Fusion Test Reactor and the European Community's Joint European Torus are operating and are ultimately intended to reach breakeven conditions with D–T fuel. Japan's JT-60 tokamak, also operational, will not use tritium fuel; it is intended to generate a "breakeven-equivalent" plasma using ordinary hydrogen and deuterium. The Soviet Union's T-15 experiment has just been put into operation. Experimental results are anticipated over the next few years. In addition to these major devices, each of the programs operates several smaller fusion experiments that explore the tokamak and other confinement concepts. Each program is also developing aspects of fusion technology.

3b. Inertial confinement

The second major approach to confining a fusion plasma is inertial confinement. In inertial confinement fusion, laser, or particle beams — called *drivers* — compress a pellet containing deuterium and tritium to a density many times that of lead and then heat it to ignition. At that density, about 10 billion times the density of a magnetically confined plasma, the confinement time needed is less than one-billionth of a second—short enough so that the ensuing fusion reactions can generate net fusion power before the pellet blows itself apart. The fuel's own *inertia* serves to confine it.

Inertial confinement already has been demonstrated on a very large scale in the hydrogen bomb, an inertially confined fusion reaction whose input energy is provided by a fission (atomic) weapon. The challenge of laboratory-scale inertial confinement research is to reproduce this process on a much smaller scale. In a hypothetical inertial confinement reactor, microexplosions with explosive yields equivalent to about one-tenth of a ton of TNT would be generated by bombarding fusion pellets—called targets—with laser or particle beams; these explosions would be repeated several times a second.

The milestones of "breakeven" and "ignition" used to mark the progress of magnetic fusion are also applicable to inertial fusion, but they occur in the opposite order. An inertial confinement fuel pellet can ignite—support a self-sustaining fusion reaction in its center—without generating enough energy to "pay back" the energy originally put into the driver. Therefore, ignition does not imply breakeven. To reach breakeven, the pellet must be large enough for the central fusion reaction to propagate outward into the surrounding fuel. The reaction must release enough energy not only to match what is initially delivered to the pellet, but also to compensate for inefficiencies in the lasers or particle beams that generate the incident energy. These drivers are expected to be only 10% to 25% efficient. Moreover, an additional factor of 4 to 10 beyond breakeven is required to produce substantial net output. In a commercial reactor, therefore, each pellet explosion would have to generate as much as 100 times the energy it was driven with.

Inertial confinement fusion (ICF) has not been studied for as long, or at as high a level, as magnetic fusion energy in the United States. Although it has the long-term potential to serve as a civilian energy source, its near-term applications are military: inertial confinement experiments are used to validate computer models, collect data, and develop diagnostic instruments for use in the nuclear weapons program. In the longer term, a "single shot" ICF facility capable of producing significant pulses of fusion energy (greater than about 100 MJ or 1/40 ton of TNT) could simulate the effects of nuclear detonations without requiring underground testing of an actual nuclear weapon. Such a facility would be fired every few days.

Due to its military applications, most ICF research is funded under the nuclear weapons activities portion of the Department of Energy's (DOE's) budget. It is conducted largely at nuclear weapons laboratories, and a substantial portion of this research is classified because of the close relationship between inertial confinement fusion target design and thermonuclear weapon design.

3b.1. Applications. At present, many of the near-term requirements for military and civilian applications of ICF coincide. Both will require development of an ICF facility with high gain, capable of generating fusion output several times greater than the driver input. However, military and civilian needs diverge once a high-gain facility has been demonstrated, with a commercial ICF reactor requiring still further research and development.[8] Whereas military ICF tests can be worthwhile even if they are expensive and infrequent, commercial energy production stresses cost effectiveness, reliability, and high efficiency and requires a facility capable of detonating several pellets per second. According to a 1986 review of inertial confinement research by the National Academy of Sciences,

> The optimum ICF targets from the point of view of theory will probably be very difficult to fabricate. . . . If ICF is ever to provide commercial energy, a revolution in target fabrication methods will be needed to provide sufficiently inexpensive fuel pellets for a reactor.[9]

3b.2. Driver development. Four principal driver candidates are now being studied in the U.S. inertial confinement research program. Two of them—solid-state or glass lasers and light-ion[10] accelerators—have by far the largest facilities; the other two principal candidates—gas lasers and heavy-ion accelerators—are in lesser stages of development.

Major U.S. glass lasers include NOVA, the highest power laser in the world, at the Lawrence Livermore National Laboratory (LLNL) in California, and the OMEGA laser at the University of Rochester Laboratory for Laser Energetics in New York. The highest power light-ion accelerator in the world is the PBFA II at Sandia National Laboratory in New Mexico. The krypton–fluoride gas laser, the third driver candidate being studied, is being developed at Los Alamos National Laboratory. Additional contributors to the laser and light-ion inertial confinement programs are the Naval Research Laboratory and KMS Fusion, Inc., the only private corporation significantly involved.

A fourth driver candidate—the heavy-ion accelerator—is much less developed than the laser or light-ion drivers. Unlike light-ion and laser research, heavy-ion accelerator research is a non-military program funded by the Office of Energy Research, the same DOE office that funds the magnetic confinement fusion program.[11] Heavy-ion experimental work is limited to accelerator technology, and in the United States it is concentrated at Lawrence Berkeley Laboratory in California.

3b.3. Direct and indirect drive. Drivers are focused on fusion targets in two different ways. In direct drive, energy is focused directly on the target, and it must be deposited with very high uniformity to ensure that the ensuing compression produces a small, spherical hot spot in the target center that can reach ignition. In the indirect drive approaches, driver energy is transformed into x rays inside a cavity containing the target. These x rays are then absorbed in the outer surface of the target, driving an implosion. Due to the intermediate conversion step, the driver energy in indirect drive experiments can be considerably less uniform than it must be in direct drive. However, indirect drive requires more energy. Direct drive research in the United States is generally unclassified; indirect drive research is conducted at the nuclear weapons laboratories and many of its specifics are classified.

3b.4. Schedules and budgets. The ICF program in the United States began in the 1960s, a decade after the magnetic confinement program. During the 1960s it was funded at about an order of magnitude below the magnetic fusion effort. ICF budgets caught up somewhat in the 1970s, and ICF has been funded at between one-third and one-half the level of the magnetic fusion since then. Laboratory experiments have not yet reached either ignition or breakeven. According to the National Academy ICF study,

> It seems unlikely that even the largest currently existing laser driver, the Nova laser at LLNL [that can generate up to 60 kilojoules of energy, depending on wavelength], can ignite thermonuclear burn in a target. There is some possibility that the more energetic (1–2 MJ) PBFA II light-ion facility [at Sandia National Laboratory] can lead to ignition . . .[12]

In March 1988, the *New York Times* reported that two years earlier, researchers had ignited fusion reactions in hydrogen fuel pellets that were "triggered by a blast of radiation from an exploding nuclear weapon" at the Government's underground nuclear test site in Nevada. The Department of Energy has had no detailed comments on any such tests, although it has released a fact sheet describing a program called HALITE/CENTURION in which "underground nuclear explosions at the Nevada Test Site have been used to extend the range of research carried out in inertial fusion . . . . Experiments conducted by this means have allowed demonstration of excellent performance, putting to rest fundamental questions about basic feasibility to achieve high gain."[13]

ICF research has progressed sufficiently for the Department of Energy to complete a major study to define characteristics for a next-stage inertial confinement fusion device. Such a device, called the Laboratory Microfusion Facility (LMF), would cost on the order of a billion dollars. It would, for the first time, generate laboratory fusion reactions producing many times the energy needed to drive them, and would therefore provide an important contribution to ICF's civilian applications. Moreover, the LMF would have considerable military significance, providing for the first time enough ICF power to simulate some effects of nuclear explosions. DOE has not yet decided to propose to Congress that such a facility be funded.

3b.5. Comparison to magnetic confinement. The military applications of inertial confinement fusion have no parallel in magnetic confinement. For energy applications, inertial confinement eliminates the large and expensive magnets and other systems associated with heating, confining, and controlling a magnetically confined plasma, and replaces them with large and expensive drivers (lasers and/or particle beams) and target fabrication/positioning facilities. ICF has the significant potential advantage that these drivers can be located some distance away from the reaction chamber, permitting them to avoid the radiation, neutron-induced activation, and thermal stresses from microexplosions.

The systems that support and maintain the plasma in a magnetic confinement reactor cannot be isolated from the high radiation, high neutron-flux environment surrounding the plasma. A second potential advantage of ICF is that the reaction chamber can incorporate neutron absorbing materials such as liquid lithium inside the structural walls. This results in walls that last the lifetime of the reactor, with lower induced radioactivity. Such an approach would be inconsistent with the high vacuum requirements for magnetic confinement.

Inertial confinement also has disadvantages compared to magnetic confinement. Whereas some magnetic confinement concepts are steady-state, and others might have pulse lengths greater than 1 hr, systems within ICF reactors have to withstand explosions equivalent to a few hundred pounds of TNT several times a second. Inertial confinement reactors must also focus high-power driver beams precisely and repetitively on target in this environment. Targets must be cheap. Furthermore, the energy gains needed for ICF facilities must be much larger than those required in magnetic confinement to make up for driver inefficiencies.

On balance, the inertial confinement program involves scientific and technological issues that are quite distinct from those relevant to magnetic fusion, although the two approaches do share some technological aspects in tritium breeding, energy conversion, and materials. Scientific progress in magnetic confinement, in terms of what is known or expected to be known with current or near-term facilities compared to what must be known to determine fusion's feasibility, is probably ahead of that in the unclassified inertial confinement program. However, the DOE Fact Sheet on Inertial Fusion[13] refers to—but does not elaborate upon—considerable progress that has been made in classified inertial fusion research.

3b.6. Status of world programs.   The Soviet Union and Japan both conduct inertial confinement research, but at a significantly lower level of effort than their magnetic fusion programs. International collaboration in inertial fusion is much more restricted than it is in magnetic fusion due to U.S. national security constraints. Japanese researchers, e.g., have published papers that would have been classified in the United States.

4. Cold fusion

The extraordinary technical complexity of both the magnetic and the inertial confinement approaches results directly from the extremely high temperatures required to overcome the mutual repulsion of the electrically charged fuel nuclei. Therefore, reports by two research groups in 1989 stunned the scientific community and captured world-wide attention. If such *cold fusion* reactions indeed could take place in appreciable numbers, hundred-million-plus degree plasmas would no longer be necessary and the obstacles to developing fusion power would be enormously reduced. Unfortunately, the claim that fusion is responsible for heat production many times greater than the electrical energy input[14] has almost certainly been proven baseless. The verdict is less clear regarding production of neutrons at an extremely low level[15], but even if those reports are ultimately confirmed, there is as yet no reason to believe this process will have any significance as a potential energy source.[16]

An entirely different process that has also been termed cold fusion has been observed and understood since the 1950s. A subatomic particle called the muon, having an electrical charge equal to but opposite that of a hydrogen nucleus, can serve to shield the nucleus's electrical charge enough to permit it to approach another nucleus without being repelled electrostatically. The muon binds so tightly to the nucleus that the shielding remains effective down to distances where a fusion reaction can take place. (Electrons, which are also negatively charged and can also screen the positive charge of the nucleus, bind more loosely to nuclei and do not provide screening at the ranges needed for fusion.)

If a negative muon induces a fusion reaction and is subsequently freed, it can become captured by another nucleus to repeat the process. In this way it serves as a catalyst, enabling fusion energy to be released without itself being consumed. However, the muon is unstable, and muon catalysis can be practical only if each muon generates more than enough energy during its 2.2 µs lifetime to make its own replacement. Therefore, the feasibility of this approach depends on how many fusion reactions a typical muon can catalyze.

In the reactions observed in high-energy physics experiments in the 1950s, muons were rarely observed to induce more than one fusion reaction each before decaying, compared to the hundreds of reactions per muon that would be necessary to make the process worthwhile. However, more recent experimental and theoretical work has shown that the number of fusion reactions that can be catalyzed by a single muon depends on parameters such as the density and temperature of the deuterium–tritium mixture into which the muon is injected. Muons are capable of catalyzing many more reactions during their lifetime than had been thought many years ago.

Whether this process can ever yield net energy production depends on increasing the number of fusion reactions per muon. The number is not yet high enough for the process to be scientifically feasible, and fundamental limits may prevent it from ever being so. Muon-catalysis research currently focuses on understanding the limits to how many fusion reactions can be induced by a single muon. If muon catalysis proves to be feasible in principle, a substantially increased level of effort and a more detailed comparison of its potential benefits and liabilities to those of the other fusion approaches may be warranted.

Should a muon-catalyzed fusion reactor be feasible, it would not require plasmas or the associated confinement systems, high-energy drivers, or target fabrication facilities. However, it would require a large and expensive particle accelerator to serve as a muon source. Tritium breeding and energy recovery systems would still be needed, and the problems of induced radioactivity due to neutron activation would remain. Like an ICF reactor, a muon-catalyzed reactor should be able to use shielding techniques unavailable to magnetic confinement. The operating temperature and pressure of such a reactor would be determined by the chemistry of the atoms and molecules created when muons bind to hydrogen atoms and molecules. Present research is conducted using pressurized deuterium–tritium at thousands of atmospheres pressure and hundreds of degrees C.

5. Fusion as an energy program

The following discussion of possible characteristics of a fusion reactor is based on the magnetic fusion system study literature, which is more extensive than that for ICF studies. Muon–catalysis fusion, still in the process of resolving basic scientific uncertainties, awaits any significant analysis of reactor technology and characteristics.

Several of the points discussed below, such as resource requirements, provision for tritium breeding and handling, and the inability of fusion reactions to run away, would be common to all D–T fusion approaches. Other aspects, such as the degree of neutron activation of the reactor structure and consequent radioactive waste generation, have been specifically analyzed below with respect to magnetic confinement. The other approaches must also contend with neutron activation. They may be able to incorporate more effective shielding techniques, but these shielding materials themselves may then become an additional source of waste.

5a. Pure fusion vs fission/fusion hybrid reactors

The long-term goal of the fusion programs in the United States—both magnetic and inertial—is to produce electricity.[17] Fusion reactor designs attracting the most attention today would produce energetic neutrons, whose energy would be recovered in the form of heat and would be used to generate electricity. An alternate approach—the *fission/fusion hybrid reactor*—would use fusion neutrons to breed fissionable fuel or to induce fission reactions in fuels contained within the reactor. Most of the energy ultimately converted to electricity would come from fission reactions, with fusion supplying neutrons that would significantly increase the fission cycle's efficiency. Since the fusion power output of a fission/fusion hybrid would be less than that of a pure fusion reactor generating the same amount of electricity, the technical requirements on the fusion core would be less demanding.

A hybrid reactor combining the fusion process with fission, however, could combine their liabilities. After all, many of the potential environmental and safety advantages of fusion stem from the absence of fissionable fuels and fission products. Reintroducing fission might therefore be seen as counterproductive. Consequently, hybrid reactors are not fusion's primary intended application in the United States, Western Europe, and Japan. (The Soviet fusion effort does—or at least did prior to Chernobyl—appear oriented towards producing fuel for fission reactors.)

The impetus for developing fission/fusion hybrid reactors depends on the price of uranium fuel. Therefore, analysis of the hybrid concept should take into account not only possible advanced converter fission reactors that would use fissionable fuel much more efficiently than present-day reactors, but also pure fission breeder reactors.

5b. Safety and environmental characteristics

If magnetic confinement fusion development is successful, it may be possible to ensure that fusion reactor accidents due to malfunctions, operator error, or natural disasters could not result in immediate public fatalities. This safety would depend on passive systems or on materials properties, rather than on active systems that could fail or be overridden. A number of attributes of the fusion process should make environmental and safety assurance easier for fusion reactors than for fission reactors[18]:

• Fusion reactions cannot run away. Fuel will be continuously injected, and the amount contained inside the reactor chamber will only operate the reactor for a short period of time. Energy stored in the plasma at any given time can be dissipated by the vacuum chamber in which the fusion reactions take place.

• With appropriate choice of materials, the amount of heat produced by the decay of radioactive materials in the reactor after the reactor has been shut down should be less for fusion reactors than for fission reactors. Fusion reactors should therefore require simpler post-shutdown or emergency cooling systems, if any such systems are required at all.

• The radioactive inventory of a fusion reactor, in terms of both the total amount present in the reactor and the fraction that would be likely to be released in an accident, should be smaller than that of a fission reactor. Fusion wastes may have a greater physical volume than fission wastes, but they should be substantially less radioactive and orders of magnitude less harmful. Fusion will not generate long-lived wastes such as those produced by fission reactors. Depending on fusion reactor design, material selection, and the parameter being compared, potential safety hazards from fusion reactors can range from being comparable with fission designs to orders of magnitude lower. Except for tritium gas, the radioactive substances present in fusion reactors will generally be bound in metallic structural elements.[19]

• With the exception of tritium, practically none of the radioactive materials which could be released from a fusion reactor in the event of an accident would tend to be absorbed in biological systems. Tritium is an inherent potential hazard, but the risk it poses is smaller than that of the gaseous or volatile radioactive by-products present in fission reactors. Active tritium inventories in current fusion reactor designs are small enough that even their complete release should not produce any prompt fatalities off-site. Nevertheless, handling large tritium flows and maintaining or replacing radioactive and tritium-permeated portions of the reactor structure will be an important issue for fusion reactor design.[20]

This discussion does not imply that fission reactors are unsafe. Indeed, efforts are underway to develop fission reactors whose safety does not depend on active safety systems. However, the potentially hazardous materials in fission reactors include fuels and by-products that are inherent to the technology. While the tritium fuel required by a D–T fusion reactor is a potential hazard, the byproducts of fusion are not in themselves hazardous. Since there is much greater freedom to choose materials that minimize safety hazards for fusion reactors than there is in fission reactor design, a higher degree of safety assurance should be attainable with fusion.

This safety assurance will not come automatically. It will take considerable effort to realize the advantages that are potentially available with fusion, and there is no guarantee that these advantages can in fact be attained in an economically viable reactor.

5c. Nuclear proliferation potential

The ability of a fusion reactor to breed fissionable fuel could increase the risk of nuclear proliferation. A pure fusion reactor would not contain materials usable in fission-based nuclear weapons, and it would be impossible to produce such materials by manipulating the reactor's normal fuel cycle. However, fissionable material could be produced by placing other materials inside the reactor and irradiating them with fusion neutrons. This procedure, in effect, would convert a pure fusion reactor into a fission/fusion hybrid reactor. If such modifications were easily detected or were extremely difficult and expensive, pure fusion reactors would be easier to safeguard against surreptitious production of nuclear weapons material than existing fission reactors, and fusion reactors would therefore pose less of a proliferation risk.

D–T fusion reactors will produce tritium, a material that could be used in nuclear weapons. However, nuclear weapons cannot be made with tritium alone. Fissionable materials such as enriched uranium and plutonium are also required, and these materials are considered to pose a greater proliferation danger than tritium. Facilities able to produce tritium would, however, pose problems for arms control regimes based on control of nuclear materials.[21]

5d. Resource supplies

Shortage of fuels will not constrain fusion's prospects for the foreseeable future. Enough deuterium is contained in the Earth's waters to satisfy energy needs through fusion for billions of years at present consumption rates. Domestic lithium supplies should offer thousands of years worth of fuel, with vastly greater amounts of potentially recoverable lithium contained in the oceans.

Materials required to build fusion reactors may pose more of a constraint on fusion's development than fuel supply will, but at this stage of research it is impossible to determine what materials will eventually be developed and selected for fusion reactor construction. No particular materials other than the fuels appear at present to be indispensable for fusion reactors.

5e. Cost

It is currently impossible to determine whether a fusion reactor, once developed, will be economically competitive with other energy technologies. The competitiveness of fusion power will depend not only on successful completion of the remaining research program but also on additional factors that are impossible to predict—e.g., ability to license the plant, construction time, and plant reliability, not to mention factors less directly related to fusion technology such as interest rates. Fusion's competitiveness will also depend on technical progress made with other energy technologies.

Bibliography

Many texts in plasma physics appealing to readers with various degrees of preparation can be found. Texts or general works specifically discussing fusion research and energy applications are less common. Among these, most focus on magnetic confinement. A few are listed below, along with other useful sources of information regarding fusion energy.

U.S. Office of Technology Assessment, *Starpower: The U.S. and the International Quest for Fusion Energy*, OTA-E-338 (U.S. Government Printing Office, Washington, DC, October 1987). GPO stock number 052-003-01079-8.

The primary reference for this chapter; this report concentrates on magnetic confinement research but briefly addresses inertial confinement and muon catalysis as well. Written for the interested lay reader, it covers the history of fusion research, fusion science and technology, fusion as an energy program, fusion as a research program, fusion as an international program, and future paths for the magnetic fusion program.

National Research Council, *Physics Through the 1990s: Plasmas and Fluids* (National Academy Press, Washington, DC, 1986), Chapter 4, "Fusion Plasma Confinement and Heating," pp. 144–242.

The chapter covering fusion physics in this volume, one of the comprehensive Physics Through the 1990s series summarizing the state of physics in the late 1980s, provides a readable discussion of the physics underlying magnetic confinement and – to a somewhat lesser extent – inertial confinement fusion. Technologies required to create, heat, and maintain fusion plasmas are also described. However, other technology issues that will be required for reactor design but that have less physics content – heat recovery and conversion; fusion materials research; tritium breeding, recovery, and handling; etc. – are not covered.

Argonne National Laboratory, Fusion Power Program, *Technical Planning Activity: Final Report*, commissioned by DOE, Office of Fusion Energy (OFE), ANL/FPP-87-1, 1987.

A detailed analysis of the technological issues that must be resolved before fusion can be evaluated as a source of energy. Over 50 scientists and engineers from throughout the fusion community identified the milestones to be reached before fusion's potential can be assessed.

John P. Holdren *et al.*, "Exploring the Competitive Potential of Magnetic Fusion Energy: The Interaction of Economics with Safety and Enviromental Characteristics," Fusion Technology **13**, 7–56 (1988).

This article is the executive summary of the report of the Senior Committee on Environmental, Safety, and Economic Aspects of Magnetic Fusion Energy. Using consistent economic and safety models, the committee evaluated eight proposed fusion reactor designs, two fission/fusion hybrids, and four fission designs, with respect to safety, environmental characteristics, and economics. The cases chosen spanned a wide range of materials choices, operating parameters, fuel cycles, and power conversion cycles. Fusion Technology, the journal in which this summary is published, prints much of the literature on fusion reactor system studies and technology assessments.

Weston M. Stacey, Jr., *Fusion: An Introduction to the Physics and Technology of Magnetic Confinement Fusion* (Wiley-Interscience, New York, 1984).

A general introduction to magnetic confinement fusion research covering both physics and technology issues. Targeted for the advanced undergraduate or first-year graduate student.

Weston M. Stacey, Jr., *Fusion Plasma Analysis* (Wiley-Interscience, New York, 1981).

A graduate plasma physics text that concentrates on the role of plasma physics in the development of fusion power. This book takes an engineering physics perspective and covers non-physics aspects of fusion technology only to the extent that these other factors interact directly with the physics or provide important constraints.

Fusion, Edward Teller, ed. (Academic Press, New York, 1981).

Volume 1 (which itself is in two parts) of this work is a comprehensive and detailed text aimed at graduate students and practitioners in the field of magnetic confinement fusion research. It concentrates on the theory and current experimental status of every significant confinement approach; however, it also has chapters on plasma heating and a 217 page chapter on Magnetic Fusion Reactors by Robert Conn that can qualify as a text in its own right.

References and notes

1. This paper is based on U.S. Congress, Office of Technology Assessment, *Starpower: The U.S. and the International Quest for Fusion Energy*, OTA-E-338 (U.S. Government Printing Office, Washington, DC, October 1987). However, the views expressed here are those of the author, who directed the OTA study, and do not necessarily represent those of OTA or the Technology Assessment Board.

2. One electron volt (eV) is the energy that a single electron can pick up from a 1 V battery. It is equal to 1.6×10^{-19} J, 1.52×10^{-22} BTUs, or 4.45×10^{-26} kWhr. One thousand electron volts is called a kiloelectron volt, or keV.

3. One reaction attracting considerable interest now is the one between deuterium and helium-3, which does not produce neutrons and therefore could greatly reduce radioactivity and ensuing environmental and safety concerns. (Side reactions accompanying the main deuterium–helium-3 reaction do produce neutrons; however, these can be minimized by using a fuel mixture rich in helium-3.) Helium-3, although extremely rare on earth, can be found on the moon; commercial deuterium–helium-3 fusion reactors would require economical exploitation of lunar helium-3.

4. The fusion reactions that power stars are confined by a different mechanism—the star's own gravity. Stars are so massive that the temperatures and densities reached in their cores are sufficient to ignite fusion reactions. Although of intense interest to astrophysicists and indirectly to the rest of us, this approach cannot be utilized on earth. Even the planet Jupiter, which is over 300 times more massive than Earth, does not have sufficient mass to gravitationally confine a fusion reaction.

5. Four-fifths of the energy produced by the D–T fusion reaction is carried out of the plasma by the neutrons and therefore has no effect on plasma behavior. Only the fraction that is imparted to the alpha particles—20%—will be retained within the plasma. At breakeven, fusion reactions in the plasma produce as much energy as is injected into the plasma from outside sources. However, most of the energy that is produced escapes with the neutrons, and the plasma behavior is still dominated by the external heating. In order for fusion reactions to generate enough self-heating to dominate other sources of heating, total fusion power generated by a plasma must be many times greater than the total input power. Consequently, ignition is a much more meaningful milestone than breakeven from a plasma physics perspective.

6. As of June 1989, the construction cost for CIT was estimated at $455 million dollars, with an additional $132 million for related R&D, diagnostic equipment, and certain operation costs. Although some construction money has already been appropriated for CIT design in fiscal years 1988 and 1989 construction funding was withdrawn from the fiscal year 1990 budget, and its status is still up in the air. With the original request, CIT had originally been scheduled for completion in 1996; now, however, its schedule is indefinite.

7. Argonne National Laboratory, Fusion Power Program, *Technical Planning Activity: Final Report,* commissioned by the Department of Energy, Office of Fusion Energy (OFE), ANL/FPP-87-1, 1987.

8. In addition to providing energy, ICF reactors could also be designed for military application in producing fissile fuel and tritium for nuclear weapons. A third potential long-term application might be thrust for space propulsion.

9. *Review of the Department of Energy's Inertial Confinement Fusion Program,* Committee for a Review of the Department of Energy's Inertial Confinement Fusion Program, National Academy of Sciences, March 1986, p. 21.

10. Light ions are ions of light elements such as lithium.

11. The heavy-ion research program is far smaller than the magnetic confinement program and is managed by a different part of the DOE Office of Energy Research.

12. *Review of the Department of Energy's Inertial Confinement Fusion Program, op cit.,* pp. 23–24.

13. U.S. Department of Energy, "Inertial Fusion Fact Sheet," September 30, 1988. Quoted material was also presented by Sheldon Kahalas, U.S. Department of Energy, in a talk at the Fusion Power Associates Symposium "Energy and the Environment," June 1–2, 1989, Washington, DC.

14. M. J. Fleischmann and S. Pons, J. Electroanalyt. Chem. **261**, 301–308 (1989).

15. S. E. Jones *et al.,* Nature **338**, 727–740 (1989).

16. For a recent summary of the status of these cold fusion claims, see Robert Pool, "Cold Fusion: Only the Grin Remains," Science **250**, pp. 754–755, 9 November 1990.

17. As discussed earlier, inertial confinement has nearer-term military applications not associated with power generation.

18. Much of the following is based on calculations summarized in Holdren *et al,* "Exploring the Competitive Potential of Magnetic Fusion Energy: The Interaction of Economics with Safety and Environmental Characteristics," Fusion Technology **13**, 7 (1988). This paper summarized the report of the Senior Committee on Environmental, Safety, and Economic Aspects of Magnetic Fusion Energy, chaired by John Holdren.

19. Some fusion reactor designs use liquid lithium as a coolant and tritium-breeding materials. Lithium burns when exposed to air and it also reacts vigorously with water. Even so, calculations of the maximum plausible fraction of the radioactive inventory released by a lithium fire in a fusion reactor are significantly less than the equivalent fraction calculated to be released in fission reactor accidents.

20. Fusion reactors operating on advanced fuel cycles would not require tritium. Some tritium might be produced in side reactions, but in smaller amounts than those found in tritium-fueled reactors. See note 3 earlier.

21. The nuclear superpowers are thought to have enough plutonium and/or enriched uranium on hand to meet a range of future needs for nuclear weapons even in the event of a materials production cutoff. Tritium, however, must be continously generated due to its 12.3 year half-life. To the extent that tritium is helpful or essential to superpower nuclear weapon design, control of tritium could impose significant controls on superpower stockpiles. Since substantial tritium flows are inherently associated with D–T fusion reactors, fusion reactors could be difficult to reconcile with tritium controls unless these flows could be monitored with confidence.

Photovoltaics

Gary Cook

1. Introduction

From 1950 to 1987, the world increased its energy consumption by 300%, with the electricity sector growing by 1000%.[1,2] Many projections expect growth trends to continue in the coming decades, although not at such steep rates.[3-5] Yet, the world must decrease its dependence upon fossil fuels, the basis for much of this growth, or face the consequences of possibly severe environmental degradation, including that of climate change resulting from the accumulation of greenhouse gases. Photovoltaics (PV), which uses solid-state technology to turn the photon energy of sunlight directly into electrical energy, could make an important contribution in helping the world meet the demands of energy and the environment. Not only can PV accommodate centralized and dispersed electrical needs, large and small applications, and the diverse electricity requirements of all sections of the world, it is also one of the most environmentally benign energy technologies available. The operation of photovoltaic systems requires no fuel and no cooling water and generates no fumes, noise, or other pollutants. However, there is some concern for the land area needed for large-scale photovoltaic power production and possible toxic chemicals released to the environment during manufacture of photovoltaic materials.

Considered strictly in terms of the energy (sunshine) and material (such as silicon) resources available, the potential that photovoltaics holds for energy and the environment is enormous. But the impediments to approaching that potential can also loom large. These include the cost of PV systems, a giant energy infrastructure that resists change, a fledgling photovoltaic industry that lacks financial strength and whose production capacity is small, minimal government financial support for research and development, and a lack of coherent national policies that would recognize the value of nonpolluting technologies and that could accelerate the technology into the market. Also because photovoltaic systems generate electricity only when the sun shines, reliance on photovoltaics as a major producer of electricity will require either a backup power source or extensive energy storage (see *Energy Storage Systems*, by John G. Ingersoll).

Over the next two or three decades, research and development will not only help photovoltaics overcome many technological problems, but will also help reduce costs, increase production capacity, and open up markets. Enabling PV to become one of the world's preferred technologies for generating electricity, however, will also take a shift in government policies and incentives.

2. Evolution of PV technology

The French physicist, Edmund Becquerel, first recorded the PV effect in 1839 when he noted the appearance of a voltage when illuminating two identical electrodes in a weak conducting solution.[6] Nothing noteworthy became of this discovery until the late 19th and early 20th centuries when others found that selenium also exhibited the effect. Selenium was quickly adopted by the field of photography for photometric devices.

The field then lay fallow until the advent of improved doping techniques and a method for growing crystals in the 1940s and 1950s. Shortly thereafter (1954), Bell Telephone Laboratories made practical silicon PV cells that reached sunlight-to-electricity conversion efficiencies as high as 6%.[7] Since 1958, PV cells have powered most U.S. satellites in space.

Because of their expense, however, PV systems were not considered for terrestrial applications until after the 1973 oil crisis, when the National Science Foundation (NSF) organized a conference at Cherry Hills, N.J. to lay the groundwork for a national research and development (R&D) program in terrestrial PV.[8] Since then, scores of new PV materials and many new cell designs have been explored with the objective of developing long-lasting, reliable, low-cost PV systems for terrestrial use. As a result, dozens of new efficiency records have been set and reset.[9] Today's photovoltaic modules typically provide more than 20 years of relatively maintenance-free operation.[10] And the cost of modules has dropped by as much as 20-fold, to between $4 and $4.50/W, which over the lifetime of a system can translate to electricity costs as low as 30 cents/kWh.

Manufacturing processes are becoming automated and semicontinuous. Cell and module designs now incorporate transparent conducting oxides, flexible substrates, laser-scribed connections, microgrooved surfaces, point contacts, multijunctions, and light-capturing techniques. Japan, European nations, China, India, and others have established and enlarged their PV programs. The number of organizations involved in manufacturing and R&D has grown from less than a dozen in the early 1970s to more than 200 today. Some companies are building new facilities to produce 10 MW of modules per year — a scale that could reduce the cost of PV modules to $1.15/W,[11] which may translate to approximately 12 cents to 15 cents/kWh over the lifetime of the module. Others are closing deals to provide systems for utilities.[12]

The photovoltaic industry has grown substantially, with module shipments increasing by 47% from 1987 to 1989, to a record 42.1 MW.[13] Industry revenues are also growing, with some predicting they will reach nearly $800 million for 1990 — a figure that includes equipment for balance of systems.[14] Much of this growth has occurred in the consumer products market, where small PV systems (a few milliwatts to a few watts) are used to power things like calculators, radios, and walk lights, and where PV is competitive with the primary alternative, batteries. But the largest market for PV continues to be for stand-alone applications, uses where PV is not only competitive with many conventional alternatives, but more convenient and reliable. Today, tens of thousands

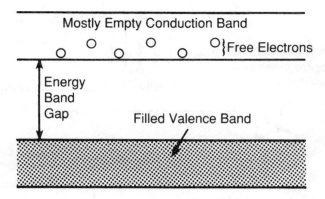

Figure 1. Under normal conditions a semiconductor has a filled valence band and a nearly empty conduction band.

of PV systems are used to power remote homes and cabins, run communications, pump water, provide electricity to entire communities, and much more.

This growth and advancement bodes well for the future of photovoltaics. By the turn of the century, markets may grow by as much as an order of magnitude,[15] the industry will have become stronger and healthier, and manufacturers will have incorporated the research advances of the 1990s into their designs. The resulting higher efficiencies and lower cost modules should help spur the market, with PV electricity even becoming competitive with peak, and some base load, utility power from conventional sources.

3. Progress and status of photovoltaic technology

3a. The basic technology

Perhaps the principal reason that PV is making such rapid progress of late is that it is an integral aspect of today's solid-state revolution. The key to the PV effect is the semiconductor. Under normal circumstances a semiconductor acts like an insulator, inhibiting the flow of electrical energy. This is because in a semiconductor there are very few free electrons in the conduction band where they can move through the crystal. Most are bound in the valence band where they cannot carry electric current (Figure 1). An external stimulus of energy equal to or greater than the band gap—the energy difference, characteristic of the material, between the bottom of the conduction band and the top of the valence band—releases electrons from the valence band to the conduction band, enabling the semiconductor to conduct electricity. For a transistor, the external energy is provided by a pulse of electric current. For a PV cell, the energy comes from the photons, or light particles, in sunlight. As long as a PV cell is exposed to sunlight, electrons are moved to the conduction band, and the cell conducts electric current.

The PV cell is more than just an absorber of sunlight. It converts the electromagnetic energy in sunlight to electric current. This is so because of the electric field at the junction formed between dissimilar semiconductor materials in the

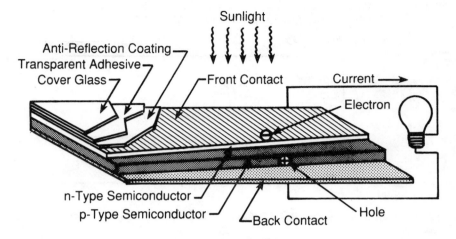

Figure 2. In a typical PV cell, sunlight generates electron-hole pairs that are separated at the junction. This creates an internal voltage that drives current through an external circuit.

cell. Light shining on a cell continuously knocks electrons into the conduction band where they act as free electrons. The absence of an electron in the valence band acts like a positive charge and is called a hole. Holes effectively "travel" through the semiconductor material when they are filled by electrons from the valence band which in turn leave a hole in a different place. The electric field, the magnitude of which is characteristic of the cell material, separates the electrons and holes formed near the junction and sweeps them to opposite sides of the junction. Electrically connecting the cell to an external load completes the circuit and allows electric current to flow. The power produced by a cell is equal to the product of its current, which is a function of the number of charge carriers generated, and its voltage, which is related to and limited by the electric field (Figure 2).

There are three common designs used to form the electric field for a PV cell: homojunction, heterojunction, and p-i-n junction. The homojunction (Figure 3) is formed by doping, diffusing impurities into a single material, most

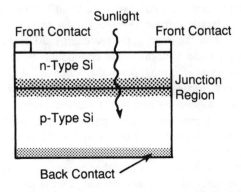

Figure 3. In a homojunction cell the junction is formed by doping a single material so that one side is p-type and the other is n-type.

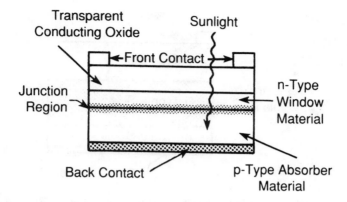

Figure 4. Heterojunction devices use a high band gap window material that allows almost all of the sunlight through to a low band gap absorber material. The junction is formed between the dissimilar materials.

commonly crystalline silicon. One side of the junction is p-type, that is, it contains an impurity with one less valence electron than silicon so it easily forms holes in its crystal structure. The other side of the junction contains an impurity with one more valence electron than silicon so it readily produces free electrons and is called n-type material. The junction is formed so that maximum light is absorbed near the p-n boundary, optimizing the probability that charge carriers, electrons and holes, will diffuse across the junction. In the normal configuration, the p-type layer, which is relatively thick compared to the n-type layer, lies beneath the n-type layer and absorbs the great majority of photons generating most of the free carriers.

In a heterojunction (Figure 4), two different semiconductor materials form the junction — a "window" material with a high band gap (generally greater than 2.5 eV) and a lower band gap, absorber material. The band gap of a junction is the energy difference between the valence and conduction bands. Although either material can be doped p-type or n-type, generally the window material is n-type while the absorber is p-type. This structure is most often used in thin-film cells where both semiconductor materials are very thin (on the order of 1μm). The window material allows almost all of the light through to the highly absorbent partner material where most of the light is absorbed near the junction. The most notable material combinations for heterojunction devices are cadmium zinc sulfide on copper indium diselenide ($CdZnS/CuInSe_2$), cadmium zinc sulfide on cadmium telluride (CdZnS/CdTe), and aluminum gallium arsenide on gallium arsenide (AlGaAs/GaAs).

Most amorphous silicon (a-Si) devices and some cadmium telluride (CdTe) devices use a p-i-n structure (Figure 5), which employs an undoped intrinsic (i) layer interposed between a p-type layer and an n-type layer. Most of the charge carriers are generated in the i layer and are separated by the electric field set up by the p and n layers.

The typical cell produces a small amount of power, equal to its current times its voltage. Connecting many cells together in a module, the building block of

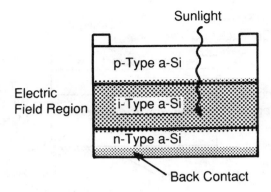

Figure 5. A p-i-n structure employs an undoped intrinsic layer (i) interposed between a p layer and an n layer.

PV systems, produces more power output and provides protective packaging for the cells. When the cells are connected in series, their voltages add, while connecting them in parallel adds the currents they produce. Modules can also be made by depositing amorphous or polycrystalline semiconductor layers over a large area and then encapsulating the layers in protective coatings. Modules, in turn, can be combined into arrays, and finally into systems. The dc electricity thus generated is often passed to a power conditioner to produce ac power. In this modular fashion photovoltaic systems can be designed to meet *any* requirement of power, voltage, and current.

There are two types of module systems: flat plate and concentrator. A flat-plate system builds its modules on a flat surface to capture unconcentrated sunlight. Flat-plate systems are relatively uncomplicated and can use both the diffuse and direct components of sunlight.

Concentrator systems use lenses to focus the sunlight on the cells. Because of the large amount of sunlight that can be focused on the cells, a concentrator system uses sophisticated, small-area cells, such as highly efficient, single crystal, multijunction devices (although multijunction devices, see below, are also being designed for flat-plate use and sophisticated single-junction devices are being used in concentrator systems). This drastically cuts the cell area required to produce a given amount of power, eliminating a large portion of system cost. Concentrator systems, however, can only utilize the direct component of sunlight, which can be less than 80% of the total. They also have lens losses and must depend upon mechanisms that track the sun precisely along two axes: daily east to west and seasonally north and south.

3b. Efficiency and cost

The ultimate goal of PV technology is to produce significant amounts of electric power for large segments of society. To do this, PV must be a cost-effective option competitive on the open market. Since the PV module can constitute more than 50% of a system's cost, it is imperative to make modules, and hence the cells that comprise them, cost-effectively. (Balance of systems—support structures and foundations, tracking mechanisms, conversion devices, storage, and so on — can also make up about 50% of the cost of a system. The possibilities for cost reduction here, however, are not as great because options are more confined, technologies and methods are older and already have been refined more towards their limit, and some costs, such as labor for installation, may even rise. Consequently, most efforts for making PV more competitive are concentrated on the solid-state end of things: materials, manufacturing processes, and device refinement.) To do this, whether for flat-plate or concentrator systems, there are three basic options: to make them efficient, inexpensive, or both. In each case, the modules must operate reliably for 20 to 30 years.

3c. Efficiency

If you can raise the efficiency of a given material/device design while holding or decreasing other costs, you effectively lower the cost of the electricity produced. This is because the device will produce more electric power, decreasing unit electricity costs. Plus, for a given amount of power, a more efficient system will take up less space, decreasing costs associated with land, site preparation, and balance of systems.

The conversion efficiency is the ratio of the incident power of sunlight to the output power of the device. It is determined by the product of the open-circuit voltage, the short-circuit current, and the fill factor (V_{oc} x I_{sc} x ff). The fill factor is the ratio of the product of the current and voltage obtained at maximum power to the product of the open-circuit voltage and short-circuit current ($I_m V_m / I_{sc} V_{oc}$).

No PV device can be 100% efficient. The highest theoretical efficiency of a single-junction cell under unconcentrated sunlight, for example, is about 33% (for gallium arsenide, GaAs).[16] Single-crystal silicon has a theoretical maximum almost as high as that of GaAs. Materials not made as single crystals, especially the thin-films, have lower theoretical limits.

Single-junction cells are thus limited because they do not capture the solar spectrum efficiently. Sunlight reaching the Earth is composed of a spectrum of photons whose energy content ranges from greater than 4 eV in the ultraviolet to less than 0.5 eV in the infrared (Figure 6). Whereas most of the sunlight energy is concentrated in the visible region (from about 1.6 eV to 2.8 eV), a significant amount falls in the infrared portion, and a small but important quantity is in the ultraviolet region. Photons with energy less than the band gap of the cell pass through the cell without producing charge carriers. Although photons with energy greater than the band gap create charge carriers, their excess energy is wasted as heat.

The way to overcome this limitation is to use multijunction concepts where cells employing different cell materials and hence different band gaps are layered on top of each other in descending band gap order (Figure 7). In this manner, the high band gap cells collect the high-energy photons and pass the rest of the photon energy to the lower band gap cells for collection.

Thermodynamic considerations, however, limit the efficiency of multijunctions to about 68% under unconcentrated sunlight and to 87% under concentrated sunlight, even if an infinite number of cells with descending band gaps

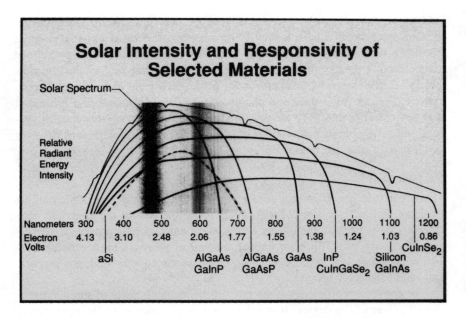

Figure 6. Sunlight reaching the Earth is comprised of a spectrum of photons whose energy content ranges from greater than 4 eV in the ultraviolet to less than 0.5 eV in the infrared.

are used.[17] Efficiencies as high as 40% for some single-crystal combinations and in the mid-20s for some thin-film combinations may be reached through the more practical approach of layering two or three cells.[18]

Most other loss mechanisms are not generally inherent limitations on the efficiency of a device. These include carrier recombination in the bulk material, the junction, the surface, and the contacts; absorption of light by defects and antireflection coatings; reflection of light from the surface; shading by the electrical grid which connects the cells in the array; and series and contact resistances. These loss mechanisms can be minimized through choice of material combinations, materials processing, sophisticated cell fabrication, and by clever device design and engineering.

3d. Cost

The cost of a PV device is also determined by several factors. These include the kinds of materials used and the amount required, the choice of substrates, the fabrication process, and the sophistication of the device design. Regarding this last point, consider multijunction devices. Although they can theoretically achieve higher efficiencies than single-junction devices, their major drawback is that they can be much more complicated to make, and hence more expensive. There are more parameters that must be carefully controlled, such as the thicknesses of cell layers, and the deposition of dopants and gases. If the device is made monolithically, as a single piece of material, the atomic spacing of the lattices of the different materials have to match closely to avoid stress-related defects. The contacts on the back of upper cells must be transparent to allow

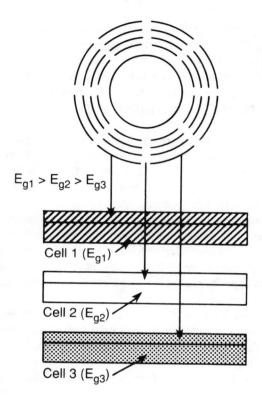

Figure 7. In a multijunction device, cells are layered in descending band gap order.

the unfettered passage of low-energy photons to the lower cells. Cells stacked in series must have matching photocurrents. And semiconductor materials for the different cells must be carefully selected for proper band gaps.

The tradeoffs between cost and efficiency, however, are not linear because of other system costs. A 20%-efficient cell, for example, can cost 7 times as much as a 10%-efficient cell of the same area and still result in the same overall system cost because of the decreased number of cells needed which in turn saves costs in manufacturing an array. For a 30% efficient cell, the cost can be as much as 14 times that of a 10%-efficient cell.[19]

4. Photovoltaic materials

The choice between higher efficiency and lower cost often boils down to a choice between crystalline and thin-film materials. Crystalline materials are generally more efficient; thin-film materials typically cost less. Many materials are being investigated, but the emphasis is on crystalline silicon, amorphous silicon, polycrystalline thin films, and III-V materials.

4a. Crystalline silicon

Crystalline silicon comes in two forms: single crystal and polycrystalline.

Single-crystal silicon has long been the workhorse of PV technology. It has the longest history, the largest technology and scientific base, and has dominated the markets since the first practical application of PV cells in satellites. The 1980s, however, saw the relative market share of single-crystal silicon slip as consumer products using thin-film cells joined the competition. But single-crystal silicon remains the dominant choice for power systems, which is the direction PV must head if it is to contribute significantly to future energy supplies.

In the early and mid 1980s, research and industry began to deemphasize crystalline silicon technology because it looked as if it would be too expensive to compete for future, large-scale, power systems. It has, for one, a low absorption coefficient, that is, it has a relatively low ability to absorb photons to generate charge carriers. Consequently, the typical cell used as much as 300 μm of material to absorb sufficient quantities of light. Furthermore, since many charge carriers are generated relatively far from the junction, the crystalline material used must characteristically have a long diffusion length, the length the average charge carrier travels before recombining. (If a charge carrier recombines with its opposite charge, hole or electron, it cannot add to the current of the device.)

Next, the method used to make modules was slow, wasteful, labor and energy intensive, and did not seem to lend itself well to automation or continuous production. Cells were made from cylindrical Czochralski ingots, a crystal growing process invented in the 1940s where single-crystal boules of silicon solidify as they are slowly drawn from a silicon melt. The ingots were then sawed, polished, doped, made into cells, and connected into modules.

Finally, although the technology of crystalline silicon was an extremely reliable technology, with modules promising to operate well for 20 years and more, advances in efficiency seemed to reach stasis. The best cells weren't much more than 16% or 17% efficient and commercial modules appeared to reach a plateau of around 12%.

But then things began to change dramatically with all facets of the technology improving. Manufacturing techniques became semiautomated and less wasteful with, for example, new, multiple-blade saws for slicing boules, and lasers for scribing cells. Module assembly began to be computerized and throughput rose dramatically.

The most dramatic changes occurred in the understanding of the physics of the devices, and in the design and efficiency of devices. By revisiting the physical phenomena involved, several researchers revised their predictions of the maximum theoretical efficiency that silicon cells could reach and suggested several ways to optimize design.[20] This new understanding led to innovative device designs.

Perhaps the most radical of these is the so-called "point-contact" cell designed by Swanson and his co-workers at Stanford University.[21] The cell employs a pyramidally textured front surface to reduce the reflection of incident light and to scatter the light into the cell at nearly random angles, which is necessary for light trapping (Figure 8). A back-surface mirror reflects more than 90% of the unabsorbed photons at angles that result in total internal reflec-

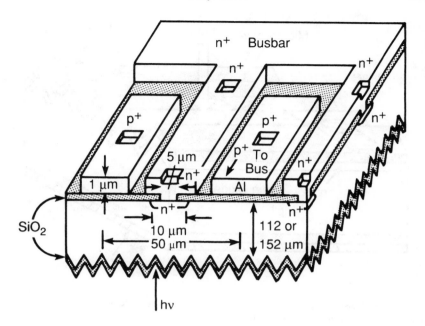

Figure 8. The point-contact cell from Stanford University. (Adapted from Ref. 35 with permission of the IEEE.)

tion. This makes it possible to absorb nearly all of the usable light with cells that are approximately 100 μm thick. The high percentage of light absorption produces a great number of charge carriers, and the thinness of the cell helps the carriers reach the terminals for collection before they recombine. The researchers also added thin, silicon oxide layers at the front and back surfaces to help reduce surface recombination. But, perhaps the most radical aspect of the design is the electrical contacts. Instead of metal fingers on the front surface, this cell incorporates extremely small (10 to 30 μm) point contacts that alternate between p-type and n-type in a polka-dot pattern on the back surface. Putting them on the back surface removes any shading or recombination that might otherwise have occurred by putting·them on the front surface. Making them small significantly reduces the contact surface area, which decreases recombination and contact resistance and improves cell performance.

An equally impressive device has been designed by Green and others at the University of New South Wales (UNSW), Australia.[22] They use a microgrooved texture on the front surface to minimize reflection and help trap light inside the cell (Figure 9). They also incorporate a thin layer of oxide at the front surface to decrease recombination at the surface and front contacts. In addition, they minimize the contact area between the front surface metal and the silicon, as they do the shading by the grid. The design has top and bottom metal contacts that allow the cell to be easily mounted in concentrator modules.

The result of these and other innovations has been little less than spectacular. New efficiency records are continuously being established, both by research devices and by production cells and modules. As of this writing, for example,

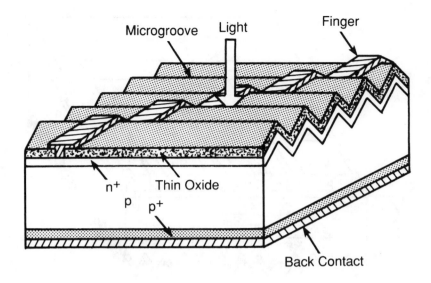

Figure 9. The innovative silicon cell designed at the University of New South Wales. (Adapted from Ref. 35 with permission of the IEEE.)

research cells have reached efficiencies of 22.8% under unconcentrated sunlight and 28.2% under sunlight concentrated 100 times.[9] Module-ready cells have gone as high as 25.5% under 150 times concentration. And things look like they can only get better. By incorporating the UNSW-type cells into 200X concentrators, for example, the U.S. Department of Energy (DOE) expects to achieve concentrator *module* efficiencies of 20% or more. And the UNSW cell can be further improved with better surface texturizing and light trapping techniques. Also, using a prismatic cover to direct light onto unmetallized regions of the top surface would circumvent problems of shading by the metal and would enhance light trapping.

The point-contact type of cell could also be improved by making the cell thinner, reducing resistance, and optimizing the texturized surface and the carrier lifetime. With such innovations, the theoretical efficiency limit of the cell could be as high as 30% under ordinary sunlight and 36% under concentration. The practical limits would have to be somewhat lower because of unavoidable losses such as band-to-band Auger recombination, some emitter recombination, and resistive voltage drops. Even so, single-junction cells could eventually reach efficiencies as high as 31% under concentration.[23] A degradation problem has been observed with the point-contact cell, but work is under way to ameliorate or eliminate it.

These results have been mostly for expensive, small-area cells that would be used almost exclusively in concentrator systems or in specialized, high-value markets such as space applications. Equally important have been the results achieved for relatively inexpensive, large-area cells for flat-plate systems. Several organizations and companies have made production-size cells that have

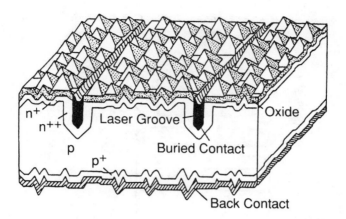

Figure 10. The laser grooved, buried contact PV cell. (Adapted from Ref. 35 with permission of the IEEE.)

reached 20% efficiency. One of the most interesting of this group is the laser-grooved, buried contact cell designed by Green and his co-workers at the University of New South Wales. This cell is notable for its size (12 cm²), its efficiency (19% to 20% under global unconcentrated sunlight), and the fact that its method of fabrication promises to be cheaper than most current commercial cells, which are appreciably less efficient. Like other modern cells, this cell uses a textured surface, an oxide coating, and commercial-grade Czochralski silicon (Figure 10). But the innovative part is that it uses standard laser scribing equipment to etch grooves 20 μm long and 60 μm deep into the front surface. When filled with plated copper, the grooves result in a large cross sectional area of high-conductivity metal for the contact fingers, while giving very little shading loss.

Meanwhile, some organizations are beginning to produce flat-plate modules with efficiencies greater than 15%, a long-sought goal of the National Photovoltaic Program. These modules are being produced with production cells that average around 17% efficiency. Incorporating cheaper, larger cells that promise to average over 20% efficiency will enhance module efficiency.

Finally, most of the new highly efficient silicon cells use pure silicon crystals grown by the float-zone technique. [In this technique, a polycrystalline ingot of silicon is placed atop a seed of single-crystal silicon and heated, causing the interface between the materials to become molten (Figure 11). The heating coils are slowly raised while the ingot remains stationary. Pure, single-crystal silicon solidifies below the molten interface and grows upward as the heating coils move up.] Scientists at the Solar Energy Research Institute (SERI) have improved the technique by adding controlled amounts of gallium and boron to the silicon material. This allows them to make low-resistivity material with long minority carrier lifetimes, four to five times as long as previously attained. (Minority carriers are charge carriers, free electrons or holes, that are in the minority in the particular material and that are the principal source of

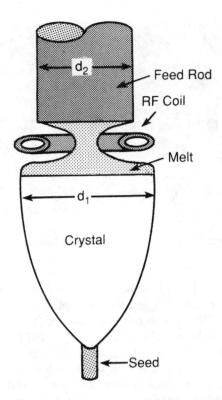

Figure 11. Researchers use the float-zone technique to grow very pure, high-quality silicon crystals.

current generation. Long lifetimes mean that the charge carriers exist freely for a long time, which helps to generate more current.)

4b. Polycrystalline silicon

High efficiency is not the only road that crystalline silicon can take. Another approach is to compromise on efficiency in the interest of achieving lower production costs. The two primary methods for this are ribbon technology, which grows entire sheets of silicon, and cast silicon, which uses simple casting processes to grow square blocks that are then sawed into wafers.

Polycrystalline silicon physically differs from single crystal silicon in that it is composed of many individual grains of single-crystal silicon. The size and quality of the grains determine their effect on cell performance. The larger and more perfect the grains the more nearly their electrical behavior resembles that of a cell made of single-crystal silicon. The main advantages of polycrystalline silicon are that it has less stringent growth requirements; uses less material, especially if prepared in ribbon rather than block form; requires less pure silicon; holds the potential for much greater throughput, more automated manufacturing, and cheaper production, especially in ribbon form; promotes more efficient use of silicon feedstock; and can be deposited on less expensive substrates. At the same time, since rectangular polycrystalline cells can be

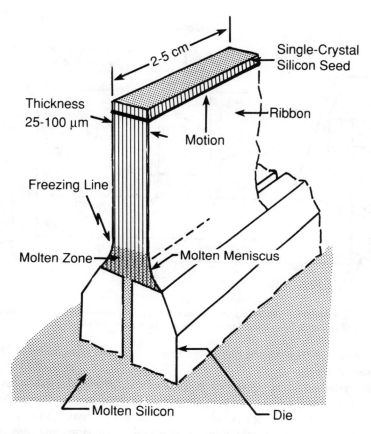

Figure 12. Edge-defined film-fed growth (EFG) uses capillary action to draw molten silicon through a narrow, slotted mold.

packed more densely into modules than round single-crystal cells, polycrystalline modules can be nearly as efficient as those of flat-plate, single-crystal silicon.[24]

Although many techniques for producing sheets of silicon have been studied, primarily under the National Photovoltaic Program, only one is a commercial reality today, edge-defined film-fed growth (EFG). Edge-defined film-fed growth uses highly pure molten silicon that is drawn upward by capillary action through a narrow, slotted mold of carbon or other suitable material, producing a continuous crystalline silicon ribbon (Figure 12).

Another technique, dendritic web growth, is nearing commercial status. This process is similar to EFG, except the mold is eliminated, thus avoiding problems of mold erosion and of contamination associated with the mold. In the dendritic web process, two single-crystal dendrites are allowed to touch and grow slightly into the molten silicon. As the dendrites are slowly withdrawn from the melt, because of silicon's surface tension, a web of crystalline silicon forms between them, solidifying as it rises from the melt and cools (Figure 13).

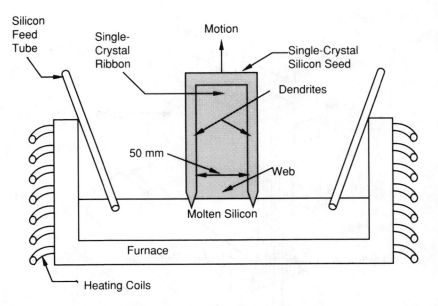

Figure 13. In the dendrite web process, two silicon dendrites are slowly drawn from silicon melt, with a web of crystalline silicon forming and solidifying between the dendrites.

The ribbons are then wound onto large reels, cut into lengths, cleaned, doped, and made into cells.

Polycrystalline silicon devices use some of the same design advances used with single-crystal silicon, such as back-surface reflectors and antireflection coatings. Polycrystalline devices also incorporate atomic hydrogen to passivate (i.e., make electrically inactive) defects and grain boundaries and thus reduce carrier recombination. Atomic hydrogen ties up dangling bonds at the grain boundaries, deactivating recombination centers.

As a result of these and other improvements, experimental cells from the dendritic web ribbon process are routinely achieving efficiencies near 17%, while some production-size cells from this and EFG are nearing 15%.[25,26] At the same time, production rates are increasing substantially, with dendritic web gaining a fivefold increase in throughput over 1984.[23]

Modules produced by EFG, however, account for only about 0.1% of the market for power modules, while ribbons of polycrystalline silicon from the dendritic web process have only been produced in pilot quantities.[13] To become commercially viable, they must overcome at least two related obstacles. First, they must substantially increase their throughput. Second, thin silicon ribbons can cool unevenly, creating stresses and crystal imperfections that can cause ribbons to buckle or crack. This can limit cell yields and performance. These obstacles are being addressed with the help of sophisticated growth models and computer aided production that constantly samples ribbon growth and feeds the information back to the process for adjustment.

Cast silicon technology, although it does not contain all the potential cost-savings of ribbon techniques, is more mature than ribbon technology. Several major PV companies emphasize cast polycrystalline silicon modules. And in 1989 about 11 MW of polycrystalline modules were shipped to customers, which was more than one-fourth of the worldwide market[13]. Some cast-silicon production cells are over 15% efficient and large-area cells approach 14%. Industry is also exploring processes that promise to reduce the cost of solar-grade silicon by 50% or more. Achieving this will substantially decrease the cost of polycrystalline silicon modules.

Finally, there is the possibility that polycrystalline cells could be made more than ten times thinner than present cells, about 30 μm, while at the same time achieving efficiencies of 19% and more. The fundamental trick to achieving such a cell would be the employment of light trapping techniques pioneered with single-crystal silicon. But the design also depends upon passivating grain boundaries and using silicon material that has a minority carrier diffusion length at least twice the thickness of the absorber layer (which is achievable since the absorber layer will be so thin). The concept also promises to be quite inexpensive because it uses a low-cost ceramic substrate and an innovative deposition process where the thin layer of silicon is deposited on the ceramic from solution. Although the concept is new, in 1988 researchers made an experimental cell that reached 14.2% efficiency.[27]

4c. Amorphous silicon

Amorphous silicon (a-Si) is today's primary alternative to crystalline silicon. It presently dominates the consumer end of the PV market. But with continued rapid improvement, it is positioning itself for the large power markets in the future.

Such an assertion would have sounded ludicrous in 1973 when the first a-Si:H cell was made—it was less than 1% efficient and appeared to many to be a mere curiosity.[28] Since that time, however, single-junction cells have reached 12% efficiency, multijunctions have gone to 13.3%, and large submodules have surpassed 9%.[9] Industry and research organizations have been quick to recognize the potential of a-Si, investing large sums of capital and effort. In fact, almost all of Japan's interest in PV to this point has been centered around amorphous silicon.

Amorphous silicon is a rather dense and disordered substance that lacks the structural uniformity of crystalline silicon. It was this disorder and the resulting multitude of defects that made many suspect amorphous silicon would be unacceptable for electronic devices. But the first conducting films of amorphous silicon were produced from silane (SiH_4) and thus contained hydrogen. The hydrogen filled the defects and restored many of the crystalline properties.

There are several advantages to hydrogenated amorphous silicon (a-Si:H). First, it absorbs sunlight 40 times more efficiently than does crystalline silicon and thus needs only 1 to 2 μm to soak up 99% of the incident light above its 1.7 eV band gap. The small amount of material required yields significant cost savings. Also, because of its disordered state, it can be deposited on low-cost substrates. These characteristics have led much of the worldwide PV effort into a major new direction: low-cost, thin-film PV, and are helping to usher in an era of automated, in-line, continuous production.

Typically, making an a-Si:H p-i-n cell begins by depositing a p+ layer onto a textured transparent conducting tin oxide film (the top electrode). This is followed by the intrinsic (i) layer, and a thin n+ layer. The cell is then given a reflective coating on the bottom, usually of aluminum or silver. Using a textured transparent conducting oxide substrate and a reflector help to trap the maximum possible amount of light in the PV cell.

Glow discharge chemical vapor deposition is the method that has been used to make the most efficient a-Si PV cells. In the most common form of this method, a stream of silane is passed between a pair of electrodes whose polarity is reversed at high frequency. This reversal of the voltage induces an oscillation of energetic electrons between the electrodes. The electrons collide with the silane, breaking it down into SiH_3 and hydrogen. Since SiH_3 is a radical, and hence unstable, it adheres to a substrate on one of the electrodes to gain stability. Hydrogen is then released from the substrate, leaving a film of amorphous silicon with about 10% hydrogen. Doping can be accomplished by adding diborane (B_2H_6) or phosphine gas (PH_3) to the silane.

Most commercial amorphous silicon PV modules are made by this method. The successive layers of the module are formed by moving the appropriate substrate sequentially to different electrode stations in separate chambers, and depositing the appropriate materials one at a time. Moving the modules through different chambers minimizes the opportunity for contamination from previous steps. This process uses little energy and is potentially able to produce modules a few square feet in area, on a variety of substrates and in an assortment of shapes.

There are two basic drawbacks to amorphous silicon. The first is the Staebler-Wronski effect, where a-Si:H devices lose efficiency after initial exposure to light. When the effect was first recognized, devices were losing up to 50% of their efficiency. And that was for devices that were relatively inefficient to begin with. Since that time, researchers have begun to understand the phenomenon and have proposed ways to mitigate it. One of the causes of the Staebler-Wronski effect is light-induced defects, known as dangling bonds, in the intrinsic layer. Most a-Si:H devices employ a p-i-n structure. (See above.) The wider the intrinsic layer, the more pronounced the effect. Consequently, some manufacturers are making modules with very thin intrinsic layers. Others go a step further, employing two or three stacked a-Si:H cells, each with an extremely thin intrinsic layer. These approaches have resulted in degradations of less than 15% for modules and 10% for cells.

Since the depositions needed to make thin-film multijunction devices do not use much energy, multijunction devices are potentially inexpensive to fabricate. Making a multilayered cell can be very similar to making one cell — just adding on one thin film after another. In fact, the trend in amorphous silicon is away from single junction and toward multijunction devices.

The other drawback is one of low efficiency. Commercial cells are generally

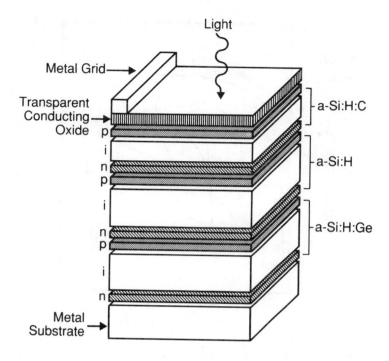

Figure 14. A three-junction a-Si device that uses a-Si:H:C for the high band gap cell, a-Si:H for the middle cell, and a-Si:H:Ge for the low band gap.

under 8%, while commercial modules reach efficiencies of only 4% to 6%. But proponents of amorphous silicon are betting that by improving fabrication and design and by minimizing both recombination in the intrinsic layer and the resistive effect of the transparent conductor on the current, the efficiency of small-area cells should reach 15% and that for large-area submodules should reach between 10% and 12%.

Even higher efficiencies are possible using alloys of a-Si in multijunction devices. A typical 3-junction device uses a high band gap cell of a-Si:H:C, a mid band gap cell of a-Si:H, and a low band gap cell of a-Si:H:Ge (Figure 14). The a-Si:H:C cell captures the blue end of the spectrum, the a-Si:H cell absorbs the middle portion, and the a-Si:H:Ge captures the red part. Theoretical calculations indicate that three such stacked cells with band gaps of approximately 2.0, 1.7, and 1.45 eV, respectively, could attain an efficiency of about 24%.

To approach such a limit requires high-quality cells. Currently, adding carbon to increase the band gap or germanium to decrease the band gap degrades cell quality with defects and voids in the microstructure. For example, a-Si:Ge:H cells with band gaps above 1.5 eV are generally of high quality. But once the band gap has been lowered to between 1.4 and 1.5 eV, the material begins to deteriorate. Researchers are attempting to understand and circumvent the degradation. If they succeed, it should be possible to produce inexpensive modules with efficiencies in the 15% to 20% range.

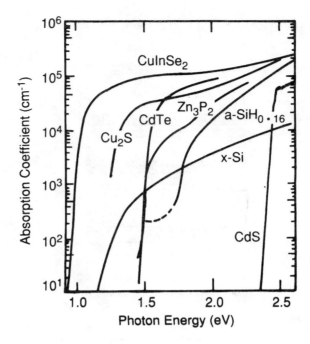

Figure 15. Absorption coefficients of some PV materials. A material's band gap is taken as the energy of the steep increase in absorption. Slopes of the polycrystalline thin films rise sharply, indicating their high absorption coefficients.

4d. Polycrystalline thin films

Polycrystalline thin film semiconductors are extremely thin (from a few micrometers to a few tenths of a micrometer) and are composed of many very small crystalline grains of semiconductor materials. The two most prominent polycrystalline materials are copper indium diselenide ($CuInSe_2$) and cadmium telluride (CdTe). The fundamental quality that makes these materials prime candidates for use in thin film devices is their very high absorption coefficients. Copper indium diselenide, in fact, has the highest absorption coefficient of any PV material known and can absorb 90% of the solar spectrum available to the cell in 0.2 μm of the material (Figure 15). Consequently, like a-Si, polycrystalline thin film devices use very little material. Also like a-Si, they can use inexpensive substrates and they are amenable to automated manufacturing processes.

They have additional advantages. For one, they do not suffer from light-induced degradation: $CuInSe_2$ is one of the most stable PV materials available. Also, the methods for deposition promise to be an order of magnitude or so faster than the chemical vapor deposition used for a-Si. Copper indium diselenide depends upon variants of physical vapor deposition to deposit the material, like vacuum evaporation, where sources of copper, selenium, and indium are separately evaporated and then deposited on a substrate; and sputtering, where thermal evaporation is replaced by physical bombardment of the elemental sources by high-energy ions. A method that promises to be scalable, low-cost, and have high throughput is a two-step process whereby copper and indium are sputtered on a molybdenum-coated glass substrate, which is then annealed in an atmosphere of hydrogen selenide (H_2Se).

Of the deposition methods being explored for CdTe, two have the potential for being the most inexpensive, in terms of capital equipment, and for having very high throughput: electrodeposition and spray pyrolysis. In electrodeposition, electricity is passed through a solution containing ions of the required elements, causing them to be deposited out of solution onto an electrode, which acts as the substrate. In spray pyrolysis, solutions of the salts of the necessary elements are sprayed onto a hot substrate. They react under the elevated temperatures to form the required layers, while the solvent evaporates. Both of these methods are already widely used industrial processes with spray pyrolysis, for instance, being used to coat large areas of plate glass. These low-cost, "low-tech" alternatives need merely be successfully adapted to photovoltaics. And the prognosis looks good for both, with cells made with electrodeposition achieving efficiencies of 11% while those from spray pyrolysis have reached 7%.

Polycrystalline thin films also pose problems. Cadmium telluride, for example, undergoes performance degradation because of difficulties contacting p-type CdTe. Fortunately, researchers may have solved the contact problem with CdTe by using a p-i-n device structure with an n-type CdS window, an intrinsic CdTe layer, and a p-type ZnTe layer, rather than an n-CdS/p-CdTe heterostructure (which requires metallization contact to the p-type CdTe layer). This and other improvements have led to new record high efficiencies as well as successful long-term stability.

Copper indium diselenide has had a history of lower open-circuit voltage. The open-circuit-voltages in cells made from most other compound semiconductors tend to be at least 50% of the band gap, with many approaching 66%. With a band gap of 1 eV, however, copper indium diselenide cells have only reached voltages between 0.4 and 0.45 V. To overcome this problem, researchers have added gallium to the compound, raising the band gap to 1.1 eV and hence the open-circuit voltage to at least 0.55 V. Of course, if no other strategies are used, increasing the band gap lowers the current, since those photons with energy between 1 and 1.1 eV no longer create charge carriers. This loss of current can be surmounted to a large degree by increasing the band gap of the window by replacing the CdS window with a CdS/ZnO window.

By employing some of these and other strategies, $CuInSe_2$ has achieved what many consider a crucial breakthrough for thin films—square-foot submodules that are more than 11% efficient. Moreover, new, small-area cells are reported at over 14% efficiency. With these kinds of results, and with still a great potential for further increasing the open-circuit voltage, and improving morphology and device design, it looks as if single-junction copper-indium diselenide modules have the potential to break the touted 15% module efficiency barrier, and at low cost.

Beyond that lies the ample promise of multijunction concepts, of which there are two general avenues being pursued. The first is to use a-Si:H as the top cell in a two-junction device and CuInSe$_2$ as the bottom cell. With this concept, the top cell will absorb photons with 1.7 eV or greater and pass the rest through to the bottom cell, which will soak up photons with energies between 1 and 1.7 eV. One organization already reports a 15.6% efficiency for a multijunction device using this structure.

The second approach is to use CdTe, which has a band gap of 1.45 eV, as the top cell and CuInSe$_2$ as the bottom cell. Several organizations are pursuing this concept, which has the potential to reach efficiencies greater than 20%.

4e. III-V materials

The highest efficiencies may be obtained with III-V materials, most notably gallium arsenide (GaAs) and its alloys such as aluminum gallium arsenide (AlGaAs), indium gallium arsenide (InGaAs), and gallium indium diphosphide (GaInP$_2$). To many, GaAs represents the ideal PV material. It has the optimum band gap (approximately 1.42 eV) for single-junction cells, a very high absorption coefficient, the highest theoretical efficiency for any known PV material (approximately 39% for single-junction cells under concentrated light of 1000 suns), and can be alloyed with many different materials to alter the band gap as desired for high efficiency multijunction configurations.[16] These attributes have enabled GaAs-based cells to establish efficiency records for single-junction cells under one-sun illumination (25.7%), single-junction cells under concentrated sunlight (29.2%), and the all-time record for all cells, a GaAs/Si two-junction device that attained an efficiency of 31.0% under concentrated sunlight.[9,29]

The cells used to establish this last record are worthy of mention. The top cell is a slight variation of the record-holding single-junction GaAs cell cited above. As the top cell in the two-junction stack, this single-crystal device absorbed the blue end of the spectrum to register an efficiency of 27.2%. The bottom cell was a state-of-the-art, point-contact, single-crystal silicon cell (described above). By itself, this particular cell had registered an efficiency of 26%. But in the two-junction stack it added only 3.8% to the efficiency of the whole package because only a small percentage of light made it through to the bottom part of the cell. This illustrates one of the difficulties with multijunction devices: stacking cells does not add their efficiencies; most of the light is absorbed by the top cell or by the interface between cells. That is why some argue that it would be wiser to make less complicated, less expensive, but highly efficient single-junction devices.

There are other ways to make multijunction devices, however. Rather than make the cells separately and then physically stack and connect them, you can grow the entire device monolithically, in one system. With this approach you can either grow the device from the "top-down" or from the "bottom-up." To grow it bottom-up, you start with a substrate, grow the bottom cell on the substrate, deposit interface layers and connects, then grow the top cell. Either way, it is a difficult process. There are many variables that must be controlled simultaneously such as elemental sources, and heat. And you must be careful to use

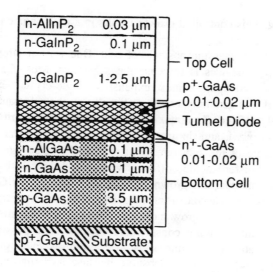

Figure 16. Device structure of the SERI record-setting GaInP₂/GaAs multijunction cell.

cell and interface materials whose atomic spacing match well; otherwise growing one cell on top of the other will cause defects resulting from stress.

Even with all the difficulties, the eventual result may well be worth the effort, as is being borne out by recent tests on a device fabricated by the Solar Energy Research Institute. This device, which uses a GaInP$_2$ top cell, a GaAs tunnel diode interconnect, and a GaAs bottom cell (Figure 16), recently reached efficiencies greater than 27%, a record for the materials and for monolithically grown devices. (A tunnel junction allows charge carriers to move between the top cell and the bottom cell through a very thin insulating layer.) The materials used exhibit many of the characteristics desirable for monolithic multijunction devices: their lattices match and the materials are rather insensitive to wide variations in growth conditions. Furthermore, with band gaps of the GaInP$_2$ and GaAs at 1.9 and 1.42 eV, respectively, this combination has a theoretical efficiency of 34%.

Multijunction devices that use III-V materials could reach practical efficiency even higher than 34%. Employing GaAs as a middle cell in a three-junction device, and alloys of GaAs as the top and bottom cells, efficiencies as high as 40% could be attained.[18]

Growing GaAs (and the related alloys used in multijunction devices) layers, whether monolithically or in stacked configuration, is very often accomplished with metalorganic chemical vapor deposition (MOCVD). Here, a heated substrate is exposed to gas-phase molecules gallium and arsenic (and other materials, such as aluminum or phosphorous), which react under high temperatures, freeing atoms to adhere to the substrate. This technique grows single-crystal layers epitaxially — that is, the new atoms deposited on the substrate continue the same crystal lattice structure as the substrate, with few disturbances in the atomic ordering. This controlled growth results in a high degree of crystallinity and high cell efficiency.

III-V devices still face the obstacle of cost. One way of circumventing the high cost of III-V devices is to reuse the substrates. Most GaAs cells have historically depended upon a thick, expensive, single-crystal GaAs substrate. Because the cell was generally grown epitaxially on the substrate, the substrate was permanently affixed. (In epitaxial growth one crystal is grown on the surface of another crystal, and the growth of the deposited crystal is oriented by the lattice structure of the substrate.) But the CLEFT (cleavage of lateral epitaxial films for transfer) method offers a way around this. With CLEFT a thin film of single-crystal GaAs is grown atop a thick single-crystal GaAs substrate. The thin GaAs is then cleaved from the substrate for incorporation into a cell, and the substrate is used to grow more thin GaAs. The substrate can be reused as many as ten times, saving considerable expense. The possibility of using inexpensive, reusable germanium substrates promises even greater savings. Germanium may be a candidate substrate because its lattice constant closely matches that of GaAs.

The CLEFT process is made possible by using a mask that exposes the thin layer of GaAs to the substrate only at select openings. Epitaxial lateral growth is initiated at the openings and a continuous single-crystal film can be grown to any desired thickness. Using the mask creates a weak plane. Growing the thin film along the principal cleavage plane of GaAs allows the film to be cleaved from the substrate without damage to either (Figure 17).

The CLEFT method has enabled GaAs cells to recently reach an efficiency of 22.4%,[9] and an experimental GaAs flat-plate module to achieve greater than 20% efficiency. The cleaved thin films may be used in any compatible device configuration. One possibility under consideration is to use it as a top cell in a two-junction device with $CuInSe_2$ as a bottom cell. This combination could have advantages for space applications since both materials can withstand radiation damage quite well.

Some properties of gallium arsenide and its alloys (such as fast response time and direct band gaps) make them highly efficient light emitters as well as light absorbers. This makes them prime candidates for incorporation into optoelectronic applications. In fact, some companies and R&D organizations are exploring GaAs for this use as well as for PV. They are finding that their work in PV fabrication, device design, and thin-film techniques is applicable to optoelectronics technology. And, as they progress with their optoelectronic research, they will find that advances there will enhance progress in PV technology. Such is the symbiotic relationship between PV and other aspects of the solid-state revolution. Advances in the technology of one begets advances in the other.

5. Photovoltaics and the greenhouse effect

The obvious incentive for proceeding with PV R&D and commercialization is that PV has the potential for being one of the world's most capable and widespread energy technologies, with the capacity to supply the world with electri-

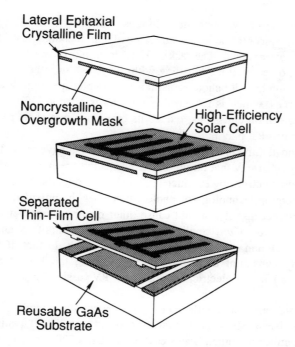

Figure 17. In the CLEFT technique, the substrate is reused to grow other thin films of GaAs.

cal power as long as the sun shines. But recently an equally strong incentive has begun to emerge: for an energy technology, PV is quite environmentally benign (although it has such environmental side effects as use of a large land area and possible release of toxic materials during the manufacture of cells). This is especially true *vis-à-vis* climate change due to the greenhouse effect, the gradual warming of the Earth because of the continuous influx of anthropogenic greenhouse gases. Climate change due to the greenhouse effect just might become the supreme environmental challenge to face the world in the 21st century.

Excess carbon dioxide (CO_2) released into the atmosphere, primarily from the burning of fossil fuels, may be responsible for as much as one half of the projected worldwide rise in temperature in the next century. A major source of CO_2 is the generation of electricity from coal, petroleum, and natural gas. The problem is, reliance on electricity is growing faster than that of any other form of energy. Moreover, this growth is taking place in almost every nation on Earth. And as Third World countries continue to expand their populations and develop their economies, their electricity consumption is bound to grow steeply with many, like China, turning increasingly to vast coal reserves.

The Environmental Protection Agency, on the other hand, suggests that the world must cut its generation of CO_2 by at least 50% *from present levels* by the year 2100, to reach an acceptable state of stability in the atmosphere.[30] An "acceptable state" means the built-in increase in the global average temperature should be from 1.5 °C to 3 °C.

Can the world achieve such an ambitious goal? It is hard to say. There are just too many uncontrollable variables involved: technological, economical, political, and international. The task is Herculean and is one which diplomats, scientists, and others will be wrestling with for years. So rather than tackle this involved question here it behooves us to ask a simpler question: how much can the United States, which is responsible for approximately 23% of the world's anthropogenic carbon emissions, reduce its emissions in a short period of time and what role can PV play in this?

Technologically, the problem is not insuperable. We have many technologies for producing electricity that do not contribute to the greenhouse effect. Among these are hydroelectric power, nuclear, wind, and photovoltaics. And among these, the emerging technology of photovoltaics has the potential for becoming the most facile because it does not cause other major environmental problems, can be designed to fit *any* application, relies on a ubiquitous and ample resource (sunlight), and can be used anywhere on Earth. In fact, if the resource were the only constraint to deploying PV systems, the United States could easily derive all of its electrical needs from PV, assuming adequate storage capacity.

To illustrate this point, let's take a look at a recent study performed by resource scientists at SERI. Their analysis showed that 55,000 quads of sunlight fall on the continental United States each year; that's 700 times as much energy as the United States annually consumes. Assuming that each state could set aside 1% of its land area upon which it would install 10%-efficient systems, the study concluded that all but ten states could supply all the energy (not just electrical energy) they needed. And those ten states could import their extra required energy from neighboring states.

The question is not so much what we could physically do, but what our infrastructure and our economy could reasonably be expected to accomplish in a short period of time. There are a number of projections available on how much energy PV could supply within the next couple of decades. The DOE Office of Policy, Planning, and Analysis, for example, projects that the United States will install approximately 1 quad of PV energy by the year 2010.[31] This projection is based on the assumption that the PV industry and the National Photovoltaic Program will reach its goal of 6 cents/kWh by the year 2000.

Some in the PV industry feel that this is a very conservative number and that, given the investment and incentives, the United States could install as much as 10 GW of capacity per year by 1995.[32] If such were the case, it would result in an installed capacity of at least 150 GW by 2010. If we assumed that the installed systems operated at about a 30% capacity factor, the installed PV capacity would produce approximately 4 quads of energy per year (where 1 quad is the equivalent of approximately 96.6 billion kilowatt-hours).

Still another study, prepared for the DOE's Energy Research Advisory Board in 1984, arrives at a different conclusion. It states that, by 2010, PV has the upper-bound potential to supply the United States with 12.5 quads of energy. This upper bound, or ECP (energy contribution potential) is defined as the maximum amount of energy that can be economically extracted from the renewable energy resource base on an annual basis. This upper bound may be approached a few years after PV becomes competitive with conventional forms of electricity production, reaches the status of a preferred option, and begins to penetrate the electricity market substantially. (We must note that 12.5 quads of PV energy is a very large amount of energy. Currently, for example, the United States has 0.003 quad of PV in place.[32] To get 12.5 quads of energy from PV, the United States would have to install 450 to 500 GW of capacity, assuming that the installed capacity generated electricity at a 30% capacity factor. For comparison, the United States presently has a total installed electric generation capacity of about 700 GW.)

For our purposes we will assume the most conservative of these estimates and let the reader conclude what could follow under the more liberal projections.

How much carbon emissions could 1 quad of PV save? If we assume that PV electricity displaces electricity generated by coal, then 1 quad of PV would save almost 25 million tons of carbon annually. (A quad of coal in the United States produces an average of about 25 million tons of carbon.) In 1986, the United States emitted approximately 1.3 billion tons of carbon into the atmosphere. Thus, 1 quad of PV displacing coal would eradicate 1.9% of 1986 U.S. carbon emissions.

There are, however, at least two things wrong with this calculation. The first is that we cannot assume that PV will displace only coal. It may also equally displace gas, oil, hydroelectric, nuclear, and others. Second, in 2010, the United States will consume considerably more energy than it did in 1986, and some of the increase is bound to be in the forms of coal, oil, and gas. Let's explore each of these problems in turn.

To a first approximation we could guess that PV would displace other sources of electricity in rough proportion to the relative contribution the electric technologies currently make. In 1986, the major technologies contributed the following proportions of America's electricity: oil (5.4%); gas (10.1%); coal (54.2%); nuclear (16.8%); and hydroelectric and other (13.5%).[33] So, if these approximate ratios were to hold for future generation mixes, and if PV were to displace each in like proportion, then in 2010 PV systems could displace 0.054 quad of oil, 0.101 quad of natural gas, 0.542 quad of coal, 0.168 quad of nuclear, and 0.135 quad of hydroelectric and other. This would translate into a savings of 1.1 million tons of carbon from oil, 1.48 million tons of carbon from natural gas, and 13.47 million tons of carbon from coal. (Nuclear and hydroelectric do not emit carbon.) This is a total of 16.05 million tons of carbon, which is equal to 1.2% of the total carbon emitted in the United States in 1986.

Even with this figure we must throw in a few more caveats. First, it is unlikely that PV would displace technologies in the assumed proportion. Technologies will change by 2010, needs will change, and the mix of electric generation will change. Besides, much of PV will be used for remote and specialized applications where it may predominantly displace diesel fuel and batteries. Additionally, PV is presently limited by the fact that it is only an intermittent power source; PV systems can produce power only when the sun shines. So, until we develop efficient, reliable, and inexpensive means to store PV electricity for use during periods of no sunshine, PV systems for central utilities will be limited to supplemental and peaking power, and not for base power. This implies that PV systems for utilities would mostly be in competition with gas or oil-powered units, not coal, nuclear, or hydroelectric, which are used primarily for base power. Of course, the next couple of decades should see ample progress made in the ongoing research in storage technologies, so we should see this limitation of PV systems disappear.

Finally, 20 years from now the United States will undoubtedly consume considerably more energy that it does now, much of which will come in the form of electrical energy. The DOE Office of Policy, Planning, and Analysis, in fact, projects that the United States will consume 108.2 quads of energy by 2010, under the assumptions that present energy consumption trends will continue and that we do nothing actively to stem the growth in carbon emissions.[33] This is nearly 32 quads more energy than we used in 1986.

Some of this increase will be due to consumption of renewable energy, which the DOE forecasts to rise to 12.7 quads per year by 2010. But most of the increase will come from an increased use of fossil fuels: coal, oil, and natural gas. Of these, the DOE predicts that the consumption of coal will increase the most, from 17.3 quads in 1986 to 36 quads in 2010. This, along with the increased use of natural gas and oil will drive U.S. carbon emissions up a dramatic 38.6%, from a present level of 1.32 billion tons of carbon to 1.83 billion tons. Obviously, the consumption of energy under this "business as usual" projection will exacerbate the greenhouse effect.

There are, however, things we can do to stem this trend. Using the DOE projections as a base, SERI performed an analysis that indicated we could adopt several strategies to mitigate the rise in carbon emissions, at no extra cost to our economy.[34]

First, we could adopt cost-effective conservation and energy efficiency technologies, such as enacting efficiency standards for automobiles, appliances, and lighting. By increasing our energy efficiency sufficiently, by about 1.9% per year, we could reduce our expected energy consumption for 2010 to 91.8 quads. (Increasing our energy efficiency by 1.9% per year is not unprecedented; between 1970 and 1985 the United States increased its energy efficiency at a rate of 2.1% per year.)

Second, we could displace 5 quads of coal with renewable energy sources like hydropower, biomass, geothermal, and PV. And we could displace another 4 quads of coal with gas, which emits about half the carbon that coal does, per unit of energy.

Third, we could help produce and then import another 1.5 quads of Canadian hydropower.

Fourth, we could reforest federal lands and plant trees on one-half of the current inventory of crop set-aside lands. This strategy would fix an additional 3 quads of coal-related carbon emissions over a 20-year period.

These strategies, which include the 1 quad of PV projected by the DOE, would have renewable energy technologies supplying America with as much as

21% of its energy by the year 2010, a total of about 19.5 quads, counting the imported hydroelectric power. All of these strategies together would result in carbon emissions of 1.19 billion tons. This is 640 million tons, or 35%, less than that entailed by the DOE projections. Moreover, we have to add to this the possibility of using conservation and renewable energy to practically eliminate the use of chlorofluorocarbons. And all of this is at no extra cost to the economy.

It is still, however, only 130 million tons, or 10%, less carbon emissions than the 1986 level. But we must note that, like the DOE projections, the SERI analysis assumed economic neutrality, i.e., that renewables would have to be economically competitive with other technologies. It also assumed no major initiatives in R&D or commercialization. It becomes apparent, then, that if the government also substantially increases its efforts in R&D, commercialization, setting stringent energy and environmental standards, and providing incentives and subsidies, the contribution from renewable energy technologies could accelerate even more.

The point of this sobering view is that no one technology or single-faceted strategy can meet the challenge of reducing greenhouse gas emission to an acceptable level. If we are to do it, we must use a multifaceted approach that includes all possible technologies, governmental policies and incentives, intensified R&D, curtailment of population growth, and high levels of international cooperation.

In this milieu, PV can play an initially small but a growing and significant role.

Bibliography

Solar Energy Research Institute, *Basic Photovoltaic Principles and Methods* (Van Nostrand Reinhold Company, New York, 1984). This book presents a nonmathematical overview of the theory and design of photovoltaic cells and systems. The book uses the silicon cell as a model since silicon is a widely used cell material and provides a good basis for understanding the photovoltaic cell. It is written for those who desire an introduction to the field. Each chapter concludes with a short bibliography listing sources of additional information.

M.A. Green, *Solar Cells: Operation Principles, Technology and System Applications* (Prentice Hall, Englewood Cliffs, New Jersey, 1982). Clearly written and well illustrated, this text covers many aspects of photovoltaics but is particularly strong in cell theory and design. The mathematics included here requires a college math background.

PV News (PV Energy Systems, Inc., 2401 Childs Lane, Alexandria, Virginia 22308). This newsletter reports the latest research, industrial, and international developments in photovoltaics. Issues regularly contain editorial commentaries and bibliographic citations for new PV publications.

Solar Cells (Elsevier Science Publishers, 52 Vanderbilt Avenue, New York, New York 10017). This periodical is devoted to the physics and chemistry of solar cells, their technology, applications, and economics. In-depth review articles and original papers are abstracted and illustrated. Book reviews and conference announcements are included.

Photovoltaic Technical Information Guide, Second Edition, SERI/SP-320-3393 (Order Number DE88001184, National Technical Information Service, 5285 Port Royal Road, Springfield, Virginia 22161, August 1988). This guide is designed to help you find information about photovoltaics. It contains history, technology basics, descriptions of current research and development activities, and references as helpful sources for further information.

References and notes

1. United Nations, *Energy Statistics Yearbook, 1987* (Publishing Division, United Nations, New York, New York, 1989).
2. United Nations, *Yearbook of Energy Statistics, 1980* (Publishing Service, United Nations, New York, New York, 1982).
3. U.S. Department of Energy, Office of Policy, Planning, and Analysis, *DOE Long Range Energy Projections to 2010, Report DOE/PE-0082* (U.S. Government Printing Office, Washington, DC, 1988).
4. Chevron Corporation, *World Energy Outlook* (Chevron Corporation, Corporate Planning and Analysis, San Francisco, CA, 1987).
5. J. Goldemberg *et al.*, Annual Reviews of Energy, **10**, 613–688 (1985).
6. E. Becquerel, Comptes Rendus **9**, 561 (1839).
7. D.M. Chapin, C.S. Fuller, and G.L. Pearson, J. Appl. Phys. **25**, 676 (1954).
8. *Proceedings of the Conference on Photovoltaic Conversion of Solar Energy for Terrestrial Applications*, Cherry Hills, N.J., 23 to 25 October 1973 (Publication JPL-SP-43-3, Jet Propulsion Laboratory, Pasadena, CA, 1973).
9. U.S. Department of Energy, *Report DOE/CH10093-40* (Solar Energy Research Institute, Golden, CO, 1989).
10. R.G. Ross, Jr., in *Proceedings of the 17th IEEE Photovoltaic Specialists Conference*, Kissimmee, Fl, 1 to 4 May 1984 (IEEE, New York, 1984), pp. 464–473.
11. D.E. Carlson, in *Proceedings of the Eighth International E.C. (European Community) Photovoltaic Solar Energy Conference*, Florence, Italy, 9 to 13 May 1988 (Kluwer Academic, Norwell, MA), pp. 635–641.
12. Paul Maycock, ed., PV News **7**, 1 (October 1988).
13. Paul Maycock, ed., PV News **9**, 1 (February 1990).
14. Personal Communication, Sandia National Laboratories.
15. Paul Maycock, ed., PV News **9**, 2 (March 1990).
16. G.L. Araujo and A. Marti, in *Proceedings of the 20th IEEE Photovoltaic Specialists Conference*, Las Vegas, NV, 26 to 30 September 1988 (IEEE, New York, 1989), pp. 672–678.
17. A. DeVos, J. Phys. D **13**, 839 (1980).
18. J.C.C. Fan, B.Y. Tsaur, and B.J. Palm, in *Proceedings of the 16th IEEE Photovoltaic Specialists Conference*, San Diego, CA, 27 to 30 September 1982 (IEEE, New York, 1983), pp. 692–702.
19. D.L. Bower and M. Wolf, IEEE Trans. Components, Hybrids Manufacturing Technol. **ChMT-3**, 464 (1980).
20. P. Campbell and M.A. Green, in *Proceedings of the 18th IEEE Photovoltaic Specialists Conference*, Las Vegas, NV, 21 to 25 October 1985 (IEEE, New York, 1986), pp. 39–42.
21. R.M. Swanson, in *Proceedings of the 21st Intersociety Energy Conversion Engineering Conference: Advancing Toward Technology Breakout in Energy Conversion*, San Diego, CA, August 1986 (American Chemical Society, Washington, DC, 1986), pp. 2144–2146.
22. M.A. Green *et al.*, in *Proceedings of the Eighth International E.C. Photovoltaic Solar Energy Conference*, Florence, Italy, 9 to 13 May 1988 (Kluwer Academic, Norwell, MA, 1988), pp. 164–168.
23. R.M. Swanson, in *Proceedings of the 8th International E.C. Photovoltaic Solar Energy Conference*, Florence, Italy, 9 to 13 May 1988 (Kluwer Academic, Norwell, MA, 1988), pp. 156–163.

24. P. Maycock, *Photovoltaic Technology Performance, Cost, and Market Forecast to 2000; a Strategic Technology and Market Analysis* (Photovoltaic Energy Systems, Casanova, VA, 1988).

25. U.S. Department of Energy, *Report SERI/SP-320-3382* (Solar Energy Research Institute, Golden, CO, 1988).

26. Int. Solar Energy Intell. Rep. **14**, 307 (1988).

27. A.M. Barnett *et al*, in *Proceedings of the 8th International E.C. Photovoltaic Solar Energy Conference,* Florence, Italy, 9 to 13 May 1988 (Kluwer Academic, Norwell, MA, 1988), pp. 149–156.

28. D.E. Carlson and C.R. Wronski, Appl. Phys. Lett. **28**, 671 (1976).

29. R. Pool, Science **241**, 900 (1988).

30. Environmental Protection Agency, *Policy Options for Stabilizing Global Climate,* Draft Report to Congress (February 1989).

31. U.S. Department of Energy, Office of Policy, Planning, and Analysis, *Report DOE/PE-0082* (U.S. Government Printing Office, Washington, DC, 1987).

32. H.M. Hubbard, Science **244**, 297 (1989).

33. Energy Information Administration, *Report DOE/EIA 0383 (89)* (Energy Information Administration, Washington, DC, 1989).

34. Testimony of Thomas Bath Before the United States Senate, Committee on Energy and Natural Resources, March 14, 1989.

35. *Proceedings of the 19th IEEE Phtotvoltaic Specialists Conference* (IEEE, New York, 1987).

Solar thermal energy

John G. Ingersoll

1. Introduction

Energy from the sun reaching the surface of the earth is mostly converted to heat, while a small fraction of the incident radiation, perhaps one percent, is used in photosynthesis by plants on the land and in the oceans. A fraction of that incident solar flux can be used by man to meet at least a portion of his societal energy needs. It is estimated that the rate at which solar energy arrives at the surface of the earth is about 1.72×10^{11} MW, of which two-thirds is absorbed by the surface or in the atmosphere and one-third is reflected back into space.[1] Assuming that no more than one percent of this solar input is ultimately captured to generate power for human activities, a utilization rate which will not disturb the environment, an upper limit of solar power availability of 1.1×10^9 MW can be established. This power is more than one hundred times greater than the present power generation on earth. Thus the solar input resource is immense. Nonetheless, it also happens to be intermittent and diffuse in nature.

The utilization of solar energy began at least a few thousand years ago. The ancient Greeks and Romans practiced passive solar heating of buildings. Solar evaporation of salt brines to recover the salt has been used for centuries as has solar drying of crops. In the last hundred years, periods of intense activity producing significant progress have been followed by periods of limited development. Since the early 1970s, the increasing costs of energy from conventional sources and mounting concern over environmental issues have renewed interest in the utilization of solar energy.

The design and analysis of solar processes and systems present unique challenges because the resource is intermittent in occurrence and is diffuse in intensity compared to conventional sources (fossil fuels, etc.). The present chapter summarizes the utilization of solar energy by thermal processes only; that is to say, processes in which solar radiation is absorbed by a surface and converted into heat. This heat is then stored and/or used directly for water heating, building heating and cooling, crop drying, water desalination, and electricity generation. The direct conversion of solar radiation into electricity by photovoltaic devices and biomass via photosynthesis are treated in *Photovoltaics* and *Energy from Biomass* of this volume.

2. Water heating

Dwellings need hot water for washing and other domestic purposes. Domestic hot water heating consumes close to 1% of the annual energy used in the U.S. Moreover, large volumes of hot water at low temperatures, i.e., less than 100°C, are used for process heat in industry. As a result, solar water heating constitutes an extremely important application of solar thermal energy.

The main component of a solar water heating system is a flat plate collector. This term is somewhat misleading for it is used to describe the generic geometry of the collector without attention to the detailed construction of the collector itself and the methods of transferring the heat absorbed from solar radiation to the water. Nonetheless, the term implies that both direct and diffuse radiation are absorbed in contrast to focusing collectors which make use of only the direct solar radiation. While simpler collectors hold all the water that is to be heated, the more advanced designs heat only a small portion of water at a time in order to minimize energy losses from the system. The heated water is accumulated in separate storage tanks. The majority of flat plate collectors consist of five main components:[2]

    i. a transparent cover of one or more layers of glass or plastic that transmits visible and near infrared (wavelength $<2 \times 10^{-6}$ m) and absorbs far infrared (wavelength $>2 \times 10^{-6}$ m) radiation;

   ii. an absorber plate made of a high thermal conductivity material such as a metal coated with a selective material having a high solar absorptivity and a low thermal emissivity;

  iii. tubing or channels through which the heated fluid passes and which are made of metal and are coated with the same material as the absorber plate;

  iv. a casing or container enclosing all components;

   v. insulation on the back of the absorber plate/container and sides of the container to minimize heat losses.

In general the performance of a flat plate collector can be increased by reducing first, the convective heat transfer between the absorber plate and the outer cover with the addition of extra transparent layers, and second, the radiative loss from the absorber plate with the use of selective surface coatings having a high ratio of solar absorptivity to thermal emissivity.

The heat transfer fluid (usually water) in a water heater with separate storage must move from the collector to the storage tank and back to the collector in order to transport energy. Forced circulation systems use a single speed pump, while in thermosiphon systems, circulation is driven naturally by the density difference between hot and cold water. Typically, the water circulates at about one storage volume per hour in the forced system and one storage volume per two hours in the thermosiphon system. In forced circulation systems, the pump must be activated by sensing a positive temperature difference between the surface of the collector and the water in the storage tank.

A special class of water heaters employs evacuated collectors to further reduce heat losses and attain a fluid temperature approaching 100°C. The collector of an evacuated system typically consists of two tubes, one inside the other with the fluid going through the inner tube and 1 Torr or less pressure between the inner and the outer tubes. The inner tube is made of a material of high thermal conductivity and its exterior surface is coated with a material of high absorptivity and low emissivity as described above. The exterior tube is transparent to solar radiation (visible and infrared) and opaque to thermal radiation.

The efficiency of a flat plate collector can be expressed as the product of two separate efficiencies: the efficiency of the absorber in capturing the incoming solar radiation and the efficiency of transferring heat from the absorber to the heating fluid. Uniform standards for independently testing the performance of solar water heaters have never been developed. (This is indeed a very unfortunate state of affairs for it leads to the marketing of substandard designs. It needs to be remedied, if the use of solar water heaters is to become more widespread.) However, based on the figures quoted in numerous publications, we can estimate those efficiencies for a good solar water heater design. The typical value of the absorber efficiency for capturing solar radiation lies in the range between 50% and 70%, while the heat transfer efficiency from the absorber to the fluid may be as high as 85%.[2,3] This implies an overall flat plate collector efficiency in the range of 40% to 60%.

The collector efficiency depends on the desired temperature rise between ambient and hot water as well as the available insolation. Clearly, the higher the temperature rise, the lower the efficiency as the losses increase. The total efficiency of the solar water heater will be decreased by additional heat losses of the fluid going through the pipes and being stored in the tanks. The pump power must be also taken into account. Thus, for a water temperature rise of about 20°C to 30°C and a total horizontal solar flux of over 750 W/m^2, the solar water heater efficiency may be as high as 40% (the fraction of the energy of incident solar radiation stored in the water).

Besides low performance, the other major potential problem with solar water heaters is reduced system life expectancy due to corrosion. This results from the use of mixed metals in contact with the heat transfer fluid, which is typically water, as well as the presence of dissolved oxygen in the heat transfer fluid. Solutions for the corrosion problem have been worked out over the years and must be incorporated in the design of the solar water heater.[2]

3. Heating and cooling of buildings

The space conditioning of buildings, both heating and cooling, with the use of solar energy is an old idea which has received renewed interest in the last fifteen years. In general, one can distinguish passive and active means of solar space conditioning. The distinction is primarily based on the degree to which mechanical systems and hardware are used to attain the end result and is not an absolute one. For example, the use of fans to move air is a legitimate part of a passive solar space conditioning system, although not a fundamental one. On the other hand, the use of air heating collectors and solar driven heat pumps and air conditioners are characteristic of active systems. Another way of distinguishing between passive and active means of solar space conditioning is that the former constitute, by and large, an integral part of the building design, while the latter can be viewed mostly as add-on features.

Based on cost considerations and degree of technical development, passive features are generally less costly and better understood, and therefore can more easily penetrate the new building market. Hence, we will concentrate on passive means for solar space conditioning of buildings and cover in passing some of the more recent developments in active means.

Passive means of solar space heating in buildings have been traditionally classified according to the main mechanisms for collecting and storing solar heat. The corresponding classification for passive solar space cooling is not as well developed at the present time as interest in that area is much more recent. Over the last decade several research organizations including the Solar Energy Research Institute and the Lawrence Berkeley Laboratory Building Energy Data Group have collected data on the performance of all types of solar buildings. The major conclusion is that properly designed and constructed solar buildings perform well. In addition, passive designs have outperformed active systems.[4]

A good passive solar building design starts well outside the physical boundaries of the building, a very important but frequently overlooked detail. This means that the first step in a passive solar building design is the proper subdivision and site layout of parcels of land since the objective is to incorporate passive solar designs in all buildings rather than a selected few. For maximum solar gain in the winter and minimum solar gain in the summer, a building must be constructed with an E-W axis aligned within about 30° in either direction of true E-W. Lot frontage must increase in the east-west direction and lot depth must decrease. More E-W streets will be required so that the width of each street has to be slightly reduced. For example, in a typical single family home subdivision, lots have a 28 m front, a 36 m depth and streets are 20 to 21 m wide. Lots are divided about equally between north or south entry on E-W streets and east or west entry on N-S streets. In a solar subdivision with the same density of housing (10 per hectare), streets are reduced in width to about 17 m, lot frontage is increased to 38 m, and lot depth is reduced to 27 m. Moreover, almost all lots have north or south entry on E-W axis streets.[5] Thus, buildings have an increased solar gain in the winter from higher southern exposure and a reduced solar gain in the summer from lower western exposure.

A proper solar subdivision also requires the availability of solar access to the buildings. Standard sunpath charts can be used to assess the impact of obstructions on the collection of solar energy.[6] Consideration of the legal aspects of solar access have shown that good zoning bylaws and site layout, rather than attempts to define "sunshine rights" best protect solar access.

A passive solar design attempts to maximize the amount of useful solar energy that enters the building in the winter and minimize it during the summer. The easiest way to accomplish this is to increase window area on the south facing side of the building and decrease the window area on the east and west sides of the building. South facing windows receive maximum amounts of solar energy during the winter and are easy to shade in the summer with simple overhangs. The sun elevation above the horizontal is low in the winter (e.g., noon midwinter elevation above horizontal at a 40°N latitude is 27°) and high in the summer (e.g., noon midsummer elevation at 40°N latitude is 73°). Moreover, the sun moves more slowly across the southern sky in the winter than it does in the summer. Hence, sun rays strike a vertical south facing window in the winter at a more normal angle for relatively longer periods of time, so that more solar radiation gets through the window than when the sun is high in the sky. East and west windows should be avoided, because they are ineffective in the winter and receive maximum solar radiation in the summer, when it is not needed.

In addition to standard windows, other types of glazed area such as the clerestory and the skylight, may be used in passive solar buildings. The clerestory is a wide strip of windows projecting up above the roof line, while the skylight is an opening in the roof. Both are traditionally used to admit light but both can be properly designed, particularly the clerestory, to also admit winter heat in the sunless part of the building.

A passive solar building requires the presence of added thermal storage or thermal mass, as it is commonly called, in order to be able to utilize effectively the solar heat gains in the winter and minimize their impact in the summer. *Thermal mass* refers to the ability of materials to store heat either by raising their temperature (sensible) or by changing their phase (latent). Every building has thermal mass represented by its construction materials. For example, a one story insulated timber-framed house with a 180 m^2 floor area and lined on the inside with a 15 mm thick dry wall (walls and ceiling) has a heat capacity of about 10 MJ/°C.[7]

The temperature of a building's interior changes with some delay due to its thermal mass and does not follow the pronounced diurnal ambient temperature variability experienced both in the winter and the summer. In the winter, excess solar heat is stored in the thermal mass during the day time and is released at night to keep the building interior warm. In the summer, the unnecessary solar heat is likewise absorbed during the day time so as to keep the building interior at a lower temperature and then released at night and exhausted to the outside. Obviously, attaining the proper time delays to maintain comfortable interior temperatures with a minimum use of auxiliary heating and cooling is the essence of a passive solar building design.

The major questions regarding thermal mass are:
i. how much is it needed?
ii. where should it be placed? and
iii. what materials should be used?

Extensive research in the last fifteen years has provided answers to the above questions, although some uncertainty still remains. In general, a building with window area in excess of 7% of the floor area requires extra thermal mass beyond that of a conventional building.[8] Traditionally, designers assumed that sunlight must fall directly on the thermal mass for heating purposes. Moreover, the surface area of the thermal mass, if not directly exposed to the sun, ought to be about four times as much as when directly exposed.[9,10] More recent research indicates that direct sunlight on thermal mass has no significant effect on long-term thermal storage (i.e., after a day or so).

The accessibility of thermal mass in terms of heat transfer has proved more important than direct light on thermal mass.[8] In other words, a thin masonry wall or floor of large surface area is much more effective in eliminating interior temperature fluctuations in the winter and the summer than a thick wall or floor with the same volume.[7] Thus, ideal designs distribute thin layers of thermal mass over the entire interior surface of a passive solar building. For example, increasing the thickness of the drywall or plaster is an ideal technique.

Typical materials used to increase thermal mass include concrete, brick, rock, and water. The thermal capacity of the first three is on the order of 1500 $kJ/m^3°C$, while that of water is about 4200 $kJ/m^3°C$. Most common building materials have a heat capacity similar to that of masonry products. Iron, iron oxides, and steel constitute an exception having a thermal capacity over 3900 $kJ/m^3°C$. A number of materials which can store heat by changing phase have become commercially available in the last few years. They are normally packaged in translucent pods, rods, steel cans, and plastic trays or pouches. One particularly appropriate material for heating and to some degree for cooling is specially formulated calcium chloride hexahydrate with a phase change at 27°C and a latent heat capacity of about 270 MJ/m^3. The manufacturer claims that the life expectancy of the material is several thousand cycles. (In the past, a common failure of hydrated salts has been separation of phases after many cycles.) For a temperature rise of 10°C, even water has a heat content per volume of an order of magnitude less than that of a hydrated salt phase change material. Consequently, it may be overall cost effective to use the more expensive phase change materials in a passive solar building rather than water which requires an expensive structure to contain it.

The generic designs of passive solar heated buildings are traditionally divided into the following six categories: direct gain; attached sunspace; convective loops; Trombe (mass) or water walls; roof ponds; and roof space collectors. These can be combined to produce a variety of alternative designs.

Of the various passive solar building designs, the direct gain solar building appears to be the preferable design because of a relatively lower cost and a better understood performance that can be readily modified to accommodate local climatic conditions. A direct gain building has a larger than usual south-facing glazing and has extra thermal mass to store the enhanced solar gains. Thus, window area may vary between 15% and 25% of floor area depending on the type of climate (hot or cold, respectively). The thermal mass may be as much as 40 MJ/°C and is distributed over all the walls, ceiling, and portion of the floors in layers of no more than 6 cm thickness (using mostly extra thick dry wall). The building envelope is very well insulated with average wall and ceiling thermal transmittance values as low as 0.1 $W/m^2°C$, in the colder climates, and an average infiltration rate of about 0.2 air changes per hour.

Since the excess window area is likely to result in significant heat losses during a winter night, movable insulation is necessary in most climates in addition to multipane glazing. Moreover, the windows must also be shaded during a summer day to avoid building overheating. Unfortunately, no single method of either movable insulation or shading has emerged as convenient and economic for all situations. Recent advances in the development of superglazings that include heat mirror and electrochromic windows appear more promising. Heat mirror windows are designed to transmit solar radiation and reflect thermal radiation instead of absorbing it as ordinary glass does. These windows are typically double-paned and make use of low thermal emissivity films or coatings applied on the surfaces between the glass panes. Electrochromic windows consist of an optically variable layer in the glass which is activated by an external voltage (with a power consumption of less than 1 mW/cm^2) to reduce transmission of solar radiation. A combination of these two types of glazing can have a revolutionary impact on the widespread adoption of direct gain and other passive solar building designs. However, certain technical details regarding the longevity of such windows remain to be solved and the cost must also be reduced.

A sunspace, appearing under a variety of names such as solarium or greenhouse or atrium, can sometimes be combined with, or be used to modify, a direct gain building. The sunspace is not conditioned but is used to temper the main building in the cold winter nights and hot summer days. The utilization of an atrium becomes a necessity, however, in commercial passive solar buildings where a direct gain design is generally not suitable.

The remaining passive solar designs do not appear to have a universal applicability for different technical reasons in addition to higher costs. The roof pond cannot be used in climates with subfreezing temperatures. A mass wall, except in mild climates, does not utilize solar energy very effectively, because the hot outer wall surface radiates much of the heat collected back out through the glazing. Convective loops with or without associated rock thermal storage bear similarities to the thermosiphoning designs for heating domestic hot water and share many of the problems inherent in active collectors. These include high temperatures in the event of stagnation as well as high absorber plate temperatures resulting from poor heat transfer to the air and thus low efficiency. Moreover, the collector is an add-on piece of hardware. Of course the principle of convective loops can be applied without formal acknowledgement to many situations such as transferring heat from a greenhouse into an adjacent living space. Finally, the roof-space collector that consists of a south facing glazed roof with air flowing under it has not been tested adequately. There are practical concerns with leakage and roof timber damage due to high temperatures and cycling. In addition, it should be noted that the major drawbacks of a direct gain design may be excessive glare on a bright day as well as danger of overheating, but both of these can be diminished with proper design.

Passive solar cooling of buildings comprises the following four generic designs: landscaping, ventilation, roof radiators, and earth tubes. Landscaping changes the cooling load of the building in two different ways. First landscape features like trees can directly shade the building. The selection of plants and their placement with respect to the building is very critical in order to accomplish the desired shading effect in the summer but not obstruct the sun in the winter. Secondly, plant evapotranspiration directly reduces the ambient temperature. Vegetation tends to lower the air temperature by a few to several degrees centigrade while potentially increasing humidity, as studies at Lawrence Berkeley Laboratory have shown in the last few years. Quantifying the impact of landscaping on the cooling load of buildings is, however, a very involved task and will require additional research.

Ventilation of buildings occurs predominantly at night. The placement of the windows and interior partitions of the building must be such as to allow natural air flow from the outside to the inside and vice versa. Fans can be used to assist this air flow. In the summer ventilation cooling at night uses the lower ambient air temperature to reduce the thermal mass temperature established earlier in the hot day. Also during the day, appropriate ventilation can capture prevailing winds and direct them through portions of the building.

The remaining two generic designs, i.e., roof radiators and earth cooling tubes are at best in a research stage of development. Roof radiators consist of high thermal conductivity and high thermal emissivity flat plate collectors that cool down below the ambient air temperature at night by radiation to the clear sky. Earth cooling tubes consist of plastic tubes buried one to two feet below the soil surface where the temperature during the summer is such that ambient air passing through the tubes loses part of its moisture by condensation. Not enough is currently known about these designs to enable us to use them effectively.

Several active heating and cooling systems have been tried in the past, but none has proved successful either because of performance limitations or high cost or both. One recent effort involves solar-fired desiccant cooling systems. These systems require regeneration temperatures of about 110°C, but apparently use of lower temperatures of 70 to 90°C, more compatible with low-temperature solar collectors, incurs only a small penalty in performance with proper design of the silica gel dehumidifier.[11] Numerous other research efforts in the area of solar assisted space heating and cooling are taking place around the country. Practical designs and products, however, lie several years away at best.

4. Thermal electric power

Solar radiation can produce high enough temperatures to operate a heat engine which can generate electricity with reasonable efficiency. In order to attain the required temperature range, collectors must use concentrators. Depending on the concentrating method of the collectors, we can obtain intermediate temperatures (100°C to 250°C) or high temperatures (over 250°C). For the practical generation of electricity, i.e., with reasonable conversion efficiencies, we need high temperatures. Although there is no strict definition of what constitutes a reasonable conversion efficiency, we can assume that it has to be over ten percent from incident solar radiation to the generator electricity output. In this section we will consider the various concentrating collector systems and types of heat engines appropriate for the task. We will emphasize designs that are proven viable alternatives in electricity generation at the present time.

4a. Solar concentrators

A concentrating collector includes a receiver, where the incident solar radiation is absorbed, and a concentrator, which is the optical system that directs the beam portion of the incident radiation onto the receiver. In general, the concentrator moves continuously tracking the trajectory of the sun from sunrise to sunset so as to intercept the maximum possible radiation. The aperture of the concentrator is its projected area in a plane perpendicular to the direction of the beam (direct) radiation. The concentration ratio of the collector is defined as the ratio of the area of the aperture of the concentrator to the area of the receiver. The concentration ratio ideally equals the ratio of the solar flux density at the receiver to that at the collector. In reality, however, the flux density is not constant across the receiver of a practical system. Moreover, the temperature of the receiver cannot be increased indefinitely by increasing the concentration ratio. In fact, the receiver temperature cannot exceed the equivalent black body temperature of the sun in accordance with Kirchhoff's law. In addition the sunrays subtend a finite half-angle of 0.25° at the surface of the earth. Thus, the maximum achievable concentration ratio is limited to about 43,000 (the square of the ratio of the sun-earth distance to the sun's radius).

Two typical concentrator designs are the parabolic trough and the parabolic bowl.[12,13] The former consists of a parabolic mirror with the receiver running parallel to its axis through the parabola focus so it concentrates in only one direction. Tracking is relatively simple as the mirror rotates about an east-west axis following the sun's elevation. Maximum concentration in one dimension cannot exceed about 200 with a corresponding maximum receiver temperature of about 1200°C. Practically obtainable temperatures are less than that for three main reasons:

  i. practical troughs are not perfectly parabolic so that the solar image subtends an angle greater than 0.25°;
 ii. heat is removed from the receiver by passing a fluid through it;
iii. some heat is lost to the environment by conduction/convection even if the receiver is properly shielded.

A good practical parabolic trough may attain a maximum receiver temperature of 700°C, corresponding to a concentration ratio of about 150. This maximum temperature assumes negligible conductive/convective losses from the receiver (e.g., an evacuated tube down to 10^{-2} Torr surrounding the receiver tube).

A parabolic bowl concentrator requires a much more complicated tracking arrangement as it follows both the sun's tilt (elevation) and azimuth. The concentrator itself is shaped like a paraboloid of revolution. The sun, considered as a black body, has a surface temperature of about 5800 K. Even discarding any convection/radiation losses, the temperature of the receiver of a parabolic bowl concentrator is unlikely to exceed 4200 K. At the present time, imperfections in the tracking and the shape of the mirror limit realistic upper receiver temperatures to the order of 3500 K. Even then, the size of the high-temperature area is exceedingly small to have practical value for power generation.

Collectors with concentrators can attain temperatures in the range of 300 to 700°C or even higher. Thus they can be used to operate a heat engine which generates electricity with reasonable efficiency. It is impractical from an engineering point of view to build large tracking collectors with apertures in excess of about 500 m². Even a collector of that size would receive a peak thermal power that is obviously too small for use by a typical centralized utility unless several such collectors are used together. Multiple collectors must be configured as a solar thermal-electric power station large enough to be considered "central" from the utility point of view. Based on experience gained by previous developments, we can assume that a typical central station size will range from a minimum of a few MW of electric output upwards to a maximum an order of magnitude larger. Two approaches have been investigated so far: 1. the distributed collector system; and 2. the central tower system.

The *distributed collector system* consists of a large number of small individual collectors, typically parabolic troughs rather than parabolic bowls. It is easier to collect the heat from the receiver of a parabolic trough and transport it to a central heat engine. Moreover, parabolic troughs require a less complex, hence a less expensive, tracking mechanism and have an adequate concentration ratio to generate high enough receiver fluid temperatures (i.e., well over 300°C) for an efficient heat engine operation. In order to reduce end-effects at off-noon hours, it is necessary to string a number of collectors in line so that the length of the array is several times the focal length. This arrangement makes it difficult to mount the collector string in any other way except to have its axis horizontal.

A horizontal axis is best suited to orientation with the axis in the east-west direction. However, in this case there is a loss of performance at off-noon hours because of the large angle between the direction of the sun's rays and the normal to the aperture. Nonetheless, the daily output remains fairly constant throughout the year. If the axis lies in the· north-south direction, the performance is strongly enhanced during the summer months because the array aperture surface normal is not far from being parallel to the direction of the sun rays at moderate latitudes (35 to 45°N) in midsummer. The performance of a north-south horizontal axis parabolic trough falls below that of the east-west axis in the winter months. For example, the performance of an east-west horizontal axis parabolic trough at 40°N latitude on a clear day is 2646 W/m² on December 21, 2811 W/m² on March 21 or September 21, and 2645 W/m² on June 21.[13] The corresponding performance of a parabolic trough collector with a north-south axis but identical otherwise is 840 W/m² on December 21, 2738 W/m² on March 21 or September 21, and 4280 W/m² on June 21.[13] Thus, if a strong summer peaking load exists, a north-south axis of orientation is appropriate. Otherwise, an east-west orientation axis has a more desirable annual performance.

Besides the added complexity of tracking in two directions, bowl concentrators have more severe heat transfer losses due to their higher operating temperatures. In a distributed collector field, heat transfer losses can be avoided by mounting a heat engine (e.g., Stirling or Brayton cycle) at the receiver. This eliminates an extended network of pipes, but integration of the electrical output may be a problem. An alternative solution uses a single very large dish in a fixed configuration with a tracking receiver. However, it has never been demonstrated that such approaches have viable performance and cost characteristics.[2]

A *central power tower* system consists of a large field of fixed mirrors with adjustable orientation which reflect solar radiation to a receiver located atop a tower. A power station is located on the ground directly under the tower or adjacent to the field. The heat transfer fluid is pumped up to the receiver where it is heated. The hot fluid returns below to the heat engine for conversion to electricity. Each of the fixed mirrors is called a heliostat. The heliostats must be capable of rotation about two perpendicular axes. The reflective area of each heliostat is typically in the range of 40 to 70 m^2. Some designs use metallized plastic reflectors encased in a plastic dome for protection from the weather. The heliostats are steered by computer-controlled servo-motors according to stored geometrical algorithms with periodic checking to ensure accuracy.

The field layout requires specification of tower position and height, the field shape, and the density of heliostats. In general, the higher the tower the more horizontal the mirrors can be, thereby reducing losses. However, the spreading of the reflected beam becomes worse. For a power plant of several MW, a tower height of about 100 m is deemed appropriate, while tower heights of over 350 m are envisioned for large plants in excess of 100 MW. For locations in the region of 30 to 40°N latitude, the power tower is located near the south end of the field because the prevailing angles of solar incidence lie mainly in the northern sky. This allows the mirrors to remain, on the average, more perpendicular to the direction of the sun's rays than if the tower were located at the center of the field. The fill factor, or fraction of the field covered with mirrors, appears to have an optimum in the 40% to 50% range.

The design of the receiver is very critical and is still evolving. In general receivers are classified as external, if there is an external network of piping containing the heat transfer fluid, and as internal, if a cavity is used to trap radiation onto an internal network of tubing for heat transfer. Whether the receiver is external or internal, it must be such as to absorb a large fraction of the incident radiation. A novel idea in the case of the internal receiver is the utilization of air containing carbon dust particles as the heat transfer fluid so that over 90% of the incident radiation is trapped.[14] The hot air with temperatures as high as 1400°C can then be used to drive a heat engine such as a Brayton-Joule cycle turbine where the air is introduced into the receiver at about 600°C.

Besides the hot-air Brayton-Joule cycle engine, another power cycle considered for the central tower system is a Rankine steam cycle. In this case, the heat transfer fluid can be either sodium, heated to 600°C and then used to produce 500°C superheated steam at 13.8 MPa (137 atm) by using a heat exchanger, or water, boiled at 10 MPa and superheated to 500°C at the receiver. The central receiver system is ideal for high-temperature operations in excess of 550°C. It is also possible to use a two-stage system, where the heat transfer fluid is preheated by a less sophisticated collector system and then its temperature is boosted at the tower receiver.[12] Typically, the simpler system may supply 40%–45% of the total heat energy input, and the remainder is supplied by the central power tower system.

4b. Heat engines

As the name implies, a heat engine is a device that converts heat to mechanical work and ultimately electricity. For the last 100 years, the standard heat engine used in the generation of electricity has been a steam turbine. A turbine provides a very efficient means of converting the internal energy of the hot working fluid, such as steam, into rotational energy of a shaft. The shaft is in turn coupled to a generator which produces electricity. The increased internal energy of the turbine's working fluid is supplied by either burning fossil fuels or by the heat generated in a nuclear reactor or, in the present case, by the incident solar radiation. The boiler of a conventional steam power plant is replaced by the receiver of the solar concentrating collector.

The working fluid enters and leaves the turbine in a steady flow with an amount of mechanical work being produced. The motion of the fluid in a turbine is rather complex. For water or an organic fluid as the working fluid, the process can be described in an idealized fashion by the *Rankine cycle*. This is a closed cycle and entails the phase change of the working fluid in order to convert heat to work. Solar energy absorbed at the receiver of the concentrating collector delivers heat at a high temperature to the liquid fluid passing through the receiver. The fluid, which is pumped-up to a high pressure prior to entering the receiver, is vaporized at the prevailing high pressure. The high-pressure vapor is used to drive the turbine. The exhaust from the turbine, usually consisting of a mixture of low-pressure vapor and liquid droplets, is condensed to a liquid by heat removal at a cooling tower. This liquid is then pumped-up in pressure before it enters the receiver to repeat the cycle.

The efficiency of the Rankine cycle heat engine in converting heat to mechanical work is close to the corresponding Carnot cycle efficiency operating between the same heat source and heat sink temperatures. For example, the Rankine cycle efficiency using water as the working fluid is about 0.33 for a 230°C heat source, a 38°C heat sink, and a fluid pressure of 2900 kPa. The corresponding Carnot cycle efficiency is 0.385.

One can improve the efficiency of the Rankine cycle by injecting heat at as high a temperature range as possible.[15] One method of adding heat at a higher temperature is to superheat the steam at constant pressure above the boiling point at that pressure. The reheat cycle employs a multistage turbine to superheat the exhaust from each successive stage at a reduced pressure before it enters the next stage. The regenerative cycle also uses a multistage turbine, but in this case some of the exhaust from a higher pressure stage preheats the condensate returned to the boiler or receiver. Thus, the regenerative cycle increases the temperature at which heat is added to the working fluid at the boiler or receiver. Of the aforementioned three methods, the last one is best suited for solar systems.

In addition to water, solar systems can operate with organic fluids. The pressure of vapor entering the turbine should be at least 700 kPa and preferably greater than 2100 kPa. To attain those pressures with water, one needs temperatures in the range of 150 to 200°C. Many organic fluids have a much higher vapor pressure than water in the 75 to 175°C temperature range. Hence, organic fluids are more suitable for lower temperature cycles. However, several practical disadvantages are associated with organic fluids. These include a relatively high cost, flammability, and sometimes toxicity. Moreover, the technical disadvantages of organic fluids are also significant. Because their heats of vaporization are lower than water, a larger mass flow rate is required for a given heat input. This results in increased parasitic power losses for pumping that can be as high as 20% of the generated mechanical shaft power. In comparison, the corresponding loss for water is 2% to 5%.

A number of alternatives to the Rankine cycle exist, but apparently the two most appropriate cycles for solar energy are the Stirling and Brayton-Joule cycles.[13,15] In the *Stirling cycle* an exterior combustion chamber or source furnishes heat to the engine. In the case of solar energy, either the receiver acts as the engine heat input surface or heat is stored and then transferred to the engine by a fluid. The Stirling cycle has a potentially higher efficiency than the Rankine cycle for the same maximum temperature. This is because, in a Stirling cycle, all the heat is delivered at the high temperature and rejected at the low temperature, as in a Carnot cycle. In practice, Stirling engines have difficulty obtaining good heat transfer between working gas in the engine (air or helium) and the engine heat input surface (exterior cylinder walls).

The *Brayton cycle* basically duplicates the cycle used in a conventional jet aircraft engine except that in a solar application the cycle is closed. Air is heated and compressed essentially by solar energy. The air then expands through a turbine to convert its internal heat energy to mechanical work. The Brayton cycle is most efficient for temperatures in excess of 500°C. The gas used as the working substance is compressed by a compressor mounted on the same shaft as the turbine and the electrical generator. Solar radiation at the receiver of a concentrating collector then heats the compressed gas to a high temperature. The hot compressed gas expands through the turbine thereby producing enough work to drive the compressor and the generator. The turbine exhaust air passes through a heat rejection unit and is recycled back to the compressor.

In practice the Brayton cycle undergoes two modifications.[13,15] First, a regenerator (heat exchanger) is introduced between the turbine exhaust and the heat rejection unit for preheating the compressed gas prior to the main heating step. Second, the hot exhaust gas from the turbine supplies heat to a Rankine cycle system. For example, assuming air as the working gas, a maximum cycle temperature of 825°C, a compression ratio 5:1, and an inlet (environment) air temperature of 33°C, we can calculate a 36% gas turbine efficiency vs a Carnot efficiency of 72%. Introduction of a regenerator boosts the efficiency of the turbine under the same operating conditions to 57%. Similarly, introducing a Rankine steam cycle at the exhaust of the gas turbine results in a combined system efficiency of over 50%.

4c. Practical systems

Examples of the distributed collector system and the central power tower system have been built and operated for some time. After extensive studies funded by the U.S. Department of Energy, the central power tower system was selected for construction as a demonstration project more than ten years ago. A prototype plant was built and is still operating near Barstow, California. The plant has a 10 MW electric power output and consists of 1818 plane mirrors, each 7 m by 7 m, and a central tower 101 m tall.[3] The reflected radiation is concentrated on the raised boiler atop the tower. Southern California Edison purchases the electricity generated. The cost of the plant, totally subsidized by the government, was $10,000 per kW electric. It was expected that the cost per kW would be significantly reduced if several units were to be built. The U.S DOE was initially planning the construction of a 200 MWe plant following the completion of the 10 MWe plant. However, the unexpectedly high cost of the first demonstration unit resulted in the abandonment of any future construction. Based on the experience of the 10 MWe plant, solar thermal electric power generation was judged prohibitively expensive compared with electricity generation using nuclear fission or coal, notwithstanding the environmental costs associated with both of these.

The introduction of a distributed collector system by Luz International, Ltd. has reversed the bleak outlook for solar thermal electricity generation. In 1978 Luz began research and development of a line focus parabolic trough concentrator.[16] The objective was to develop a superior solar thermal electricity generation system in terms of efficiency and cost. After several design iterations and exhaustive testing, both in Israel and the United States, the first commercial unit with a 30 MWe output was completed in 1986. A second unit of the same capacity was completed in 1987. Additional units have been added at a faster pace bringing the total installed capacity to 285 MWe as of January, 1990. Plans call for an additional capacity of 400 MWe over the next 5 years. The first two units were built near Kramer Junction, California on land leased by Southern California Edison, which also purchases the generated electricity.[16] More recent units have been built on leased land in the same general area of the Mojave Desert.

Each 30 MWe unit consists of 202,000 m^2 of line focus parabolic trough collectors with the axis on an east-west direction. The fundamental building block is the solar collector assembly with a total length of 52.3 m and an aperture opening of 2.45 m. Sixty-four separate mirrors make up a single solar collector assembly (SCA). The mirrors are manufactured in Germany to Luz specifications and are hot formed from white glass with low iron content to decrease the glass absorptivity. Each mirror is designed with an 81° rim angle and has a concentration ratio of 61 (which is the ratio of the aperture width to the receiver tube diameter). The receiver of each SCA consists of 16 connected segments of stainless steel pipe with a selective black chrome coating, each segment having a 3.2 m length. The entire tube assembly is encased in a Pyrex glass tube which has been treated with antireflective coating. The glass tube is plasma welded to the stainless steel pipe flanges at each end of the pipe using expansion bellows to accommodate the differential expansion between glass and steel. A reduced pressure down to 1 Torr is established in the annular space between the steel pipe and the glass tube. This low vacuum not only decreases the heat losses from the receiver, but even more importantly protects the selective coating from oxidation.

A specially developed oil by Chevron, the working fluid, is circulated from a cold tank at 245°C through the receivers and into a hot tank at 310°C. Both oil storage tanks are well insulated. This system allows for a shift of electrical generation to meet peak demand, if necessary. The hot oil is pumped to a steam generator where it produces saturated steam at 250°C and 3.8 MPa. The saturated steam is then superheated with the aid of an independent gas-fired superheater to about 415°C and fed to the turbine. After it leaves the steam generator, the oil is used to preheat the feedwater and subsequently returns to the cold tank. Each 30 MWe unit has its own turbine, which is a single line, flow impulse design operating at 60 revolutions per second. The efficiency of the Rankine cycle turbine is about 31.5%, while the efficiency of each solar collector assembly is no less than 66%.

Each 30 MWe unit, including collectors, storage tank, turbine, generator and other auxiliary facilities occupies an area close to 65 ha. It is estimated that each such unit can generate enough electricity to power 50,000 homes. These units are modular in construction to allow for a variation in size according to land, economic, or other considerations. Thus the latest added unit has an 80 MWe capacity, and according to Luz, this is the ideal unit size in terms of economics. The development of the entire Luz concept has been privately financed with no government subsidies whatsoever. The capital cost of the latest commercial unit is about $2,900 per kW electric. This is significantly less than the estimated cost of nuclear power plants of $4,000 per kWe at the present time. Moreover, the average cost of electricity for the latest Luz unit has been estimated to be $0.08 per kWh (capital cost plus operation and maintenance costs amortized over the 30 year lifetime of the unit). Consequently, the Luz design of a distributed collector solar thermal power plant is economically competitive with other conventional electricity generating systems.[17]

5. Toxic waste destruction

One of the latest practical applications of solar thermal energy is the destruction of hazardous chemical wastes. The U.S. annually generates some 290 million tons of wastes of which only 4 million tons are incinerated or recycled. The rest is stored on site, placed in landfills or dumped in waterways. A cooperative effort involving researchers from the government and academia has demonstrated the feasibility of using concentrated sunlight to destroy hazardous wastes. Moreover, this method has a number of advantages over conventional incineration techniques. Photons cause chemical reactions that destroy wastes more effectively, at lower operating temperatures, and in shorter periods of time.[18] Concentrated sunlight systems can reach temperatures in excess of 3000°C which allow destruction of some wastes not readily affected by conventional techniques.

Recent laboratory experiments have shown that the presence of photons in the concentrated sunlight system reduces the destruction temperature by about 800°C compared to conventional methods. Destruction of toxic wastes may become less capital intensive as reduced operating temperatures relieve the harsh temperature requirements on reactor materials. Furthermore, the operational costs of destroying wastes decrease as the energy required for the destruction process is supplied from the sun and there is no need for additional conventional power generation. This could help the U.S. increase the quantity of hazardous wastes destroyed every year.

6. Water desalination

The arid regions of the Southwestern U.S. obtain their water from underground aquifers or transportation from reservoirs in the north. In both instances, the supply of water is limited and may be insufficient to meet the future needs of human activities and agriculture. Since the Southwest has a very high insolation (over 2200 kWh/m^2 yr) and an extensive coastline nearby, it is reasonable to use solar energy to distill sea water. The most straightforward approach, the so-called single effect still, makes use of a shallow basin or pond with a transparent vapor-proof cover over it. The cover slopes towards a collection channel at the edges of the basin. Water evaporates from the surface of the basin, rises as vapor due to thermal convection, and condenses at the cover surface. The condensed water slides down the cover and into the collection channel. Assuming a typical insolation of 20 MJ/m^2 day and a 100% efficient solar still, we can estimate an output of about 8 kg/m^2 day. A more realistic output yield is probably on the order of 3 kg/m^2 day.[19] Clearly solar desalination requires a substantial covered area to produce enough fresh water for human consumption, let alone agricultural and other uses.

The output of the single effect still can be significantly improved by utilizing a multiple effect system. In such a system, the heat given off by condensation of the distilled water evaporates an additional mass of water. Hence some of the heat of vaporization is recovered and less solar heat input is required per unit mass of distilled water produced. A pilot multiple effect solar distillation system, designed by the Environmental Research Laboratory of the University of Arizona and built in Sonora, Mexico, produced 19 kg/m^2 day of fresh water.[19] The plant employed 1000 m^2 of simple solar collectors and five distillation stages. The collector consisted of a series of shallow ponds covered by a single layer of plastic. The multiple effect was obtained by reusing the air stream that had been dehumidified by the inlet cooling water. After a pass through the warm water in a tower evaporator, the water vapor was condensed out by the cooling water system. This dried-out air was then passed through the warm water in a lower section of the tower and was rehumidified for subsequent condensation in the cooling portion of the system. This particular configuration used effectively one-eighth of the heat of vaporization per unit mass of fresh water. The multiple effect still can also make use of a pressure staging. If the water in any stage is condensed at a higher pressure than the next stage evaporator, the heat of vaporization from the condensing water can be trans-ferred into evaporating water in the stage at a lower pressure.

In recent years a new approach to water desalination has attained commer-cial stature, albeit it has not yet been used at the scale of the traditional solar desalination systems. This new technique is based on reverse osmosis with the sea water pumped against the osmotic pressure across special membranes which prevent the flow of dissolved material. Energy is expended in the process of pumping and the latest estimate of electric input for small systems is about 0.09 kWh/kg of fresh water.[20] However, the electric energy consumption for pumping can be probably reduced substantially for a large system. In the mid 1970s, the required electricity for the membranes then available was about three times larger due to higher pumping pressure necessary to push sea water through the membrane. It is expected that new membranes will bring the elec-tricity consumption for pumping down to less than 0.04 kWh/kg. The required electricity can be supplied, for instance, by a thermal electric plant of the Luz type described in an earlier section. The economics of the reverse osmosis process has been drastically improved in recent years as the price of the mem-branes has been reduced while their longevity has increased to several thousand hours.

With regard to the cost of water desalination systems, it is important to note that in places with an annual rainfall in excess of 0.4 m it is less expensive to build a rain catching and storage system than any desalination system.

7. Conclusions and recommendations

Solar thermal energy can be used in several other applications not covered in this chapter. Two traditional applications are crop drying and low-temperature hot water generation in solar ponds. Both of these applications can be, and are being, used in this country. However, it is unlikely that either one of them, and particularly solar ponds, can be used to a great degree in the U.S. Other more recent applications still in the developmental stage include the use of high-con-centration sunlight to develop materials with unique properties by surface treat-ment with intense photon radiation and to drive a variety of lasers and similar devices.

In the more traditional applications with large potential to reduce conven-tional energy use, several developments need to occur to accelerate the market penetration of solar thermal systems. Performance and manufacturing standards must be established for solar hot water heaters analogous to other appliance and equipment standards. Certification, similar to a UL certification, should be re-quired by local and state authorities in order to allow use of any particular solar water heater. This would weed out all substandard designs and ultimately lead to lower prices through increased volume sales by fewer manufacturers.

Less expensive and more durable selective coatings will require additional research. The development of superglazings, both of the heat mirror and elec-trochromic types, is also very critical for the widespread introduction of passive solar buildings, particularly of the direct gain design which is by far the easiest and least expensive to implement on a large scale. Developers, home builders, and contractors must become familiar with the specifics of direct gain passive solar buildings for their introduction in the market. Consequently, additional research must be carried out to determine how to use public education to introduce passive solar buildings. Moreover, extensive research is needed with regard to passive, and perhaps active, cooling systems for buildings. The generation of electricity from solar heat seems to be underway. However, there is room for improving the efficiency and optimizing the Luz design configuration.

The three areas where solar thermal energy can have a significant impact in reducing our dependency on conventional energy sources are domestic hot water heating, space heating and cooling with passive systems, and thermal electricity generation. These three areas then ought to become the focus of solar thermal energy research and development.

References and notes

1. B.K. Ridley, *The Physical Environment* (Ellis Harwood Ltd., Sussex, UK, 1979).
2. J. C. McVeigh, *Sun Power: An Introduction to the Applications of Solar Energy,* 2nd ed. (Pergamon Press, Oxford, UK, 1983).
3. J. Twidell and T. Weir, *Renewable Energy Resources* (E & F Spon, London, UK, 1986).
4. M. Holz, R. Frey, R. Bishop, and J. Swisher, Solar Age **10**, 49–56 (1985).
5. J. Hix, *Subdivisions and the Sun,* Report M55 1 Z8 (Ontario Government Bookstore, Toronto, CN, 1979).
6. M. Jaffe and D. Erley, *Protecting Solar Access for Residential Development: A Guidebook for Planning Officials,* Report HUD-PDR-445 (U.S. Department of Housing and Urban Development, Washington, DC, 1979).
7. J. Litter and R. Thomas, *Design with Energy: The Conservation and Use of Energy in Buildings* (Cambridge University Press, Cambridge, UK, 1984).
8. C. Carter and J. de Villers, *Passive Solar Building Design* (Pergamon Press, New York, 1987).
9. E. Mazria, *The Passive Solar Energy Handbook* (Rodale Press, Emmaus, PA, 1979).
10. *Passive Solar Design Handbook,* J.D. Balcomb, ed. (U.S. Department of Energy, Washington, DC, 1980).
11. T. Penney, "Geometry Improves Dehumidifier Performance," *SERI Sci. and Tech. Rev.,* No. 3 (Solar Energy Research Institute, Golden, CO, 1988), p. 8.
12. A. Rabl, *Active Solar Collectors and Their Applications* (Oxford University Press, New York, 1985).
13. D. Rapp, *Solar Energy* (Prentice-Hall, Inc., Englewood Cliffs, NJ, 1981).
14. S.M. Schoenung, P. DeLaquil III, and R.J. Loyd, "Particle Suspension Heat Transfer in a Solar Central Receiver," *Proceedings, ASME-JSME-JSES Solar Energy Conference, Honolulu, HI, March 1987* (Am. Soc. Mech. Eng., New York, 1987), pp. 333–338.
15. D.W. Devins, *Energy: Its Physical Impact on the Environment* (John Wiley & Sons, New York, 1982).
16. C.H. Miller, "Solar Thermal Trough Technology: The Luz SEGS-I Project," *Proceedings, Eleventh Energy Tech. Conf., Washington, DC, March 1984* (Government Institute, Inc., Washington, DC, 1984), pp. 1453–1458.
17. J. Hogan, "The Finance of Fission," Sci. Am. **200**, 82–83 (1989).
18. B. Gupta, "Concentrated Sunlight and Its Vast Potential," *SERI Sci. and Tech. Rev.,* No. 4 (Solar Energy Research Institute, Golden, CO, 1988), pp. 2–5.
19. A.P. Meinel and M.P. Meinel, *Applied Solar Energy* (Addison Wesley Co., Reading, MA, 1977), pp. 554–555.
20. Recovery Engineering, Inc., private communication, Minneapolis, MN, 1989.

Hydroelectricity

John Dowling

1. Overview

The development[1] of hydroelectricity in the United States started in the 1880s, Niagara Falls was tapped in the 1890s, and many of the then existing hydro-mechanical sites were converted to hydroelectric facilities. Hydropower developed rapidly during the first half of the twentieth century, and by 1930 40% of U.S. electricity was supplied by hydropower. But the development of hydro (hereafter, hydro is synonymous with hydroelectric and hydropower) could not match the rapidly increasing energy demands of the 1950s and 1960s, cheap fossil fuels, and the optimistic outlook for nuclear energy. However, the oil crises of 1973 and 1979, the rising costs of new fossil fuel plants, and the battles over nuclear power sparked new interest in hydropower. In 1980 hydro generated nearly 14% of U.S. electricity[2] and accounted for 4.6% of U.S. energy consumption.

Hydroelectric facilities can be run-of-river operations, but more generally are hydroelectric dams. Run-of-river operations extract energy from running water and are mostly used to supply peaking power or to supply isolated sites which require low power outputs. A hydroelectric dam extracts energy from the release of the stored water. Hydroelectric dams can be used solely to generate electricity, as a pumped-storage facility or as a combination generator/pumped-storage facility.

As expected, there are pros and cons to hydropower. The main advantages are a clean, nonpolluting, technologically mature, renewable energy source which does not contribute to the greenhouse effect. The primary disadvantages are land use, environmental problems and high initial cost.

2. Hydroelectric potential

In December 1988 the U.S. had about 90,000 Megawatts of hydroelectric generating capacity with approximately 3,000 Megawatts more under construction.[3] Hydro production varies from season to season with the amount of water in rivers and streams. In regions where it rains heavily during the winter, capacity is greater during the winter. In regions where water forms snow during the winter, hydro capacity will be greater in the spring and summer months.

In 1988, hydro generated 223 TWhr for a total of 8.2% of U.S. electrical generation. In 1973, hydro generated 272 TWhr which was 14.6% of the U.S. electrical generation. Clearly although the electric power generated by hydro

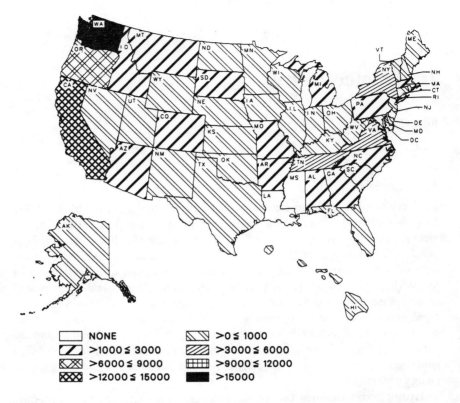

	NONE		>0 ≤ 1000
	>1000 ≤ 3000		>3000 ≤ 6000
	>6000 ≤ 9000		>9000 ≤ 12000
	>12000 ≤ 15000		>15000

Source: Energy Information Administration, Form EIA-860, "Annual Electric Generator Report."

Figure 1. Geographical distribution of hydroelectric generating capacity (Megawatts) in the United States as of December 31, 1988. See Reference 3.

has decreased only slightly from 1973 to 1988, hydro's share of electricity production has decreased dramatically. For the world, hydro showed a small gain in its share of total electrical production. In terms of primary energy consumption, hydro has consistently provided around 4% of U.S. energy.[4]

Figure 1 shows the geographical distribution of hydroelectric generating capacity in the United States. Not surprisingly, this capacity is concentrated in the wet, mountainous regions of the Pacific coast. Large-scale use of hydroelectric generation requires the presence of suitable sites along rivers with large water flows. Thus the hydroelectric generation is geographically limited. Like other renewable energy technologies, increased exploitation of hydro will depend on developing technologies for storage and electric power transfer.

A simple order of magnitude calculation gives an upper limit to the potential for energy production from hydro in the U.S. Energy available from water power is proportional to the average height (h) of a country and the average rainfall in meters (h_R). The potential energy of the water equals the product of the density of water (δ), the area covered by the water area (A), and the acceleration of gravity (g). For a back of the envelope calculation, consider that the area of the U.S. is 9.16 x 10^{12} m^2 and the density of water is 1000 kg/m^3. Assume an average height of 150 m and an average annual rainfall of 0.33 m. Then,

$$E = 150 \text{ m}(0.33 \text{ m})(1000 \text{ kg/m}^3)(9.16 \times 10^{12} \text{ m}^2)(9.8 \text{ m/s}^2)(\text{hr}/3600 \text{ s})$$
$$= 1,234 \text{ TWhr}.$$

The 223 TWhr generated annually represents roughly a 20% use of what is available. This agrees well with a 36% estimate of the hydropower exploited for North America, given by Deudney and Flavin.[5]

The power that can be developed at a site depends primarily on the rate of flow of water multiplied by the effective head, or height through which the water falls, at the turbine. This water flow and the head must both be determined to see if the site is worth developing. The amount of power generated can be calculated using

$$P = e \, \delta \, g \, dQ/dt \, H,$$

where H is the average vertical distance between the upper and lower reservoirs, dQ/dt is the flow rate (in m^3/s), δ and g as defined above, e is the system efficiency. About 85% of the potential energy stored in water can be extracted to perform work. The remaining 15% is mostly lost to friction and to turbine efficiencies.

Hydropower depends on the consistent operation of the hydrological cycle to ensure its success since flow rate in rivers and streams depends on rainfall. The severe drought of the summer of 1988 brought this home. Although the drought caused significant declines in hydroelectric production, the adequate margins of reserve capacity and the diversity of the fuel supply eliminated any serious power shortages. Droughts have relatively little impact on the electrical grid because hydropower is only 8% of the total capacity, and droughts are seldom nationwide.

Most of the logical large hydro sites have been developed in the United States. Those that have not been developed are not likely to be developed either because of the costs or for environmental reasons. But there is room for expansion: by increasing the output at many existing hydro dams, converting existing nonhydro dams to power-producing units, raising the height of existing hydro dams, and building new small hydroelectric facilities. While problems do exist for the first two cases they are relatively minor since a dam already exists and the resulting reservoir has occupied the requisite land area. Conversion to hydroelectric generation may damage local fish populations or change water flow down stream. Even raising the height of existing hydro dams may not be feasible. The existing dam may not allow for retrofitting. The new, higher dam will increase the area of the reservoir raising environmental concerns. The utility must purchase expensive waterfront property and bear the costs of relocating people, roads, utilities, and all the other facets of waterfront communities.

The development of new sites is contingent on many factors: proper precautions being taken with all the environmental considerations, surmountable relocation problems, and a favorable legislative and financial environment. Federal legislation has been enacted that (1) encouraged small-scale hydro by applying tax credits for plant development; (2) exempted small plants from some environmental regulations; and (3) forced utilities to purchase power at favorable rates to the small plant operators. Interested readers should examine the following legislation: Federal Power Act (PL 66-280), Water Resources Planning Act (PL 89-80), Wild and Scenic Rivers Act (PL 90-542), National Environmental Policy Act (PL 91-190), Clean Water Act (PL 92-500), Endangered Species Act (PL 93-205), Public Utility Regulatory Policies Act (PL 95-617), Crude Oil Windfall Profit Tax (PL 96-223), Energy Security Act (PL 96-294), and the Fish and Wildlife Coordination Act (PL 96-336). Cliffe[6] also discusses these issues in the article "Hydro: Past or Future?"

Loeb[7] estimates that if existing dams were refurbished then small-scale hydro could add 7% to the electric capacity of New England. This 7% is based on a list of about 8200 existing or former dam sites in the New England states: 300 were rated at over 2000 kW, 1400 over 50 kW, and 6500 at less than 50 kW. Total power generated would be approximately 1,500,000 kW. If new small-scale hydro dams could match this nationally, it would mean an approximate 14% increase in current hydro generating capacity or a 2% increase in electric generating capacity.

3. Mechanics of hydroelectric power

Many people associate water power with a rustic old mill and its old-fashioned water-wheel: a large diameter, slow turning device, driven by the flow of the water. Their slow turning makes them more useful for producing mechanical power for grinding and pumping than for electrical power. Waterwheels provide small amounts of electric power, but only via a complex "gearing up" to an electric generator. Most electricity is generated by specially designed hydroelectric dams, the main features of which are shown in Figure 2.

The forebay is the water storage area created by a dam. The dam regulates the flow of water and adds to the elevation of the water. This usually raises the effective head to operate the wheel or turbine, increases the supply of water available for generating power and contributes some stability in supplying electrical generating capacity.

Water from the forebay is conveyed to the turbines by a large pipe called a penstock. The type of turbine used depends on the flow rate and the hydraulic head. The turbines may be of three general types: propeller, curved vane, or bucket. The turbine drives a generator which is connected to an electrical grid system that distributes the power. The hydraulic head and the quantity of water available determine the total energy of any given hydropower site. The gross hydraulic head is the approximate difference in elevation between the headwater (the water from the forebay) and the turbine. The effective hydraulic head is the gross head with the frictional losses taken into account. The tailwater is the water leaving the turbine and exiting the plant.

Some forebays have a concrete enclosure with underwater holes which allow water to pass through, but not floating debris. Even if there is not such a structure, water from the forebay passes through screened intakes called trashracks

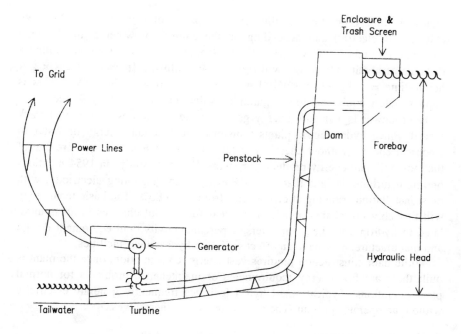

Figure 2. Pertinent features of a typical hydroelectric plant.

(sometimes automated) into the penstock. Fish screens can also be installed to help prevent fish from entering the penstock. The penstock conveys the water to the turbines, which can be several kilometers downstream, particularly if there is an advantageous drop in elevation from the forebay to the turbines. Penstocks, usually steel, can be either buried or run along the surface.

The heart of the hydroplant is the turbine (for an excellent discussion of turbines see *Energy Alternatives*[8]). Turbines are classified as either impulse or reaction turbines. In an impulse turbine a jet of water hits the cup-shaped blades at their ends, forcing them to rotate. These impulse turbines are called "Pelton wheels" and require high velocity jets associated with heads greater than 300 m. A reaction turbine is totally embedded in the water stream and is powered by the water pressure on the vanes or blades of the device. Reaction turbines can operate efficiently with 6 m heads since they do not require a high velocity jet. Some reaction turbines can adjust the pitch of the blade to accommodate variable water flow rates and different demand loads. These characteristics make reaction turbines viable in relatively flat-terrain and in some run-of-river operations.

The mechanical energy from the spinning turbine is converted to electrical power by the generator. An electrical generator relies on a basic phenomenon of electromagnetism: when a properly oriented electrical conductor is moved through a magnetic field, current flows in this conductor. Large power plants use a synchronous generator: the rotation speed of the generator is related to the frequency of the current produced. In these generators the conductor is stationary and the magnetic field is rotated. Each turbine usually has its own generator. Turbine-generators usually have a capacity of from 100 to 400 MWe (Megawatts electric) and depending on the amount of water available, may have several units installed. Typically a kilogram of water per second falling a distance of about 100 meters will produce one kilowatt [power = mass × g × height/time = 1 kg(9.8 m/s^2)(100 m)/(1 sec) = 980 watts ≈ 1 kW]. This is approximately equivalent to one gallon of water per second falling 92 feet.

Hydropower is a mature energy generation technology with a proven track record. Since hydroelectric plants transform the potential energy of water to electrical energy, without chemical or thermal transformations, they are one of the most efficient energy-producing systems. Doland's study[9] in 1954 mentions optimum turbine designs that are 94% efficient in generating electricity, with most installations running at efficiencies from 75 to 80%. The basic technology will probably not change much in the near future. But changes in areas unrelated to hydropower, i.e., computers, superconductivity, and equipment design and manufacture, may have an effect on future hydropower facilities.

While dam construction requires vast energy consumption, once the plant is built there are few energy requirements or maintenance problems for normal plant operations. Typical small hydro facilities frequently generate electricity without an operator present. The plants almost run themselves.

4. Pumped storage

A pumped-storage facility is a hydroelectric plant that uses water that has been pumped to an elevation higher than the powerhouse. Usually the water is pumped to the higher elevation by the utility company during times of low demand (late night and early morning) and then released during peak demand periods. The power to pump the water may be from a hydroelectric facility or from any conventional electrical generating facility. When power is needed the water is released and a pumped-storage plant operates much like a normal hydro plant.

The generators in a pumped-storage facility are reversible; that is, they act as motors when supplied with electricity. During periods of low demand electricity is supplied to the generator-motors to rotate the turbines (the reverse of normal operations) to pump the water to the upper storage area. During periods of high demand the water is returned to the turbines, recapturing about two-thirds of the energy used to pump the water uphill. The losses are incurred due to friction of the water in the conduits, evaporation, and seepage, and because the turbines need to be kept spinning to allow a quick response to changes in demand.

Pumped storage is currently the best method available for generating large amounts of electrical energy quickly, that is, it acts like a battery with the stored mechanical energy released as electrical energy. The largest pumped storage facility in the world is on Lake Michigan's eastern shore at Ludington, Michigan. The average head is 85 m and the six pump/turbines can generate more than 2000 MW of power. Lake Michigan is the lower reservoir, while the upper reservoir is an artificial lake. When the plant is fully discharged, it generates 15 MWhr.

At present some 2% (one-seventh of the hydro energy) of all U. S. electric energy is obtained from pumped-storage facilities. It is unlikely that this small fraction will increase by much. Expansion is limited by the lack of suitable sites near large electric-generating plants, most of which are typically located close to large metropolitan areas. Other difficulties associated with pumped storage are the environmental issues which arise including land-use problems and damage to local flora and fauna. Consequently, it has been suggested that underground caverns be used as an alternative to above ground pumped storage. In the underground cavern scheme, water is stored in a small upper reservoir and passes through the pump/turbine to the lower underground cavern, which can be as deep as one thousand meters. The higher head means a lesser volume of water is needed. The cavern-excavation technology is available as is the technology for high-lift pump turbines. While suitable formations exist in the northeastern, north central, and western United States, no underground pumped storage plant has ever been built.

The Kinzua Dam in northwestern Pennsylvania is a typical pumped storage dam which is mainly used to supply peak power to Cleveland, Ohio (240 km to the west). Kinzua has a storage pond which is about 20 m deep and 800 m in diameter, and provides a head of about 220 m. At full capacity Kinzua can generate 870 Megawatts. At Kinzua the change over from generator to motor normally involves a down time of about an hour. Five years ago, using off-peak power at Kinzua to pump water to storage cost about 2 cents/kWh, while the power generated by the stored water during peak demands was sold for 5 cents/kWh.

The combination of a nuclear power plant and a pumped-storage facility permits a nuclear plant to run at a consistent operating load. It can thereby use off-peak excess electricity to pump water which is then used to supply overload requirements during peak demands. Obviously, the combination of pumped-storage and nuclear facilities is only feasible with a substantial water supply and a large difference in elevation.

Pumped-storage facilities are not as efficient as conventional hydro plants because the generator/motor must both generate electricity and pump water. This dual purpose compromises the efficiency of the facility.[10] Pumped-storage plants operate with efficiencies between 50 and 92%, most plants running around 70%.

5. Environmental aspects

Obviously there are both pros and cons associated with a typical hydroelectric dam. The pros are a nonpolluting, renewable energy source for the generation of electricity, flood control, irrigation, and recreational use. The cons are the many relocation problems, possible failure of the dam, power lines emanating from often scenic areas, the damming of wild rivers, and the multitude of problems associated with the storage and release of the water. Local and migrating fish, in particular, die when they are drawn into the turbines of hydroelectric plants. Fish ladders have proven effective at several sites but construction of fish ladders can add as much as 50% to the cost of hydro projects.[11]

The environmental problems associated with development of hydro depend on the situation of individual projects. These pros and cons can be best examined by looking at several sites: the Salt River Project in Arizona, the Tioga-Hammond Dams in Pennsylvania, the Glen Canyon Dam in Arizona, and Falls Creek, in Ithaca, New York.

Case I : The Salt River is impounded by Roosevelt Dam, creating Roosevelt Lake, about 90 kilometers upstream of Phoenix, Arizona. This hydroelectric facility supplies electrical power to Phoenix and the surrounding area. Downstream of Roosevelt Dam are two smaller nonhydro dams which form Apache and Canyon Lakes. These dams are in an unpopulated, rugged, mountainous, and desert wilderness region near the famed Superstition Mountains. It is important to note that these dams were built well before present day concerns about the environment. These three lakes are oases in the desert, each one a mecca for fishing, swimming, boating, camping, and hiking. After leaving the lakes the bed of the Salt River winds through the desert and passes to the south of Phoenix. But only on rare occasions does water flow in the Salt River (today many streets traverse the dry river bed). The water is nearly all channeled into irrigation canals for the Phoenix environs. So the Salt River Project essentially terminates the Salt River, sets up three good-sized recreational lakes, produces electricity, and irrigates local farm lands. The population of Phoenix almost never complains about the Project. Life in Phoenix would be very different without it.

Case II : Such a complaint-free environment is not usual for hydroelectric dams. There are two general choices for a hydroplant: a site where a dam can impound the water or a site where there is a large head (often a scenic waterfall). There are almost always relocation problems with any dam. In Mansfield, Pennsylvania's sparsely populated Tioga County, two flood control/recreational use dams (not hydrodams) were planned in the late 1960s and built in the 1970s. Part of Tioga, PA was taken over, a major north-south U.S. highway relocated, several other roads rerouted, many local farms flooded (some farmed for generations), houses and businesses moved or abandoned, and utilities reestablished. This relocation was a heart-rending experience for those involved. During the time these dams were being built, over 20 families who were being relocated by the dam came to the new owner of a farm just outside the flood plane to ask what he paid for the farm, how many acres in production, whether there were animals, what improvements had been made, etc., in order to establish a fair price for the farms which they were being forced to abandon. None was happy at being evicted from a lifelong home. Any dam in a populated area has the same problem, it is just multiplied many times over.

Case III : Glen Canyon Dam in Arizona is yet a different situation. Davis[12] summarizes the problem:

> "Dam builders failed to take most of these ecological disruptions into account when they designed the structures. Most dams went up long before 1970, when the National Environmental Policy Act first required environmental impact statements for new, major water projects. Dam builders' major mistake was to assume that a dam could serve several functions simultaneously without slighting any of them."

At the construction site in 1964, the mood was excitement: jobs and money were being brought in, the Colorado was being tamed, the impounded water was providing insurance against droughts, and a huge lake for recreation was being created.

Today Davis says, "Thanks to the Glen Canyon Dam, little that lies along the Colorado River is truly natural." The dam has lowered water temperatures downstream, drowned Glen Canyon, affected the ecological systems along the river, speeded erosion of beaches, etc. Davis also cites positive changes. While some native fish have been destroyed, others, like trout, thrive, trees grow along the river's banks, etc. Glen Canyon Dam is having a major impact on the area. Whether for good or for bad depends on each individual's views.

Case IV : A final example of the problem of building a hydroelectric dam is found at Cornell University in Ithaca, New York. Falls Creek passes through the campus of Cornell and has two hydro sites: the upper one is about 600 m downstream from Beebe Lake, the lower one another 600 m further downstream. Beebe Lake is created by Triphammer Dam built in 1832 by Ezra Cornell, founder of Cornell University. Both sites were used as sources of waterpower (to turn water wheels) from the early 1800s. The upper site, owned solely by Cornell was used as a source of hydroelectricity from 1904 until 1969. By 1969 the upper site was in need of major repairs, but the cheap price of electricity did not warrant refurbishing the plant. The lower site was located just above Ithaca Falls, a popular fishing and swimming spot. An old dam above Ithaca Falls diverted water to the Ithaca Gun factory, again, for waterpower, later for hydroelectricity, and was abandoned in the late 1950s.

The energy crises of the 1970s made Cornell reevaluate the situation.[13] The university decided to refurbish the upper site and by 1981, the plant was operational. While the application to restart was pending they installed new turbines, cleaned out and rebuilt part of the penstock, installed new valves, and reworked the forebay and trash racks. The cost of the turbines and renovation was around $1,250,000. Operating costs on the fully automated plant run around $25,000/year. In 1978 the federal government's National Energy Act required states to pay reasonable rates for electricity produced by independent small-scale hydroelectric facilities. Perhaps even more significant was the Public Utilities Regulatory Policies Act which required utilities to buy power from independent power producers for the utilities' avoided cost. In 1982, the first full year of operation of the refurbished plant, New York State paid Cornell approximately 6 cents/kWh for the electricity the hydroplant generated, while Cornell was paying only 4.2 cents/kWh for electricity Cornell was using. The reason for this is the following: When you buy power you pay the average cost of all power produced. When you sell power as an independent producer, you receive the avoided cost, which is the cost to the utility of adding needed capacity or the cost of displacing the utility's most expensive existing capacity. The plant can generate up to 1.3 MW (approximately 5% of Cornell's daily use) and normally sells around $250,000 worth of electricity/year to New York State Power Authority. Thus the refurbished plant paid for itself in about five years.

The only problem Cornell had with the license application was a complaint by New York's Department of Environmental Conservation. There was a concern about whether salmon would be endangered by reduced flow of water. So flow sensors were installed to ensure sufficient flow over Triphammer dam. The hydro facility is shut down if there is insufficient flow. Normally the plant generates electricity for about two-thirds of the year.

Cornell University was interested in seeing the lower site developed and filed an application to develop the site in 1982. Problems arose and the application was held up. All that had to be done at the upper site was to replace the generators in the existing building and to repair the existing penstock. But the lower site requires a dam and a new building. Although Cornell is now no longer interested in developing the lower site, an independent developer is. In the November, 1988 election in Ithaca a referendum barely passed which proposed to reactivate the lower site. Since no one lives in this section of Falls Creek, there are no real relocation problems. However, local environmentalists actively fought against development on the grounds that fishing and swimming would be endangered, and that the gorge's scenic beauty (including the lovely Ithaca Falls) and a natural resource study area would be destroyed.[14] Currently, the project is in limbo.

Two of the environmental problems of the lower site at Falls Creek are typical of any hydro site: killing fish and maintaining sufficient water flow. Both can be alleviated somewhat. Fish screens can be placed to mimimize killing fish by diverting them to the natural river run (in some dams fish ladders are constructed so fish can find their way around the dams). Restrictions on water use would be enforced when river flow is below certain minimums. These two measures, if properly handled, would ensure minimal effects on both fishing and swimming. The effects on the flora and fauna and scenic beauty are not so open to interpretation or solution.

It should be pointed out that gross violations do occur: the Spokane River passes through Spokane and during parts of the year it is literally turned off; the water is needed to pass through the turbines. A hydro plant near Munising, MI is sited alongside a beautiful falls, and a bridge to the plant has been thrown over the middle of the falls.

The choice at Falls Creek, as for any hydroelectric site, ultimately has to be a compromise. Unfortunately, the compromise is very complex, taking into account the environmental aspects, the scenic beauty, the need for electricity, and where and how the electricity would be obtained if the site were not developed. No one who has wandered through the gorges in Ithaca would want them to be destroyed. The simple fact is that water-power was taken from both sites in Falls Creek for almost a century and a half. One site has been refurbished with minimal new effects. The remnants of the second site are there today. The present scenic beauty of the gorge was not destroyed; it coexists with the refurbished upper site and the remnants of the lower site.

Another drawback of hydroelectric dams is associated with the storage and release of the water. Storing the water generally raises its temperature. Many aquatic ecosystems flourish only within relatively small temperature ranges.

The increased heat in the released water may be enough to disrupt delicate environments. The famed Tennessee Snail Darter is only one example. The Snail Darter held up the construction of the Tellico Dam on the Tennessee river for two years because it was thought to exist only in that section of the river. Happily, the Snail Darter was found elsewhere.[15] Unfortunately, this is not always the case. Occasionally the increased temperature in the storage area serves as a habitat for ecosystems that did not exist there previously. There is also the problem of silting over of the dam. Any obstacle in a river serves to impound the silt carried off by the river. Eventually the dam fills up and reservoirs can become big silt swamps.

The use and release of the water can cause problems downstream associated with the fluctuating height of the river. People like to build houses on stream banks and care has to be taken that houses, bridges, and roads can accommodate changes in the river level.

The last problem to be dealt with is the failure of the dam. Risk estimates have to be made on the proper siting of the dam, materials used in the dam, construction of the dam, and the size of the associated reservoir. (See the chapter on risk assessment in this book.) While all these factors are fairly well understood, close attention must be paid during all phases of construction to ensure that all is done properly. This gives rise to an unquantifiable problem — the fear of dam failure by downstream inhabitants. Often these people are not reassured by risk assessments. They know that there are large dam failures almost yearly. The result is that there can be a large group of people downstream who will fight any proposed hydrosite.

6. Cost-effectiveness

Economically, hydropower has inherent advantages[16] over thermal plants. Hydropower consumes no fuel (except in the case of some pumped-storage plants) and is a renewable energy resource. The lifetime of a hydro plant is often two to three times that of a thermal plant. The efficiency is usually twice that of thermal processes. Maintenance costs are very low, and hydropower can respond quickly to increased load demands.

These inherent advantages are often offset by the fact that initial capital costs are higher for hydropower than for thermal plants. This cost disadvantage is being narrowed by (1) the sharp increases in investment costs associated with the siting and construction of fossil fuel and nuclear plants, (2) the cost increases associated with equipment and operating costs for air pollution control, and (3) rising fuel costs. However, over the lifetime of a hydro plant costs generally become economically feasible. To make hydro projects operational, developers usually make arrangements with local utilities to compensate for a project's lack of economic feasibility in the early years (called front-end loading). Utilities pay developers rates based on the replacement costs of fuel and purchased power credits. Purchased power credits allow developers to achieve positive cash flows from the start, while utilities accumulate purchase power credits to draw on once the replacement cost of fuel exceeds the cost for hydro.

In addition, economic feasibility is heavily dependent on the rates charged by utilities. For instance, the economics of developing a hydro project would be better in Central Vermont than near Niagara Falls, where hydro does not have to compete with long-established hydro. Overall, however, the economic attractiveness of hydro projects will most likely improve as the price for fossil fuels begins to increase again.

The impact of the two oil crises of 1973 and 1979 created a favorable climate to refurbish old hydro sites and to develop new ones. This was primarily due to tax credits and an Accelerated Costs Recovery System. The latter allowed hydro producers a five year depreciation instead of the standard 30 year term. This costs recovery system had the greatest impact on the small, independent producers, which is where most of the remaining hydro potential is in this country. But this interest has since diminished because both the credits and cost recovery programs were eliminated.[5]

Costs for new plants are difficult to assess because land and relocation costs are very site dependent. Here are a few estimates: Loeb[6] mentions $3000 per installed kilowatt; the Army Corps of Engineers[17] gives an average between $1000 to $2000/kW; the refurbished Falls Creek site was about $960/kW; while the Oklahoma[7] study cites $168/kW for a new dam in Oregon, from $200 to $400 for average costs per installed kilowatt, and $150 to $220 per installed kilowatt for pumped-storage facilities.

Little development of hydro sites is currently being undertaken by the electric industry. Most growth in hydro will come from the kind of small developers typified by the Falls Creek project in the case studies presented above. If small developers can overcome the initial problems of funding construction, small-scale hydro offers considerable potential as one of many local energy sources that can make substantial contributions to the U.S. energy supply.[18]

7. Summary and conclusions

To recapitulate, the advantages of hydropower are the following:
- a clean and renewable energy source;
- no fuel costs;
- long-lived efficient plants with low maintenance costs;
- mature technology with few surprises in store;
- one of the few means to store energy on a large scale;
- no mining and transport costs;
- flexibility in output;
- quick start capability eliminating the need for expensive oil or gas generators for peak demands;
- much less severe pollution problems;
- no radioactive waste;
- no harmful combustion products and their accompanying wastes;
- no contribution to the greenhouse effect.

The disadvantages are:
- the high initial costs;
- the attendant extensive relocation costs;

- the environmental aspects (large construction projects, ecological changes, fish problems, reduced water flow, etc.);
- the risk of dam failure;
- the large demands on land;
- the destruction of scenic sites.

Resistance to a new large-scale hydroelectric facility would probably be akin to that for a new nuclear facility. However, there appears to be room for expansion; for both new and refurbished old small-scale sites, enlarging present hydrodams, and the conversion of normal dams to hydrodams (this assumes the environmental problems and the relocation aspects are not insurmountable). The fact that using hydro means not using the much more hazardous, nonrenewable sources such as coal, nuclear, or oil fuels should be an important factor in the future development of hydroelectric power.

Acknowledgements

Gary Shaw, a student at Mansfield University, did original research on this paper. John Anderson and Alan Chachich did a careful reading of the initial draft and were very helpful in supplying information and in structuring the paper. John Ingersoll supplied information on pumped-storage facilities. George Hinman provided helpful comments. Henry Doney supplied information on the Falls Creek Site.

Bibliography

Energy Alternatives: A Comparative Analysis (United States Government Printing Office, Washington, DC, 1975). This is an excellent, quite readable overview. It was particularly helpful in bringing together all the concepts dealing with the mechanics of generating hydropower.

Daniel Deudney and Christopher Flavin, *Renewable Energy* (Worldwatch Institute, Norton, NY, 1983). Chapter 9 provides an overview of hydropower.

Energy Primer: Solar, Water, Wind, and Biofuels by the Portola Institute (Fricke-Parks Press, Fremont, 1974). The chapter on water gives many technical details and an overview of hydroelectric generation of electricity.

References and notes

1. U.S. Army Corps of Engineers, "Policy Issues in Hydropower Development," *National Hydroelectric Power Study* (Institute for Water Resources, Fort Belvoir, February, 1981).
2. U.S. Army Corps of Engineers, *Preliminary Inventory of Hydropower Resources* (Hydrological Engineering Center, July, 1979).
3. Energy Information Administration, *Inventory of Power Plants in the United States 1988*, DOE/EIA-0095(88) (U.S. Department of Energy, Washington, DC, August 1989).
4. See Tables 5, 6, 7 and 8 in Bodansky, of this volume.
5. Daniel Deudney and Christopher Flavin, *Renewable Energy* (Worldwatch Institute, Norton, NY, 1983), p. 168.
6. Sarah Cliffe, "Hydro: Past or Future?" Technology Review 15–18 (Aug/Sept, 1986).
7. William A. Loeb, "How Small Hydro Is Growing Big," Technology Review 51–59 (Aug/Sept, 1983).
8. Oklahoma University Science and Public Policy Program, "The Hydroelectric Resource System," *Energy Alternatives: A Comparative Analysis*, Chapter 9 (United States Government Printing Office, Washington, DC, 1975).

9. James J. Doland, *Hydro Power Engineering* (Roland Press, New York, 1954), pp. 13 & 27.
10. Federal Power Commission, *1970 National Power Survey* (U.S. Government Printing Office, Washington, DC, 1971), pp. IV-1–81.
11. Michael D. Devine, Michael A. Chartock, David A. Huettner, Elizabeth M. Gunn, Steven C. Ballard, and Thomas E. James, Jr., "Small-Scale Hydropower," *Cogeneration and Decentralized Electricity Production* (Westview Press, Boulder, CO, 1987) p. 224.
12. Tony Davis, "Managing to Keep Rivers Wild," Technology Review 27–33 (May/June, 1986).
13. D. Fixell, Cornell Alumni News, 23–25 (February, 1982).
14. The Grapevine **9**, 9 (13 Oct 1988).
15. U.S. News & World Report **6** (29 Aug 1983).
16. United States General Accounting Office, *Hydropower—An Energy Source Whose Time Has Come Again* (United States Comptroller General, Washington, DC, January 11, 1980).
17. U.S. Army Engineer, *Identification of Selected Small Scale Dams Suitable for Hydroelectric Power Development* (Institute for Water Resources, Fort Belvoir, VA, July, 1978).
18. Devine *et al.*, pp. 201–240.

Geothermal energy

Ruth H. Howes

1. Background

Geothermal energy is the energy produced from the heat trapped in the earth's core and the decay of radioactive isotopes in the crust, mantle and core. Heat is conducted to the atmosphere through the earth's crust with an average flow rate of 0.06 W/m^2 which produces a change in temperature of approximately 30°C/km in the crust.[1] This might be compared to the average solar flux for the United States which is about 200 W/m^2. The heat flux through the upper 10 kilometers of the crust differs from place to place and is influenced by such factors as the volcanic history of an area and the specific geology of the region. For example, the Wairakei thermal field in New Zealand has a heat flux of 30 W/m^2.[2] Geothermal heat sources that can be used for energy production must produce a substantial heat flux at the surface and be large enough to sustain the flux over time. A heat source, such as a magma body, can produce a sufficiently large local heat flux. The heat is transferred to a large volume of hot rocks lying above the heat source, called a *reservoir*, and is trapped there by a cap of rocks which are both insulating and impermeable to fluids so that heat is stored in the rocks and their temperature rises. Finally the reservoir must contain fluid heated by the rocks which serves as a medium for the transmission of heat to the surface once the cap rocks have been pierced.

Local concentrations of geothermal energy have long been exploited. Bathing in hot springs dates back into prehistory. In the United States, the first downhole heat exchanger for a domestic heating system was installed in 1931 at Klamath Falls, Oregon — and functioned for 25 years.[3] Large-scale, modern geothermal space heating projects began in Reykjavik, Iceland in 1930.[4] Despite these early and successful local uses, by the end of 1985 the global use of geothermal energy had only the capacity to generate 4500 MWe of electric power and provide 7500 MWt of thermal power for space heating and other direct applications.[5]

Research on technologies for exploiting geothermal energy since 1973 has defined at least three major areas where work is needed: locating geothermal fields suitable for exploitation and increasing the lifetime of productive geothermal reservoirs to commercially acceptable lengths; producing materials and instrumentation able to function in the hostile environments of geothermal hot brines; and developing technology to exploit the energy trapped in hot dry rocks to supplement the geographically limited number of naturally occurring geothermal reservoirs which contain trapped fluids.

This review will briefly describe existing geothermal reservoirs and the technologies used to exploit them for the generation of electricity and for direct heating applications. Various approaches to the engineering problems of maintaining serviceable reservoirs, including reinjection of cold geothermal fluid, and the multiple materials problems encountered will be presented, as will environmental problems in the exploitation of geothermal energy. Finally, attempts to create artificial geothermal reservoirs by injecting water into hot dry rocks close to the surface will be discussed.

2. Characteristics of geothermal reservoirs

Geothermal reservoirs vary widely in the ease with which their energy can be extracted. The easiest to mine are superheated, vapor dominated reservoirs where the heat energy of the reservoir is released by discharge of pressurized steam that can be used directly to drive turbines, as at Larderello, Italy and The Geysers, California. More common reservoirs release mixtures of steam and water. Finally, hot water reservoirs release energy in the form of liquid water under pressure. If the temperature of the reservoir is above the boiling point of water at atmospheric pressure, the water will "flash" into steam as pressure on it is reduced.

The value of a geothermal reservoir is measured by the heat energy stored in it as well as in the fluid that releases heat for exploitation at the surface. By calculating the heat balance for The Geysers steam field, Economides[6] clearly demonstrates that the heat stored in the reservoir cannot be replaced by natural heat flow nearly as quickly as it can be withdrawn by production of geothermal electricity. Wells at The Geysers drain a surface area of 11 acres or 45,000 m^2. On the basis of geological evidence, it is reasonable to assume that the rock surface area drained is ten times the reservoir's above ground surface area. Thus the effective area of the reservoir is 453,640 m^2. The steam from The Geysers produces 18×10^7 watts of heat or a reservoir outward heat flux of 3.9 W/m^2.

The conduction of heat into a reservoir from the magma body or other source is governed by Fourier's law of heat conduction which states that the heat flow per unit area (Q/A), the heat flux, is proportional to the temperature gradient (dT/dx) between the source and the reservoir:

$$Q/A = -k \, dT/dx,$$

where k is the conductivity of the reservoir rocks.

Since untapped reservoirs have come to nearly stable equilibrium over geologic time, the heat flow into the reservoir is balanced by the heat flow out through the cap rock. If enough heat is withdrawn from the reservoir to reduce its temperature, the heat flow into the reservoir should increase. The net heat flow into a reservoir is the difference between the heat flow in and the heat flow out. Assuming temperatures of the magma body and the reservoir characteristic of The Geysers, Economides calculates that the net heat flux from the heat source into the reservoir should increase by 46 percent of the original heat flux as the reservoir temperature decreases by 100°F (56°C). For The Geysers reservoir, the heat flux into the reservoir is calculated to be 0.04 W/m^2 at equilibrium. Thus, should the temperature of The Geysers reservoir fall by 56°C due to production, heat flow into the reservoir could supply only about 0.5 percent of the 3.9 W/m^2 removed by production. Clearly geothermal reservoirs are not renewable in short periods of time.

Geothermal reservoirs are essentially finite, exhaustible resources that are used up over time. When pressure and temperature in the reservoir drop below certain limits, the reservoir is no longer useful for heat production. One of the major problems in assessing geothermal resources has been to assess their probable productive lifetimes. For example the Ahuachapan geothermal field in El Salvador, which has been exploited for electric power production since 1975, has shown unexpected depletion in spite of the reinjection of geothermal fluid. Current methods for evaluating reservoirs can only estimate their useful lifetimes.

The fluid in the reservoir must be able to travel through the rocks of the reservoir sufficiently easily so that geothermal wells do not run dry quickly. The fluids themselves contain a variety of suspended solids, dissolved gases, and dissolved minerals. During production, the hot thermal brines may attack drilling and production equipment. As they cool and are placed under reduced pressure, the gases and minerals in the brines come out of solution and produce scaling on equipment and in the pores of the rocks feeding the geothermal well. The exact problems depend on the specific chemical composition of the fluid at a particular site, which varies widely. The chemical behavior of hot brines is complex and poorly understood, but it is important to consider the physical and chemical characteristics of the geothermal fluid itself in assessing the value of a geothermal reservoir.

The first step in exploiting sources of geothermal energy is to survey the region of interest and to characterize it according to the extent of the geothermal resources, the chemical and physical characteristics of the fluid that can be used to extract the heat, the mobility of the fluid in the rocks of the reservoir, and the enthalpy of the reservoir as a unit. Because the original source of heat for a geothermal reservoir must be a temperature anomaly in the earth's crust, geothermal resources are generally distributed in regions of geologically recent volcanic activity. They follow the ring of volcanic activity around the Pacific Ocean, the mid-Atlantic ridge, along the Mediterranean, across the Alpine-Himalayan Range, down the east coast of Africa, and in the interior of Asia. The tectonic activity that has led to the presence of geothermal reservoirs frequently makes these areas physically rugged, and many of them are not industrially developed. The widespread use of geothermal energy has been limited by the fact that its sources are far from the people who need it.

Hot geothermal fluids cannot be transported hundreds of miles from their sources so use of geothermal energy to provide direct heat is limited to applications close to the source of the fluid. On the other hand, electric power can be transported over long distances with acceptable losses, but its production requires higher enthalpy geothermal resources than does the exploitation of geothermal resources for direct heat. Utilization of geothermal resources divides itself naturally into electric power production and direct utilization of the resource.

3. Electric power production from geothermal energy

In all geothermal electric power production plants, hot gases produced by the heat of the geothermal fluid are used to turn a turbine which in turn drives a generator for the production of electric energy. Figure 1 shows simplified schematics of the major cycles used. In the simplest case, dry steam is extracted under pressure from a geothermal well and used directly to drive a turbine.[8] Dry steam fields are in successful commercial use at The Geysers in California and Larderello, Italy. After delivering power to the turbine, the steam is then condensed isothermally. *Enthalpy*, which is defined as the internal energy of the vapor plus the product of its pressure and volume, is traditionally used to describe geothermal reservoirs since it measures the phase of the geothermal fluid as well as its internal energy.

Dry steam fields are easy to exploit for the production of electricity but are unfortunately rare. Far more common geothermal reservoirs contain pressurized liquids or liquid-steam mixtures. For pure pressurized liquid systems, the high-temperature, high-pressure geothermal fluid is flashed into steam by reducing the pressure on it. The flashed steam may then be used to drive a turbine, as in the case of the dry steam field. The power which can be delivered to the turbine depends on the temperature at which the geothermal liquid is flashed.

The actual temperature of the fluid may be well above the optimal flashing temperature which results in a waste of power several times the power actually delivered to the turbine. To avoid this waste, working plants generally use multiple flash systems where the vapor of the first flash, conducted under pressure, is fed to the turbine system and the pressurized liquid residue is then flashed again at lower temperature and drives a second stage turbine. For geothermal systems which produce mixtures of vapor and liquid, the pressurized vapor is separated from the liquid and used to drive the first stage turbine while the residual liquid is flashed and used to drive the second stage turbine. Because of the wider availability of two phase and pressurized liquid reservoirs, flashed cycles are used for practical power production at many sites in the United States, the Philippines, Mexico, Italy, Japan, New Zealand, El Salvador, and Kenya, among others. The advantages of flashed cycles over direct steam cycles are that the flashed steam is usually less contaminated by dissolved gases and other contaminants or suspended solids than is the direct steam and thus produces less corrosion and scaling in the turbine and condensation systems. The major disadvantage of flashed steam cycles lies in their requirement for reservoirs containing fluids at high pressures and temperatures of tens of degrees above 100°C.

Experimental cycles referred to as total flow concepts are designed to use the expansion of vapor fluid mixtures to drive turbines using the liquid along

Geothermal Cycles for Generating Electric Power

A. Dry Steam Cycle

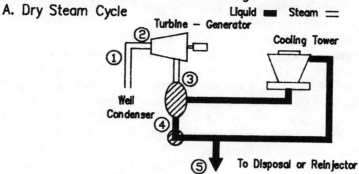

B. Dual Flash Cycle

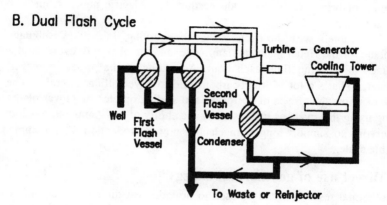

C. Organic Cycle

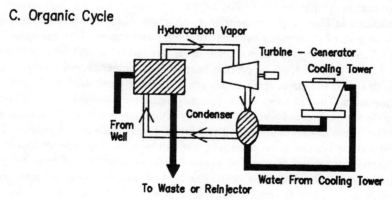

Figure 1.

with the vapor. In principle, total flow systems should be more efficient than flashed liquid systems since none of the energy of the geothermal fluid is lost during the flashing process. The concept has been field tested but not placed in commercial use.

In order to exploit lower enthalpy geothermal resources for the production of electric power, organic Rankine cycles use the heat of the geothermal fluid to flash an organic working fluid into vapor which then drives the turbine and is condensed and recycled. Various modifications of the basic cycle involve superheating the fluid for more efficient conversion. A demonstration 45 megawatt, organic Rankine binary cycle electric power plant is under construction in Heber, California. This is the largest geothermal electric power plant based on the organic Rankine cycle that has been constructed, although the technology has been used successfully on a smaller scale for industrial applications. The advantage of such a plant is that it is both more efficient than conventional steam cycles and that it will use lower temperature, more widely distributed geothermal fluids for its operation. The Heber plant will use geothermal brine whose wellhead temperature is 182°C and will require only two thirds the amount of brine needed to produce equivalent electric power from a conventional cycle.

The problem with the cycle is that the working fluid (90% isobutane–10% pentane for the Heber plant) is flammable, and the plant will cycle eight million pounds of working fluid each hour. According to the project manager, John Bigger, "That's a lot of hydrocarbon fluid to have rolling around."[9] The plant will examine the effects of corrosion and scaling on electric power plants operating using this cycle as well as study the ability of low-temperature geothermal reservoirs to support long term electric power production on a commercially viable scale.

4. Direct use of geothermal energy

Geothermal reservoirs can be used to provide heat directly instead of converting heat to electric energy.[10] The technology involved is generally simpler than that used in electric power generation although there are still problems of the interaction of geothermal brines with piping and pumps. In its simplest form, hot fluid from the geothermal reservoir is circulated through pipes and its heat is used to provide space or process heating. The temperatures of the geothermal fluids useful for direct applications are lower than those used in electric power generation. Thus in some cases, pumps driving the circulation can be installed inside the geothermal well. In other cases, the pumps are on the surface which makes them easier to service. The hot fluid must then travel through a pipe distribution system to the point where it will be used. The pipe system must be thermally insulated in order to prevent unacceptable loss of heat before the fluid arrives at its desired destination. Most working systems also provide access to the pipes for maintenance such as removal of silt which settles out of the hot brines as they cool and are under less pressure.

The hot fluid can provide heat directly or be passed through heat exchangers to heat a working fluid. In some cases, heat exchangers are actually lowered into geothermal wells so that the working fluid produces less silting, corrosion, and scaling in the pipe and pump systems.[11] Table 1 lists temperatures of geothermal reservoirs which can be usefully exploited for a variety of applications.

The most traditional use of geothermal heating is in balneology, a traditional cure for a variety of ailments. A second widespread application of direct geo-

Table 1. Uses of geothermal energy.

Temperature °C	Application	Production method
20	irrigation to warm soil fish hatching animal husbandry	compressor type heat pump
50	soft drink production balneology synthetic rubber production meat processing residential heat pumps	heat exchangers and heat pumps + exchangers
80	domestic hot water pharmaceutical production greenhouse heating grain and fodder drying	heat exchangers
110	electric power (Rankine cycle) air conditioning plastic production oil regeneration desalination municipal heating	absorption heat pumps
140	electric power (single flash) aluminum oxide (Bayer process) paper production textile manufacture	direct fluid extraction
170	electric power (dual flash) refrigeration heavy water production evaporation processes	direct fluid extraction
240	electric power (direct steam) electric power (dual flash)	direct fluid extraction

thermal energy is in space heating. Projects in Iceland, Japan, New Zealand, Hungary, the United States and elsewhere have long exploited geothermal energy for space heating. More recent systems such as the one at Creil in France use geothermal energy to drive heat pumps.[12] Most space heating systems require input of electrical energy to drive pumps as well as direct use of geothermal heat and some have backup conventional heating systems. Other recent uses of direct geothermal energy include a variety of agricultural applications such as heating greenhouses,[13] heating livestock barns and fish farm ponds, refrigeration, and industrial process heat. Industrial uses generally require higher temperatures than space heating and agricultural uses and in many cases require steam for processes such as dehydration or distillation. It is frequently profitable to cascade the use of geothermal brines so that, for example, the condensate from a steam driven industrial process is used to provide space heating for the buildings in the vicinity. Such cascading is not always practical due to the geographical distribution of users and resources or the fact that all users require maximum energy at the same time.

In general, direct uses of geothermal energy are specific to the region where they are employed. Because they require relatively unsophisticated technology, they are ideal for developing nations and are beginning to be exploited where they can be found. Direct use must occur close to the geothermal reservoir, so the areas where such exploitation is possible are more limited than areas that can be served by the conversion of geothermal heat to electric power.

5. Problems in exploiting geothermal energy

One problem confronting developers of geothermal resources is that the heat of the reservoirs is essentially finite. As hot brine or steam is pumped out of wells sunk into the geothermal reservoir, the pressure in the well may decrease abruptly or sufficient fluid may not enter the well to replace that which has been withdrawn. Once a well fails to produce adequate fluid, it becomes worthless. Since drilling geothermal wells and constructing power plants is expensive, producers generally figure that a well should produce for times on the order of thirty years in order to repay its initial costs.

The first obvious step in producing geothermal energy is to carefully survey resources before any production facilities are constructed. Relating field data to the future productivity of geothermal wells is not trivial. Early geothermal projects were begun by drilling in areas where geysers and naturally occurring hot springs seemed to indicate the presence of heat sources near the surface. At best, this proved a hit or miss proposition, particularly near the edges of a thermal region. Drilling test wells is also much more expensive than other geological techniques, such as seismic, gravity, or resistivity surveys. A literature on surveying geothermal reserves using techniques other than full scale drilling has borrowed from the petroleum industry.[14] Even the best technological evaluations may easily lead to wells with poor production, and research is needed in the field.[15]

In order to replenish the fluid withdrawn from the well and to dispose of the warm brine without damage to the environment, many geothermal projects reinject the cooled geothermal fluid into the reservoir in hopes that it can extract heat from the hot rocks of the reservoir and maintain the production level of the wells. Reinjection must be conducted with care so that the brine, which may contain dissolved toxic materials such as arsenic as well as salt and dissolved silica, does not contaminate surface water, especially that used as drinking water for men and livestock. Reinjection must be done far enough from the production wells so that the cool water does not cool fluid in the production well. The brines contain suspended solids which tend to silt up the well used for reinjection. Silica in the cooled brine may stop up the pores which drain reinjection wells. In other cases, the cold brine may open new passage ways through the rocks as thermal stress opens existing faults. Designers guard against such openings which bring cool brine into contact with the drawing area of the production well, a process known as thermal breakthrough.

The circulation of reinjected fluid in geothermal reservoirs has been extensively studied using chemical and radioactive tracers added to the reinjected fluid and collected at the production well. In many cases, the reinjection strategy appears to recirculate the geothermal fluid through the hot rocks of the reservoir, effectively mining more heat from the reservoir. In other cases, reinjection has reduced the production of geothermal wells apparently due to thermal breakthrough. At present the long-term effects of reinjection are under study around the world at sites using producing geothermal wells.[16]

6. Chemical problems with geothermal fluids

Geothermal brines do not provide an easy environment for instruments and materials. Many of the problems associated with the use of geothermal energy arise from these adverse effects. Problems associated with materials can be divided into three basic groups: *silting*, the buildup of suspended solids in flashers or pipes; *scaling*, the crust or scales of dissolved minerals that form on pipes, wells, or equipment as the dissolved solids come out of solution in cooled, depressurized brines; and *corrosion*, the destruction of materials by the action of the hot brines. Silting problems are similar to those encountered in other mining operations and can be handled by filtering the geothermal fluid. The filter material may be subject to blocking by scale and corrosion.

Prevention of scaling and corrosion by geothermal fluids is complicated by the fact that the fluids vary widely in composition so that the problems differ from site to site. The behavior of water-soluble materials at high temperatures has been little studied. In many cases, the basic chemistry and thermodynamics of the geothermal brine are not known. Both processes are strongly influenced by the pH of the solution as well as its composition.[17] Corrosion can be prevented to some extent by careful selection of materials for pipes, pumps, instrument cables, and well casings. Unfortunately materials which best resist corrosion are expensive. The search for inexpensive materials that will resist corrosion forms an important part of research on using geothermal energy.

Scales form from such natural contaminants as silica, sulphates, sulphides, carbonates, silicates, and the halides as well as substances, such as clay, that are introduced during the process of drilling. These substances precipitate from the brine as it is cooled and depressurized and adhere to pipes and other equipment. Attempts to prevent scaling include adding substances such as HCl to the geothermal fluid to change its pH, adding scale inhibitors, usually organic compounds, to the fluid, and adjusting the amount of CO_2 dissolved in the solution. Techniques such as flash crystallization collect the materials which would cause scale in a portion of the handling equipment where they may easily be removed.[18] In addition the fluids may be subject to physical treatment such as desilication in fluidized beds,[19] flotation or filtering before scale can adhere. As in the case of corrosion, the major problem with such methods is their expense. The search for cheap additives or techniques to remove scale-forming materials from geothermal brines is ongoing and includes such materials as combustion residues.[20]

Instrumentation for drilling, assessing, and operating geothermal wells presents special problems because the wells are necessarily hot. The weight and density of drilling muds are difficult to regulate and the drill bits themselves tend to wear out quickly in the hot environment and the presence of corrosive brines. Many advanced techniques can be borrowed from the petroleum industry but the temperatures within geothermal reservoirs present unique drilling problems.[21]

Assessing even so basic a problem as the temperature within a geothermal well has proven difficult and a major research effort has been made to find reliable methods for making such measurements by adding indicators to the drilling mud used in the well[22] and the development of high-temperature electronics. Conventional electronics fail to operate at temperatures over 150°C, and a research effort is underway to produce electronic probes that can function at temperatures up to 450°C. Electrical cables must be armored to protect them against the corrosion of the geothermal fluids and strong enough to support their own weight over distances of 6–8000 meters as well as rugged enough to be handled in the environment of a drilling rig. Data collection from geothermal wells is being computerized and automated in order to increase understanding of these complex systems.[23]

Pumps which can function reliably at temperatures of 400°C do not exist, and work on such systems as well as the development of new materials for turbines and other components of geothermal production systems is underway. The recent burst of activity throughout the materials science community holds promise for the development of materials for economic exploitation of geothermal energy.

7. Environmental impact of geothermal energy production

The environmental impact of geothermal development has just begun to be studied.[24] Clearly, dumping of toxic or merely highly saline geothermal brines into local surface waters such as lakes and streams may be expected to have a deleterious effect on local wildlife. Research is underway to study the detailed effects and to quantify their seriousness. A particular concern is that geothermal brines often carry significant quantities of dissolved toxic elements. The 180 MW Wairakei geothermal power station in New Zealand dumps 156 tonnes of arsenic and 0.006 tonnes of mercury into the Waikato River each year in addition to the 6500 tonnes of hot water dumped every hour. This must be viewed in light of the fact that these toxic chemicals are already discharged by natural sources at a rate one fourth that of the power station. Other concerns include contamination of ground water and hence drinking water by the brines and the environmental impact of the heat generated by the geothermal brines and exhausted to the atmosphere. At the time of this writing, CBS news reports studies are underway to determine whether Crater Lake may be affected by exploitation of a geothermal reservoir outside of the national park.

The major gaseous discharge to the atmosphere from geothermal projects has been H_2S which is smelly in low concentrations (below 10,000 ppb) and toxic at high concentrations. A concentration 600,000 ppb of H_2S is fatal after 30 minutes. The quantity of H_2S released depends on the chemistry of the local geothermal reservoir and varies widely. Estimates are available for a few well-studied reservoirs, but many reservoirs are too poorly studied to permit quantitative estimates or the release.

At The Geysers in California, various measures have been tested to trap the estimated 50 tons/day of H_2S released by the geothermal project. They have proved expensive and technically difficult. The first efforts involved reacting H_2S with oxygen in the presence of an iron catalyst (a modified Claus process) which failed because both moisture and CO_2 tended to reduce the efficiency of the process and the recovered elemental sulphur was too impure to market and difficult to remove. The use of iron sulphate and sulphuric acid followed by addition of caustic soda to the condensers and hydrogen peroxide into the cooling tower sumps produced a corrosive sludge which destroyed the efficiency of the cooling tower and contaminated the surroundings. At present the Stretford process, which uses an expensive surface condenser to reduce contamination of water and must be followed by a secondary treatment with hydrogen peroxide and acetic acid, is being used, but it is too expensive to provide a long-range solution. Geothermal plants also produce significant quantities of CO_2, which is dissolved in the brines.[25]

A very troubling problem with exploiting geothermal heat has been the noise made by the escaping steam and water. At Wairakei, the water is passed through twin cylindrical towers where it is swirled around to lose its kinetic energy to friction before it is expelled. Steam escapes from the top of the towers and most of the noise is carried up, reducing the noise found in the environment. Similar designs are used at most geothermal installations since the shriek of escaping geothermal fluid is intolerable over large areas. The plants are still noisy, but the towers provide at least a partial solution.

Because geothermal fluids are stable only when they are under pressure and liable to flash into steam, breaks in pipes can lead to serious accidents and a variety of safety systems must be installed. For example in 1960 at Wairakei, a casing break 600 feet below the surface formed a small steam volcano which collapsed a hillside but caused no serious injuries. Other potential environmental problems include subsidence of the land as geothermal fluids are drained. Local geysers may cease to erupt as reservoirs are exploited, hurting tourism. It should be noted that geysers may naturally change their behavior in time. Scenery may be damaged if trees are killed by the discharge from a geothermal plant. There has also been concern that reinjection might trigger seismic shocks, but to date there has been little data to support this fear.

8. Artificial geothermal reservoirs

The use of geothermal energy is limited to a relatively small number of suitable geothermal reservoirs. In many areas of the world, hot rocks lie near the surface of the earth but they are dry so that their energy cannot be extracted. Successful use of the energy stored in hot dry rock for production of electricity would greatly increase the areas around the world where geothermal energy could be used commercially and the energy stored in commercially viable geothermal reservoirs. In the United States alone, hot rocks within reach of drilling could provide approximately 10 times the heat energy of known U.S. coal deposits. It

Hot Dry Rock Geothermal Production

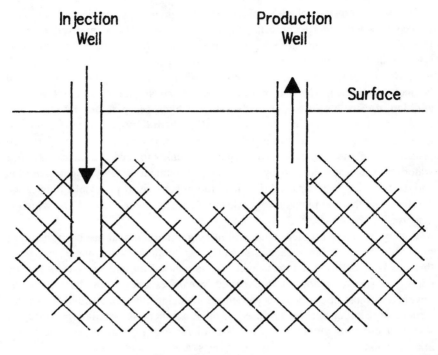

Figure 2.

should be noted that most hot rocks lie near the surface in volcanic regions such as Hawaii, Alaska, and the western U.S. In some regions, notably the mid-Atlantic coasts and the United Kingdom, sediments blanket and insulate granite intrusions which are heated by radioactive decay. If sufficiently deep drilling techniques can be developed, any site on earth might be a prospective site for the production of geothermal energy. Current state-of-the-art in drilling is about 10 kilometers and economic wells are usually limited to 6 km.[26]

Figure 2 shows a schematic drawing of the wells used for production of geothermal energy from hot dry rocks. Designs involve drilling two wells, an injection well and a production well. The injection well is drilled at an angle some hundreds of meters below the production well. The rock between the wells is fractured by many small cracks which provide a path between the two wells through which water can flow. Water is pumped down the injection well, traverses the fracture path between the two wells and is then collected at the production well after it has been heated on its journey between the two wells. It is important that the flow between the two wells not be too rapid in order to allow the water time to mine heat from the hot dry rocks of the reservoir. The fracture system must also be limited in scope so that at least a substantial portion of the water injected into the reservoir can be recovered and used as would the fluid from a traditional geothermal reservoir.

In addition to the difficulties of drilling through hot rocks, the major problem encountered in constructing a geothermal reservoir from hot dry rock is the fracturing of the rock to connect the injection and production wells. Hydraulic fracturing, used in the petroleum industry, opens fractures in the rock by forcing water into the rock. However the mechanism of fracturing is not well understood because in petroleum production it had not proved a critical issue. The petroleum drained through the fractures to the well regardless of whether new fractures were created by the pressure of the water or whether the water merely increased the size of existing fractures. For geothermal production, it is important that the pattern of fractures occur in such a way as to link production and injection wells while the water flows over surface areas on the order of a million square meters.

During the 1970s, research on using hot dry rocks for production of geothermal energy was undertaken at an extinct volcanic caldera near Los Alamos, New Mexico in the United States and at a quarry in Cornwall, Britain. Initial results at the U.S. site were very encouraging. Two test wells were drilled to about 3 km at a temperature of about 200°C and connected by a fracture system. The fracture system was explored at a variety of production conditions and studied extensively with tracers and seismic monitors. Water losses gradually decreased to about 1% of the input. Reservoir size was increased from 8000 m^2 to 50,000 m^2 by further hydraulic fracturing and the water loss increased only 30%.[27]

In 1980, the first electric power was generated from a geothermal system connected to a hot dry rock reservoir by connecting the flow from these wells to a binary cycle electric plant. The test wells produced 3 megawatts of thermal power. Encouraged by these results, the Los Alamos group drilled two new wells reaching a depth of 4.66 km and temperatures of 320°C with the aim of creating a sufficiently large, hot reservoir to produce electric power on a commercial scale. Unfortunately it proved very difficult to connect the two wells by a fracture system. Research demonstrated that fracturing occurs along joints which follow the stresses in the rock. The stress patterns in hot rock are complex because they depend on the location of the thermal source providing heat to the rock. In 1985, the production well was drilled in a new direction to intersect the fracture pattern that had been created and flow tests are under way on the new system, which has produced 9 megawatts of thermal power at a temperature of 240°C, well below the 30 megawatt goal of the well.

The British research has focused on lower temperature reservoirs (80°C) created at depths of 2.1 km. The British group also experienced problems linking the wells by a fracture system but finally managed to link them using a gel to fracture the rock. They have produced 5 megawatts thermal power with a water loss of 20%. The West Germans, French, and Japanese are currently working with the U.S. and the U.K. to develop hot dry rock reservoirs.[28]

To date, creation of geothermal reservoirs in hot dry rock has proved technically more difficult than was expected during the early experiments in the late 1970s. On the other hand, results are sufficiently encouraging so that continuing research in the area seems justified. Basic research on the mechanisms by which rocks fracture and on the stresses in rocks near a thermal source must be conducted in order to predict the fracture patterns around geothermal wells. Techniques for creating larger and more controlled fracture patterns need to be developed. Experience will be needed to determine the lifetime of producing geothermal wells in hot dry rocks. The chemical interactions of injected fluid and the rocks must be explored. Drilling in hot rock is still expensive and difficult despite the substantial advances made in drilling technology. The water supply for geothermal production must be studied including the rate of water loss and the fate of the water that is lost as well as its potential environmental effects. It is possible that techniques will have to be developed to clean the water before it is reinjected into the reservoir since it may contain silt and dissolved minerals. The commercial viability of geothermal energy production from hot dry rock reservoirs will depend on the results of these research efforts.

9. Economic and political issues

Where favorable natural reservoirs exist, geothermal energy has proved itself an economically viable alternative to fossil fuel. Construction costs and thirty year operating expenses (not counting steam costs) of a 110 megawatt electric power plant for the reservoir at The Geysers, California, are estimated at $1.48 billion in 1985 dollars. A 50 megawatt plant for a hot water reservoir whose temperature is 100°C will cost $1.9 billion in 1985 dollars including operation and maintenance for a 30 year lifetime. Costs for direct use systems must be estimated on a case by case basis and the economic viability of such systems determined by comparison with the cost of conventional heating systems.[29] In regions where convenient reservoirs exist, geothermal heating systems are currently viable and will become more attractive should the price of fossil fuel increase. The cost of hot dry rock production systems will depend on the results of basic research now in progress.

A final issue in exploiting geothermal energy is the legal system governing the use of water resources. Water and mining laws, particularly in the western United States, are noted for their complexity, the passion with which they have been developed, and the individuality of the state systems. Geothermal resources are regulated according to mining laws in some states and according to ground water laws in others. Some states claim ownership of geothermal resources and lease them to developers while in other cases ownership belongs to the person who holds title to the surface. The federal government (and many states) limits the acreage of geothermal leases. Federal law forbids any one developer to lease more than 20,480 acres in any one state. Developers claim these limits restrict exploration and development. Lease terms and conditions vary from state to state. Both the federal and state governments have a variety of loan, tax incentive, and grant programs to encourage the development of geothermal energy. Discussion of a federal insurance program to insure developers against early depletion of geothermal reservoirs is underway.[30]

Although the legal system governing geothermal resources is complicated by the fact that geothermal fluid is considered ground water in some states and a mineral resource in others, the leasing and legal system is probably not a major obstacle to development of these resources.

10. Conclusions

Although geothermal energy is the largest source of unconventional power in the United States, it represents only a tiny fraction (0.2% in 1987)[31] of the U.S. electrical generating capacity. This comes from the single development at The Geysers. There is little indication that this share will substantially increase in the near future. The technology in use at The Geysers requires naturally occurring dry steam, a commodity in very short supply. Other technologies for generating electrical energy from geothermal energies have encountered greater technical difficulties, and the high enthalpy geothermal reservoirs which they require are also in very limited supply. Major examples of the use of this technology are found in other countries. Small pilot plants in the United States have used a variety of approaches but have not proven economically attractive. In the event of a steep climb in oil prices, geothermal generation of electricity might become economically attractive but would be severely limited by the small number of reservoirs in the United States which are suitable for exploitation.

Efforts to exploit hot dry rocks for a more widely distributed energy source have been beset by technical difficulties and slowed by budget cuts. As long as cheap oil is available, there seems little chance that such technologies would be economically feasible. The use of geothermal energy for the production of electricity will continue, at least for the short term, to be limited to those sites fortunate enough to produce natural dry steam.

The exploitation of geothermal energy as direct heat seems more likely to prove useful in the near future. Its applications require relatively unsophisticated technology, so it may prove useful on an individual basis. For example, the Electric Heating Institute of Indiana is advertising ground-source heat pumps to reduce electrical consumption in individual homes. These applications of geothermal energy are similar in many ways to conservation measures and must be considered in terms of payback periods. Direct application of geothermal heat may also prove valuable in developing countries with available geothermal resources and technologically unsophisticated populations.

Bibliography

Van Thanh Nguyen, *Geothermal Energy: Resource and Utilization* (American Association of Physics Teachers, College Park, MD, 1983) provides a short, simple and general introduction to geothermal energy in a discussion designed for high school physics teachers.

H. Christopher H. Armstead, *Geothermal Energy* (E. & F.N. Spon, New York, 1983) provides a detailed introduction to geothermal energy that is readable and accessible to a general audience. It is probably the best single text on the subject.

Applied Geothermics, M. Economides and P. Ungemach, eds. (John Wiley and Sons, Ltd., Chichester, 1987) is a collection of review articles on all aspects of the use of geothermal energy. These essays are comprehensive and up-to-date. They are designed for a technical audience and provide details which other general texts neglect.

H. Christopher H. Armstead and Jefferson W. Tester, *Heat Mining* (E. and F.N. Spon, New York, 1987) is a basic text on the exploitation of hot dry rock and is probably the best general source on this technology.

References and notes

1. R. Cataldi and C. Sommaruga, "Background, Present State and Future Prospects of Geothermal Development," Geothermics **15**, 359–383 (1986).
2. H. Christopher H. Armstead, *Geothermal Energy* (E. & F.N. Spon, New York, 1983), p. 31.
3. D.H. Freeston and H. Pan, "The Application and Design of Downhole Heat Exchanger," Geothermics **14**, 343–351 (1985).
4. J.S. Gudmundsson and G. Palmason, "The Geothermal Industry in Iceland," Geothermics **16**, 567–573 (1987).
5. Pierre Ungemach, "An Introduction to Geothermal Energy," in *Applied Geothermics*, M. Economides and P. Ungemach, eds. (John Wiley and Sons, Ltd., Chichester, 1987), p. 1.
6. Michael J. Economides, "Engineering Evaluation of Geothermal Reservoirs," in *Applied Geothermics*, Economides and Ungemach, eds., pp. 9–109.
7. Malcom A. Grant, Ian G. Donaldson, and Paul F. Bixley, *Geothermal Reservoir Engineering* (Academic Press, New York, 1982), pp. 231–233.
8. Reviews of the production of electric power from geothermal sources are presented by: Pierre Ungemach, "Electric Power Generation from Geothermal Sources," in *Applied Geothermics*, Economides and Ungemach, eds., pp. 137–187.
   Sergio Siena, "Small-Medium Size Geothermal Power Plants for Electricity Generation," Geothermics **15**, 821–837 (1986).
   Van Thanh Nguyen, *Geothermal Energy: Resource and Utilization* (American Association of Physics Teachers, College Park, MD, 1984).
9. John Bigger, project manager for EPRI at Heber, as quoted in Nadine Lihach, "Heber: Key to Moderate-Temperature Geothermal Reserves," Geothermal Energy **13**, 11–16 (1985).
10. Reviews of direct uses of geothermal energy are given by:
    J.S. Gudmundsson and J.W. Lund, "Direct Uses of Earth Heat," in *Applied Geothermics*, Economides and Ungemach, eds., pp. 189–219.
    V.N. Kruzhelnitsky, "Space and Process Heating: State-of-the-Art and Prospects," Geothermics **14**, 165–173 (1985).
    Direct uses are also treated in Nguyen as cited above.
11. Freestone and Pan, as cited above and Robert Harrison, "Design and Performance of Direct Heat Exchange Geothermal District Heating Schemes," Geothermics **16**, 197–211 (1987).
12. P. Jaud, "Geothermal Heating Systems Using Heat Pumps," Geothermics **14**, 189–196 (1985).
13. K. Rafferty, "Some Considerations for the Heating of Greenhouses with Geothermal Energy," Geothermics **15**, 227–244 (1986).
14. See Grant, Donalson and Bixley, and also Economides, pp. 96–104 for general discussion. See M.Y. Corapcioglu and N. Karahanoglu, "Simulation of Geothermal Production," in *Alternative Energy Sources II* T. Nejat Veziroglu, ed. (Hemisphere Publishing Corporation, Washington, DC, 1981), Volume 5, pp. 1895–1918 for discussion of modeling problems. See Tuncer Esder and Sakir Simsek, "The Relationship between the Temperature-Gradient Distribution and Geological Structure in the Izmir-Seferihisar Geothermal Area, Turkey," in *Alternative Energy Sources II*, Veziroglu, ed., Volume 5, pp. 1919–1940 for typical summary of field evaluation data.
15. Lee Dye, "Geothermal Plant May Prove Worthless," reprinted from the Los Angeles Times in Geothermal Energy **12**, 8–11 (1984).
16. See Economides as cited above and R.N. Horne, "Reservoir Engineering Aspects of Reinjection," Geothermics **14**, 449–457 (1985) for general reviews of reinjection strategies and

Z.F. Sarmiento, "Waste Water Reinjection at Tongonan Geothermal Field: Results and Implications," Geothermics 15, 295–308 (1986) for a specific example.

17. Reviews of materials problems are found in G. Culivicchi, C.G. Palmerini, and V. Scolari, "Behavior of Materials in Geothermal Environments," Geothermics 14, 73–90 (1985) and O.J. Vetter and V. Kandarpa, "Chemical Thermodynamics in Geothermal Operations," in *Applied Geothermics*, Economides and Ungemach, eds., pp. 125–135.

18. Riccardo Corsi, "Scaling and Corrosion in Geothermal Equipment: Problems and Preventive Measures," Geothermics 15, 839–856 (1986).

19. R.C. Axtmann and D. Grant-Taylor, "Desalination of Geothermal Waste Waters in Fluidized Beds," Geothermics 15, 185–191 (1986).

20. E. Pernklau and E. Althaus, "Desalination of Geothermal Brines by Means of Combustion Residues?" Geothermics 15, 193–195 (1986).

21. Samuel G. Varnado and Alexander B. Maish, "Geothermal Drilling Research in the United States," in *Alternative Energy Sources II*, Veziroglu, ed., Volume 5, pp. 1949–1964.

22. J.V. Gaven, Jr. and C.S. Bak, "Temperature Indicators for Geothermal Use," in *Alternative Energy Sources II*, Veziroglu, ed., Volume 5, pp. 1941–1948.

23. N. Adorni, L. Ceppatelli, R. Papale, and A. Parmeggiani, "Instrumentation for Geothermal Wells: Present Status and Future Prospects," Geothermics 14, 287–307 (1985).

24. Good reviews of environmental problems of geothermal energy are given in Armstead, as cited above and Louis J. Goodman and Ralph N. Love, *Geothermal Energy Projects: Planning and Management* (Pergamon Press, New York, 1980). Case studies include Vincent H. Reske, Gary A. Lamberti, Eric P. McElroy, John R. Wood, and Jack W. Feminella, *Quantitative Methods for Evaluating the Effects of Geothermal Energy Development on Stream Benthic Communities at The Geysers, California* (California Water Resources Center, Davis, CA, 1984) and *The Tenth Annual Interagency Geothermal Coordinating Council Report for Fiscal Year 1985* (U.S. Government Printing Office, Washington, DC, 1984/85).

25. Specific examples given here are taken from Armstead.

26. Good discussions of the exploitation of hot dry rock reservoirs include H. Christopher, H. Armstead, and Jefferson W. Tester, *Heat Mining* (E. & F.N. Spon, New York, 1987.); Richard A. Kerr, "Hot Dry Rock: Problems, Promise," Science 238, 1226–1228 (1987); and A.S Batchelor, "Hot Dry Rock Exploitation," in *Applied Geothermics*, Economides and Ungemach, eds., pp. 221–234.

27. Z.V. Dash, H.D Murphy, R.L. Aamodt, R.G. Aguilar, D.W. Brown, D.A. Counce, H.N. Fisher, C.O. Grigsby, H. Keppler, A.W. Laughlin, R.M. Potter, J.W. Tester, P.E. Trujillo, Jr., and G. Zyvoloski, "Hot Dry Rock Geothermal Reservoir Testing: 1978–1980," J. Volcanology Geothermal Res. 15, 59–99 (1983).

28. Numbers here come from Kerr, and discussion is based on Kerr and Batchelor as cited above. A discussion of technology for hydraulic fracturing is found in Donald S. Dreeson and Robert W. Nicholson, "Well Completion and Operations for MHF of Fenton Hill HDR Well EE-2," Trans. Geothermal Resources Council 9, 89–94 (1985). An optimistic review of early work in New Mexico is given by P.R. Franke, "Hot Dry Rock Geothermal Energy Development Program," in *Alternative Energy Sources II*, Veziroglu, ed., Volume 5, pp. 1993–2007.

29. A.D. Stockton, "What Do Geothermal Projects Cost?" Geothermal Energy 14, 8 (1986) and Charles V. Higbee, "Life Cycle Cost Analysis for Direct Use Geothermal Systems: An Introduction," Geothermal Energy 12, 7–8 (1984).

30. R. Gordon Bloomquist, "A Review and Analysis of the Adequacy of the U.S. Legal, Institutional and Financial Framework for Geothermal Development," Geothermics 15, 87–132 (1986).

31. See Chapter 1 for a discussion of the relative importance of different energy sources.

Energy from the oceans

M. M. Sanders

1. Introduction

Solar energy is deposited in the oceans and the atmosphere at an average rate of approximately 80 trillion kW — about 8000 times the rate of current world energy consumption. This flux drives the weather systems over the oceans, whose winds in turn drive ocean waves; it feeds the river systems of the earth through cycles of evaporation and condensation; it deposits sensible heat into the ocean waters to establish stable thermal gradients and to drive ocean currents. In addition to the solar energy income, energy is dissipated in the oceans at the rate of about 3 billion kW through the tides.

Each of these processes has attracted adherents who claim that it can be exploited to serve human energy needs. Some of the ideas are old; references to tidal mills are found in the Domesday book. Some of the ideas, such as ocean thermal energy conversion (OTEC) and schemes to exploit salinity gradients near the estuaries of major rivers, are recent. Typically the notions of using conspicuous ocean energy resources (i.e., tides, currents, waves) have a longer history than schemes involving inconspicuous resources (thermal gradients, salinity gradients). However, the amount of energy stored in inconspicuous form is estimated to be several orders of magnitude greater than that stored in conspicuous form. Consequently, the development of technologies which can utilize inconspicuous energy resources will probably have a greater impact on the world energy future.

This chapter will review some major proposals for the utilization of conspicuous and inconspicuous ocean energy resources. In particular, it will deal with OTEC technology as this is the most promising ocean technology in the near term owing to both the extent of the resource and the maturity and flexibility of the technology, but in addition will discuss wave energy conversion and tidal power. The utilization of salinity gradients as an energy source has not been as yet realized in any practical sense and so it will not be discussed.

2. Ocean thermal energy conversion

2a. Principles and background

Ocean Thermal Energy Conversion designates a simple heat engine whose thermal cycle operates between the warm temperature found in tropical waters near the ocean surface and the near freezing temperatures of the ocean's depths.

OTEC converts the solar thermal energy in the ocean into the mechanical energy of a spinning turbine and then to electric power.

The solar flux onto the surface of the oceans and the atmosphere provides around 40,000 billion kilowatts which maintains a thermal gradient in the tropical waters that could be harnessed to produce as much as 10,000 GW of electric power worldwide by some estimates.[1] The ocean's great thermal mass damps the daily fluctuations which plague other solar power schemes. OTEC derived energy may prove cheaper than that produced by land based solar plants.[2] The thermal power cycle upon which OTEC is based is similar to that found in conventional power plants except that the temperature difference which drives OTEC is much smaller. The thermodynamic efficiency of an OTEC cycle ($\approx$ 5%) is much lower than that of a fossil fuel power cycle ($\approx$ 30%). However, the energy which maintains the thermal gradient in OTEC comes for free, in contrast to the fuel for conventional plants. All the fluids involved in an OTEC thermal power cycle are at much lower temperatures and pressures than those found in conventional units. Consequently, the efficiency of heat exchangers can be significantly increased at the expense of structural strength without any penalty. In fact, OTEC shows promise of reliably providing a large portion of the world's energy requirements at a favorable real cost.

In the OTEC cycle the warm water found near the ocean surface evaporates a working fluid. The vapor expands through a turbine which is typically coupled to an electrical generator. Vapor leaving the turbine is condensed by thermal contact with cold water pumped to the surface of the ocean through a large pipe extending to an optimal depth of 500–1000 m. A schematic diagram of the process is shown as Figure 1.

The OTEC idea was first suggested by d'Arsonval (1881)[3] and first implemented by his friend and student Georges Claude (better known for the invention of the neon sign) off the coast of Cuba in Matanzas Bay in 1929. He lost a 2 km cold water pipe to heavy seas before he was able to successfully install a unit which generated 22 kW of electric power from a 14°C temperature difference.[4] Warm sea water was flash evaporated into a partial vacuum to provide the vapor for the turbine. The spent vapor was condensed by direct contact with the cold water and returned to the ocean. Unfortunately, more than 22 kW was required to maintain the partial vacuum of the flash evaporator. The use of water as the working fluid in the thermal cycle is termed open-cycle operation or the Claude cycle.

An alternative to the Claude cycle is closed-cycle operation which incorporates a different working fluid, such as ammonia, freon, or propane, in an entirely self-contained thermal cycle. The closed-cycle working fluid is evaporated by heat exchange with warm seawater to drive the turbine, condensed by heat exchange with the cold water, and returned to the evaporator. Open-cycle operation is attractive in that it obviates the need for indirect contact heat exchangers which are both expensive and inefficient. Open-cycle operation can also yield fresh water as a byproduct (although fresh water production requires an indirect contact heat exchanger on the cold water end to segregate the condensing vapor from cold seawater).

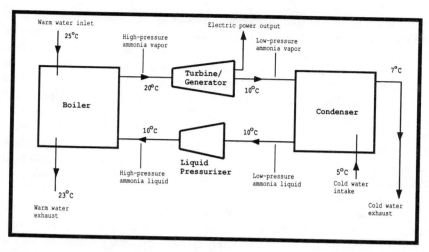

Figure 1. Schematic diagram of OTEC closed-cycle operation. Each MW of electrical capacity requires a flow of 4 cubic meters per second of water on both the warm and the cold side.

In open-cycle operation, technical difficulties are posed by the turbine to utilize the low-pressure steam and by the need to extract the noncondensible dissolved gases before the vapor passes through the turbine. In contrast, all the components necessary for closed-cycle operation are more highly evolved, and a large low-pressure turbine, suitable for open-cycle operation, could have a prohibitive cost. However, a study by the Westinghouse Corporation[5] indicates that advances in turbine blade technology might give open cycle schemes parity in cost.[6] More recent work by the Solar Energy Research Institute (SERI) indicates that near-future, open-cycle (OC-OTEC) plants in the 5–15 MWe range at onshore tropical sites will be technically and economically feasible[7] — in fact, economically superior — for the production of electrical energy, desalinization of water, and provision of nutrients for local mariculture.

While most of the work in the OTEC field to date is based around either the closed- or open-cycle heat engine scheme as described above, other techniques have been suggested. Variations on the open-cycle theme have been proposed, largely involving schemes for using heat in the surface water to lift the water either by cavitation,[8] or by foam.[9] Lifting transforms a thermal head into a hydraulic head which could drive a hydraulic turbine. These untested schemes would eliminate the gigantic low-pressure turbine. Another alternative proposal incorporates the thermoelectric effect[10] for direct conversion of a thermal gradient into an electrical current.

2b. Siting concerns

The thermal resource of a proposed site, the type of product, the end use of the product, and effects both on and by the local marine environment due to the presence of an OTEC station must all be weighed in choosing a site. If the station will transmit electricity to an established land-based power grid, it needs to be close to shore. Alternating current transmission is economical for distances less than about 30 km while direct current transmission can be feasible for distances up to 400 km. Finding sites for direct transmission can prove to be difficult since the continental shelf generally prevents access to deep water at offshore sites within acceptable transmission range. It seems likely that the first commercial OTEC powerplants will be located near population centers on island nations in the South Pacific where the combination of high thermal resource with the high price of oil and the need for desalinized water provide a favorable economic environment for OTEC.[11] Alternatively, OTEC installations could be stationed far from land in tropical waters where the OTEC electrical power could be used on site to produce hydrogen or ammonia which could be delivered ashore via pipeline, or by ship or barge. If no umbilical connections to shore exist, the powerplant is no longer constrained to maintain its station and can graze the oceans in search of energy.[12]

The intrinsic thermodynamics of the OTEC process and the mean temperature of different ocean locations set limits on the geographical location of favorable powerplant sites. The thermodynamic efficiency improves as the temperature difference (ΔT) between the hot water source and the cold water sink increases. As indicated in the thermal map of the world's oceans (Ocean Data Systems, see Figure 2), the average temperature difference between surface waters and water 1000 m deep is greatest in the equatorial waters of the South Pacific, about 24°C, and diminishes as the latitude increases. A temperature difference of 24°C corresponds to a thermodynamic efficiency of about 7%. Estimates indicate that an OTEC power plant operating over this temperature difference would have an actual efficiency of about 2.5% with the discrepancy due to temperature differences across heat exchangers, energy losses incurred by pumping deep sea water against its own density differences, and irreversible losses due to friction in the pipes.

On the order of half, the cost of a closed-cycle OTEC station depends on the area of the heat exchangers. For practical purposes, the relationship between heat exchanger size and ΔT is more critical than the relation between the Carnot efficiency and ΔT. Consequently, OTEC is not considered feasible if ΔT is less than about 17°C. The best OTEC sites are located in the tropics and subtropics, roughly between 25°S latitude and 32°N latitude, typically far from human habitation and in general far from dry land.

Other noteworthy ideas concerning optimal siting for thermal energy conversion devices include transporting icebergs to warmer waters and using their meltwater as the cold water supply to the condenser (ICETEC?).[13,14] Another proposal would involve installation of closed-cycle devices on icecaps in arctic regions. The water under the icecap would supply heat to the thermal cycle evaporator while the condenser would operate by heat exchange with the frigid polar air above the icecap (ARCTEC?).[15]

2c. Recent developments: Closed cycle research and development — 1970s

The development of OTEC technology stagnated after the initial work of Claude, although France maintained a sporadic interest in the idea which culminated in the design of a 7 MWe facility to be located on the Ivory Coast in the 1950s. The plan was abandoned when a hydroelectric plant was constructed

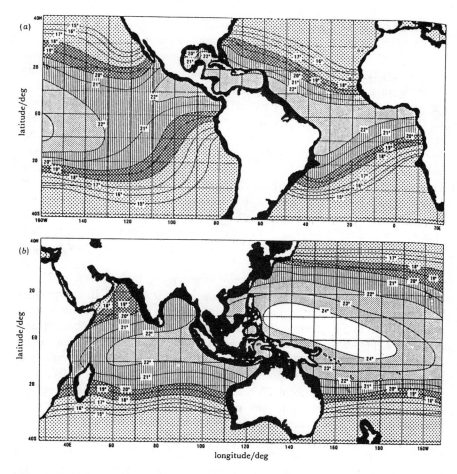

Figure 2. Annual average monthly temperature differences (°C) between the Ocean Surface and 1000 m. Black areas indicate depths less than 1000 m (from Ocean Data Systems).

nearby. In the United States, in the mid 1960s, James Anderson and his son conceived the idea of closed-cycle OTEC, and they formed a company, Solar Sea Power, Inc. (SSPI), which produced an engineering design for an installation in the Caribbean.[16] In the early 1970s, the National Science Foundation (NSF) funded two studies of closed-cycle operation, each of which resulted in the design of a closed-cycle OTEC powerplant. Two industrial teams used this work to produce a design for a floating closed-cycle OTEC power station which had a net capacity in excess of 100 MWe.

i. Floating CC-OTEC station designs. The major components of a floating closed-cycle OTEC station consist of a hull, a cold-water inlet pipe, evaporators, condensers, pumps, turbines, generators, and station maintaining apparatus in the form of a mooring or some dynamic propulsion system. Except for the turbines and generators, which are standard, each of these components presents a particular engineering challenge in adapting it to an OTEC facility and the design efforts of the 1970s displayed substantial variation.

The SSPI design, which pioneered the adaptation of the closed Rankine cycle to OTEC, features external plate-fin evaporators and condensers, using the refrigerant, R-12/31, as working fluid, with the evaporators mounted below the condensers in order to allow natural convection to assist the flow of working fluid and to balance the pressure across the evaporators. The thermal cycle hardware is mounted on a dynamically positioned barge type hull. One of the design studies funded by the NSF concluded that ammonia would be the optimal working fluid in an OTEC installation.[17] The second study proposed propane as the thermal cycle working fluid which would flow through plate-fin design copper-nickel alloy (which is incompatible with ammonia) heat exchangers mounted on a catamaran hull. They considered use of the unit to provide ac power directly to Miami, Florida, or to run an electrolysis plant which would transmit H_2 gas via pipeline to produce electricity for the New England power grid.[18]

One industrial design was a spar-buoy arrangement, built around a 1220 m fiberglass cold water intake pipe and producing 100 MWe.[19] It incorporated four separate thermal cycles; each utilizing ammonia as a working fluid with shell-and-tube design titanium heat exchangers acting as the evaporators and condensers (Figure 3). Stationing was provided by steering jets of spent warm and cold side ocean water. The other industrial design was also based on a spar buoy, incorporating a 305 m prestressed concrete cold water pipe in five telescopic sections, ranging from 31.1 to 38.4 m in diameter and designed to deliver 160 MWe. There were four modular power units which each feature titanium shell-and-tube heat exchangers with ammonia as a working fluid.[20]

ii. OTEC factories — Alternative products. In the mid-1970s, a consortium of U.S. labs and industries produced a design for a floating chemical refinery which relies on OTEC derived electricity to convert seawater and air into ammonia.[12] The ammonia would be used as fertilizer in tropical countries which are located close to ocean waters attractive for OTEC. Ammonia serves as a safe and effective carrier of hydrogen which could be shipped from a floating OTEC station to a land-based plant that would catalytically decompose it to N_2 and H_2 and feed the gas to H_2-O_2 fuel cells to produce electricity. Use of ammonia could prove much less difficult than liquifying hydrogen gas on site and maintaining cryogenic conditions during shipping. The group concluded (in 1975) that OTEC plant-ships of this nature would produce ammonia at a competitive price with land-based processes employing natural gas as an energy source in the mid to late 1980s.

iii. Heat-exchangers for closed cycle OTEC. Although none of these closed-cycle OTEC proposals of the 1970s have been deployed, they did identify the major areas where further work needed to be done. The heat exchangers are the most significant component in a closed-cycle OTEC design, representing approximately half the cost of an installation. Due to the relatively low tempera-

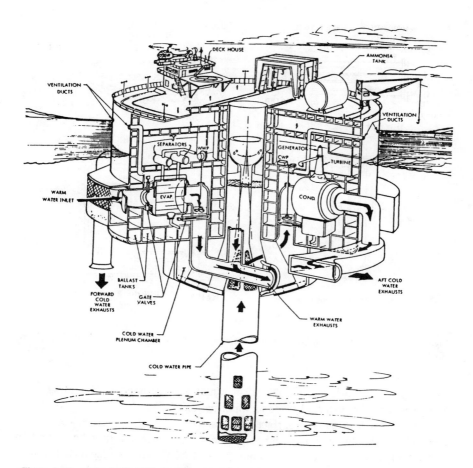

Figure 3. Concept of 100 MWe baseline OTEC power plant from TRW, Inc. Platform diameter is 100 m. One of the 25 MWe power modules is shown in the cutaway.

tures involved in the process and the large flows of ocean water, the total heat exchanger area in a closed-cycle OTEC power cycle is much larger than that which would be necessary to operate a Rankine cycle in an oil fired power station of the same output. Consequently there have been major efforts towards optimizing their design in terms of physical arrangement, material, and operation. Much of the current information derived from the operation of the engineering test facility, OTEC-1, in 1980–1981 off the Kona Coast of Hawaii.

The basic design configurations for closed-cycle heat exchangers, illustrated in Figure 4, are divided into the shell-and-tube category and the plate-fin category. The more traditional shell-and-tube design is found in fossil-fuel and nuclear power plants. The plate-fin or extended surface design represents more of a departure from current engineering practice, but its use may be warranted for closed-cycle OTEC. Other design variables are the recommended material which might be used (aluminum, titanium, copper-nickel alloys, and plastic[21] have been suggested), thickness of the heat transfer surfaces, surface preparation of the heat transfer surfaces (see below), proposed working fluid (am-

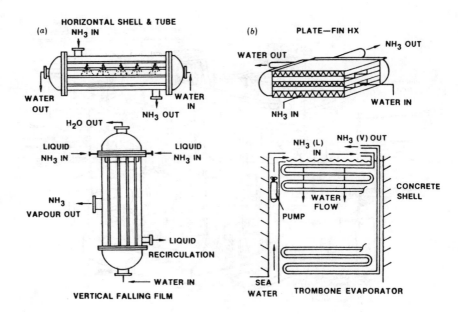

Figure 4. Examples of shell-and-tube (a) and plate (b) closed-cycle heat exchanger designs (from [Coh82]).

monia, propane, various freonlike refrigerants), and methods of biofouling control (mechanical devices, ultrasound, chlorination).

The area of heat transfer surface necessary to produce a given power output depends on the thermal impedance of the heat transfer surface between the water and the thermal cycle working fluid. This thermal impedance itself is composed of three partial impedances for the flow of heat between the water and the heat exchange surface, for the flow of heat between the heat exchange surface and the working fluid, and for the flow of heat from one side of the heat exchange surface to the other with the direction of heat flow depending on whether the heat exchanger is acting as an evaporator or a condenser. In either case the partial impedances depend on the composition of the heat exchanger, its geometry, the roughness of the surface and, in the case of the partial impedance between the working fluid and the heat exchange surface, the composition of the working fluid itself. Typically, the thermal impedance associated with the working fluid side is much higher than the thermal impedance of the water side both for evaporators and condensers. The high thermal impedance from a smooth wall to an evaporating fluid is due to nucleating bubbles while the high thermal impedance from a smooth wall to a condensing fluid is due to the buildup of a film of the condensate on the heat exchange surface.

The thermal impedances associated with the working fluid side of the heat exchangers can be diminished by the augmentation of the actual heat exchange surface. On the evaporator side, permanent nucleation sites can be formed by almost closed channels[22] on the surface[2] or by coating the inner surface with a layer of spongy copper.[23] Corrugating or fluting the heat exchange surface alleviates formation of the film of condensate.[24]

Heat exchangers in OTEC applications tend to build up a layer of slime on the seawater side, a process known as biofouling. The biofouling action can be inhibited by the use of materials such as stainless steel or titanium, but these materials are quite expensive. Copper-nickel alloys are cheaper and have good resistance to biofouling but their use is incompatible with ammonia which is the most attractive working fluid for closed-cycle operation from a purely thermodynamic standpoint.[12] The present consensus favors the use of aluminum, as it possesses favorable heat transfer characteristics, and is relatively inexpensive although some positive action needs to be taken to keep it clean. Commercial devices used for the manual cleaning of heat exchangers include the M.A.N. brush and the Ampertamp sponge rubber ball. Biocides such as chlorine may be used to prevent the buildup of the slime layer. Newer proposals include the use of ultraviolet radiation and ultrasound.[25]

iv. Cold-water pipes. The largest physical feature of an OTEC power station is the cold water pipe which must extend hundreds of meters into the depths of the ocean and have sufficient diameter to move cold ocean water to the power cycle condensers at the rate of roughly 4 m^3/sec for each MWe of power output. In addition, the pipe must be outfitted with pumps that are sufficiently powerful to support this flow against the losses due to friction in the pipe, density differences between deep water and surface water, and pressure drops in the condenser. Pumps of this capacity do not represent a significant departure from current pump technology although they may represent about 10% of the cost of a closed-cycle plant.[26] The pipes must be constructed so as to be able to withstand the stress of ocean currents. Various materials have been proposed: aluminum, steel, fiber-reinforced plastic, and reinforced concrete.

The deployment of the cold water pipe presents a considerable challenge in itself. It may be that the technologies developed by the petroleum industry for the deployment of offshore drilling rigs apply in the deployment of OTEC installations[12] as OTEC power replaces oil power — an observation which contains a hint of irony.

v. Closed-cycle installations. Although the 100 MWe scale closed-cycle installations have never been deployed, smaller-scale versions have been constructed and tested. A 50 kW (2 MWt) test unit, Mini-OTEC, operated off the coast of Hawaii during 1979; DOE ran a 1 MWe (40 MWt) installation, known as OTEC-1, off Hawaii in 1980; in 1981 the Japanese built a 100 kW (4 MW) installation on the Pacific island of Nauru.

Mini-OTEC[27] was deployed on August 2, 1979, on a barge moored 2 km off Keahole Point. It was the first closed-cycle OTEC plant ever built and the first OTEC plant of any kind to run with a net energy gain, developing a net power output of 10–15 kWe out of its gross capacity of 50 kWe. It used titanium shell-and-tube heat exchangers and the cold water pipe was 0.6 m in diameter and took water from a depth of 650 m. For a total of four months of operation, continuous chlorination of 0.1 mg l^{-1} was used and no biofouling problem on either the evaporator or the condenser was observed. The heat exchanger performance was satisfactory. The power that mini-OTEC developed was dissipated on board.

DOE operated an engineering test facility known as OTEC-1,[28] located about 30 km off the Kona Coast of Hawaii. The facility tested the performance of candidate heat exchanger designs rated at 1 MWe (40 MW). No electric power was generated. The platform for the experiment was a T-2 tanker equipped with shell-and-tube heat exchangers. The condenser shell, made of carbon steel, was about 15 m in length by 3 m in outer diameter and contained 5526 titanium tubes 12.6 m long with a 2.54 cm outer diameter and wall thickness of 0.7 mm. The seawater flowed through the titanium tubes while the working fluid (ammonia) flowed around them on the shell side. The evaporator was of a similar design except that the insides of the titanium tubes were coated with a Union Carbide High Flux surface in order to enhance the heat transfer. The polyethylene cold water pipe (actually a bundle of three pipes) had an effective diameter of 2 m and extended to a depth of 700 m.

The object of OTEC-1 was to obtain data on heat transfer, biofouling, corrosion, and fouling countermeasures. It became fully operational on December 31, 1980 and the project was terminated prematurely on April 15, 1981 due to budget constraints. The lifespan of OTEC-1 was sufficient to obtain good heat transfer data but not long enough to yield conclusive results on biofouling, corrosion, and fouling countermeasures.

OTEC-1 continued to function in winds gusting to 84 km h^{-1} with subsurface currents of 5.6 km h^{-1}. During one storm on March 4, 1981 the cold water pipe came into contact with the gimbal stops.[29] On another occasion, the installation was allowed to drift free of its mooring in order to ride out the storm. These events provided a degree of confidence in the ability of OTEC installations to survive in open sea conditions and furnished information for the design of better mooring and piping schemes.

The Japanese government Ministry of International Trade and Industry (M.I.T.I.) commissioned the construction and operation of a 100 kWe OTEC installation on the shore of the island of Nauru.[30] Ocean water was supplied to the plant through three aqueducts. The cold water pipe extended 900 m to a depth of 540 m. The heat exchangers were shell-and-tube design with stainless steel tubes in the evaporator and exterior fluted titanium tubes in the condenser with freon as the working fluid. It operated intermittently between October of 1981 and September of 1982, during which time it had a 240 h period of continuous operation. It developed a net power output of 35 kWe which was more than double the expectation for the installation. The power output was fed into the island's commercial power grid. The Nauru operation was the last complete OTEC installation anywhere in the world as of this writing.

2d. Open-cycle research and development: Overview

Georges Claude's effort off the Cuban coast was the first—and last—open-cycle OTEC deployment anywhere in the world as of 1989. Despite inherent problems, OC-OTEC has a greater fundamental potential as a profitable energy technology than its closed-cycle counterpart due to the basic efficiency advantage in direct contact heat exchange. In direct contact heat exchange the working fluid vapor (steam in this case) mixes directly with the fluid, evaporating it on the evaporator side and condensing it on the condenser side. Therefore, the temperature of the steam entering the turbine is essentially the same as the temperature of the warm water exhaust, and the condensing steam temperature is that of the exhausted cold water. In a closed-cycle mode, the evaporating working fluid vapor is at a lower temperature than the exhausted warm water while the condensing working fluid is at a higher temperature than the cold water exhaust, thus reducing the temperature difference for the cycle and its thermodynamic efficiency. Some recent system studies have indicated that for onshore OTEC installations in the 5–10 MW range, OC-OTEC plants are superior to closed-cycle designs.[31] For this reason, over the past five years, the major focus of the international OTEC research and development community — in particular, the DOE's Ocean Energy Technology Program — has shifted from development of viable closed-cycle technology to an effort to solve the problems of open-cycle installations.

In terms of major system components, OC-OTEC is similar in many respects to CC-OTEC. For example, the technology associated with the flow of the warm and cold seawater through the system would not be substantially different. However, there are two major component systems in OC-OTEC which are particularly demanding of attention. One of these is the heat exchange system which must evaporate and condense seawater while removing atmospheric gases which were dissolved in the seawater. The other is the turbine which must be rotated by the low-pressure water vapor flow to drive a generator in the MW range.

i. OC-OTEC turbines. Because the low-pressure steam that drives OC-OTEC turbines carries a low power density, the necessary turbine size is immense. An OC-OTEC installation producing 11.7 MWe of net power would require a turbine diameter of nearly 12 m.[32] The stresses on the blades of a turbine this size are far beyond the capacity of present designs and materials. As of this writing, the upper limit of power production of an OC-OTEC turbine is in the 2–5 MW range,[7] although it is possible to operate more than one turbine in parallel in order to obtain a higher net power output from one OTEC installation. Crossover technology from the aerospace industry[5] and studies of alternative turbine geometries such as 3–6 m squirrel cage designs, vertical axis units, and radial inflow designs are under consideration.[11]

ii. Direct contact heat exchange for OC-OTEC.[32] Direct contact gas (or vapor) to liquid heat exchange can occur in four, more or less distinct, ways: simple gas cooling, gas cooling with vaporization, gas cooling with partial condensation, and gas cooling with total condensation. Each of these processes is described by a separate set of complex relations. The hardware for direct contact heat exchange falls into five classes: packed towers, spray towers, cross-flow

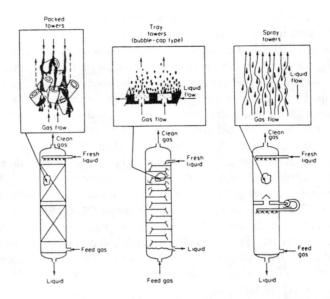

Figure 5. Schematics of gas-liquid direct-contact transfer devices (from [KB88]).

tray towers, baffle tray columns, and pipeline contactors with the most common designs being the baffle tray columns and the spray towers (Figure 5).

In a typical OC-OTEC thermal cycle, the entering warm water is admitted to a partially evacuated chamber so that the temperature of the incoming stream is at or above the flash point of the water. The resulting steam passes through the turbine where it is condensed by direct contact with the cold water stream. The condensation process is mediated to some extent by the presence of noncondensible atmospheric gases which were originally dissolved in the warm seawater. These gases must be pumped out of the cycle by external means so that the pressure on the downstream side of the turbine does not increase. The engineering challenge of designing a heat and mass transfer system to accomplish all these tasks lies in holding the flash evaporation and the condensation processes as close to their thermodynamic ideal as possible while minimizing the parasitic power required both to pass the material through the heat exchange devices and to pump out the noncondensible gases.

iii. Evaporators. The heat transfer in flash evaporation depends on the rapid renewal of liquid surface regions in order to work efficiently. The flash evaporation process cannot be characterized by a normal heat transfer coefficient because the area over which the heat transfer takes place is not really well defined. The effectiveness of the flash evaporation process is the ratio of the temperature difference between the liquid influx and the liquid exhaust to the temperature difference between the liquid influx and the vapor temperature corresponding to the chamber saturation pressure. An effectiveness of 1 is the maximum value allowed by thermodynamics.

Studies have shown that the most effective flash evaporation system for

OC-OTEC is the falling planar jet system. The falling planar jet evaporator shoots the water through one or more thin slots arranged so that a flat jet of water sprays directly downward from each slot. The jets shatter into a droplet spray upon entering the evaporator. The falling planar jet design was observed to have an effectiveness of 0.7–0.8 for countercurrent vapor flow[33] and an effectiveness as high as 0.95 for crosscurrent vapor flow.[34] Effectiveness was independent of initial jet width and could be further improved by passing the jet through one or more metal screens placed at various locations in the flow path.

The vertical spout evaporator in which the inflow water sprays upward out of one or more narrow vertical pipes maximizes the surface breakup in the flow without introducing a severe pressure drop penalty in the incoming liquid flow. The spray may be passed through a screen to further encourage the breakup of the flow. This design yielded an effectiveness of 0.97 in some trials.

iv. Condensers. On the downstream side of an OC-OTEC turbine, the design goal for a direct contact condenser is an arrangement with high heat transfer efficiency in conjunction with small volume and low parasitic power loss as a consequence of moving material through it. Current thinking finds that packed towers using ordered packing material such as corrugated sheets or wire gauze are most efficient. The use of ordered packing material improves the performance characteristics of packed tower condensers in the sense of diminishing the pressure losses associated with the vapor flow, although the ordered packing designs are more expensive to manufacture.[35]

System studies show that the optimal condenser will incorporate a two-stage process. In the first stage, the steam is directed downward in a co-current flow with the coolant. Most of the condensation occurs in the first stage. The second stage passes the exhaust vapor from the first stage into a counterflow with the coolant. The exhaust from the second stage becomes enriched in noncondensible gases and so the energy involved in the final exhaust pumping process is minimized. Calculations indicate that 70% of the steam is condensed in the first stage with the rest condensing in the countercurrent stage.[36] The vapor entering the countercurrent flow contains about 2% noncondensible gas while the exiting vapor is enriched to about 40% noncondensibles. The concentration of noncondensible gases which is performed by the counterflow stage reduces the exhaust pumping power by about 80%. It may prove that performance can be improved further by using several condensation stages instead of just two.

v. Indirect contact heat exchange for OC-OTEC: Desalinization. One of the potential benefits of OC-OTEC is the production of desalinized water from the the spent vapor. For obvious reasons, this cannot be done if all the vapor is condensed by direct contact with seawater. The solution to the problem is to install an indirect contact heat exchanger in tandem with the multistage direct contact condenser and divert enough steam through it to produce the desired volume of fresh water. Unfortunately, this plan leads back to facing the inherent inefficiency of indirect contact heat exchange.

2e. OTEC economics

The future of Ocean Thermal Energy Conversion as a viable energy technology is essentially an economic issue. The viability of OTEC in a particular region is strongly linked to the local price of oil, since oil fired power plants are the most common competing technology. It appears certain that OTEC is technically possible right now, with all indications that the price of developing it will fall as we progress along the learning curve.

The primary reason why industry has not pursued OTEC seems to be that OTEC power plants are capital intensive relative to fossil fuel plants although they do not have fuel costs after construction. A calculation shows that if a 100 MWe OTEC installation were to be installed at a cost of about $6k/kWe with an interest rate of 10%, it would take nearly ten years to pay off the capital investment if the payments were made at the rate of the operating costs of a 100 MWe oil fired power plant where the initial price of oil was $40/bbl and rose with inflation.[37] This calculation assumed a higher price of oil (by a factor of 2) than currently exists and perhaps a slightly low estimate of the installation cost of OTEC.[38,39]

i. Federal support. In 1980, at the end of the Carter Administration, two OTEC specific laws were enacted by Congress. One of these, U.S. Public Law 96-310, established targets for federally sponsored OTEC demonstration plants of 100 MWe by 1986 and 500 MWe by 1989. This law also set a commercial industry target of 10,000 MWe for 1999. The companion law, U.S. Public Law 96-320, was enacted for the purpose of making OTEC investment attractive to industry. It specified that OTEC plants and plant-ships were to be considered as vessels and therefore they were to be qualified for Federal mortgage loan guarantees — up to near $1500 M for construction of up to five OTEC power plants. In addition, OTEC would fall under the provisions of the Public Utilities Regulatory Policies Act which specifies that utility companies must pay for electricity purchased at the same rate they would pay to get the same electricity from oil.

The economic climate brought about by cheap oil in the 1980s has done away with the installed capacity goals specified by PL 96-310, so the time scales over which the law is to take effect have been expanded. The responsibility of administrating PL 96-310 lies with the Federal Ocean Energy Technology Program (FOETP), a branch of DOE. The basic strategy of the program is to perform the high risk research and development activities necessary to design an OTEC installation in the 5–15 MWe range. The interim (five year) goal of FOETP is to bring the estimated cost of both CC-OTEC and OC-OTEC to below $7200/kW, at which level it would be competitive with fossil fuel power on Pacific islands. The long-range FOETP goal is to bring the cost of both technologies below $3200/kWe, at which cost it would become competitive with conventional power sources located on or near the U.S. Gulf Coast.

There are a number of potential spin-off benefits associated with an onshore OC-OTEC installation which could enhance the economic feasibility of OTEC power production. The production and sale of desalinated water could lessen the effective cost of power production. In certain locations, the entire cost of the station could be subsidized by the sale of fresh water alone.[32] The cold seawater that is pumped to the surface during the OTEC process is rich in nutrients. Natural upwellings of cold water cover 0.1% of the ocean surface, yet produce 44% of the world's fish. The artificial upwellings caused by OTEC could spawn large-scale mariculture operations.[28,40] In addition, the exhausted cold water could be piped to nearby buildings in order to provide air conditioning and refrigeration.

2f. Environmental consequences

The questions that arise in regard to OTEC's effect on the environment are: What are the effects of OTEC on the ocean's temperature? What is the effect on local ocean chemistry? Are there any effects on marine life? Will it affect local weather patterns? Also, considering the possible locations of OTEC stations, it is important to assess the impact of the environment upon OTEC and ask what effect the local meteorology will have on the OTEC process.

One very simple global estimate[2] of the effect of OTEC on ocean temperature assumed an equilibrium between the absorption of solar energy by the oceans and the loss of heat due to vaporization. The equilibrium changes when OTEC plants are introduced. The conclusion was that the average surface temperature of the tropical oceans would be lowered 1°C due to production of 60 billion kWe via OTEC—an amount of power that would provide each member of (the 1970 estimate of) the world's population in 2000 with the amount of electrical power that was enjoyed by a U.S. citizen in 1970. The analysis went on to assert that the cooling of the surface waters in the tropics by pumping surface water below and bringing cold water up would eventually set up global convection currents in the ocean which would have the effect of warming the waters of northern latitudes. On a more local scale, one analysis by Bathen[12] concluded that a 20 MWe OTEC plant off Keahole Point in Hawaii would drop the local (e.g., 1 km^2) water temperature only 0.02–0.07°C, well below the natural seasonal fluctuation. The effects of large ocean temperature changes on marine life, ocean chemistry, and meteorology is an issue that should be seriously considered before large-scale deployment of OTEC plants begins.

According to Henry's law, the solubility of gases in water increases as the water temperature drops and water pressure increases. For these reasons, the ocean water below the thermocline is a copious reservoir of dissolved atmospheric gases. When this deep water is brought to the surface and its temperature is raised, these dissolved gases escape to the atmosphere. The largest fraction of the released gas is CO_2; thus it turns out that OTEC could contribute to the greenhouse effect although an OTEC cycle is estimated to release one-third of the carbon/MWe that a fossil fuel cycle would produce.[41] The release of CO_2 also raises the pH of the ocean water, normally at a pH of 8. Above pH 9, $CaCO_3$ precipitates and may not redissolve, leading to a turbid, alkaline effluent.

Closed-cycle OTEC power has the potential for a spill of the freon, propane, or ammonia used as the working fluid in the power cycle, or pollution occurring as a consequence of the use of biocides in cleaning the heat exchangers.

The potential for large-scale pollution is magnified, of course, if OTEC is powering a floating chemical refinery.

2g. Present and near future

At present, there are no working OTEC power plants. There are, however, facilities where various components of the OTEC process are installed and operating and there are near term plans to construct a fully operational facility in Hawaii. The Natural Energy Laboratory of Hawaii (NELH) is conducting a number of experimental investigations of aspects of the OTEC process at their Seacoast Test Facility (STF) located at Keahole Point. They have installed a water supply system consisting of a 6000' long (40" dia.) polyethylene cold water pipe that can bring up to 13,000 gallons of water per minute from a depth of 2200' and a 500' (28" dia.) warm water pipe that can draw up to 9600 gallons per minute from a depth of 60'. This is the only continuous supply of deep ocean water in the world. Plans call for this experimental system to evolve into a fully operational OC-OTEC power plant that will produce 165 kWe gross power sometime in the 1990s.

2h. Summary and conclusions

There is a tremendous amount of thermal energy stored in the earth's oceans — in theory, enough to satisfy the total world's energy requirement. The problems of converting this energy into an immediately useful form and transporting that energy to locations where it is needed are difficult to solve, but not insurmountable. Over the last twenty years, interest in the OTEC process has been growing worldwide and a substantial engineering knowledge base has been growing along with it. Although, it is true that OTEC can be accomplished through an integration of currently existing technology, it is clear that more innovation is necessary in order to increase the efficiency of the process to the level where it becomes economically competitive with current energy producing technology. The high capital cost of OTEC makes investment in it seem risky. Accurate long-range prediction of future cost of energy is necessary to be able to assess the success of OTEC investment since OTEC's payback period is long.

It is unfortunate that the level of commitment to OTEC is not as high as originally mandated by the OTEC specific laws of 1980. The amount of activity in this regard seems to be inversely related to the price of oil. Nevertheless, the next five years should see the construction of some small-scale OTEC demonstration plants, followed by development of OTEC power in island communities in the Pacific. The next step would be the development of OTEC in the Gulf Coast region. These predictions all depend on a rising price of oil coupled to a lowering of real cost of OTEC as its technological learning curve is traversed.

OTEC is demonstrably a working technology which is in need of some improvement in certain areas. It has enough intrinsic merit that its further research and development should be vigorously pursued in the near future so that a mature and viable OTEC technology will exist if it is needed to alleviate a significant portion of the energy needs that will arise in the 21st century.

3. Tidal power

3a. Basic principles

Both the sun and the moon exert gravitational forces on the earth's oceans so that the oceans develop a slight bulge along an axis which is approximately in line with the moon. Since the earth rotates with respect to the tidal bulge axis, a given point on the seacoast will experience a fluctuation in sea level over a slightly more than twenty-four hour period. The tidal period is slightly longer than twenty-four hours because the moon has moved slightly ahead in its orbit around the earth during the period of the earth's rotation. Local coastal geographical features complicate the situation.

The tidal fluctuation comprises two high tides alternating with two low tides each day. The extent of the tidal fluctuation depends on the relative positions of the sun, moon, and earth. The highest tides occur when all three celestial objects are aligned. The tidal amplitude along the seacoast also depends strongly on the local geography. Certain tidal basins are approximately in resonance with the rhythm of the tidal fluctuation and are able to act as tidal amplifiers. Canada's Bay of Fundy, for example, has tidal fluctuations of as much as 16 m due to the size and shape of the Gulf of Maine.[42]

Most serious schemes to exploit the energy in the tides involve the hydraulic head available between high tide and low tide. A dam at the mouth of a tidal bay is equipped both with sluice gates, which allow the free passage of water between the ocean and the tidal basin, and with turbine gates which shunt water through turbines which generate electricity. In the simplest mode of operation, the sluice gates are opened as the tide is rising, then closed at absolute high tide. Approximately six hours later, the tide has fallen once again and the high water can be let down out of the basin through the turbines until the action of the returning tides and the draining of the basin degrade the hydraulic head.

Using this scheme would allow the generation of electrical power in spurts that would occur at different times during each day and about twelve hours apart. The amount of energy generated depends on the number of turbines that are installed. The maximum possible energy that could be extracted per tide from a specific tidal basin would be derived in the case where all the water in the tidal basin could be drained instantaneously through turbines that could extract all of its gravitational potential energy. This ideal program cannot be realized, of course. Nevertheless, the maximum energy case can be approximated by real projects.

Designs that maximize energy production generate tidal power at times which don't often correspond to times of peak demand for electricity. A number of schemes have been proposed which diminish energy to obtain increasingly levelized power output. One option for power levelling is to allow the high tide to fill the basin, then close the gates at high tide, but continue pumping water into the basin as the tide is ebbing to raise the head in the tidal basin

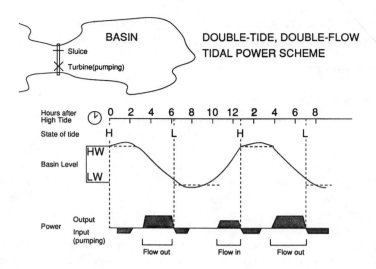

Figure 6. Illustration of a double tide double effect tidal power plant scheme.

above the high tide level. After the tide has ebbed for an hour, the pumps are shut off. After another hour the basin water is allowed to flow through the turbines at a rate that maintains the amount of head between the basin and the ocean at a constant level. Consequently the power output is constant for about four hours on an ebbing tide. It is also possible to close the gates at low tide, pump water out of the basin for some time in order to maximize the head between the basin and the ocean so that the returning tide can generate uniform power for a few hours as it fills the basin (as shown in Figure 6).

Other ideas for promotion of uniform power output involve dividing the tidal basin into two or more separate chambers, and managing the tidal flow so that there is always a tidal head between either a pair of adjacent chambers or a chamber and the ocean. Current wisdom in the tidal power engineering community holds that all designs should maximize energy output, since tidal power will be one of many options available on a local power grid. Load management would be accomplished through integration of many diverse resources.

3b. Tidal resource – sites

It has been estimated that 3,000 GW of power is continuously dissipated on the whole earth due to the action of the tides.[43] Of this, only about 2% is considered usable. The extent of the usable resource is a consequence of the coastline geography.[44] The potential tidal energy available depends linearly on the area of the tidal basin and on the square of the tidal range. The actual energy that can be derived from a tidal power project is lower than the potential energy by a utilization factor which depends on the nature of the tidal power scheme and falls in the range of 0.22 to 0.34.[45]

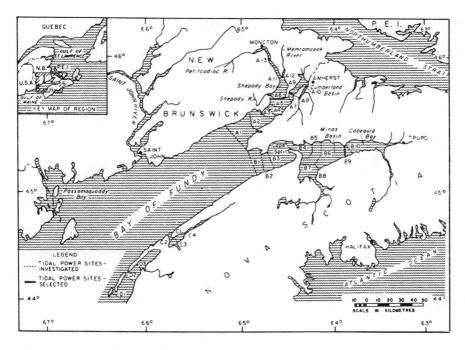

Figure 7. Proposed tidal power sites in the Bay of Fundy.

Another figure which governs the feasibility of a tidal power project at a particular site is Gibrat site ratio, the ratio of the length of the dam in meters to the annual energy production in kWh. A smaller site ratio is more desirable. Some examples are La Rance: 0.36; Severn: 0.87; Passamaquoddy: 0.92.[46]

Ideal basins for installation of a tidal power unit possess large surface areas with a narrow channel connecting them to the open sea. Moreover, the basin size must be in resonance with the tidal period so the tidal amplitude is large enough to make development worthwhile. In the continental United States, the only site that has legitimate merit is the Passamaquoddy-Cobscook Bay area in eastern Maine, an area with the potential for a 1 GWe tidal power installation. The Cook inlet area, near Anchorage, Alaska, offers the potential for a 2.6 GWe tidal power installation.

The Government of Canada and the provincial governments of New Brunswick and Nova Scotia authorized the formation of the Bay of Fundy Tidal Power Review Board (BFTPRB) to assess in detail the economic and technical feasibility of development of tidal power projects in the Bay of Fundy. The Board examined twenty three different proposed sites and concluded that three of them, Cobequid Bay (B9), Shepody Bay (A6), and Cumberland Basin (A8) would provide the best prospects for development. (See Figure 7 and Table 1.)[47] Tidal power from these sites could be fully absorbed into the power grids of the maritime provinces and New England. Without government subsidies, the initial investment costs for development of these sites would place an inordinate financial burden on utility customers. None of these sites has as yet

Table 1. Proposed tidal power sites in the Bay of Fundy.

Site	Net Plant Capacity MW	Average Annual Output GWh	Costs in $Millions (June 1976 dollars)			Projected in-service cost[a] (1990)
			Tidal Plant	Transmission Link	Total	
Cobequid Bay (B9)	3800	12,653	3637	351	3988	9290
Shepody Bay (A6)	1550	4,533	2160	37	2197	–
Cumberland Basin (A8)	1085	3,423	1197	37	1234	3120

[a]Projected total in-service costs assume, for the purposes of the study, commissioning by 1990 and include escalation and interest during construction: development of Site A8 would require about seven years to complete and that of Site B9, about 11 years: in-service cost of A6 was not computed since a development at that site is not economic. (Bay of Fundy Tidal Power Review Board.)

been developed, although a 20 MWe pilot plant was constructed at Annapolis Royal, Nova Scotia.[48] When the Annapolis Royal Project opened in 1984, it had the the distinction of being the first (and is, as yet, the only) tidal electric power station to operate in North America.

Other countries have been more successful. The Rance river in the northwest of France is the site of a 240 MWe tidal power plant that has operated there since 1966.[49] At Kislaya Bay, on the Barents Sea north of Murmansk, the USSR has operated a 400 kWe tidal generating station since 1969[50] with plans to build a couple of more units nearby around the White Sea—one site having a potential for producing 6 GWe. The UK has extended extensive scrutiny to the proposal to construct a tidal power system in the Severn estuary on the west coast of England, near Wales. The Severn site could be developed to produce as much as 12 GWe of power according to the latest studies but the most economically favorable development scheme involves construction of a 7.2 GWe facility.

3c. Existing technology

Both tidal power plants and hydroelectric plants require the construction of dams to create and maintain a reservoir of water at a large hydraulic head; both require the installation of turbines and generators that can transform the flow of water, driven by the hydraulic head, into electrical power. Tidal power generation differs from hydroelectric in that (1) the hydraulic head produced by the tides is less than that found in a typical hydroelectric project, (2) the flow is bi-directional, and (3) salt water is more corrosive than fresh water.

The hydroelectric industry spawned the development of low head hydraulic turbines which are suitable to tidal applications. The petroleum industry pioneered the use of prefabricated structures for permanent stationing at sea, such as oil drilling platforms, a construction technique that is predicted to be very useful in future tidal power development. For example, the components of the Soviet tidal project at Kislayaguba were prefabricated onshore, floated to site, and sunk in place. In terms of economics, this is generally considered to be

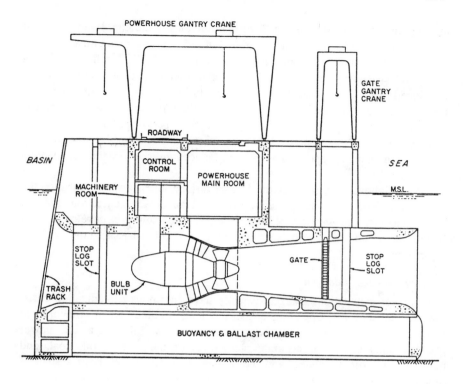

Figure 8. Illustration of modular tidal powerhouse caisson containing a Kaplan Bulb Turbine. (From reference 50.)

the optimal construction technique.[51] Floating caissons which can house either a turbine or sluiceway that would be integrated into the powerhouse of the tidal power plant are constructed out of steel or concrete. Each caisson (see Figure 8) can then be floated to the site of the tidal barrage, positioned, and sunk in place.

The existing tidal power projects both at La Rance and Kislayaguba use horizontal shaft bulb turbines where the hydraulic rotor is placed in a plane perpendicular to the tidal flow. The generator is located in a large steel bulb directly behind the turbine itself directly in the tidal flow. This design allows a large hydraulic efficiency owing to the straight flow of water through the turbine. The pitch of the turbine blades of the bulb units are adjustable to change the rotation speed of the turbine in a given flow. In addition, the generators in the bulb units can also function as motors, so the turbines can also serve as pumps. The disadvantages of the bulb designs are that in order to accommodate the stator ring of the generator, the bulb must be quite large, which obliges the hydraulic passageway to be large enough to contain the bulb. In addition, cooling, maintenance, and repair become more of a problem. Each of the bulb units at La Rance has a rated capacity of 10 MWe with a rotor diameter of 5.6 m. As a group, they have been in use 95% of the time they could conceivably be used and have achieved 91% of their design output over fifteen years of use.[44]

The pilot plant at Annapolis Royal uses one 20 MWe Straflo Turbine, a horizontal shaft rim-type design, 7.6 m in diameter. The generator stator is positioned around the circumference of the turbine rotor. The Straflo design preserves the desirable flow characteristics of the bulb designs while eliminating the bulk of the generator bulb from the flow path. Special care must be taken in these turbines to prevent the water that rotates the turbine from leaking into the generator. It appears that this can be accomplished by the use of pressurized seals.

Certain innovations, such as the Straflo turbine, may prove to enhance the overall efficiency and economic benefits of tidal power generation, but the basic principles are well established. Whether tidal power is developed on a large scale around the world is an economic and political issue, not a technological one.

3d. Economics of tidal power

Tidal power installations characteristically have enormous capitalization costs, low maintenance and operating costs, and long lifetimes when compared to other energy technologies. To assess the economic viability of a tidal power plant requires a host of accurate predictions of the cost of oil, coal, and uranium, the cost of money, and the market for electric power for a span of time on the order of a century. This is not to mention the possibility that other emerging energy technologies could evolve to the point that they would provide cheaper energy.

The important factors that determine the cost of tidal power are the site ratio of the intended location of the power station and the design of the power plant, i.e., how much capacity will be installed, single or double effect, how many pools, etc. These physical and engineering considerations determine the capital cost of the facility. In addition to the capital costs are the prevailing interest rates, taxes, insurance, and operating and maintenance costs.

To determine benefits requires an assumption about the nature of the energy technology that would come to exist in lieu of the tidal power project. Having defined that scenario, it remains to assess a cost or value to it over the lifetime of the proposed tidal power project.[47,52]

3e. Environmental consequences of tidal power — Bay of Fundy model

The alteration of the tidal flow by building a barrage across the mouth of a tidal estuary will have an impact on the local environment and the human activities in the region. Specific environmental concerns include the effects of tidal power exploitation on fish, shellfish, and other marine biota, the effect on local shorelines and wetlands due to altered tidal amplitude, and the impact on transportation and recreation in the location of the installation.

All of these concerns need to be quantified in terms of their magnitude and extent in order to assess the relative impacts in any meaningful fashion. The key issue is the accurate prediction of the changes in tidal amplitudes in the region of the tidal barrage. It appears there is no rule of thumb estimate of the change of tides that a region would experience as a consequence of a tidal power project; it is a very complicated problem, and its solution will vary with each tidal basin under consideration since the tidal dynamics are so strongly dependent on local geography. The experience with the tidal project at La Rance suggests that the local tides changed only slightly as a consequence of the installation of the tidal barrage there, but that minimal impact is not thought to be typical of any tidal basin.[53]

A number of computational models have been developed in order to predict tidal dynamics in various regions around the world. The Bay of Fundy/Gulf of Maine (BOF/GOM) region is perhaps the most extensively modeled tidal system in the world today. The BOF/GOM basin is bounded on the west by Cape Cod, on the Northwest by the Maine Coast and on the East by New Brunswick and Nova Scotia. The region is bounded by the open Atlantic Ocean on the south and southeast. Although the Gulf of Maine is apparently open to the Atlantic Ocean there are two undersea features—the Scotian Shelf and George's Bank that serve to mediate the flow of water from the open ocean into the gulf. The ocean depths over George's Banks and the Scotian Shelf are typically a few tens of meters in contrast to depths of hundreds of meters that are found closer in to shore in the Gulf of Maine. Water enters and leaves the gulf via the Great South Channel (depth 140 m) which separates George's Banks from Cape Cod, and also from the Northeast Channel which lies off the Nova Scotian coast and attains a depth of 360 m. The Northeast Channel plays the dominant role in water transport in and out of the gulf. The continental shelf lies just to the south of George's Banks and the Scotian shelf and the water depth drops precipitously to 2000 m.

In a typical tidal model of the BOF/GOM, the region is divided into a finite grid of square cells of varying sizes. Equations of fluid dynamics that apply to the basin are solved in a discrete fashion on the lattice of cells by a finite difference method. The fluid equations are set up to preserve both conservation of mass and momentum over the grid and take into account the pressure gradients acting on the fluid, the frictional or viscous forces on the water, as well as the "Coriolus force" due to the earth's rotation. Once the initial values of current and water level in each cell are provided, the values of these quantities on all the boundary cells are specified for all time. Then numerical integration of these equations provides a prediction of the water level in each cell and the current flowing through each cell as a function of time.[54]

In order to model the effect of a tidal power plant in the Bay of Fundy, such a plant was installed in the model by specifying a new boundary condition at the location of the intended tidal barrage to have the flow characteristics that a single-acting, single tide tidal power plant would impose by its operation. The results of such a calculation indicated that the amplitude of the tide would be reduced by 34 cm at the tidal barrage in the Minas Basin (site B9 in Fig. 7), but that as one moved South along the New Brunswick and Maine coast, the tide would be amplified. At the mouth of the Minas Basin, no change in tidal amplitude was predicted. At St. John, New Brunswick, the predicted amplitude in-

Table 2. Amplitudes (in centimeters) and phases of the M₂ tides predicted at various locations in the Gulf of Maine by the Greenberg Model for the separate cases of development of the tidal power installations at sites A6 and B9. Immediately below each figure in parentheses appears the deviation from the actual measured value at each location. From Ref. 53.

Location	A6		B9	
	Amplitude	Phase	Amplitude	Phase
St. John	304	339°	321	338°
	(+1)	(+2°)	(+18)	(-3°)
Digby	306	330°	319	330°
	(+1)	(-2°)	(+14)	(-2°)
Yarmouth	163	309°	170	311°
	(+2)	(-1°)	(+9)	(+1°)
Portland	138	346°	148	349°
	(+4)	(-1°)	(+14)	(+2°)
Boston	139	353°	150	356°
	(+4)	(-1°)	(+15)	(+2°)
At Tidal Barrage	454	349°	558	358
	(-24)	(-1°)	(-34)	(-17°)

crease was 18 cm, 16 cm at Bar Harbor, 15 cm at Portland, Maine, and a 15 cm increase in tidal amplitude at Boston (Table 2).

Whether or not this predicted variation in the tidal amplitudes in the Gulf of Maine would have a large environmental impact depends in part on the natural variability of the tidal amplitudes in this region as the ocean water is driven one way or the other by meteorological events in the Gulf of Maine. Strong onshore winds can cause a rise of water level along the coast by as much as several meters and this rise can be accentuated by a drop in atmospheric pressure that is often associated with foul weather. A 3 kPa pressure drop can cause a 0.3 m increase in the mean sea level[53] and storm driven sea level rises of up to one meter have been observed in the Gulf of Maine. Also geological evidence indicates that Boston is sinking at a rate of between 15 and 30 cm per century due to crustal movement. In relation to all these other influences, the 15 cm potential change in tidal amplitude doesn't seem to be very significant. There is no clear consensus that human intervention in the tidal flow would have a large impact on the coastline ecology although there is some concern that the altered tides would threaten salt marsh regions and some sandy areas of the Maine coast. The shoreline flora and fauna seem to be quite robust to changes in local sea level due to the large natural fluctuations found in the region.

The alteration of the tidal amplitude in the BOF/GOM region can also enhance mixing in coastal waters. The issue of whether local waters are mixed or stratified bears strongly on the nutrient pathways in that environment and on the local fish population.[55] The Bay of Fundy area is an important spawning ground for herring and also for scallops. It is generally acknowledged that alteration of tidal currents in the region will have an effect on these creatures but no consensus exists on the nature of that effect. It has also been suggested that intensified mixing in some regions of the Gulf of Maine stimulates the growth of the red tide organism, *Gonyalaux excavata*, which causes paralysis in shell-fish.[56]

3f. Conclusions

There is well-developed technology for the production of electricity from tidal energy in favorable coastal areas. The development of this resource worldwide and nationally has been minimal so far, with the 240 MWe tidal power plant at La Rance being the largest single effort. The single effect/single basin mode of operation appears to be the most economical method to exploit tidal energy. Any power plants that are designed in this fashion will have to operate in conjunction with a number of other baseload technologies since the power output of such a tidal configuration is not steady in time and cannot be made to fit the patterns of local demand.

Tidal power installations characteristically have high capital costs, low operation and maintenance costs, and very long lifetimes. Their lifetime costs compare favorably with the costs of fossil fuel and nuclear energy production, although, in order to make these comparisons, one has to make assumptions about the cost of petroleum and uranium for a long time into the future.

The environmental consequences of the operation of a tidal power plant in a region largely depends on how much the facility will alter the tidal dynamics in the immediate area. The magnitude of this effect will depend quite strongly on the geography of the local region. The alteration of tidal patterns has an impact on the ecology of the intertidal zone, affects coastal erosion patterns and the nutrient pathways for fish and shellfish in the region. There is some environmental risk involved with tidal power, but it does not appear to be catastrophic.

4. Energy from ocean waves

4a. Ocean wave resources —A brief primer

Ocean surface waves are the consequences of some dynamic disturbance of the ocean surface. The origin of the disturbance may be (1) seismic, giving rise to tsunamis or "tidal waves" (a misnomer), (2) due to true tidal forces, or (3) due to the action of the wind playing over large expanses of the ocean surface. Waves of seismic origin are relatively rare and will not be considered as a dependable energy resource. Waves of tidal origin are predictable and their merits as an energy resource are discussed above. This section of the chapter on ocean energy resources will be concerned with the third category of waves — those seas and swells that are caused by the force of the wind on the ocean surface.

The theory of water waves was first worked out in detail by G.G. Stokes.[57] Stokes's theory is derived from basic principles of fluid mechanics under the assumptions that water is both incompressible and inviscid. These basic assumptions are incorporated into a nonlinear mathematical expression for the wave motion of the water surface. A linear water wave transfers energy through space. The energy transfer can be characterized by the energy flux or wave power.[58] The expressions for simple, linear, harmonic water waves form a basis for quantitative understanding of the energy transport of real ocean waves.

In general, the wave motion encountered on the surface of the deep ocean cannot be simply described as harmonic wave motion of a simple wavelength and period, but rather as a superposition of waves from a wide spectrum of wavelengths. From the standpoint of energy use, it is desirable to be able to characterize the *spectral density* of the seas and swells that appear in an ocean region, to assess the energy flux of the entire spectrum and to relate these quantities to observed wind loading over the expanse of ocean, or *fetch*, over which it blows. The actual form of the spectral density function must be determined by empirical observation of the local ocean wave climate.[59]

The energy flux that is carried by the waves in a random sea can be calculated by integrating the spectral density over all possible wave periods. This integral can be expressed as a relationship between the average wave period in the sea and the significant wave height, both easily and directly observable quantities.[60] The significant wave height is defined as the average height of the highest one-third of all the waves in the distribution. It is the best representation of what a human observer would report as the wave height, judging by sight. The average wave period can be determined by looking at the zero-crossings of a wave recording device. It is noteworthy that the energy flux increases like the square of the significant wave height.

Some values for the average annual energy fluxes in the ocean waves found at various nautical locations off the coasts of Canada and the United States are presented in Table 3. These values are taken at sites of varying depth and distance from shore. Typically, the energy flux diminishes as the water depth decreases, i.e., as the offshore distance diminishes. The wave power of the open ocean is degraded as the incoming waves encounter the continental shelf and the energy content continues to attenuate as the waves approach the shoreline. The information in Table 4 indicates that the ocean waves in the Pacific Northwest region carry much more power than those found in the Atlantic Northeast. Nevertheless the energy fluxes reported are large enough that the prospect of harnessing this energy is economically feasible.

4b. Extracting the energy of the waves

i. History. The first patent for a device to extract useful energy from ocean waves was issued in Paris on July 12, 1799 to the Girards, father and son. They proposed to attach a long lever to a floating ship. The lever would run to the shore where its wave-induced oscillations would run a pump or other mechanical device. Since that time there have been issued more than 1,000 patents for wave energy conversion devices in North America, Japan, Western Europe, and the United Kingdom.

The forms of wave energy conversion can be categorized; essentially, all the designs of wave energy conversion devices are based on one of a half dozen or

Table 3. Wave energy climates observed in some locations off the coasts of Canada[a] and the continental United States.[b]

Location	Average Annual Energy Flux (kW/m)	Ocean Depth (m)	Distance from shore (km)
Nova Scotia			
Osborne Head	7.9	20–40	5
Western Head	8.8	40–50	5
Cape Roseway	6.7	80–90	12
Tofin, British Columbia	24	40	10
Weather Station Papa, lat. 50°N, long. 145°W	100	4200	1000
Northern California			
Coastal	5–15	10–20	1–5
N. of Point	25–30	40–150	5–20
Reyes	30–35	500–1000	20
S. of Point	20–25	40–150	5–10
Reyes	25–30	500–1000	10
North Carolina	5–6	20	10
	12–15	150	30

[a] In "An Evaluation of the Energy of Wind Generated Water Waves in the Coastal Areas of Canada," Sept. 1980, LTR-HY-82, Hydraulics Laboratory, National Research Council, Canada.
[b] Data provided by George Hagerman, Seasun Power Systems.

so basic principles—each encompassing a particular aspect of wave motion. These categories include devices that incorporate the *heaving* or up and down motion, the *pitching* or rocking motion, *pneumatic principles* (also known as *oscillating water column* such as the Bouchaux-Praceique device, described below), *surging* motion such as in the Mauritius breakwater, presented below, and *underwater pressure variations*.

In 1910, a Frenchman, M. Bouchaux-Praceique, constructed a wave energy conversion device that was able to deliver 1 kW of power to his home. The first large-scale wave energy conversion project was conceived on the island of Mauritius in 1959 by Walton Bott. Bott's design augmented one of the natural coral reefs that surround the island with concrete so that waves above a certain amplitude would break over the concrete wall and fill a reservoir. In this fashion, a small hydraulic head would develop which could be let down through Kaplan turbines in much the same manner as in a tidal power installation. Hydraulic ram pumps in the reservoir would use the action of draining the reservoir to pump some seawater to another reservoir on much higher ground. This was largely an economic measure so that one high-head turbine could be installed instead of several low-head units to provide the same amount of electrical power. The 5 MW project was abandoned, before construction, in 1966 due to a drop in the price of oil delivered to the island.

In 1965, Y. Masuda designed a 60 W air turbine powered buoy. This design,

Table 4. Twelve near-term wave energy conversion technologies (after Ref. 69).

No.	Device Name	Developer	Largest-Scale Physical Test
1	KN System	B. Hojland Rassmussen A/S Copenhagen, Denmark	1 kWe prototype in the Oresund
2	Hose Pump	Gotaverken Energy Gothenburg, Sweden	30 kWe prototype (1/3 scale buoys) off Vinga Islands
3	Wave Energy Module	NORDCO, Ltd.[a] St. John's, Newfoundland	1 kWe prototype on Lake Champlain
4	Contouring Raft	Sea Energy Corporation New Orleans, Louisiana	1/15 scale model in wave tank
5	Nodding Duck	University of Edinburgh Edinburgh, Scotland	1/10 scale model on Loch Ness
6	Tandem Flap	Q Corporation Troy, Michigan	20 kWe prototype on Lake Michigan
7	Neptune System	Wave Power International Sydney Australia	1/12 scale model in wave tank
8	Multi-Resonant Oscillating Water Column	Kvaerner-Brug A/S Oslo, Norway	500 kWe demonstration (full-scale plant) at Tofestallen
9	Backward Bent Duct Buoy	Ryokuseisha Corporation Tokyo, Japan	1/10 scale prototype in the Sea of Japan
10	SEA Clam	Sea Energy Associates, Ltd. and Coventry Polytechnic Coventry, England	1/10 scale model on Loch Ness
11	Tapered Channel	Norgrave A/S Oslo, Norway	350 kWe demonstration (full-scale plant) at Tofestallen
12	DELBUOY	ISTI-Delaware Lewes, Delaware	250 gpd prototype off Magueyes Island

[a] Licensed by U.S. Wave Energy, Inc., Longmeadow, Massachusetts, United States.

illustrated in Figure 9, was patented by the Ryokuseisha Corporation which manufactured over one thousand of the devices for use as navigational aids. These are in use in oceans around the world. The operating principle of the Masuda buoy was that it contained a vertical cylinder whose bottom was underwater and open and whose top was above sea level and capped by an air turbine. The bobbing of the buoy causes the water column in the cylinder to oscil-

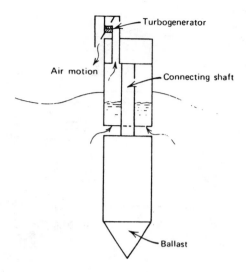

Figure 9. Diagram of Masuda's oscillating water column buoy. From McCormick (1981).

late up and down therefore pressurizing or rarifying the air above it. The air is either sucked into or expelled from the top of the buoy through a system of flap valves which ensure that the turbine spins in one direction regardless of whether the air is entering or exiting the top of the cylinder.

ii. United Kingdom — 1970s. Over a ten year period from 1974 to 1983, the British Government expended nearly $20 million on a national program of wave energy research and development. It is estimated that an average of 120 GW of wave power is dissipated on the west coast of the UK — about five times the British electrical demand.[61] The first stage of the British wave energy development program was the parallel development of four distinct wave energy conversion devices. Figure 10 illustrates each of the four technologies that were the basis of the early British effort.

Salter's Duck[62] is a pitching device called a nodding duck because the moving elements of the machine resemble the head and beak of a duck. This choice of shape induces a very favorable conversion efficiency to the device — around 90% for two-dimensional sinusoidal waves. In normal operation, several of these ducks are threaded onto a stiff shaft or spine about which they are free to rotate. The entire assemblage is moored in deep water, positioned so the spine is parallel to the incoming waves, with all the beaks facing the incoming waves. As the waves impinge on the beaks, they surrender some of their energy to rotatory oscillations of the ducks about the spine. The rotations of the ducks are coupled to a hydraulic pumping system which drives the electrical generators. Later innovations on this basic design have been the installation of gyroscopes in the beaks of the ducks. The rotations of the ducks cause precession of the gyroscopes and the precessions drive ring-cam pumps which pressurize the hydraulic oil.[63] This measure apparently reduces the bending moment of the spine — making the entire device less prone to mechanical failure.

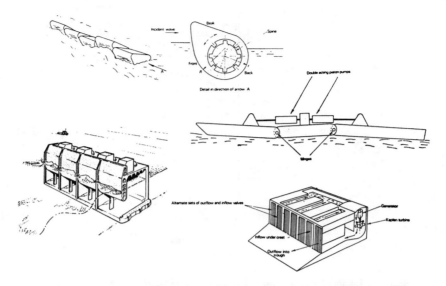

Figure 10. Four designs from the early stage of the British wave energy program. Clockwise from top left: Salter's Duck, Cockerell's wave contouring raft, NEL Breakwater, HRS Rectifier.

Cockerell's Wave Contouring Raft. In 1971, Sir Christopher Cockerell, the inventor of the Hovercraft, suggested attaching several rafts together in line by hinges. The rafts would be moored at sea with the hinges roughly parallel to the incoming wavefronts. As the waves rolled under the rafts, the linked rafts would undulate. By attaching double-acting hydraulic pistons across the hinges of adjacent rafts, the undulatory motion could be hydraulically coupled to a generator.

HRS Rectifier. The HRS (Hydraulic Research Station) Rectifier, introduced in 1975, is a surge device which would sit on the ocean floor, relatively close to shore. It is a large rectangular concrete caisson outfitted with several compartments that would face the waves. Each compartment has a set of one way flap valves so that a surging wave can open the valves and fill the chamber, but the valves cannot spontaneously open to let the water drain out of the chamber. The front to back distance of each chamber must be less than one quarter wavelength in order to prevent valve closure due to reflections from the back of the chamber. The design of the Rectifier causes the waves to fill its chambers to a water level higher than the mean sea level. The excess water drains into a holding chamber through a hydraulic Kaplan turbine which is coupled to the electrical generator. The holding chamber has flap valves that only open when the trough of a wave passes, allowing the water to drain back to the sea.

Under ideal conditions in a monochromatic wave environment, the maximum efficiency of the HRS Rectifier is about 20%.[58] On top of this fairly mediocre figure, it would also require a large capital expenditure to construct. It is not one of the more promising options for wave energy conversion technology.

NEL Breakwater-Oscillating Water Column. The NEL (National Energy Laboratories) Breakwater is a rectangular concrete structure which would sit in shallow water and parallel to the wave crests. The seaward side of the break-water has several chambers open to the ocean as in the Russell Rectifier, except that none of the apertures are blocked by any gates and the apertures themselves are totally submerged. There is a volume of air trapped above the water in each chamber which is alternatively blown and sucked through a turbine. In the early designs of this device, flap valves were installed to ensure that the flow of air through the turbine was always in one direction only. Later designs incorporated special turbines that rotate in only one direction regardless of the direction of the air flow.[64]

iii. United Kingdom — 1980s. The first generation of wave energy conversion designs that were investigated by the British DOE all proved to be rather expensive. Consequently, the DOE shifted its emphasis and funding to a set of newer ideas which held the promise to deliver cheaper power.

Lancaster Flexible Bag. The Lancaster Flexible Bag, shown in Figure 11, consists of a rigid buoyant spine that is moored so that it is perpendicular to the wave crests. Mounted atop this spine is a series of flexible membranes or bags which are filled with air. Each of these bags has two ducts, one of which connects to a high-pressure air manifold and the other is connected to a low-pressure manifold. The ducts are fitted with gate valves which prevent the flow of high-pressure air from the manifold to the bag and also prevent the flow of high-pressure air from any bag to the low-pressure manifold. As a wave crest travels along the spine, it compresses each bag and causes the bag to act as an air pump to the high-pressure manifold. When the wave passes by, the bag expands and draws air into itself from the low-pressure manifold. Power is extracted by an air turbine that operates between the high-pressure manifold and the low-pressure manifold.

Bellamy's SEA Clam. The Clam design called for a mooring scheme in which the spine of the unit would be held at an angle of approximately 35° to the wave crests with the air bags facing the open sea and bearing the full brunt of the oncoming waves. The wave action would make the bags alternatively draw and expel air through a self-rectifying turbine, one that spins in one direction regardless of the direction of air flow. Of all the first and second generation British wave energy conversion devices, the Clam was considered to be the most cost effective.[65]

One design drawback of the Clam was that the spine of the device is subjected to large bending moments. Use of a circular spine with the bags on the outside avoided this problem. Calculations and tests have shown that the circular design has several additional unforeseen benefits which include a 4-fold increase in device efficiency and a 2-fold decrease in cost of electrical energy (based on 1–2 MW capacity).[66] Scale models of the circular SEA Clam are currently being tested in Loch Ness, Scotland.[67]

Bristol Cylinder. The Bristol Cylinder, shown in Figure 12, is based on the theory that a submerged horizontal circular cylinder which is forced to revolve about an axis parallel to its own, will produce surface waves that will travel in one direction only—that being in the direction of the motion of the cylinder at

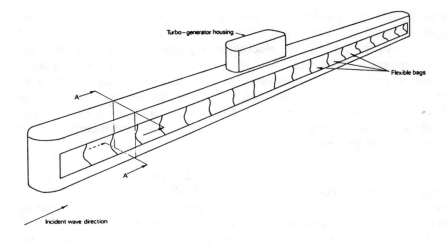

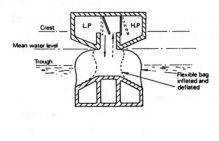

Figure 11. Illustration of the Lanchester Flexible Bag wave energy converter. From Shaw (1982).

the top of its revolution. It can be shown that a loaded submerged cylinder will absorb the energy of surface waves that are parallel to the cylinder axis. Calculations and tests on this device have shown that it is possible to achieve conversion efficiencies of 90% for small amplitude waves of wavelengths typically found in ocean climates.[68]

4c. Economics of wave power: Comparison of near-term technologies

Following the analysis of Hagerman and Heller,[69] one can pick out twelve forms of wave energy conversion technology as being representative of the fledgling industry. These technologies are listed in Table 4 and schematically illustrated in Figure 13, where they are grouped in terms of their respective principles of operation.

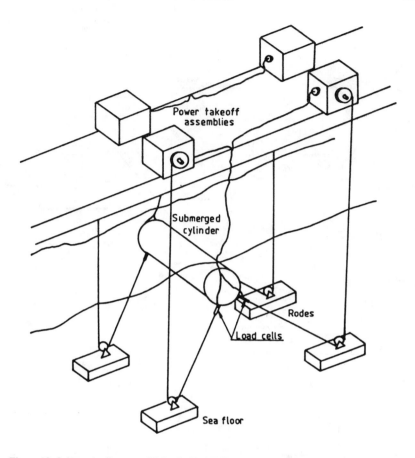

Figure 12. Schematic diagram of Bristol offset cylinder operation. From Davis *et al.* (1981).

The heaving float technologies (1,2,12) operate in modes that have not been described here yet. The KN system (1) and the DELBUOY (12) both are tightly anchored to the ocean floor by a hydraulic cylinder which is the mechanism that provides the power. The KN system is designed to produce electrical power while the DELBUOY is designed for the desalinization of seawater. Gotaverken's device (2), on the other hand, is a "hose pump." The buoy on the surface is connected by an elastic hose, suitably reinforced, to either an anchor or a large submerged semibuoyant plate. When the buoy is elevated by the crest of a passing surface wave, the hose is stretched and its interior volume decreases and pumps water at high pressure to a pelton wheel type of hydraulic turbine. When the crest passes and the buoy sinks into a trough, the hose sucks in a new charge of seawater. These devices are intended to be stationed in an array that will maximize their total power absorption capability and allow several pumps to share one turbine.

The combined heaving and pitching float technologies (3,4) are somewhat dissimilar. The NORDCO Wave Energy Module (3) is similar in principle to the Gotaverken hose pump except that the pumping action can be excited by

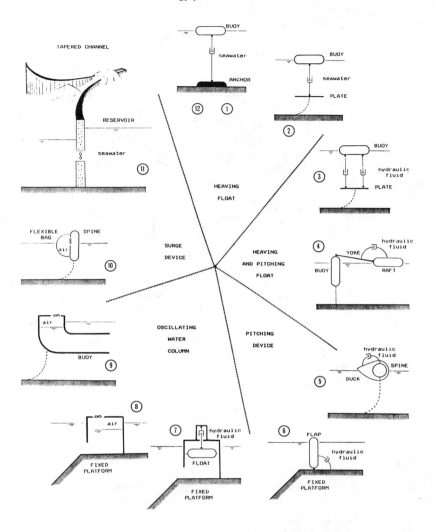

Figure 13. Illustration of twelve near-term wave energy conversion technologies. From Hagerman and Heller (1988).

pitching motion as well as heaving motion. The contouring raft (4) is designed along the lines of Cockerell's rafts except that the front of the device, the yoke, is held fixed by a mooring, and it connects to the aft section, the raft, by only one hinge.

Of the pitching devices (5,6), one is the familiar Salter's Duck (5) while the other, the Q Corporation Tandem Flap (6) is constructed on a platform which is fixed to the ocean floor. Two buoyant vertical "flaps" are attached to the platform via a hinge at the bottom of each and the flaps are positioned one behind the other with the hinges parallel to the wave crests. Numerical modeling shows that the two-flap design can absorb twice as much power from the incoming waves as a one flap design of equivalent total width. Power is obtained by hydraulic cylinders that are pumped as the flaps rotate on their hinges.

The oscillating water column devices (7,8,9) comprise both fixed and floating designs. The NEL Breakwater is an example of a fixed design, as is the Kvaerner-Brug Multi-resonant Oscillating Water Column (8) which is a pneumatic device that has a 500 kWe capacity at Tofestallen, Norway. The Neptune Systems device (7) is a fixed oscillating water column design which uses a float on top of the oscillating water column to drive a hydraulic piston instead of relying on a pneumatic conversion system. The Backward-Bent Buoy was designed by Masuda. It is a floating oscillating water column which has a novel shape that is intended to maximize the power absorption capabilities.

One of the surging devices (10,11) is the SEA Clam, which has been discussed. The other, the tapered channel (11), is currently in use in Norway at Tofestallen. The structure of the tapered channel is designed to capture and amplify the incoming waves so they will efficiently fill a holding pond which can be emptied back into the ocean through a hydraulic turbine. The Tofestallen installation has a capacity of 350 kWe and plans exist for a 1 MW installation in Indonesia.

In the study by Hagerman and Heller, the cost of power for full scale versions of each of these technologies was assessed in the following manner:

• All costs associated with power transmission ashore were ignored.

• When capital cost breakdowns were available, the capital cost of a particular technology was assessed according to prevailing construction costs in the U.S. If the cost breakdown was unavailable, the developer's estimated cost was used instead.

• Developer's estimates of capacity factor and annual operating and maintenance costs were used if provided.

• Total cost of energy was computed according to the guidelines set forth in the EPRI Technical Assessment Guide.

The capitalization costs for each different technology, for differing wave climates in some cases are presented as Table 5. The cost of energy for each of these projects as computed by the EPRI TAG formula is presented in Figure 14 as a logarithmic plot comparing wave energy costs in cents per kWh against the average annual incident wave power in kW/m. Most of the data points approximate a trend line that connects 10 cents/kWh at a resource of 50 kW/m and 20 cents/kWh at a resource of 20 kW/m and the other points are equally dispersed above and below the trend line. For each of the designs of similar technologies operating in different climates, the figure shows that the logarithm of the cost of power decreases linearly as the logarithm of the incident wave power increases.

The trends that are illustrated in the graph indicate that some wave energy conversion techniques, particularly Gotaverken's hose pump, can produce power that is almost cheap enough to compete with conventional technology. This news is encouraging, but it is important to bear in mind that the data displayed is calculated based on assumptions made concerning technologies that are in widely varying states of development—from scale models up to

Table 5. Estimated capacity factors and capitalization costs for full-scale installations of eleven near-term wave energy conversion devices. From Hagerman and Heller (1988).

No.	Plant Location	Average Annual Incident Wave Power	Plant Size (Capacity Factor)	Offshore Capital Cost	Annual O&M Cost
1	North Sea (Denmark)	9.1 kW/m	750 MWe (13%)	$ 875/kWe	0.8%
2	Baltic Sea (Sweden)	5.5 kW/m	25 MWe (37%)	$1090/kWe	6.8%
	NE Atlantic (Norway)	45 kW/m	64 MWe (32%)	$ 580/kWe	7.9%
3	NW Atlantic (Newfoundland)	21 kW/m	1.0 MWe (30% assumed)	$2080/kWe	4.0% (assumed)
4	Southern Ocean (Australia)	49 kW/m	0.5 MWe (60%)	$2280/kWe	3.0%
5	NE Atlantic (Scotland)	48 kW/m	2000 MWe (38%)	$1920/kW/e	1.8%
6	Low energy (generic)	20 kW/m	4.4 MWe (24%)	$1640/kWe	4.0%
	Medium energy (generic)	35 kW/m	4.4 MWe (35%)	"	"
	High energy (generic)	50 kW/m	4.4 MWe (43%)	"	"
7	Central Pacific (Hawaii)	15 kW/m	1.0 MWe (30%)	$2320/kWe	5.0%
8	NE Atlantic (Norway)	25 kW/m	0.5 MWe (41%)	$1600/kWe	1.0%
	SW Pacific (Tonga)	21 kW/m	2.0 MWe (33%)	$4000/kWe	1.0%
9	Sea of Japan (Japan)	13 kW/m	0.5 MWe (16%)	$4830/kWe	4.0% (assumed)
	Central Pacific (Hawaii)	23 kW/m	1.0 MWe (20%)	$3210/kWe	"
	NE Atlantic (England)	51 kW/m	3.5 MWe (20%)	$1320/kWe	"
10	NE Atlantic (Scotland)	48 kW/m	2.0 MWe (31%)	$1040/kWe	5.0%
11	NE Atlantic (Norway)	25 kW/m	0.35 MWe (57%)	$3870/kWe	0.4%

fullsize demonstration plants. Construction of real power plants may hold some surprises. As an example, the darkened circle in the figure, number 8, is based on Kvaerner-Brug's estimated cost of constructing their 500 kWe plant at Tofestallen — around 9 cents/kWh in a 25 kW/m climate. The open circle bearing the same number 8 is based on the same company's estimate of the cost of a similar plant in a similar wave climate off the island of Tonga — close to 25

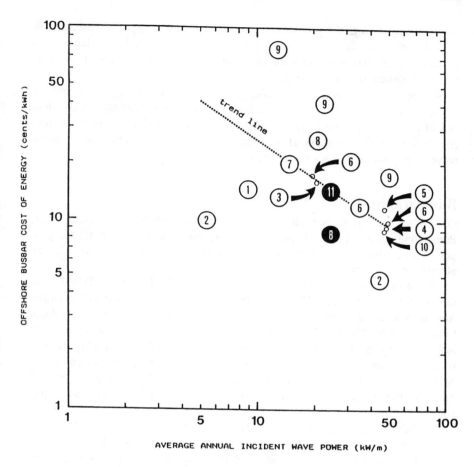

Figure 14. Cost of energy from eleven different wave conversion devices as a function of annual average wave power. Dark circles indicate preconstruction estimates for the demonstration power plants at Tofestallen, Norway. From Hagerman and Heller (1988).

cents/kWh in a 21 kW/m wave climate. In the second case the estimate was made after the Tofestallen plant was completed and operating (but before the existing plant was destroyed by storm driven seas in January, 1989!) and the new cost estimate was approximately double the old one. On the other hand, the simple innovation of converting the straight spine of the SEA Clam to a circular one had the added benefit of cutting the cost by a factor of two. The ideas and techniques are all new and no economic estimates are, as yet, firm.

4d. Conclusion

Wave energy impinging on the continental United States represents only a small fraction of the present electrical energy requirements of the United States but it is nonetheless deserving of development. Although suggestions for harnessing wave power have existed for 200 years, it has only been in the last 15 years or so that any concerted national efforts have been launched to find an economic solution to the problem. At the present time, it appears reasonable to state that electrical energy at a cost of 9–15 cents/kWh could be produced from a relatively modest wave climate of 5 kW/m—which is typical of the seas found off the coast of Cape Hatteras.

At the stage of today's technology, wave energy conversion is not an economic alternative to conventional fossil fuel technology. However, the field is still in its infancy and so far the trend has been to produce more cost-effective designs as time goes on. If this trend can be induced to continue, then the wave energy industry has a good future.

Bibliography

G.L. Wick and W.R. Schmitt, *Harvesting Ocean Energy* (UNESCO Press, Paris, France, 1981). This book gives a very nice description of all the proposed renewable energy sources in the oceans. It is very informative at an elementary technical level.

Two articles provide a good introduction to OTEC. The article "Power from the Sea" by Penney and Bharathan in the January 1987 issue of *Scientific American* is a good introduction to OTEC. It is very well illustrated. Also, Cohen's article, "Energy from the Ocean," in *Philosophical Transactions of the Royal Society of London*, **307**, 405–437 (October, 1982) is highly recommended. In addition to these sources, there is an excellent review paper by J.G. McGowan that was published in *Solar Energy* **18**, 81–92 (1976) under the title "Review Paper: Ocean Thermal Energy Conversion—A Significant Solar Resource."

For tidal power, the best sources are David Greenberg's article, "Modeling Tidal Power," in the November, 1987 *Scientific American*, and R.H. Charlier's book, *Tidal Energy* (Van-Nostrand Reinhold, 1982) which is quite comprehensive.

For a complete introduction to the theory and practice of ocean wave energy conversion, see either *Ocean Wave Energy Conversion* by Michael McCormick (John Wiley and Sons, 1981) or *Wave Energy: A Design Challenge* by Shaw (Halsted Press, 1982).

References and notes

1. G.L. Wick and W.R. Schmitt, eds. *Harvesting Ocean Energy* (UNESCO Press, Paris, France, 1981).
2. Clarence Zener, "Solar sea power," Phys. Today 26, 48–53 (January, 1973).
3. Jacques Arsene d'Arsonval, "Utilisation des forces naturelles. Avenire de l'électricité, Revue Scientifique 17, 370–372 (1881).
4. Georges Claude, "Power from the tropical seas," Mechanical Engineering 52, 1039–44 (December, 1930).
5. Westinghouse Electric Corp., *OTEC 100 MWe Alternate Power Systems* (Westinghouse Electric Corp. Final Report, 1979).
6. B. Coffay, *Commercialism Potential of Open Cycle OTEC Plants* (Technical Report, Westinghouse Electric Corp., 1980).
7. Andy Trenka, OTEC project director, SERI, September 1988, private communication.
8. Earl J. Beck, "Ocean thermal gradient hydraulic power plant," Science 189, 293–294 (1975).

9. Clarence Zener and John Fetkovich, "Foam solar sea power plant," Science **189**, 294–295 (1975).

10. T.S. Jayadev, D.K. Benson, and M.S. Bohn, *Thermoelectric Ocean Thermal Energy Storage* (Technical Report, Solar Energy Research Institute, Golden, CO, 1979).

11. *Federal Ocean Energy Technology Program: Multi-year Program Plan, FY 85-89*, Technical Report DOE/CH/10093-100 (U.S. Department of Energy, Washington, DC, December, 1985).

12. G.L. Dugger, E.J. Francis, and W.H. Avery, "Technical and economic feasibility of ocean thermal energy conversion," Solar Energy **20**, 259–274 (1978).

13. R.T. Heizer, in *Iceberg Utilization*, A.A. Husseiny, ed. (Pergamon Press, New York, 1978), pp. 657–673.

14. D.M. Roberts, in *Iceberg Utilization*, pp. 674–689.

15. J.D. Isaacs and R.J. Seymour, "The ocean as a power resource," Int. J. Environ. Stud. **4**, 201–205 (1973).

16. J.H. Anderson and J.H. Anderson, Jr., *Large-Scale Sea Thermal Power*, Technical Report 65-WA/SOL-6 (ASME, December, 1965).

17. A. Lavi, *Final report–solar sea power project*, Technical Report NSF/RANN/SE/GI-39114/PR/74/6 (Carnegie Mellon University, Pittsburgh, PA, January, 1975).

18. J.G. McGowan and W.E. Heronemus, *Gulf stream based ocean thermal power plants*, Technical Report 75-643 (AIAA, April, 1975).

19. TRW Systems Group, *Ocean Thermal energy conversion research on an engineering and test program*, Vols. 1–5, Technical Report (NTIS, Springfield, VA, June, 1975).

20. Ocean Systems of Lockheed Missile and Space Co., *Ocean Thermal Energy Conversion (OTEC) power plant technical and economic feasibility*, Vols. 1,2, Technical Report USF/RANN/SE/GI-C937/Fr/75/1&2 (NTIS, Springfield, VA, April, 1975).

21. DSS Engineers, Inc., *Development of plastic heat exchangers for sea solar power plants*, Technical Report NSF/RANN/AER74-13030A01/FR/75/1 (NTIS, Springfield, VA, March, 1975).

22. U.S. Patent No. 3,454,081, July 8, 1969.

23. R.M. Milton and C.F. Gottzman, Chem. Eng. Progr. **56** (September 1972) and *Ocean Thermal Power Plant Heat Exchangers*, Technical Report 644 (Linde Division, Union Carbide Corporation, 1971).

24. D.F. Dipprey and R.H. Sabersky, Intl. J. Heat Transfer **6**, 329 (1963).

25. Hawaii Natural Energy Institute, *Annual Report 1985* (University of Hawaii at Manoa, Honolulu, Hawaii, 1985).

26. T.E. Little, *OTEC Seawater Pump Study* (Technical Report, Westinghouse Electric Corporation, Oceanic Division, 1978).

27. W.L. Owens and L.C. Trimble, "Mini-OTEC operational results," ASME J. Solar Energy Engineering **103**, 233–240 (1981).

28. R. Cohen, "Energy from the ocean," Philos. Trans. R. Soc. London **307**, 405–437 (October 1982).

29. V.W. Harms in *Eighth Ocean Energy Conference, 1981*, E.M MacCutcheon, ed.

30. F. Ito and S. Yutaka, *Present Situation and Future Outlook of OTEC Power Generation* (Tokyo Electric Power Company, Tokyo, Japan, November, 1982).

31. J. Valenzuela et al., *Thermo-Economic Analysis of Open Cycle OTEC Plants*, Technical Report 84-WA/SOL-24 (ASME, December, 1965).

32. F. Kreith and D. Bharathan, "Heat transfer research for ocean thermal energy conversion," J. Heat Transfer **110**, 5–22 (February, 1988).

33. H.J. Green et al., *Measured Performance of Falling Jet Flash Evaporators*, Technical Report SERI/TP-631-1270 (Solar Energy Research Institute, Golden, CO, June, 1981).

34. D. Bharathan and T. Penney, "Flash evaporation from turbulent water jets," ASME J. Heat Transfer **106**, 407–416 (1984).

35. D. Bharathan, B. Parsons, and J.A. Althof, *Direct-Contact Condensers for Open-Cycle OTEC Applications: Model Validation for Structured Packings Using Fresh-water*, Technical Report SERI/TP-252-3108 (Solar Energy Research Institute, Golden, CO, 1984).

36. D. Bharathan, *Open-Cycle OTEC Thermal-Hydraulic Systems Analysis and Parametric Studies*, Technical Report SERI/TP-252-2330 (Soalr Energy Research Institute, Golden, CO, 1984).

37. D.E. Lennard, "Ocean thermal energy conversion progress and prospects," in *Energy Options:*

The Role of Alternatives in the World Energy Scene, R.H. Taylor *et al.,* eds., fourth international IEEE conference, London, 1984.

38. Another study of comparative economics is Alfred M. Perry, Warren D. Devine, Jr. *et al., Net Energy Analysis of Five Energy Systems,* Technical Report 77-12 (Oak Ridge Associated Universities-Institute for Energy Analysis, Oak Ridge, TN, 1977).

39. Data for comparison of the energy economics of OTEC with fossil fuels includes the use of the report, *Transition: A report to the Oregon energy council prepared by the office of energy research and planning* (Technical Report, State of Oregon, January, 1975).

40. T.R. Penney and D. Bharathan, "Power from the sea," Sci. Am. 86–92 (1987).

41. M.S. Quinby-Hunt, P. Wilde, and A.T. Dengler, "Potential environmental impacts of open-cycle thermal energy conversion," Environmental Impact Assessment Review **6**, 77–93 (1986).

42. C.L. Amos and B.F.N. Long, "The sedimentary character of the Minas Basin, Bay of Fundy," in *The Coastline of Canada: Littoral Processes and Shore Morphology,* S.B. McCann, ed. (Geological Survey of Canada, Paper 80-10, 1980).

43. R.H. Charlier, *Harnessing the Energies of the Ocean* (Marine Technology Society, Washington, DC, 1969).

44. R.H. Charlier, *Tidal Energy* (Van Nostrand Reinhold, 1982).

45. L.B. Bernshtein, *Tidal Energy for electric powerplants* (NTIS, Springfield, Virginia, 1965). Translated from the Russian by Israel Program for Scientific Translation.

46. R. Gibrat, *L'energie des marées* (University of France Press, Paris, 1966) in French.

47. *Reassessment of Fundy Tidal Power* (Technical Report, Bay of Fundy Tidal Power Review Board, November, 1977).

48. Diane Kaylor, "Fundy tidal power projects reach milestone," Mechanical Engineering **106**, 65–66 (April, 1984).

49. J. Cotillon, "La Rance tidal power station review and comments," in *Tidal Power and Estuary Management,* R.L. Gregory, ed., 30th Symposium of the Colston Research Society (Scientechnica, Bristol, April, 1979).

50. Charles Simeons, *Hydro Power: The Use of Water as an Alternative Source of Energy* (Pergamon Press, 1980).

51. E.M. Wilson, "Tidal power," in R.H. Taylor *et al.,* as cited above pp. 163–165.

52. *Tidal Power Study in Cobscook Bay, Maine: Preliminary Report on the Economic Analysis of the Project* (U.S. Army Corps of Engineers, New England Division, 1979).

53. Judith Spiller, "Scientific Issues in the Assessment of the Transboundary Effects of Fundy Tidal Power," in *Environmental Decisionmaking in the Transboundary Region,* Judith Spiller, Alison Rieser, and David Vander-Zwaag, eds., (Springer-Verlag, 1986).

54. David A. Greenberg, "Modelling tidal power," Sci. Am. **257**, 128–131 (November, 1987).

55. J.R. Keeley, C.J.R. Garrett, and D.A. Greenberg, "Tidal mixing versus thermal stratification in the Bay of Fundy and Gulf of Maine," Atmosphere-Ocean (1978).

56. A.W. White and C.M. Lewis, "Resting cysts of the toxic red tide dinoflagellate, gonyaulax excavata, in the Bay of Fundy sediments," Can. J. Fish Aquat. Sci. **39**, 1190 (1982).

57. G.G. Stokes, "On the theory of oscillatory waves, " Trans. Cambridge Philos. Soc. **8** (1847).

58. Michael E. McCormick, *Ocean Wave Energy Conversion* (John Wiley and Sons, 1981).

59. W.J. Pierson and L. Moskowitz, "A proposed spectral form for fully developed wind seas based on the similarity theory of S.A. Kitaigorodski," Trans. Am. Geophys. Union **35**, 747–757 (1964).

60. G.N. Crisp and N. Scott, "The spatial distribution of wave power on the western UK coast," in *Wave and Tidal Energy,* H.S. Stevens and C.A. Stapleton, eds., (BHRA Fluid Engineering, Cranfield, Bedford, England, Proceedings of the Second Annual Symposium on Wave and Tidal Energy held at Cambridge, England September 23–25, 1981), p. 7.

61. David Ross, *Energy from the Waves* (Pergamon Press, New York, 1979).

62. Steven Salter, "Wave Power," Nature **249** (June, 1974).

63. R. Shaw, *Wave Energy: A Design Challenge* (Halsted Press, 1982).

64. G. Eliot, "NEL breakwater–Britain's proposed first wave power station," in R.H. Taylor *et al.,* as cited above pp. 163–164.

65. N.W. Bellamy and A.M. Peatfield, "Futher development of the sea-clam wave energy converter," in R.H. Taylor *et al.* as cited above p. 163–165.

66. George Hagerman and Ted Heller, "Wave energy technology assessment for grid-connected utility applications," presented at RETSIE/IREC 88, Santa Clara, California, 9 June, 1988.

67. *Compendium of International Ocean Energy Activities*, prepared for the Wind/Ocean Technologies Division of the U.S. Department of Energy by Science Applications International Corp. and The Meridian Corp.

68. D.V. Evans, G.P. Thomas, W.H. Askew, J.P. Davis, R. Finney, and T.L. Shaw, "Some hydrodynamic characteristics of the Bristol Cylinder," in H.S. Stevens and C.A. Stapleton as cited above.

69. George Hagerman and Ted Heller, "Wave energy: a survey of twelve near-term technologies," in *Proceedings of the International Renewable Energy Conference, Honolulu, Hawaii, 18–24 September, 1988.*

Energy from biomass

J. W. Ranney and J. H. Cushman

1. Introduction

Biomass could reasonably provide 15 EJ of U.S. energy needs within 20 years, provided biomass research continues at mid-1980s levels and oil prices gradually rise in real costs. The U.S. Department of Energy's goal for energy crops alone is 26 quads of biomass per year within 40 years. Plant productivity and the availability of sufficient and appropriate land are central issues. Significant recent breakthroughs have occurred in biomass and biofuel research. Species selection, genetic selection and plant breeding, biotechnology, and cultural techniques have greatly increased plant productivity and reduced growing costs. The discovery of yeasts capable of fermenting pentose sugars, cheaper production of better enzymes, membrane filter development, and discovery of new catalyst-temperature-pressure combinations have resulted in greatly reduced conversion costs and higher efficiencies. The challenges are still significant and include improved conversion systems; better separation technologies; improvement of biomass harvest, collection, and storage systems; application of advanced genetic techniques in biomass production and conversion; and the economic recovery and upgrading of spent catalysts and chemicals. Efforts in all areas are necessary to make biofuels competitive in today's economy.

2. Why biomass for energy

While there have been significant breakthroughs in producing energy crops and converting them to biofuels over the past decade, the continued low price of conventional fuels has limited the adoption of biomass energy technologies in the United States. Other limiting factors have been low levels of technology awareness, technology readiness, and unfavorable agricultural and energy policies. Biomass feedstocks at $1.58 to $3.68/GJ ($1.50 to $3.50/million Btu) must compete with coal costing around $1.41 to 1.58/GJ ($1.35 to $1.50/million Btu). Biomass-derived liquid fuels appear to hold great promise for the future, but the conversion and production technologies are still developing and are not yet economically competitive.

Almost all of the energy now produced from biomass comes from direct combustion of wood wastes and residues. Such fuels now generate about 1.7 EJ (1.7×10^{15} Btu) annually in the United States in industrial applications. Another 1.3 EJ are derived from wood burned in stoves for space heating (6.5 million households),[1] agricultural residues, municipal solid waste, and corn used in ethanol production.[2] In total, biomass-fueled systems produce about 4% of total energy used in the country.

Why is the future of biomass energy important to the United States? The answer involves energy security, the balance of payments, regional and global environments, and the outlook for improvements in biomass energy technologies. Biomass is the only form of renewable energy that can be converted to a liquid transportation fuel using present technology, and we estimate that within 20 years biofuels could supply annually a net 15 EJ of fuel in liquid, gaseous, and electrical form. This number could go much higher in later years. This holds true provided that biofuel research continues at mid-1980s levels and imported oil prices rise. Because the transportation sector depends almost totally on petroleum, replacing gasoline and diesel fuel with biofuels could greatly reduce U.S. vulnerability to disruption in the supply of foreign petroleum.

Fifteen EJ equal 15% to 20% of current U.S. energy needs and is nearly equivalent to yearly oil imports in the mid-1980s. To supply this as liquid or gaseous fuels or electricity will require the growth or collection of raw biomass feedstocks containing nearly 40 EJ. We estimate that this amount could be produced without serious adverse environmental impacts, and its use could have several positive impacts. Recent estimates by the Biomass Energy Research Association[3] are in general agreement with this figure.

Table 1 lists potential U.S. sources of biomass feedstocks, i.e., biological material that forms the input to industrial processes which convert it to higher-valued energy forms, i.e., liquid and gaseous fuels and electricity.[4-15] The amount of biomass that can be made available for any purpose clearly depends on the price it will bring. Crops grown for energy production, e.g., compete with crops grown for food and which are often subsidized by price support policies. The estimates in Table 1 are based on the assumption that the selling price of the fuel product, usually an alcohol, will be about $0.32/1iter ($1.20 per gallon) (1986 dollars). Reaching the levels of production shown in Table 1 will require improvements in both the production of biomass feedstocks and their conversion to higher-valued energy forms.

Biomass, referred to here as vegetation and organic wastes from animals, can be defined as carbon-based material generated within historic time by plants. Wood, grass, algae, manure, garbage, and plant products such as starches, sugars, lipids, and lignins fall into this category. Peat is excluded from this definition although it can be classified as a biomass fuel.

Viewed as a chemical feedstock for energy, biomass is quite oxidized and less energy-rich than coal, ethanol, and gasoline. Table 2 compares some relevant values for biomass and other fuels. Biomass can be converted to more energy-dense forms through thermochemical and biochemical processes. Thermochemical treatment produces fuel gases, including a methane intermediate for methanol, or a "biocrude" oil convertible to gasoline. Biochemical treatment can produce either ethanol, through fermentation processes, or methane (biogas) through anaerobic digestion. In all processes, the major challenges in convert-

Table 1. Potential costs, amount, and yields from various biofuel feedstocks (t = trace).

Feedstock	Cost		Biomass/biofuel contributions	
	Present	Goal	Raw biomass (*Gross − loss*) × *Effic.* = *Net*	External energy cost
Commercial forest wood	<$2/GJ	<$2/GJ	(6.0–1.5) × 50% = 2.3	0.2
Improved forest mgmt. ($^1/_3$ for energy use)		<$2/GJ	(6.0–1.5) × 50% = 2.3	0.2
Logging residues	>$3/GJ	<2/GJ	(3.8–3.0) × 50% = 0.4	*t*
Urban wood wastes and land clearing	$2/GJ	$2/GJ	(1.6–0.4) × 50% = 0.6	0.2
Forest manuf. residues	$1/GJ	<$1/GJ	(2.4–0.3) × 50% = 1.1	*t*
Environmentally collectable agricultural residues	$1–2/GJ	$1/GJ	(2.2–0.2) × 60% = 1.2	0.1
Agric. oil seed	$0.38/L	$0.30/L	(0.3–*t*) × 95% = 0.3	*t*
Municipal solid waste and indust. food waste	$2–3/GJ	<$1.50/GJ	(2.6–0.2) × 50% = 1.2	0.1
Animal wastes	<$4/GJ	$3.50/GJ	(0.5–*t*) × 60% = 0.3	*t*
Wood energy crops[a]	$3/GJ	$2/GJ	(4.0–0.4) × 60% = 2.3	0.3
Herbaceous energy crops Lignocellulosics[a]	$4/GJ	$2/GJ	(5.6–0.5) × 60% = 3.1	0.4
New energy oil seed	$0.66/L	$0.30/L	(0.4–*t*) × 95% = 0.4	*t*
Aquatic energy crops Microalgae	$2.60/L	$0.30/L	(0.3–*t*) × 95% = 0.3	0.1
Macroalgae	$3.50/GJ	$2/GJ	(1.5–0.4) × 65% = 0.7	0.1
SUBTOTAL			(37.2–8.5) 16.5	1.7

Net potential energy output primarily as liquid fuel = 14.7 EJ
Additional energy and by-product contributions are about 15% as excess process steam, electricity sold to grid, and chemical by-products

a. Energy crop production could potentially be two to three times higher.

ing biomass to fuels are its high oxygen content and its complex and varied chemical composition.

Thermochemical processes are often viewed as the strong-armed approach to producing biofuels. Heat energy, and sometimes pressure and catalysts, are applied to the complex hydrocarbons in biomass. These act on all the different components in the feedstock, driving off a certain amount of oxygen while forming simpler, more valuable, and more homogeneous products. These products can be liquids or gases. The challenges in developing thermochemical processes are making them energy-efficient, determining the operational conditions required to make the desired products, and separation technologies for the product stream.

Biochemical processes use lower temperatures and pressures and are much more sensitive to feedstock composition. They differentiate among the various chemical components of the feedstock, some of which are much more easily digested or fermented than others. The challenges for biochemical processing

Table 2. A comparison of the oxygen content and higher heating value of biomass feedstocks, coal, and liquid fuels.

Feedstock research or fuel type	Oxygen content % by wt MK/kg	Higher heating value	Octane number RON
Woody and herbaceous biomass	28–45	19–21	. . .
Illinois bituminous coal[a]	14	28	. . .
Ethanol[b]	35	30	111
Unleaded gasoline[b]	0	28	92

a. D. L. Klass, "Fuels From Biomass" in Kirk-Othmer, *Encyclopedia of Chemical Technology*, Volume II, Third Ed. (John Wiley & Sons, New York, 1984), p. 42.
b. D. L. Klass, "Alcohol Fuels" in Kirk-Othmer' *Encyclopedia of Chemical Technology*, Supplement Volume, Third Ed. (John Wiley & Sons, New York, 1984), p. 42.

are forming useful products from all parts of the feedstock, simplifying the processing steps, reducing processing times, and separation technologies.

Thermochemical processes for producing liquid fuels are more fully developed than biochemical processes, and are currently more nearly competitive with conventional fuels. In the long run, however, the biochemical processes probably offer more opportunities for improvement and cost-effective fuel production.

Another approach to liquid biofuels is the direct production of fuel products by plants. Seed oils from higher plants and algal storage lipids appear to hold the greatest potential because of high productivity rates and ease of collection. This approach, which relies on the crop species to do much of the processing, has inherent advantages and disadvantages. The proportion of the plant biomass that appears in the fuel product is limited by the amount that is stored as oil. It is relatively easy to get a pure, energy-rich product, but most plant storage oils are triglycerides and require a treatment like cracking or transesterification before they are suitable for long-term use in diesel engines.[16] Small research programs are currently addressing the production of algal lipids and seed oils of higher plants for fuels.[17]

This chapter reviews the processes involved in using biomass for energy. Specifics of energy balances will not be discussed, but the general energy flows, recent improvements in efficiency, and new processes now under study are presented. The limitations and risks associated with these technologies are discussed to maintain an air of applied economic reality.

3. The biomass fuel pathway

Biomass moves through a number of distinct steps as it is grown, stored, and processed into liquid and gaseous fuels. For an example, using tree plantations as a source of biomass and biochemical conversion to ethanol, six steps can be described.

1. The interception of solar radiation by leaves and photosynthesis within the leaf tissue.

2. The allocation of the energy fixed in photosynthesis to respiration or the growth of roots, stems, or leaves.
3. Harvesting and transporting the biomass to a conversion facility.
4. Pretreatment specific to the conversion process being used.
5. Conversion of the biomass into ethanol.
6. Recovery, separation, and upgrading of the fuel alcohol.

The investments, energy costs, and potential biomass losses in each step represent opportunities for improving overall system efficiency. In the first step, the costs of planting material (seedlings or cutting) and plantation management (fertilization, weed control) represent investments to be minimized. A large leaf area and a high rate of photosynthesis increase the amount of solar energy captured in biomass. Both are distinct advantages when raised to optimal levels and can be improved through selection and breeding. Each operation requires an input of energy which must be balanced against the increased energy stored in usable biomass. For example, manufacture of fertilizer consumes energy and weed control requires energy input which must be counteracted by increased plant growth or else the use of the chemicals cannot be justified.

In the second step, environmental stresses and genetic controls determine how the plant stores the products of photosynthesis. Both can be manipulated somewhat to increase net production. Environmental stresses can be minimized through cultural techniques and crop management practices. People conducting screening and breeding studies are developing plant varieties that resist stresses and allocate photosynthates efficiently to the growth of harvestable stems and limbs.

The subsequent harvesting and handling steps have similar opportunities for improvement. In these steps, most improvement will come from speeding up processes and reducing the energy and materials consumed, while increasing the proportion of the feedstock convertible to fuel products.

Table 3 follows the biomass produced on one hectare (2.47 acres) of land from production through conversion to a liquid biofuel. This is another way to illustrate the investments and efficiencies in the steps described above. The table compares the values now seen in experimental systems and also shows levels expected in the future. The six steps listed above have been condensed in Table 3 to production (steps 1 and 2), harvesting (step 3), and conversion (steps 4, 5, and 6).

Table 3 projects that improvements all along the biomass fuel pathway could result in the near doubling of ethanol produced annually from a hectare of land. A larger percentage of the biomass feedstock ends up in the fuel product, and less energy is required in the processing steps. Corn is the only biomass feedstock currently used for large-scale production of ethanol. The conversion of the easily digestible starch in corn grain to ethanol is a much different process than the lignocellulosic (woody) biomass-to-ethanol pathway discussed here. The system used for corn involves a well-developed process, and studies of the corn-to-ethanol pathway do not indicate much potential for significant improvement.[18]

Table 3. Improvements along the wood-to-ethanol fuel pathway may result in a near doubling of energy production from a hectare of land over the next two decades.

| | Major pathway steps | | | | | | Output liters ethanol/ha | |
| | Production Mg/ha/year | | Handling Mg/ha/year | | Conversion Mg/ha/year | | | |
	Now	Future	Now	Future	Now	Future	Now	Future
Amount at start of step	18.0	22.0	14.0	20.0	12.0	18.0	2060	3500
Biomass loss in step	4.0	2.0	2.0	2.0	5.2	8.8	...	
External energy input in biomass equivalents[a]	0.8	0.5	0.9	0.8	0.1	1	...	
Internal energy recycled in biomass equivalents[b]	0.0	0.0	0.0	0.0	2.7	3.0	...	
By-product contribution in biomass equivalents[c]	0.0	0.0	0.2	0.2	2.3	2.5	...	

a. Conversion efficiencies for electricity and energy used to manufacture conversion facilities not included.
b. Steam generated from the combustion of wastes and residues in conversion.
c. Field residues which displace fertilizers and excess steam generation from the combustion of process wastes.

4. Progress in biomass production

Dry weight production per hectare is extremely important to the cost of producing biomass energy crops, and recent improvements in biomass production have come primarily from increasing dry weight production per hectare per year. Short-rotation woody crops on the best sites can now produce between 20 and 30 dry Mg/ha/year at an estimated cost of $35–$60 per dry Mg. This is almost double the productivity seen in such systems 10 years ago, and at half the cost. Increasing the dry weight production per hectare for fuel crops also minimizes the land which must be converted from food crops or other uses to produce a given amount of energy from biomass.

These improvements have come from the application of agricultural practices to tree and herbaceous crop production, from the selection of appropriate species for different types of sites, and from genetic improvements in the plants themselves. The general site conditions and management practices necessary to achieve high production rates in short-rotation woody crops have been defined. One hundred fifty hardwood tree species and herbaceous plants have been screened, and promising species have been identified for the Midwest, Southeast, Northeast, Great Lakes states, and the Pacific Northwest.[19] Genetic improvement programs have been started for 11 of these species.

Genetic research is most advanced in sorghum and the genus *Populus*, the aspens, poplars, and cottonwoods. Studies have examined the relationships between leaf arrangement, leaf area, and productivity. The ratio of leaf surface area to ground surface area, or leaf area index, is commonly around 6 in closely grown tree crops. The leaf area index of some newly developed poplar clones exceeds 11. These varieties also leaf out early in the spring and hold their leaves late in the fall. The high leaf area and long growing season have led to exceptionally high production rates. Poplar clones that allocate the products of photosynthesis very efficiently to stem and leaf growth have also been identified. This research on poplars should help speed the development of other species for different sites and regions.

Plant biotechnology is contributing to the rapid development of energy crops. Mass micropropagation allows the regeneration of whole plants from cells or tissues, and has proven an important research tool. The rapid multiplication of genetically identical plants makes it possible to move quickly from a single plant with promising characteristics to field tests. Every major energy species under consideration has been successfully micropropagated for study. Use of micropropagated plants for large-scale production plantings still requires considerable evaluation. The large number of plantlets or seeds needed in establishing fast-growing energy crops has placed a great deal of economic importance on cost/plant and the rapidity of propagation.

Biotechnology is also providing new ways to manipulate and breed plants that could be very effective in dealing with pests, diseases, and a multitude of site stresses such as drought, waterlogging, heat, and late-spring frosts. It is very important, but quite difficult, to control weeds in short-rotation tree crops. Biotechnology has already accomplished the engineered placement of a gene for tolerance to selected herbicides in hybrid poplar.

Some biologically based methods for improving production have barely been studied. More research is needed on the use of nitrogen-fixing species and root symbionts like mycorrhizae to reduce nutrient and water stresses. Selecting plants based on root arrangement as well as leaf and limb arrangement holds great promise. Mixed-species plantings currently present management problems, but viable mixed-species production systems could reduce pest problems and weather-related risks.

Another area important in current biomass production research is directly related to the developing conversion technologies. The chemical and physical characteristics of biomass feedstocks that affect the efficiencies of conversion processes are being investigated.[20] When desirable traits are identified, physiological and genetic studies can examine ways of manipulating plant growth and composition to produce superior feedstocks. Such manipulation will probably require a combination of biotechnology and conventional plant breeding in combination with extensive field testing. Mathematical modeling of plant growth is proving to be another important tool.

5. Harvest, transport, and storage

Harvesting, transporting, and storing biomass are energy-intensive and expensive. These steps in the production pathway may cost $12 to $14 or more per dry Mg, adding $0.75 or more per GJ to the cost of the final fuel product. Wood wastes have been readily accepted as biofuels because such costs are very low and feedstocks are already concentrated at a location.

Current research is not extensive, although some critical issues have been addressed. Studies have examined ways to reduce the costs of harvesting, collecting, drying, and transporting short-rotation woody crops. Time of harvest and storage are major issues for herbaceous forage crops, especially for thick-stemmed crops like sorghums and sugar cane. Algae are difficult to collect and dewater. Municipal solid waste must be sorted, and further treatment, both mechanical and chemical, can make handling easier and improve fuel quality.

Most research in these areas on wood energy crops has been directed toward harvesting. In general, trees less than 15 cm in diameter cannot be harvested efficiently with conventional forestry equipment. The biggest breakthrough for short-rotation woody crops has been the design and development of a tractor-mounted device capable of harvesting 1000 small trees (up to 15 cm at the base) per hour, about the rate that must be achieved to be economical. This harvesting device was developed in Canada and has been tested in the United States. As it moves through the tree plantation, it simultaneously severs trees with two side-by-side circular saws, collects them, and deposits them in bundles.

Another development is equipment for crushing or splintering small trees. The crushed material can be dried to 20% moisture (wet basis) in the field in several days. It can then be collected and baled with equipment similar to heavy-duty hay balers. The dried material can be transported far more cheaply than green chips, which are half water and rapidly lose biomass through respiration and decomposition. A form of wood comminution called chunking also shows promise. Chunking produces blocks of wood rather than chips. It requires much less energy than chipping and the chunks are more dense than chips, which offers advantages in transport.

Storage can be a major problem with both herbaceous and woody biomass. Piles of green woodchips can lose several percent of their biomass per month through decomposition. Herbaceous forage crops decompose even faster without some protection. Various methods of drying, storing, and ensiling have been considered; there have been no major improvements on conventional agricultural techniques.

Algae are very difficult to collect, dewater, and process. Algae crops are being developed to take advantage of the high productivity rates in nutrient-rich water bodies in warm sunny areas. However, the research is in its early stages and focuses on species improvement and the development of optimal systems for growing the crops. Methods for harvesting and processing the algae are under study, but at a very low level.

Matching biofuel conversion systems to biomass qualities can reduce handling problems. For example, water hyacinths have been grown on sewage effluent, producing biomass while performing wastewater treatment. Water hyacinths have a high water content and are very difficult to dry. In a demonstration project at Disneyworld the hyacinths were used without dewatering in a nearby anaerobic digester that required a wet feedstock.

6. Progress in biomass conversion

The major biochemical processes for converting biomass feedstocks into biofuels include producing ethanol from lignocellulosic feedstocks using fermentation and producing biogas (methane and carbon dioxide) through anaerobic digestion. Major thermochemical processes include gasification to a mixture of carbon monoxide and hydrogen (syngas) and the production of biocrude-derived gasoline. Syngas can be further processed to produce methanol.

7. Biochemical processes

Producing ethanol from wood and herbaceous feedstocks is much more complicated than just the fermentation of sugars and starches. As much as 75% of the total plant biomass consists of cellulose and hemicellulose, both largely polymers of sugars. In the wood-(or grass)-to-ethanol pathway these polymers are hydrolyzed to their component sugars, which are then fermented to ethanol. The hydrolysis processes itself, and the fermentation of five-carbon or pentose sugars resulting from the hydrolysis of hemicellulose, are critical steps. The fermentation of hexose sugars, whether from cellulose or hemicellulose, presents little problem.

Hydrolysis processes under study include the use of strong acids, weak acids, and enzymes. Problems with acid hydrolysis have led to more attention to enzymatic processes. Acid conditions sufficient to break cellulose down to sugars also degrade the sugars into undesirable by-products, often severely reducing ethanol yields. Enzymes are very specific, catalyzing the hydrolysis reaction but not degrading any of the sugars.[17] Biotechnology has allowed the rapid improvement in enzymes and the organisms (such as fungi) which produce them.

Recent major advances in fermentation include the ability to ferment xylose (a five-carbon sugar) and the development of organisms capable of continuing the fermentation process under high concentrations of ethanol. Unknown 10 years ago, xylose-fermenting organisms have been identified and are being developed.[17] This will allow much of the hemicellulose, which comprises over 20% of many biomass feedstocks, to be converted into product fuels. While hemicellulose fermentation is still less efficient than hexose fermentation, the technology is new, and improvements can be expected.

Future reductions in cost are expected to come from integrating the hydrolysis and fermentation steps in a process called simultaneous saccharification and fermentation or SSF. This will permit the sugar formation process to be integrated with the fermentation process as one processing step. In the past few years such systems have been improved significantly. Cellulose conversion has increased from 60%–70% to higher than 90%. Ethanol concentrations have increased from 2% to nearly 5%. At the same time enzyme requirements have been reduced 85%.[17]

In the near future, research will address several additional ways to improve overall system efficiency. A range of feedstocks will be examined to determine whether feedstock composition, if modified, could significantly affect the hydrolysis and/or fermentation processes. Membrane techniques designed to reduce distillation costs are showing very positive results, but clogging and durability are still problems that reduce effectiveness, which still needs improvement.

The gas production technology involving anaerobic digestion in the production of biogas is still in early stages of development. Tapping sanitary landfills and figuring out the complex mix of bacteria for efficient anaerobic digestion in controlled digesters are the most immediate challenges. The U.S. Department of Energy researchers have identified the anaerobic digestion process as stages or sets of bacteria populations. The first "breaks down" cellulose and other complex molecules enzymatically into simple sugars and other monomers. Then, other types of bacteria digest these products, producing organic acids and further breakdown into acetate, formate, hydrogen, and carbon dioxide. Finally, specialized bacteria, called methanogens, use these compounds to produce methane and carbon dioxide.[17] Other important challenges involve increasing the solids content in the liquid undergoing digestion (perhaps tripling present rates), improving conversion efficiencies from 50% to 90%, enhancing the stability and control of the process, and reducing material costs.[2,17]

8. Thermochemical processes

Biocrude-derived gasoline can be created through two different processes, low-pressure fast pyrolysis and catalytic high-pressure pyrolysis. In the first process biomass is heated and decomposes into vaporized biocrude containing small fragments of the biomass' organic structure. The vapor is rapidly cooled and condensed to a liquid, to prevent the organic fragments from polymerizing. Although this is a relatively low-cost process, large amounts of oxygen (up to 33%) remain in the liquid. About 70% of the energy in the biomass is converted to biocrude.

In the catalytic high-pressure pyrolysis, a more stable biocrude with lower oxygen content (10%) is produced, a form more suited to present-day refining technologies. In this process, the biomass is mixed with recycled biocrude, heated at high pressure in the presence of a reducing gas and a catalyst. This alters the chemical properties of the biocrude for simple refining.

Upgrading and refining processes are being developed for each type of biocrude. The highly oxygenated biocrude is heated to a vapor and placed in contact with a zeolite catalyst which causes the gas to form two fractions. The first is a gasoline-like molecule (octane 100) and the second is char. This process has the severe challenge of producing more gasoline and less char.

The higher quality biocrude is subjected to a high-pressure hydrogen treatment. This selectively breaks molecules apart removing oxygen and inserting hydrogen. This product has an octane number around 76, which is about the same for straight-run gasoline. Effort is underway to determine how to use less hydrogen.

Efficiencies of 60% to 80% (energy content converted to fuel) may eventually be realized, in which case a metric ton of biomass should yield 75 to 100 gallons of gasoline. The cost of producing gasoline by these methods is $2.00 to $3.00 a gallon. This cost could probably be driven below $0.90 per gallon with continued research. In all cases, the challenges are to improve reactor design, to lengthen the lifetime of catalysts, to easily upgrade catalysts after use, and to address auxiliary wastewater treatment systems. In any case, industry is becoming very interested in these processes.

The production of syngas through the high-temperature gasification of biomass is now well developed. Up to 95% of the biomass can be converted to syngas, with 75% of the original energy recovered in the gas. The challenges have been to remove tars and oils from the gas to suit various applications of the fuel.

Two new developments are significant in gasification. The first involves the high-pressure treatment of high-moisture biomass to form a mixture of methane, hydrogen, and carbon dioxide. This gas can be upgraded to be interchangeable with natural gas. While the high pressures required for this process require a great deal of energy, the concept and laboratory results to date have generated a great deal of interest. One particularly attractive aspect of the process is its ability to use moist feedstocks, because in most thermochemical processes drying consumes a significant amount of energy. The second development is the conversion of syngas to methanol. These efforts are in early stages of development with little yet to report.

9. Biomass energy and the environment

The environmental issues surrounding biofuels are significant but a great deal more favorable than for many other energy sources. The production and harvest of large quantities of energy crops would appear much more benign compared to conventional agriculture on the same land. This is particularly true for perennial woody and herbaceous crops. Energy crops would certainly provide a desired additional market for farmers. Aquatic crops may cause a much more significant change in the environment in the Southwest and in coastal areas if done on any large scale. However, a great deal needs to be researched before environmental effects can be identified more specifically.

Use of wastes for energy appears to offer a positive alternative to an otherwise negative disposal alternative. Gaseous emissions may be of greatest concern.

Use of agricultural residues, forest residues, and inventories of existing forests are of concern. Potential problems with nutrient depletion, erosion/siltation, and wildlife habitat would have to be monitored carefully, especially if large areas are involved.

Biomass offers a unique opportunity as an energy feedstock with respect to climate change. Like all fossil fuels, biomass does emit carbon dioxide when burned. However, biomass is the only fuel that takes almost an equal amount of carbon dioxide out of the atmosphere as it grows. This near-balancing of CO_2 emissions and uptake is in sharp contrast to fossil fuels, which continually add carbon dioxide and other harmful gases to the atmosphere. Other clean technologies like wind and hydropower are available for energy production, but none of these produce liquid fuels.

While the overall environmental outlook for biomass technologies is favorable, some concerns are still largely undefined. The conversion of biomass to liquid and gaseous fuels will involve the production, use, and disposal of catalysts, enzymes, phenolic wastes, acids, char and ash, and various flue gases such as carbon monoxide and other partially oxidized components of biomass. Because many of the conversion processes are not developed to the point of determining cleanup requirements, it is not clear how severe such problems will be. However, both researchers and sponsoring agencies should be very aware of the need to address such issues as soon as possible.

Energy analysts must also bear in mind that production of biomass consumes energy for cultivating crops and processing biomass feedstocks. Attention must be given to the cost in energy of producing biofuels as well as to the comparative economics of biomass and fossil fuels. Large-scale use of biofuels may require changes in the infrastructure that provides transportation fuel to consumers.

10. Discussion and summary

Biomass energy technologies encompass a variety of conversion processes and systems for producing energy feedstocks. Biomass can be a significant source of solid, liquid, and gaseous fuels, and ultimately more than one type is likely to find a permanent place in our energy mix. Biomass is the only form of renewable energy capable of producing the liquid fuels so important to current transportation systems. In the long run, the energy crop-to-fuel ethanol pathway appears to hold the most promise for meeting a serious national need. However, other forms of biomass energy will still have local and regional applications.

The very diversity of biomass energy systems is a major research challenge. Progress has been made on many fronts, and many issues remain to be addressed. This discussion has cited some costs and productivity rates that indicate that biomass will be a competitive energy source only when gasoline prices rise above $1.20 per gallon. In reality, some biofuel technologies will be viable at lower gasoline costs, some at higher costs.

The biofuels technologies are also at different stages of development. If gasoline costs $1.20 per gallon, some of the gasification technologies and possibly oilseed fuels would be viable almost immediately. The technologies for thermochemically produced biocrude, ethanol, methanol, high Btu gas, and algae-derived diesel fuel require differing amounts of additional research before they will be economically competitive with $1.20 per gallon gasoline. Where biomass technologies are already in use, they tend to be restricted to geographic areas with advantages like low-cost wastes and residues or to be employed in spin-off applications. The use of short-rotation woody crops for fiber production by the pulp and paper industry is one example.

Biofuel technologies generally appear more environmentally benign than the coal- or oil-based systems they would displace. This is particularly true at the end-use stage. Burning biofuels do create emissions, but the problems associated with emissions from biofuels appear far less severe than those associated with emissions from conventional fuels.

The production of biofuels offers a way to solve some significant environmental problems, but precautions must be taken to avoid others. One spin-off application now being adopted for short-rotation woody crops is the disposal of wastewater and sewage sludges. When directed toward tree growth, the nutrients in the waste become a resource rather than a problem. Another application is the stabilization of eroded agricultural soils with energy crops like short-rotation trees and perennial grasses. Hazards associated with the indiscriminate harvesting of existing forests for fuelwood and the production of energy crops on unsuitable sites can be avoided, but care is required.

Bibliography

Biomass Regenerable Energy, D.O. Hall and R.P. Overend, eds. (John Wiley & Sons, New York, 1987). This volume provides a technically oriented overview of biomass energy focusing on conversion.

Biomass as a Nonfossil Fuel Source, D.L. Klass, ed., ACS Symposium Series 144 (American Chemical Society, Washington, DC, 1981). An excellent integrated overview of biomass energy for the United States.

References and notes

1. P.R. Sharp, Biologue **6**:2, 3–31 (1989).
2. D.L. Klass and C.T. Sen (Institute of Gas Technology, Chicago, 1987), pp. 46–52.
3. D.L. Klass, Testimony before the Senate Committee on Appropriations, Subcommittee on Energy and Water Development (April 19, 1982).
4. Task Force on Wood Energy, J. For. 77 (1979).
5. U.S. Department of Agriculture - Forest Service, *Forest Resource Report No. 23* (U.S. Government Printing Office, Washington, DC, 1982).
6. U.S. Congress, Office of Technology Assessment, *OTA-ITE-210* (U.S. Government Printing Office, Washington, DC, 1982).
7. U.S. Congress, Office of Technology Assessment, *Energy from Biological Processes*, Vol. III, Appendixes, Part A (U.S. Government Printing Office, Washington, DC, 1980).
8. Energy Research Advisory Board, *Report on Biomass Energy* (U.S. Department of Energy, Washington, DC, 1981).
9. Doane Information Service, Doane's Agriculture Report 51 # 18 (1988).
10. Gershman, Brickner, & Bratton, Inc. and Systems Architects Engineers, Inc., *A Report to the Bonneville Power Administration* (1983).
11. U.S. Congress, Office of Technology Assessment, *Energy from Biological Processes*, Vol. III, Appendixes, Part B (U.S. Government Printing Office, Washington, DC, 1980).
12. W.J. Jewell, H. Davis, W. Gunkel, D. Lathwell, J. Martin, Jr., T. McCarty, G. Morris, D. Price, and D. Williams, *TID 27164* (Cornell University, Ithaca, NY, 1976).
13. J.W. Ranney, L.L. Wright, and P.A. Layton, J. For. **85**, 17–28 (1987).
14. J.H. Cushman, A.F. Turhollow, and J.W. Johnston, *ORNL/6369* (Oak Ridge National Laboratory, Oak Ridge, TN, 1987).
15. D.A. Johnson and S. Sprague, *SERI/SP-231-3206* (Solar Energy Research Institute, Golden, CO, 1987).
16. C.L. Peterson, Trans. ASAE **29**, 1413–1422 (1986).
17. U.S. Department of Energy Biofuels and Municipal Waste Technology Division, *Five-Year Research Plan 1988–1992*, DOE/CH10093-25/DE88001181 (U.S. Department of Energy, Washington, DC, 1988), pp. 14–16.
18. U.S. Department of Agriculture, *Ethanol, Economic and Policy Tradeoffs*, GPO 1988-201-025:60457/ERS (U.S. Government Printing Office, Washington, DC, January 1988).
19. J.W. Ranney, L.L. Wright, and P.A. Layton, J. For. **85**, 17–28 (1987).
20. R.S. Butner, D.C. Elliot, L.J. Sealock, and J.W. Pyne, *PNL-6765* (Pacific Northwest Laboratory, 1988).

Wind as a utility generation option

Jamie Chapman

1. Introduction

As a bulk power generating option for utilities in the nineties and beyond, wind turbine electric power generating systems are among the most attractive of the renewable energy generation technologies. In their current form, wind-driven power stations represent not a replacement for conventional power stations but rather a complement, forming a part of the total generation mix. In principle, wind systems could replace a part of the existing conventional generation capacity. More realistically, they will preclude the need for or delay the addition of conventional capacity in order to meet the demands of load growth.

Wind-driven power systems represent a renewable technology which may be described as mid-course in its development and maturation. It is a renewable power technology which has evolved rapidly over the last decade and has accumulated significant, large-scale, utility-connected experience, principally in California.

As a result of the impetus provided by the oil embargo of 1973, the 1978 National Energy Act, the later implementation of PURPA and by tax credit support through 1985, concerted development of modern wind turbine power systems began in the mid-seventies. Both public and private sector funding has been used to develop this technology.

The first large-scale, utility-connected installations appeared in California in 1980. Since then, private sector investment capital in excess of two billion dollars has resulted in the construction and commercial operation of wind-driven power stations whose capacity was rated at more than 1300 MW by the end of 1989. Thus in this country, by the end of 1989, grid-connected wind systems had provided capacity equivalent to a sizeable nuclear or coal-fired power station.

The early experience revealed severe shortcomings in the designs of the first-generation wind turbine components. These deficiencies can be traced largely to a lack of knowledge about the extreme dynamic range and attack times of the wind and its gust structure. Second-generation designs, introduced in the mid-eighties, generally have proven themselves able to accommodate and harness these wind forces. These systems have generated substantial quantities of electricity. For example, during 1989, these systems delivered about 2 billion kilowatt-hours of electrical energy, enough to meet the residential needs of a city the size of Washington, DC or San Francisco.

2. Utility-scale wind turbine power systems

2a. Wind-driven power stations

Large-scale, utility-connected, wind-driven power stations frequently are referred to as wind farms, wind parks, or Windplants, a trade-marked term used by U.S.Windpower, Inc.

These large-scale power generating systems are composed of a large number of arrayed wind turbines. The power from each wind turbine is collected and combined with that from others by a wiring network similar to a utility distribution system. The collected power is then delivered through a substation to the utility grid for use by consumers. A small, central control facility monitors and controls the status of each wind turbine as well as the total delivered power.

Except for the wind turbine components themselves, all other components and subsystems which make up a wind farm are the same as used in conventional utility applications. This includes the pad mount transformers used to combine the power from several wind turbines and step-up the voltage for transmission to the substation. The substation itself, with its associated monitoring, control, switching, and protective devices is purchased from the same vendors who supply utilities.

2b. Integration and compatibility

The integration of wind-driven electrical generation systems with the balance of a utility system is straightforward and is accomplished using standard utility components and practices. Control and monitoring capabilities of modern wind power systems are analogous to those used for similar functions in conventional utility systems.

For current systems, there is a limitation on the amount of wind power capacity which can be integrated with conventional power sources. Based upon the second-generation wind technology currently in use, the ratio of wind to conventional power systems is thought to be in the range 5% to 15%. This limit range arises from the time variability and other technical characteristics of the power delivered by wind systems.

Third-generation wind systems, now in their earliest stages of development, will allow wind utilization to twice this range (i.e., 10% to 30%) while still maintaining existing utility standards of power quality and system stability. In certain applications, the third-generation wind systems, when augmented with moderate amounts of energy storage, will allow the conventional power sources to be turned off.

2c. Advantages and characteristics of wind systems

Compared with either fossil-fueled or nuclear power stations, wind-driven power stations have the advantages of modularity and much shorter construction times. Due to the distributed nature of the generation sources (the wind turbine components), wind-driven power stations are inherently modular. Capacity can be added incrementally, as required, with a short planning horizon.

In addition to their inherent modularity, wind-driven power stations can be installed and placed into operation very quickly. Therefore the cost of funds used during construction is correspondingly less. For non-renewable power stations, this cost of funds can be a significant component of the installed capital cost, and therefore the cost of energy.

In their operation, wind-driven power plants utilize no fossil fuels or radioactive materials. Therefore there is no associated risk of environmental degradation or contribution to global warming. The capital and maintenance costs associated with scrubbers and other pollution control devices are eliminated. The costs and continuing risks associated with decommissioning nuclear power stations are eliminated.

Finally, the best of the second-generation wind systems currently in operation are economically competitive with newly constructed nuclear and coal-fired power plants. The Life-Cycle Cost-of-Energy (LCCoE) is in the range 7 to 9 cents/kWH. Measured on a rated *power* capacity basis, the installed, all-in capital cost is in the range $1000 to $1400 per kW. As a result of integrating research advances with the significant operating experience gained over the last decade, these cost values are expected to decrease to the ranges 4 to 6 cents/kWH and $600 to $100 per kW.

By folding in the quality of the wind resource, the cost per unit of *power* generating capacity may be transformed to the equivalent and more descriptive cost per unit of *annual energy* generating capacity. Depending upon the quality of the wind resource, the installed capital cost to produce a kWH per year is 50 to 70 cents per annual kWH of energy production capability. As before this measure of cost is expected to decrease to the range 30 to 50 cents per annual kWH of energy production capacity. It is important to note that these numbers apply to a complete power station interconnected with the grid and not just to the wind turbines themselves.

3. How wind-driven power stations differ

Wind-driven power stations differ from conventional power stations in two principal ways, each with an associated consequence. These are (1) the distributed generation sources and the associated land use, and (2) the nature of the fuel and associated characteristics of the generated power.

3a. Configuration and land use

Conventional utility power systems consist of a relatively small number of generation units each having a relatively large power capacity rating. Wind-driven power stations are comprised of a large number of geographically distributed generating units (the wind turbines) each of moderate capacity. Whereas the size of a conventional generating unit is measured in units of hundreds of megawatts and may occupy only a few acres, an individual wind turbine typically is rated in units of hundreds of kilowatts. The array of wind turbines, which when interconnected, have a capacity equivalent to that of a conventional generating unit, may occupy a land area of hundreds of acres.

Although wind-driven power stations do inherently require more geographical area or acreage than conventional stations, the land on which wind power stations are situated can continue to be used for its original purpose. This is because the individual wind turbine sites, the service roads, substations, and other supporting infrastructure actually occupy and disturb only a small fraction, typically 5% to 10%, of the total area containing the wind power station. For example, the large wind installations in northern California are located on ranchland. The ranchers continue to graze cattle on their land. As another example, the 2500 wind turbines in service in Denmark are distributed throughout the country, typically in small clusters associated with an industrial facility or group of residences.

3b. Fuel and power characteristics

The fuel for conventional power generating units typically is either a fossil fuel (coal, oil, or natural gas) or uranium. The availability of this fuel is not a function only of the vagaries of nature but of production capability, logistics, and storage capabilities.

The fuel for wind turbines comes without requiring such industrial support and without cost. However the availability of wind, while statistically predictable, is not subject to our control. It varies with time on differing scales: short-term (minutes and hours), diurnally, and seasonally.

The energy in the wind and therefore the electrical energy delivered is reasonably predictable seasonally and diurnally. It is predictable enough so that California utilities with large amounts of wind capacity factor the expected wind production into their generation planning and dispatch.

The short-term time variability is not as predictable. Thus the short-term power output from a single turbine mirrors the fluctuations of the wind. There is however some smoothing owing to inertial effects of the turbine rotor. Power electronics and adaptive control systems now being investigated for third-generation wind turbines will further smooth wind-induced fluctuations in the output electrical power.

Surprisingly, the short-term variability of the output power from an array is quite smooth relative to that from a single turbine. This is because the short-term wind time series at nearby wind turbine sites are relatively uncorrelated. When the outputs from a large ensemble of geographically distributed wind turbines are combined, the resultant, summed power delivered to the grid can be quite smooth. The short-term fluctuations are significantly reduced, but still present.

The extreme short-term time variability (minutes to tens of minutes) is the principal reason why the fraction of wind power systems delivering power to a total utility system is limited to a value currently thought to be in the range 5% to 15%. In the simplest case, in order to deliver the constant level of power required by a constant load, the time variability of the wind-derived power must be compensated for by increasing or decreasing the power from conventional generating units. Even if technically possible, frequent, short-term ramping up and down of these units increases their required maintenance and complicates the control and coordination of the entire power system.

In addition to the time variability of the real (or useable) power delivered to the grid, there is a second characteristic of the power delivered by most of the second-generation wind turbine generators currently in use. Most of the wind turbines now in use incorporate (for very good design and economic reasons) a generator which requires some electrical support from the conventional power sources also on the utility grid. It is a sort of symbiotic relationship. The particular class of generators, called induction generators, requires for its operation both an external voltage source and a source of reactive (or non-work-producing) power. The requirement for the so-called excitation voltage poses no difficulty and carries no economic penalty since the wind turbines must be connected to the grid anyway. However, to the extent that the wind turbines require reactive power, there is an economic penalty. This is because the conventional generation sources must be operated off of their most efficient operating point.

There are standard, simple ways of minimizing the reactive power required. The usual technique involves adding capacitors, a standard, passive electrical component which has the characteristic of appearing to supply most of the reactive power requirements of the wind turbines. Still, the added capacitors do not supply all of the required reactive power, represent an added system cost, and typically do not adjust for the time-varying reactive power requirements of the wind turbines.

4. Future trends and needs

4a. Geographical usage beyond California

The wind power stations installed in California generated about 1% of that state's total electrical energy needs during 1989. Two factors will increase the impact of wind energy generation in California. First, the performance of the existing plant will increase by continuing performance and reliability improvements. Second, the currently installed capacity conservatively represents development of less than half of the state's wind resource potential.

In other parts of the U.S., studies published by the Solar Energy Research Institute (SERI) and by the Electric Power Research Institute (EPRI) indicate a much larger total potential for the generation of electricity by wind systems. Regions other than California include the Great Plains, the Atlantic seaboard, the Great Lakes coastal regions, the Appalachian mountains, the Aleutian islands, and Hawaii.

Internationally, excellent wind resource regions are found in coastal and interior China, all islands and locations subject to the trade winds, coastal regions and the southern portion of South America, and several countries of Europe. As noted previously, Denmark, a leading manufacturer and exporter of wind turbines, currently utilizes about 2500 grid-connected wind turbines. These supply about 1% of that nation's energy needs.

4b. Technology trends

The principal technology improvements will be associated with increases in wind turbine size, improved knowledge of structural materials and structural responses, the incorporation of improved control hardware and algorithms, utilization of power-conversion electronics, and the development and integration of short-term energy storage systems.

The power rating of wind turbines used thus far in the California installations, on average, has doubled from about 50 to 100 kW. For wind farm applications, economic optimization studies have indicated that the optimum size lies in the range 200 to 600 kW. This corresponds roughly to rotor diameters which lie between 20 and 40 m.

Advances in airfoil performance and materials used for the wind turbine blades will allow the much wider geographic utilization of wind turbines. Changes in the electrical configuration of wind turbine generators, made possible by advances in power electronics and digital control, will further enhance the energy production performance of wind turbine power generation systems and make them even more competitive with fossil-fueled and nuclear power stations. Wind turbines will be self-controlling machines, able to adapt to changing wind and load conditions to a much greater extent than is currently implemented.

A major improvement in the wind turbine system architecture will be the incorporation of power electronics between the wind turbine generator and the load (the utility grid). As a wind system element, power electronics will bring several major benefits. It will allow the wind turbine rotor to be more efficient in converting power in the wind to delivered electrical power. It will provide a new control dimension which should permit lighter weight structural elements, thus reducing the capital cost. Finally, power electronics will completely eliminate the requirement for reactive power from conventional sources.

As discussed above, the principal characteristic which limits the use of wind systems without any support from conventional power sources is the time variability of the real power. Research in and the development of cost-effective energy storage and conversion systems will smooth out this variability. Depending upon the degree of smoothing and the time extent of the smoothing, the incorporation of wind with conventional utility systems in amounts significantly greater than the 5% to 15% or 10% to 30% values cited earlier for current and third-generation wind systems, respectively.

4c. Improved economics

Ongoing technical development and continuing large-scale operational experience are expected to yield major improvements in the economics of the next-generation wind turbine power systems. The Life-Cycle Cost-of-Energy will go from the current range 7 to 9 cents/kWH to the range 4 to 6 cents/kWH for the third-generation machines just beginning to be developed. Measured on a rated power capacity basis, the installed, all-in capital cost will be in the range $600 to $1000 per kW.

By folding in the quality of the wind resource, the cost per unit of power generating capacity may be transformed to the equivalent and more descriptive cost per unit of annual energy generating capacity. Depending upon the quality of the wind resource, the installed capital cost to produce a kWH per year is projected to decrease to values in the range 30 to 50 cents per annual kWH of energy production capability. Again, it is important to note that these numbers apply to a complete power station interconnected with the grid and not just to the wind turbines themselves.

4d. A new export market

The advantages and benefits that increasing use of renewable energy systems bring to the U.S. economy and environment apply also to other countries. Thus wind turbine power systems and services represent a sizeable potential export market for U.S. manufacturers, both to developed and less-developed countries.

The recent performance of Denmark provides an example of what can be accomplished in terms of an export market. Denmark has been quite successful at exporting wind systems to this country and elsewhere. Half of the wind turbines installed over the last decade in California were manufactured in Denmark.

Over the coming decade, the market for wind power systems in Europe alone has been estimated to be at least as large as the California market of the eighties. The individual nations of Europe together with the Commission of the European Communities have made plans for the addition of more than 4000 MW of wind capacity through the year 2000. Even if reduced by half on account of delays or overoptimism, this added capacity represents a market value of nearly two billion dollars.

5. Public policy initiatives

The current era may be characterized as one of low-energy prices coupled with uncertainty as to when (or whether) energy prices will rise. Practically all financial incentives for the further private sector advancement of renewable energy technology have been allowed to expire. These circumstances make it difficult to attract the investment of private-sector capital for further improvements and applications of the technology. Similarly, federal research budgets have been reduced drastically.

At the same time, there is increasing concern and knowledge about the effects of conventional power sources on our environment. And a case can be made that energy costs will increase significantly, whether as a result of geopolitical factors, simple depletion, or environmental concerns and the attendant costs.

These factors, together with the general perception that there is no urgent need for renewable sources of energy, represent a unique window of opportunity. Our society is presented with the time to take the next step, to combine the results of recent research and operating experience to position this country with a still-further improved renewable energy option. The fulfillment of this opportunity will, in addition, position U.S. manufacturers with a new class of export products and services. In addition to enhancing domestic employment and strengthening U.S. manufacturing, this will benefit the U.S. balance of trade.

This is an opportunity of such importance to the nation that public policy can legitimately be employed to shape the future and assist in realizing the benefits offered by these energy generation systems. Public policy initiatives to further the technical development, economics, and application of wind-driven power systems (and all renewable power systems) include:

(1) tax incentives supporting the installation of renewable power systems, but based upon delivered energy and not the installed capital cost as was done previously;

(2) accounting for and recognition in energy pricing of not only the direct costs of energy production by both renewable and conventional sources, but also the indirect, often delayed, environmental and societal costs of energy generation by the various sources;

(3) provisions in power purchase contracts (such as multiyear, levelized payment options) for assuring lenders of funds used in the construction of renewable power facilities that predictable debt service cash flow will be available over the term of the debt;

(4) policy permitting not only independent power producers but also utilities to own or co-own renewable energy generation facilities;

(5) policy enabling and facilitating the long-distance transmission of wind-generated electricity (wheeling) from a region of good wind resource (e.g., the Great Plains) to distant load centers;

(6) incentives and assistance for the export of renewable energy capital equipment and services.

Such policy initiatives would attract private sector capital to complete the needed technical development, facilitate additional renewable energy generation facilities thus mitigating environmental damage, and permit U.S. manufacturers to establish and maintain a pre-eminent position in this capital-intensive, environmentally important power generation technology.

Bibliography

California Energy Commission, *Wind Atlas* (California Energy Commission, Sacramento, CA, April 1985).

California Energy Commission, *Wind Energy: Investing in Our Future*, Revised Edition (California Energy Commission, Sacramento, CA, March 1984). See also Rashkin and Ringer below.

Electric Power Research Institute, *Proceedings of a Workshop on Prospects and Requirements for Geographic Expansion of Wind Power Usage*, EPRI report AP-4794 (Electric Power Research Institute, Palo Alto, CA, November 1986). Results of a conference which examined the potential for use of wind power stations in regions of the U.S. beyond California. See also Solar Energy Research Institute Wind Atlas entry below.

P. Gipe, *Wind Energy Comes of Age in California* (Paul Gipe & Associates, Tehachapi, CA, May 1989). A fairly complete, rational summary of the evolution, performance of the large-scale California wind power stations, including comparisons with other countries and potential.

Olav Hohmeyer, *Social Costs of Energy Consumption: External Effects of Electricity Generation in the Federal Republic of Germany*, Document EUR

11519, Commission of the European Communities [Springer-Verlag, Berlin, 1988 (English)]. Excellent treatise by a staff member of the Fraunhofer Institute of the direct and indirect costs of various classes of energy sources.

R. Lynette & Associates, Inc., *Wind Power Stations: 1984 Experience Assessment*, Interim Report, Research Project RP 1996-2 (Electric Power Research Institute, Palo Alto, CA, January 1985). One of a series of reports sponsored by EPRI which tracked the operations and maintenance performance of a large number of California wind power installations.

Janis Pepper, *Wind Farm Economics from a Utility's Perspective* (Pacific Gas & Electric Company, San Francisco, CA, August 1985). One of a series of examinations and comparisons of wind farm economics by a utility economist. Later publications appear under the name Janis Pepper-Slate, Enertron Consultants, California.

Cynthia Pollock, *Decommissioning: Nuclear Power's Missing Link* (World-Watch Institute, Washington, DC, April 1986). One of a series of papers published by the WorldWatch Institute on the environmental impact of various means of power generation.

Sam Rashkin, *Reports from the Wind Project Performance Reporting System: 1987 Annual Report* (California Energy Commission, Sacramento, CA, August 1988). One of a series of quarterly and annual reports which track the energy production performance of the California wind power stations.

M. Ringer, *Relative Cost of Electricity Production* (California Energy Commission, Sacramento, CA, July 1984).

Solar Energy Research Institute, *Wind Energy Resource Atlas of the United States*, Solar Technical Information Program, Report DoE/CH 10093-4 (SERI, Golden, CO, March 1987). Excellent treatise on the locations and quality of wind resources in the U.S. suitable for wind power installations.

Solar Energy Research Institute, *Wind Energy Technical Reading List*, Solar Technical Information Program, Report SERI/SP-320-3400 DE88001193 (SERI, Golden, CO, May 1988). Bibliography containing references to technical reports from the federal wind research program.

R.J. Sutherland and R.H. Drake, *The Future Market for Electric Generating Capacity: A Summary of Findings* (Los Alamos National Laboratory, Los Alamos, NM, December 1984). Among other topics, portrays the component of capital cost attributable to the cost of funds used during construction of a nuclear power station, prior to its becoming a performing, operational asset.

U.S. Congress, Office of Technology Assessment, *New Electric Power Technologies: Problems and Prospects for the Nineties* (U.S. Government Printing Office, Washington, DC, February 1985).

Utility Data Institute, *1987 Production Costs: Operating Steam-Electric Plants* (Utility Data Institute, Washington, DC, October 1988).

Technologies Related to End Use

Energy storage systems

John G. Ingersoll

1. Introduction

Energy is useful only if it is available when needed in a usable form. Storage keeps energy available until it is needed and in a usable form. Some forms of energy lend themselves more easily to storage than others. For example, oil can be stored relatively simply and inexpensively in a tank or container after it is extracted from the ground, while electricity requires rather sophisticated and invariably expensive means of storage such as electrochemical batteries.

The magnitude of the demand for particular energy forms varies widely. For instance, electricity is becoming the energy form of choice for more and more end-use applications because of its versatility. Yet electricity cannot be readily stored, so in the past, electric utilities have built a huge reserve power plant capacity in order to be able to meet short notice, diurnal and seasonal fluctuations in demand.

All of the most promising renewable energy sources, such as solar radiation and wind, are intermittent or variable in occurrence and can never be utilized to their full potential unless proper storage systems are developed. Likewise in the transportation sector, replacement of the gasoline internal combustion engine by a less environmentally offensive engine cannot be realized until a replacement for the gasoline-fuel tank can be found which stores an equivalent amount of energy.

Lack of proper energy storage systems results in significant inefficiencies in the utilization of the energy resources of the country. Some of the inefficiencies associated with lack of energy storage will be considered below. Then, various storage systems will be examined from a technological and economic point of view. Finally, the energy savings from implementing energy storage will be estimated and recommendations presented for the implementation of storage systems in the closing years of this century.

2. Necessity for storage

Energy storage is primarily needed to meet either a fluctuating rate of energy demand by the end users, or a mismatch between the availability of energy and its consumption. Energy storage systems must provide a specific energy concentration in a volume of space or a length of time in order to perform a certain task. Appropriate energy storage increases the efficiency with which available energy resources are used and makes new energy sources available. Timely provision of adequate energy from storage can increase productivity. In the remainder of this section we will examine three cases where availability of appropriate storage results in the aforementioned benefits.

2a. Utility load leveling

One of the major problems facing the electric utility industry is matching supply with demand. Power generation technology works best when it supplies energy at a constant rate. Unfortunately, the "load" of the utility, i.e., the quantity of electricity that must be supplied in order to meet demand during a specified time interval, is far from constant. The load cycles up and down on a daily basis so that, on the average, peak daytime loads are over 50% greater than nighttime power requirements.

Traditionally electric utilities have responded by setting up combinations of generating plants with different operating and economic characteristics so as to produce electricity at the lowest possible cost with adequate reliability.[1] Between 40% and 50% of the system's load (the base load) is typically supplied by large coal or nuclear units of the highest efficiency, operating on the fuel of lowest cost. Such base-load plants run continuously for most of the year. The broad daily peak in demand, representing another 30% to 40% of the load (the intermediate or cycling load) is met by the system's less modern and less efficient fossil-fuel units (coal, oil, and natural gas) as well as hydroelectric units and possibly gas turbines. These units operate anywhere from 1500 to 4000 hours per year and generate electricity at a higher cost than the base units. Brief peak loads are met by still older fossil-fuel units and by gas- or oil-fired turbines and diesel generators. Such units may operate from a few hundred hours to 1500 hours per year.

This traditional three-level combination of generating plants has become increasingly less attractive as sharply rising fuel costs penalize the less efficient units. In addition, the cycling of thermal plants causes maintenance problems and increases plant outage time. Moreover, cycling is not producing any return on the investment when the plants are not generating electricity. Finally, new coal-fired units require very costly pollution control equipment that represents a disincentive to replace old plants operating in a cycling or peaking mode. All these reasons clearly represent incentives to the utilities to turn to base-load plants as a source of power now generated by cycling and peaking units. The key to moving in this direction is energy storage, which makes it possible to accumulate the output of base-load plants during periods of low demand (nights, weekends) for periods of high demand.

2b. Renewable energy resources

Renewable energy resources can be most effectively utilized if the energy they supply can match the demand. Unfortunately, this is hardly the case for many renewable energy sources. The energy supply from these resources is fluctuating and, to a great extent, is unpredictable over short periods of time such as one day. The intermittent supply of renewable energy typically occurs out of phase with demand. Atmospheric phenomena whose predictability is rather limited influence most renewable resources. Consequently, exploitating renewable resources requires the availability of storage.

The most frequently encountered renewable sources nowadays include solar radiation converted to either heat or electricity, wind energy, and hydro power. Of these, hydro power lends itself to a relatively easy storage scheme as water can be kept behind a dam and released through turbines to generate electricity as needed. Storage occurs before the conversion to electricity. In the other two cases, i.e., solar radiation and wind, energy cannot be stored unless it is converted to heat or electricity. The portion of the generated heat or electricity that is not immediately usable must be stored. In the case of electricity generation either from solar radiation (photovoltaics) or wind, leeway exists for initially avoiding any storage.

If the source is connected to the utility grid, then the grid serves as the electricity storage system. However, the capacity of the grid for storage is not unlimited, albeit it is quite large. Estimates indicate that the presently existing utility grids, because of their vast interconnections, can accommodate renewable-based electricity equal to 20% of the total electricity generated (660,000 MW generating 2.5×10^{12} kWh per year). That is to say, no storage is required for up to about 20% renewable resource penetration of the utility grid at the present time. Above that penetration level appropriate storage systems become necessary.

2c. Transportation

The transportation sector in the U.S. consumes over 10 million barrels of petroleum per day, of which about 90% is used to power vehicles with internal-combustion engines. Thus over 50% of the country's oil consumption is dedicated to moving vehicles. The average fuel consumption in 1985 was 17.9 miles per gallon for passenger cars and 14.6 miles per gallon for all vehicles. Only 10% to 15% of the chemical energy stored in the fuel is converted into kinetic energy of the vehicle. Energy storage can have a major impact in the transportation sector by reducing the oil consumption in at least two different ways: (1) replace the present internal combustion engine by an engine powered with electricity; and (2) improve the miles per gallon fuel consumption of internal combustion engine vehicles.

Large savings can be realized if electric vehicles replace present day vehicles as the conversion efficiency of electricity to kinetic energy can be as much as four times higher than the conversion of fuel oil to kinetic energy. Moreover, electricity can be generated by a fuel other than oil such as coal. Consequently, the adoption of electric vehicles not only reduces fuel consumption and oil consumption, but also removes air pollution problems from large cities to remote ones where the power plants are located. Unfortunately, replacement of the internal combustion engine is a formidable task simply because of the high density of energy storage that gasoline or diesel fuel makes possible. For instance, a 12-gallon tank with a volume of 1.5 cubic feet can store 1.5 million Btu in the form of chemical energy, enough to drive a new automobile between 250 and 400 miles. In that same volume, lead-acid batteries can store only about 3 kWh (equivalent to about 10,000 Btu). Nevertheless, the inherent efficiency of an electric motor coupled with a reduced vehicle weight and an aerodynamic body design can make up in part for the low lead-acid battery specific energy density. This is the case with the recently unveiled GM prototype "Impact" which is a 2-passenger car having a 1000 kg weight, a drag coefficient of only 0.19, an energy consumption of 0.11 kWh/mi, and a range of 124 miles. Advanced batteries under development can triple or quadruple the lead-acid battery energy storage density, so that electric powered vehicles can attain the performance of gasoline and diesel powered vehicles for most applications.

Energy storage can also improve substantially the efficiency of the conversion of chemical or electric energy to vehicle kinetic energy by storing the kinetic energy of the vehicle that is normally wasted during deceleration and braking. This kinetic energy is transferred to a flywheel, which in turn powers electric accessories of the vehicle. This technique, otherwise known as regenerative braking, can increase the energy consumption efficiency of gasoline or diesel cars by 10% to 15% and an electric car by as much as 30%. For example, the Volvo prototype LCP-2000 makes use of a flywheel storage system that accounts for 7 to 10 miles per gallon of its composite urban-highway 71 mpg fuel efficiency. This is a 4-passenger car with a diesel magnesium alloy engine, continuously varying transmission, highly aerodynamic profile, fiberglass and carbon fiber components for a 50% weight reduction by comparison to the average American car.

With the advent of fiber-composite flywheels, it will be possible to drive automobiles by drawing energy from a flywheel on board. One would spin the flywheel with a motor running on power from an electric outlet, and then the energy stored in the flywheel could power the car with the motor operating as a generator to supply power to smaller motors, one on each wheel.

3. Types of storage

Most of the many techniques for storing energy have been known for some time and a few go back to antiquity. However, only a few of these techniques are in use today. Nonetheless, the concern over energy and environmental issues may provide the impetus to improve existing techniques and realize many more in the next few decades.

Energy storage systems can be characterized by several physical properties. The most prominent is a characteristic quality associated with each form of energy called *entropy*. Entropy is a measure of the degree of disorder of the particular energy form. According to the laws of thermodynamics, energy must always flow in such a direction as to increase the entropy. Consequently, we can arrange the different forms of energy in an "order of merit" the highest one being the one with the least entropy. The entropy of an energy form is significant in that it determines the efficiency with which the particular form can be stored, extracted from storage, and converted to a different form of energy.

Energy of lower entropy can be converted into the same or a higher entropy form with 100% theoretical efficiency. Likewise, the storage and extraction

Table 1. Entropy per unit energy for different forms of energy.

Form of energy	Entropy per unit energy[a]
Gravity (potential energy)	0
Energy of rotation/orbital motion	0
Sunlight	1
Chemical energy	1–10
Terrestrial waste heat	10–100

a. In units of inverse electron-volts.

efficiency of a lower entropy energy form exceeds that of a higher entropy energy form. It should be emphasized that efficiency of storage is of paramount importance in making a storage system practical and ultimately cost effective. The entropy of a few relevant energy forms is shown in Table 1.[2] The entropy of a particular energy form varies inversely with the temperature associated with that energy form.

Based on the preceding classification of energy forms, we can distinguish at least five types of storage: mechanical, electrical, chemical, biological, and thermal. Mechanical and electrical storage are treated separately, even though from an entropy per unit energy point of view, they are equivalent. Likewise, we also separate chemical and biological types of storage in order to discern the origin of the energy stored, even though the form is in both cases chemical. Thus, from the point of view of the second law of thermodynamics, three general types of storage would suffice. However, the number of storage types has traditionally been augmented to five.

Only schemes with significant potential for realization will be examined in the following sections. Significant potential here implies either a state of commercialization or else an advanced stage of development. The reader may realize that since most of the energy storage schemes are presently at a developmental stage, the possibility always exists of a technological breakthrough that may propel a storage scheme from the less developed status to a status even more developed than that of storage technologies presented herein.

4. Mechanical storage

Pumped hydroelectric storage, flywheels, and compressed air constitute three different techniques for mechanical storage, each using a different phase of matter as a storage medium.

4a. Pumped hydrostorage

A pumped hydrostorage system uses two reservoirs, an upper and a lower. During off-peak hours, electricity from base load plants is used to power pumps that lift the water from the lower to the higher reservoir. When peak demand exceeds the base load capacity, the water is returned to the lower reservoir through a turbine, thereby generating electricity. In practice the same machine (usually a Francis turbine) is used both as pump and as turbine.

The annual energy generated by pumped storage is proportional to the annual water flow, the system efficiency, and the height between the upper and lower reservoirs. Typical values of the height are on the order of 100 m. Pumped hydroelectric storage plant efficiencies are on the order of 70%. The 30% losses include 15% input power lost to friction and distribution and another 15% of input power used to keep the turbines/pumps spinning in order to allow quick response.

The largest pumped storage hydroelectric facility in the world is at Ludington on the eastern shore of Lake Michigan.[1,3] The average water head is 85 m and the six pump turbines can generate more than 2000 MW of power. Lake Michigan is the lower reservoir, while the upper reservoir is an artificial lake contained by an earth-fill dam six miles long. If the plant were fully discharged, it could generate 15 GWh. Over two-thirds of the energy expended in charging the reservoir is recovered.

At present some 2% of all U.S. electric energy is obtained from pumped-storage facilities. It is unlikely that this small fraction will be much increased because there are few suitable sites near large electric-generating plants. Pumped storage also produces environmental problems such as inundation of canyons and other scenic locations. Consequently, it has been suggested that underground caverns be used as an alternative to above ground pumped hydro-storage. In the underground cavern scheme, water is stored in a small upper reservoir and passes through the pump/turbine to the lower underground cavern, which can be as deep as 1000 m. The higher head implies a lower volume of water for a given energy storage. The cavern-excavation technology is available as are high-lift pump turbines. Suitable geological rock formations exist widely in the northeastern, north central, and western U.S. No underground pumped storage plant has ever been built, although several engineering studies have been conducted to define the costs, risks, benefits, and potential of this untested concept.

4b. Flywheels

Flywheels are rotating bodies that store mechanical energy, which can be put in and taken out. The ancient potter's wheel, which is mentioned in the Old Testament, is essentially a flywheel. In more recent times the flywheel has been employed in numerous devices from the steam engines of the Industrial Revolution to the present-day internal combustion engine of automobiles, trucks, and diesel locomotives.[4] Current research seeks to use flywheels in electricity storage and transportation. The advent of composite materials in recent years has substantially improved flywheel performance characteristics (e.g., energy density) so as to bring this technology near commercialization.

The kinetic energy of a flywheel depends on its moment of inertia and the square of its angular velocity. Obviously, the higher the angular velocity the more energy is stored in a flywheel. However, the strength of the material limits the maximum possible angular velocity as the centrifugal forces tend to

Table 2. Comparison of physical properties of flywheel materials.

Material	Density (g/cm^3)	Maximum stress (MPa)	Modules of elasticity (GPa)
Steel, stainless 301	7.80	200	190
Aluminum, 5083	2.70	124	71
Carbon fiber	1.50	2760	35–689
E-Glass	2.53	3447	72
S-Glass	2.48	4481	85
Aramid (Kevlar – 290)	1.44	2758	124
Fused silica	2.16	13,790	72
Aluminum oxide	3.96	19,305	414
Silicon carbide	3.18	11,032	841

break the flywheel apart.[5] A flywheel made with a low density, high tensile strength material, distributed in the form of a disk will have the highest energy density per mass. The maximum energy density per swept volume is directly proportional to the tensile strength of the material. The maximum energy density per unit mass is also directly proportional to the tensile strength of the material and inversely proportional to its density.

Traditionally, steel has been used as a flywheel material because of its relatively high tensile strength. Steel also has a high density which results in a lower energy density per unit mass. With the advent of lightweight composite materials, very high energy densities are possible. Moreover, the energy density can be optimized by using several materials on the same flywheel rather than only one in the so-called composite flywheels. The average radius of a thin ring will change due to centrifugal forces. This radius change is proportional to the centrifugal acceleration of the ring and the density of the ring and inversely proportional to the Young's modulus of elasticity of the material. Since centrifugal acceleration increases for an increasing radius, a flywheel can be made with a series of rings, one outside the other, such that the ratio of density to Young's modulus decreases from the center ring outwards. This implies that each ring has to be made from a different material. Table 2 shows the pertinent properties of flywheel materials.[3–5]

During the late 1970s and early 1980s, the Lincoln Laboratory of MIT under the sponsorship of the U.S. DOE designed a 40 kWh flywheel for residential photovoltaic systems.[6] A $1/10$ scale prototype was built and tested. The experimental flywheel used a 180 kg steel rotor, supported by magnetic bearings, and stored 1 kWh of energy at 15,000 rpm. The in-out storage efficiency was 82% for a one hour storage time, and the unit loss was 4W per hour or less than 0.5% of the stored energy per hour. A composite flywheel was also built using 56 kg of Kevlar with epoxy. This flywheel which was to also develop 1 kWh at 15,000 rpm suffered a catastrophic failure due to fatigue in the rotor spokes.

More recently the University of Ottawa has designed, built, and tested a composite flywheel for the Canadian Coast Guard.[7] These flywheels were developed to replace lead-acid battery storage systems connected to photovoltaic powered remote light stations. Magnetic bearing/suspension systems are used for the rotor enclosed in a hermetically sealed vacuum housing to minimize friction losses. The rotor is of a triannular ring design. From the exterior to the interior of the rotor S-glass, carbon fiber, and aluminum are used. The rotor is designed to store 8.5 kWh at 23,000 rpm (47.5 Wh/kg), has a diameter of 61 cm and a length of 49.5 cm (59 Wh/l). Its life expectancy is estimated at 20 years.

A distinct safety advantage of composite flywheels is that in the event of a catastrophic failure, the rotor material disintegrates to small pieces unlike a steel rotor.

4c. Compressed air

In this system, off-peak electricity powers a compressor system that pumps compressed air into an underground reservoir. When peak electricity is required, the pressurized air is used to power a turbine which in turn drives a generator. This storage technique has several distinct advantages over pumped hydroelectric storage. First, a wider choice of geological formations can serve as the reservoir. The storage cavern could be in either hard rock or salt. Second, air can be rapidly compressed and slowly expanded. As a result large pressure fluctuations will be smoothed out. Third, the energy storage density is relatively high and the minimum size smaller.

There is, however, a complication. During the storage period, the air gets very hot and must therefore be cooled in order to avoid fracture of the rock or creeping of the salt that comprises the reservoir walls. During the discharge period, the stored air must be preheated as the air is expanded into the turbine. The energy to reheat the air is supplied by burning of fossil fuels. Engineering studies have also been performed to determine the possibility of "saving" the heat extracted during compression by means of a recuperator and adding it back during expansion. Consequently, the need of external heat generation may significantly be reduced thereby increasing the overall efficiency of the compressed air electricity storage process.

The energy stored in the compressed air is equal to the work performed to compress the air from an initial pressure to a final one. If it is assumed that the process takes place under constant temperature and that the air behaves according to the ideal gas law, the change in entropy is zero. Under ideal (reversible) conditions the heat that has to be supplied to the gas during expansion is a minimum, but owing to irreversibility during expansion extra heat must be added. This is a simplified description of the process, but nevertheless adequate for the present study.[3]

The world's only commercial compressed air storage facility has been in operation in Huntorf, near Bremen in Germany, since 1979.[1] The facility is connected to the local utility grid. During off-peak hours electricity from the grid compresses air to about 70 atm into two caverns. The combined storage of these caverns, which have been leached out of salt deposits, is about 300,000 m^3. During period of peak demand the air is expanded, heated by the burning of fossil fuels, and fed through turbines that generate 290 MW for two hours. At the Huntorf plant each kWh of electrical output requires 0.8 kWh for air compression and 5600 kJ of heat input for air reheating at the turbines. The actual efficiency of storage is then 42.5%. In general efficiencies in excess of 50% and even as high as 70% are possible, if some of the heat extracted during compression can be saved to reheat the air during expansion. The energy density of the Huntorf storage plant is about 0.03 MJ/l (8 Wh/l) or 0.4 MJ/kg (118 Wh/kg). These energy densities are expected to be typical of compressed air storage plants. At the present time, a similar facility is under development in the state of Alabama as a joint venture between the Alabama Electric Cooperative and EPRI. The size of the facility will be on the order of 60 MW and the anticipated cost of construction on the order of $150 per kW. An advanced recuperator will be employed to recover turbine exhaust heat so that efficiency is expected to be 47.7% (per kWh of output the design input is 0.78 kWh of electricity and 4250 kJ of heat).

5. Electrical storage

Electrical energy can be stored in static electric fields, static magnetic fields, and certain reversible chemical reactions. If the generated fields or chemicals can be maintained without the addition of energy, practical electricity storage systems are obtained. Such systems include capacitors, electromagnets, and electrochemical batteries.

5a. Capacitors

The simplest such system will consist of two parallel conductive plates separated by a dielectric material. The electric energy is stored in the form of an electrostatic field built between the plates by the electric charges accumulated on the plates. The energy stored in the capacitor is equal to the work expended in bringing electric charge to the plates under an electric potential. The stored energy depends on the geometry of the plates and the dielectric constant of the material between them. Each dielectric material can accommodate a maximum field strength before it breaks down. Typical values of the dielectric constant are on the order of 2 to 3.5 for plastics such as Teflon, polyethylene, and nylon and can be as high as 100 for titanium dioxide.[3] Corresponding dielectric strengths are in the order of 50 to 60 kV/mm for the aforementioned plastics, but only 6 kV/mm for titanium dioxide. Consequently, the energy density that can be obtained in a capacitor is on the order of 10 J/l (0.0003 kWh/l). This is an extremely small energy storage density and therefore appears to be totally impractical for large systems as it would require immense volumes. Capacitors can be charged slowly and release their accumulated energy in a single pulse. Capacitor storage thus finds application in situations where single, very high energy pulses of electrical energy are required.

5b. Superconducting coils

In its simplest configuration, this system consists of a coil of wire carrying a large current. The wire is kept at very low temperatures (typically from a few to several degrees Kelvin above absolute zero) so that its electrical resistance is equal to zero. The energy is stored in the form of a magnetic field. Work is required to establish the magnetic field because initially an electromotive force is generated that opposes the current. This emf depends on the rate of change of the current which in turn depends on the physical characteristics of the winding of the coil. The ratio of the opposing emf to the rate of charge of the current is a constant for a given coil configuration and is called the inductance. In an inductor when the current is shut off the magnetic field collapses and the stored energy is fed back into the circuit — unlike a capacitor, where the charges remain on the plates, as long as the latter ones are well insulated, after the charging voltage is removed. Consequently, we can conceive of a circuit that would enable a solenoid to be charged with current from a generator, then switched to a short circuit containing only the coil and isolating the generator. When energy output was desired, the switch could be returned to the original position. This is possible because of the electrical inertia associated with an inductor. We can thus define a time constant inversely proportional to the resistance of the inductor. For superconducting coils the resistance of the circuit is extremely small. The energy densities associated with magnetic storage are on the order of 10^3 J/l. This is comparable to the energy density of pumped hydroelectric storage.

The efficiency of superconducting magnetic energy storage is about 95%, of which 3% are losses due to ac to dc conversion and 2% are consumed for refrigeration. A magnetic storage system can reverse its mode from charging to discharging in less than 1 s. Consequently, magnetic storage systems can stabilize the utility grid.

The magnetic storage technology is in its infancy. No superconducting coil has ever been built for energy storage. During 1983 and 1984 the Bonneville Power Authority successfully tested a small superconducting coil to demonstrate its feasibility as a means to stabilize the voltage on transmission lines. This coil, specified by Los Alamos National Laboratory and built by GA Technologies (formerly General Atomics) stored only 30 MJ (8 kWh) of energy. Currently, Bechtel Corporation is working on the conceptual design of a large superconducting magnetic energy storage system.[8] Funding is provided by the U.S. Department of Energy through Los Alamos. The Bechtel concept envisions a standard utility interconnected system with a 5000 MWh storage capacity. This system will be designed to charge or discharge in five hours at a constant rate of 1000 MW. The size of the coil, which is circular, is about 1000 m in diameter. It consists of 556 windings with a wire length of 1600 km and carries 200 kA of dc current. The key elements of the system are a superconducting coil, a helium refrigeration system, a thermal protection system, a vacuum vessel, and electronic controls. The cost is estimated at $900 million or about $180/kWh.

5c. Electrochemical batteries

In any chemical reaction the energy content of reaction products is different from that of the original reactants. If the energy content of the products is less than that of the reactants, we have what is termed an exothermic reaction. Normally, the liberated energy in an exothermic reaction is in the form of heat. However, many exothermic reactions will channel their energy into electricity with proper conditions and hardware.

Generation of electrical energy from chemical reactions requires (i) the separation of the reactant molecules into oppositely charged ions in a suitable medium and (ii) the utilization of the resulting potential to drive an electric current through an external circuit. The charge transfer medium is an ionic conductor called an electrolyte. An electrolyte can be an ionic liquid, a plastic membrane, or even a rigid solid such as a ceramic material. The charges accumulate on suitable electrodes, which form the interface terminals between the internal electrochemical reaction and the external electric circuit. When the chemical reaction takes place electricity flows in the external circuit. The chemical reaction can be reversed and the original condition of the reactants can be restored by applying an external voltage to the system.

The minimum system is called a cell and consists of two half-cells, each corresponding to the respective electrode. At the surface between the electrolyte and each electrode there exists a potential difference called the electrode potential. The sum of the two electrode potentials constitutes the full cell potential.

Maximum efficiency in electrochemical systems requires that the processes occur extremely slowly, implying correspondingly a small current flow. Departure from this condition results in irreversibilities that reduce the cell efficiency. Moreover, physical changes may occur in the reactants after each charge–discharge cycle leading to a finite cell lifetime.

A practical electrochemical system must meet certain technical constraints as follows:

1. The electrolyte must be a good ionic conductor and be chemically compatible with the electrodes.

2. The electrolyte and electrodes must suffer an extremely small change per charge–discharge cycle of the cell for a long lifetime.

3. The electrodes must offer a large effective surface area at their interface with the electrolyte to facilitate the process.

4. Materials used for the electrodes and electrolyte must be readily available so as to obtain a low cell (battery) cost.

5. The energy density either per unit mass or per unit volume must be high.

Of the several possible and/or available electrochemical systems, four offer the greatest promise in meeting all or most of the preceding constraints in the next decade. These systems or batteries are: (i) advanced lead acid; (ii) sodium-sulfur; (iii) lithium-iron sulfide; and (iv) zinc-bromide. Some of the general characteristics of these four systems are given in Table 3. A discussion of each system is given in the Appendix.

6. Thermal storage

A substantial portion of the energy used in the world is in the form of low-temperature heat. It is usually not sensible to meet this demand for heat from higher grade sources, which can be best saved for higher grade uses. Consequently, it would be better to use renewable resources such as solar radiation

Table 3. Characteristics of selected electrochemical storage systems.

Type (1)	Specific energy (Wh/kg)	Specific power (Wh/kg)(2)	Energy density (Wh/l)	Energy efficiency (%)	No. of discharge cycles (#)(3)	Operating temp. (°C)
Lead-acid, standard (C)	23–35	20 (Max 60)	. . .	75–80	1000	Ambient
Lead-acid, advanced (C)	35–45	25 (Max 100)	80	75–80	1500	Ambient
Sodium/ sulfur (NC)	180–200	100 (Max 200)	100	75–80	1000–1500	350
Lithium/metal disulfide (D)	100–135	50	>100	75–80	200	150–450
Zinc/ bromine (D)	25–35	40 (?)	15–20	70–75	500	Ambient

(1) C: Commercial; NC: Nearing Commercialization; D: Developmental.
(2) Maximum specific power can be attained for very short periods of time without reducing the specific energy.
(3) At 80% depth of discharge.

to meet heat energy requirements. This obviously entails appropriate storage of heat. Moreover, waste heat from other processes can also be stored to be used as a source of low-temperature heat.

Heat can be stored in two different ways: by utilizing the specific heat capacity of a material (sensible thermal storage) or by utilizing the heat associated with a phase change of a material (latent thermal storage). A third possibility of storing heat (thermochemical storage), which typically requires much higher temperatures, stores the heat as chemical energy in materials produced by the thermal dissociation of a properly selected chemical compound.

6a. Sensible heat storage

In this storage system, heat is used to raise the temperature of a large mass of material, which is thermally isolated from the environment to reduce heat losses. The heat which can be stored depends on the specific heat capacity of the material, the mass present, and the temperature increase which is allowed.

Water has one of the highest heat capacities, but most other common materials are within a factor of 2. Table 4 summarizes some of the pertinent thermophysical properties of typical sensible storage materials. In addition to the energy density attained in sensible heat storage, the rate at which heat can be stored or extracted from the storage mass — typically in a solid state — depends on the thermal diffusivity of the material. It turns out that iron and iron oxide have excellent properties in this regard. The specific heat capacity of these materials is very close to that of water. Moreover, high-temperature storage is also possible, unlike water which has a relatively low boiling temperature. The

Table 4. Pertinent thermophysical properties of sensible heat storage materials.

Material	density $\rho(kg/m^3)$	Specific heat $c(J/kg\ °C)$	Heat capacity $\rho c(J/m^3\ °C)$	Thermal conductivity $k(W/m\ °C)$	Thermal diffusivity $k/\rho c(m/s^2)$
Water	1000	4180	4.18×10^6	0.585	0.14×10^{-6}
Iron	7600	528	3.51×10^6	46.85	3.40×10^{-6}
Iron oxide (Fe_2O_3)	5200	753	3.93×10^6	20.79	5.29×10^{-6}
Granite	2700	794	2.17×10^6	2.72	1.27×10^{-6}
Concrete	2470	920	2.26×10^6	2.42	1.07×10^{-6}
Brick	1700	836	1.42×10^6	0.63	0.44×10^{-6}
Dry earth	1260	794	1.00×10^6	0.25	0.25×10^{-6}

cost of iron balls or iron oxide pellets such as taconite spherules ready for smelting is rather moderate, i.e., on the order of $0.2 to $0.9 per kilogram. Sensible heat storage densities are on the order of 0.2 to 0.05 MJ/kg and 0.2 to 0.4 MJ/l.[16] Depending on the duration of heat storage as well as the degree of insulation of the storage mass or volume, the total efficiency of sensible thermal storage can be anywhere between 50% to 90%, with 70% being typical for the larger industrial applications.

Sensible thermal storage is very common in solar buildings where the material normally used is either water or granite. Sensible thermal storage in the form of high-temperature steam during off-peak hours has been used in Germany for many years. The major problem with high-temperature steam storage is the high cost of the storage device, typically a pressurized steel vessel. Since this is impractical for use with power plants, where rather large storage volumes are required, underground caverns offer an alternative solution.

6b. Latent heat storage

The phase change of a material from liquid to solid or from vapor to liquid liberates heat at constant temperature. A phase change in the opposite direction absorbs the same amount of heat, called the latent heat of the phase change. The latent heat storage density depends on the latent heat per unit mass and the density of the material. In addition to a desired high latent heat for practical storage systems, the phase change temperature has to be relatively low, i.e., up to a few hundred degrees centigrade.

Unfortunately most materials that have high latent heats have very high phase change temperatures. Water is the most common phase change material in the ordinary temperature range, because of its high latent heat of fusion or heat of vaporization.

A class of materials, called hydrates, exhibit relatively high latent heats at ordinary temperatures. Hydrates are chemical compounds in which water molecules occupy definite lattice sites in the crystalline structure, thereby requiring a substantial amount of energy for melting. Upon melting, the water coalesces and the compound dissolves in it. Hydrates have high costs and most of them have rather limited thermal cycling expectancy. At the present time,

Table 5. Pertinent properties of latent heat storage materials.

Material	Phase change	Transition point (°C)	Latent heat (kJ/kg)
Water (H_2O)	Liquid–gas	100	2260
Water (H_2O)	Solid–liquid	0	335
Lithium hydroxide (LiOH)	Solid–liquid	462	430
Sodium hydroxide (NaOH)	Solid–liquid	318	167
Aluminum hexachloride (AL_2CL_6)	Solid–liquid	190	263
Solium sulfate decahydrate ($Na_2S_4–10H_2O$)	Solid–liquid	32	234
Calcium chloride hexahydrate ($CaCl_2–6H_2O$)	Solid–liquid	27	171

only one hydrate, calcium chloride hexahydrate, is commercially available in a number of configurations, including rods, translucent pods, and steel cans. It is claimed that the life expectancy of calcium chloride hexahydride, in a DOW Chemical proprietary formulation, is several thousand cycles. This phase change material is essentially designed for heat storage in residential and possibly small commercial buildings. Because the phase change temperature of this material is 27 °C, it is appropriate for space heating applications, where the living space temperature is ideally around 21 °C. Proper balancing of the material surface area and its mass are required in order to achieve sufficient heat transfer to the building interior to maintain a comfort temperature and not overheat the environment. Besides storing heat, "cold" storage is becoming more important.[9]

Using off-peak electricity, one can freeze water, or in some instances chill water or salts. The cooled mass is then used for air conditioning, refrigeration, or other applications during on-peak electricity demand time periods. Such systems are predominantly appropriate for medium to large commercial buildings. Several utilities support the storage of "cold" energy by supplying financial incentives to the customer. For example, the Department of Water and Power (DWP) in the City of Los Angeles provides a rebate of $250 per kW permanently shifted to off-peak up to a maximum or 40% of the installed system cost, whichever is less. The DWP will share 50% of a technical feasibility analysis up to $5000. The minimum cooling capacity is stipulated to be 25 kW. Several projects using ice or chilled water are being built around the country at the present time even without the support of the utilities. The primary incentive is the existing kWh cost differential between on-peak and off-peak hours.

Table 5 presents a summary of the latent heat storage properties of some materials primarily for comparison purposes. Typical heat storage densities for the hydrates at room temperatures and for ice are on the order of 0.25 to 0.9 MJ/kg and 0.3 to 0.6 MJ/l, while steam has densities of 2.2 MJ/kg and 0.02 MJ/l (at 20 atm pressure) at a 100 °C operating temperature.

7. Thermochemical storage

In the traditional large-scale generation of electricity from solar radiation, many concentrating collectors tracking the sun transfer solar heat to some fluid, typi-

Table 6. Candidate reactions for themochemical energy storage.

Reaction	Reaction energy per unit mass
$SO_3 \rightarrow SO_2 + 1/O_2$	1.21×10^6 J/kg
$1/2H_2O + 1/2CH_4 \rightarrow 1/2CO + 3/2H_2$	6.02×10^6
$COCl_2 \rightarrow CO + Cl_2$	1.09×10^6
$2NF_3 \rightarrow N_2 + 3F_2$	1.80×10^6
$CH_4 + CO_2 \rightarrow 2CO + 2H_2$	4.10×10^6
$CH_3OH \rightarrow 2H_2 + CO$	3.26×10^6
$NH_3 \rightarrow 1/2N_2 + 3/2H_2$	3.30×10^6

cally water, which in turn can be used directly in a turbine to generate electricity. In a thermochemical storage system, solar heat is concentrated on a fluid which undergoes a reversible chemical dissociation. The products of the dissociation can be stored and then allowed to recombine at a desirable rate so as to release heat which in turn can raise the temperature of the working fluid of a turbine to generate electricity. The advantage of the thermochemical storage system is that the heat converted and stored in the chemical is not lost between collectors and the heat engine. Consequently, transmission can be over a long distance or a long time allowing for electricity generation irrespective of solar radiation availability. Several candidate reactions for thermochemical energy storage have been identified as shown in Table 6.

The need for low-cost energy transfer components appears to favor reactions with high reaction energy per mass and high fluid densities (dependent on pressure). Moreover, desirable features of potential reactions include lower operating temperatures, use of less expensive catalysts, and less hazardous and toxic chemicals. Of the potential reactions listed in Table 6, two have been studied in some detail for a thermochemical system and are briefly described.

7a. The ammonia system[10]

Solar energy is collected by means of a large number of paraboloidal pressed steel mirrors, each 10 m² in aperture area. At the focus of each mirror is an absorber in which high-pressure ammonia gas undergoes chemical dissociation into nitrogen and hydrogen. This reaction is endothermic with the heat of reaction supplied by the absorbed solar energy. The temperature of the absorber is maintained constant at about 700 °C by properly controlling the flow of ammonia. The incoming ammonia and outgoing nitrogen and hydrogen pass through a counterflow heat exchanger. It is then possible for the nitrogen and hydrogen to emerge from the heat exchanger at almost ambient temperature.

The nitrogen and ammonia are stored and are then used to generate electricity at the required rate. The two gases are preheated at 490 °C and pass through a reaction chamber where partial recombination of ammonia takes place. The reaction is exothermic and the heat produced is approximately equal to the solar energy initially collected by the mirrors. The heat exchanger/catalyst chamber is thermally coupled to the working fluid of a turbine, or a heat engine in general, whereby mechanical and electrical energy can be extracted. The entire ammonia/nitrogen/hydrogen gas system operates at a constant pressure of about 300 atm. The thermal efficiency of the process is estimated to be as high as 79%, while conversion of heat into electricity may be on the order of 29%, for an overall conversion of solar heat to electricity of about 23% and the added advantages of energy storage and control of the electricity generation rate.

7b. The sulfuric acid system[11]

In order to increase the reaction energy per unit mass and in fact reduce the maximum operating temperature to about 800 °C for sulfur trioxide decomposition, a modified system has been proposed that centers around sulfuric acid. The heat necessary for the decomposition of sulfuric acid and sulfur trioxide is supplied either by solar energy or by heat from a nuclear power plant, assuming that the latter can also supply it at the right temperature. The process uses a catalytic bed, and the heat carried by the reaction products converts water into steam through a counterflow heat exchanger. This steam can then drive a turbine to generate ultimately electricity. The total thermal energy required for the reaction is 5.95×10^6 J/kg of SO_2, while the thermal energy released in the synthesis is 3.71×10^6 J/kg of SO_2. Hence the energy storage efficiency is a little over 62%. The operating pressure of the system is about 10 atm. The final thermal to electrical energy efficiency is about 23%. Also, the energy storage density of this system (liquified SO_2 at ambient temperature and 6 atm pressure) is 0.58×10^9 J/m^3. A typical system unit may consist of 6.4×10^6 mole of SO_2 storage (300 m^2 in volume) and require 7.36×10^6 mole of H_2O storage, 3.2×10^6 mole O_2 storage at 50 atm (1500 m^3), and 1.6×10^6 kg H_2SO_4 (1000 m^3). In one system where heat is supplied by solar energy, a central receiver system is envisioned with a peak thermal power of 74.1 MW adequate to supply the heat requirements of the aforementioned typical unit plant. Some 4233 heliostats, each 7 m × 7 m in size, and a 120° field angle and with a 17% density are required to supply the thermal energy to the tower. The tower is over 170 m high, has an aperture of 169 m^2, an optical efficiency of 59.7%, and cavity thermal efficiency of 93%. The thermal storage scheme is part of the central receiver solar system. Its function is to smooth the operation and output of the central receiver particularly during a period of highly variable insolation (i.e., passage of clouds, etc.)

8. Chemical storage

Energy can be held in the chemical bonds of many compounds and released by exothermic reactions. The most notable such reaction is none other than combustion. Sometimes it is necessary to apply heat or other catalysts to promote the desired reaction. Here we concentrate on two of the most important compounds that have been suggested as practical energy stores, methanol and hydrogen. Biological compounds which also store energy are treated in Chapter 12 of this volume. Energy from these compounds is released by means of their combustion in air. Energy is stored in them, on the other hand, through their production from abundant inorganic compounds such as water and carbon dioxide with the aid of solar energy.

8a. Methanol

Methanol is a liquid which can be used to replace gasoline in the operation of internal combustion engines or as a feedstock of a variety of chemicals, once more replacing oil. It is made from the catalytic reaction of H_2 and CO at 330 °C and 150 atm pressure. There are several ways to obtain H_2 and CO, but we will examine here a source of particular interest for it uses renewable energy and is environmentally desirable. In essence, the source of H_2 and CO can be, respectively, water and atmospheric CO_2, and the required energy input is supplied by some form of solar energy.

Traditionally, carbon dioxide can be removed from the atmosphere through its reaction with calcium hydroxide which yields calcium carbonate and water.[12] The latter can then be decomposed by heat into CO_2 and calcium oxide. Calcium oxide reacting with water will generate calcium hydroxide to complete the cycle. More recently, scientists at the Solar Energy Research Institute have developed specialized molecules that can remove CO_2 from the atmosphere in a cyclic process driven by electricity.[13] Hydrogen must be supplied from water reduced by electrochemical or thermochemical means. Heat and/or electricity may be supplied by solar and/or hydro and wind energy as appropriate.

An alternative approach to generating CO and H_2 would be the gasification of biological mass. In such a case plants act as the reducing mechanisms of CO_2 and H_2O, and solar radiation is the input energy. Unfortunately, photosynthesis has a very small yield, typically less than 1% of the incident energy, but even more importantly, plants store their energy in the form of carbohydrates, which by their chemical nature can be more efficiently converted to ethanol. Finally, methanol can be produced either from coal or natural gas but in both cases CO_2 is released to the atmosphere.

By virtue of its mode of generation, methanol will be more expensive than hydrogen. On the other hand, methanol, being a liquid at ambient temperature and pressure, lends itself to very inexpensive storage, unlike hydrogen. Nonetheless, it is estimated that the cost of methanol from atmospheric CO_2 and water will be about the same as the cost of liquid hydrogen, more expensive than hydrogen gas, but still cheaper than gasoline from coal.[12] The low vapor pressure of methanol requires that a 15% gasoline–85% methanol mixture be used to ensure cars would start in cold weather. Unfortunately such a plan provides only half as much ozone reduction as the pure methanol fuel. The main advantage of this approach will be stabilization and possibly reduction of the CO_2 in the atmosphere, thereby leading to lessening of the threat of the "greenhouse effect." Emissions from methanol powered engines contain fewer hydrocarbons and less CO with about the same amount of NO_x as those from gasoline powered engines. However, methanol leads to excess production of aldehydes. Other potential problems of a methanol fuel in engines is corrosion of synthetic gaskets and the mixing of water in it.[12] The energy density of methanol is 23 MJ/kg, on a mass basis, and 18 MJ/l on a volume basis (i.e., about 50% of that for petroleum derived liquid fuels on both mass and volume basis). Thus, twice as large a tank or twice as frequent refueling would be required for cars running on methanol compared to gasoline powered cars with the same driving range.

8b. Hydrogen

The idea of utilizing hydrogen as a universal fuel is not a new one. Since the discovery of electrolysis of water at the turn of the 18th century, hydrogen's peculiar properties and energy characteristics have been discussed with increasing frequency. In what is probably the most famous example, Jules Verne describes in *The Mysterious Island*, written in 1874, how hydrogen would one day become the world's chief fuel. During the 1920s and 1930s a flowering of interest in hydrogen as fuel took place, particularly in Germany and England. Interest in hydrogen as a major fuel was once more renewed in the late 1960s and early 1970s. Because of public awareness of air pollution due to hydrocarbon fuels and coal as well as the temporary disruptions of oil supply from the Middle East and attendant price increase, the hydrogen movement has maintained its momentum during the last 20 years. The phrase "hydrogen economy" with all its economic and environmental implications was coined in 1970 and has since been the subject of numerous investigations and proposals.[12]

Since hydrogen is not available on Earth in its free form, it has to be generated by splitting a naturally occurring compound such as water. Hence, hydrogen becomes the storage medium of the energy spent to split the water. Energy sources that may be used to produce hydrogen include coal, natural gas, nuclear fission, and solar radiation. Utilization of either coal or natural gas to produce hydrogen ultimately releases carbon dioxide into the atmosphere. Consequently, environmental problems are not improved and get worse when coal is used. Air pollution problems are moved from one location to another. The cost of hydrogen per unit energy content exceeds that of coal or natural gas thereby eliminating any economic advantage over the primary fuels.

Nuclear power may be utilized to generate hydrogen either by electrolysis or by proper thermochemical reactions.[12] If thermochemical reactions are used, high temperatures are needed. Maximum temperatures envisioned for future nuclear reactors are not expected to exceed 850 °C limiting practical nuclear reactor driven thermochemical cycles. Table 7 includes some such reactions with their maximum temperatures and corresponding thermal efficiency. These reactions are but a handful of the several that have been investigated since the early 1970s. Utilization of nuclear power to generate hydrogen does not release any harmful materials to the environment, unless there is a major nuclear accident. However, the disposal of nuclear wastes may be environmentally objectionable. The high energy particles (neutrons, protons, and alphas) emitted by radioactive substances may be used to break up water. The process, known as radiolysis, has an efficiency of about 1% due to recombination of the generated H^+ and O^- ions.

Table 7. Thermochemical cycles suggested for producing hydrogen.

Cycle title	Type	Thermochemical reactions	Max. temp. (K)	Thermal efficiency (%)
Metal oxide-sulfate	Bicyclic	$SO_2 + H_2O + (M = Metal) \xrightarrow{500K} MSO_4 + H_2$ $MSO_4 \xrightarrow{1100K} MO + SO_2 + 1/2O_2$	1100	85
Carbon-steam iron	Tricyclic	$C + H_2O \rightarrow CO + H_2$ $CO + 2Fe_3O_4 \rightarrow C + 3Fe_2O_3$ $3Fe_2O_3 \rightarrow 2Fe_3O_4 + 1/2O_2$	1673	75.5
Calcium oxide iodine	Tetracyclic	$6CaO + 6I_2 \xrightarrow{393K} Ca(IO_3) + 5CaI_2$ $5CaI_2 + 5H_2O \xrightarrow{800-1100K} CaO + I_2 + 5/2O_2$ $5CaI_2 + 10H_2O \xrightarrow{5500-800K} 5C(OH_2) + 10HI$ $10HI \xrightarrow{550-1100K} 5H_2 + 5I_2$	1100	30
Water shift-manganese-iron	Tetracyclic	$H_2O + CO \xrightarrow{600-700K} CO_2 + H_2$ $2MnO + CO_2 \xrightarrow{600-800K} 2Mn_2O_3 + 2CO$ $Mn_2O_3 + 2FeO \xrightarrow{900K} 2MnO + Fe_2O_3$ $Fe_2O_3 \xrightarrow{1100-1300K} 2FeO + 1/2O_2$	1300	?

Solar power can be employed to produce hydrogen either by electrolysis or appropriate thermochemical reactions. Solar power has the distinct advantage, over all the other energy sources, of being renewable. It is also by far the most benign to the environment for producing hydrogen. Solar power is therefore highly desirable as a source of hydrogen, but it also happens to be very capital intensive upfront and therefore results in a high-cost hydrogen. The cost disadvantage of solar generated hydrogen may, however, be greatly diminished if the environmental costs of fossil and nuclear fuel sources are taken into account.

A comparison of thermochemical hydrogen generation processes vs the electrochemical one (i.e., electrolysis) reveals that while the former are complex and can be relatively efficient, the latter is very simple but rather inefficient. For example, present efficiencies of electrolysers are on the order to 60% to 80% and can theoretically approach 100%. On the other hand, generation of electricity from any source, be it fossil or nuclear or solar, cannot presently exceed 40%. Hence, hydrogen generated by electrolysis may represent anywhere from 10% to 35% of the primary energy. Thermochemical cycles, on the other hand, appear to have much higher conversion efficiencies.

A major characteristic of hydrogen fuel is its high mass energy density of about 140 MJ/kg, about three times that of hydrocarbons—the volume density is however much smaller, about $1/3$ that of methane, for example. Depending on a variety of parameters, it is estimated that hydrogen pipeline transmission becomes more cost effective than overhead powerline electricity transmission for distances exceeding 1000 km at best or 2250 km at worst. This consideration may become important in the event of power generation, particularly nuclear or solar, far removed from population centers either because of safety or unavailability of resources.

Large volumes of hydrogen can be stored in underground caverns like natural gas. The comparatively higher diffusivity of hydrogen does not appear to present any leakage problems as experience has shown in storing "town-gas" in caverns near Paris, France and helium in a depleted natural gas field near Amarillo, Texas. However, the cost of pumping per unit energy stored is going to increase since the energy per unit volume in hydrogen is less than $1/3$ that of natural gas. The embrittlement of pipes due to the hydrogen diffusion in the pipe material can be eliminated by using steels high in nickel, chromium, and molybdenum and low in carbon and manganese. However, this is an expensive solution. Moreover, it is desirable to use the existing pipeline network of natural gas transmission and distribution. The problem of hydrogen embrittlement can be solved by mixing the hydrogen with a gaseous "impurity" such as a 200 parts per million concentration of oxygen. Other gases can be used, such as carbon dioxide, water vapor, and sulfur dioxide, but the limits of their effectiveness are not known.

Replacement of natural gas and other fossil fuels in residential and commercial applications by gaseous hydrogen does not appear to present any major technical problems. For example, the H_2-driven home concept has been tested for years by Roger Billings in Provo, Utah.[12] Hydrogen powers household appliances including the oven, range, barbecue, fireplace log, and heater. Special burners are used to accommodate the characteristics of hydrogen which differ from those of natural gas and also to reduce the temperature of the flame in order to minimize NO_x production. Before hydrogen can be used in the transportation sector, developers must select the desirable form of the fuel, i.e., whether gaseous, liquid, or absorbed in a solid matrix.[3,12]

Since the energy density of hydrogen per volume is less than 0.03% of that of gasoline at STP, a container for gaseous hydrogen equivalent to a gasoline car tank must be pressurized to several hundred atmospheres. This in itself presents severe problems in terms of hydrogen container weight which will be on the order of 300 to 600 kg. Moreover, the high pressure of hydrogen injected into the cylinder of an engine appears to cause immediate ignition and localized hot spots tending to produce cylinder distortion. Although the latter problems have been overcome in recent years, the consensus is that gaseous hydrogen is not practical as a transportation fuel. Utilization of liquid hydrogen presents no difficulties aside from the "dewar"-like fuel tank required. The specific volume of liquid hydrogen is about 14 l/kg. An 80 l (17.6 gal) tank of a standard size gasoline car can contain about 6 kg of liquid H_2 with an energy content of 800 MJ. The same tank full of gasoline would have an energy content of 2700 MJ. This would imply that it would take a 270 l LH_2 tank to give the same driving range to the vehicle as an 80 l gasoline tank. The efficiency of a hydrogen-powered car is on the average 40% higher than that of an identical gasoline-powered vehicle. The size of a LH_2 tank giving equivalent driving range to an 80 l gasoline tank would be about 190 l, or approximately twice in volume. Although this increment in fuel tank volume reduces the trunk space available in the car, it does not eliminate the possibility of liquid hydrogen as an automotive fuel. On the other hand, there is a fuel weight reduction from about 58 kg (80 l) for gasoline to about 14 kg for liquid hydrogen (190 l).

Another way to use hydrogen as an automotive fuel is to bind it within certain alloys as hydrides. Two of the most promising hydrides are those of magnesium (MgH_2) and of iron-titanium alloy ($FeTiH_{1.7}$). Magnesium hydride contains 7.6% hydrogen per weight and decomposes at 1 atm and 287 °C. Iron-titanium hydride releases 1.5% hydrogen at 5 atm and 50 °C. Due to its lower recomposition temperature iron-titanium hydride is the most desirable of the two candidates. It carries 1.5 times as much energy per volume as liquid hydrogen but it also weighs some 25 times as much. To release the hydrogen, heat from the engine radiator is cycled in a heat exchanger through the hydride tank.

Initially, designers preferred using a metal hydride, in particular the iron-titanium one, for hydrogen powered cars because of the simplicity of operation. However, significant developments in small-scale cryogenic storage and the severe weight penalties of hydrides have apparently swung expert opinion in favor of liquid hydrogen in the last ten years. The development of the hydrogen automotive fuel concept has been advanced by at least two major manufacturers to the point that production vehicles could roll off the assembly lines if hydrogen fuel would become available. Thus in 1989, two different prototype cars were unveiled, a Mercedes-Benz using a titanium hydride storage with a 70 mile range and a BMW using liquid storage with a 135 mile range.

Utilization of hydrogen fuel in transportation would have a tremendous impact on the environment as carbon dioxide, carbon monoxide, and hydrocarbon emissions are totally eliminated, while nitrogen oxide emissions can be reduced by at least an order of magnitude because of the greater ignition range of hydrogen vs gasoline and therefore the use of a leaner fuel mixture. In addition the generation of ozone also disappears along with hydrocarbon and nitrogen oxide emissions.

9. Fuel cells[14,15]

A system, different from an electrochemical battery, which converts chemical energy into electricity, has been traditionally called a fuel cell. Although the underlying electrochemical principles are the same for both batteries and fuel cells, the distinction is perhaps due to the fact that in fuel cells the fuel, whose chemical energy is converted into electricity, is external to the conversion device itself. In a fuel cell the reaction taking place can be described as follows:

$$fuel + oxygen = chemical\ compound + electricity .$$

A fuel cell, or more appropriately a fuel cell battery, consists of an assemblage of individual cells connected in series, in parallel, or in series-parallel. Each individual cell consists of an anode, a cathode, and an electrolyte. Hence a fuel cell in the customary terminology implies a system of individual cells interconnected in some fashion. The fuel is consumed at the anode and oxygen at the cathode. A chemical compound from the oxidation of the fuel is released in the electrolyte, and a flow of electrons (current) takes place between the anode and cathode if they are connected externally by a conductor. The reactions between conventional fuels and oxygen are theoretically capable of producing about 1 V per individual cell.

The theoretical voltage and current densities, based on the geometrical rather than the actual surface area of the anode, are more nearly reached by more electrochemically reactive and chemically simpler fuels. In order of decreasing reactivity, the conventional fuels include: hydrogen (in a class by itself); the so-called compromise fuels (ammonia, hydrazine, methanol); and hydrocarbons. Hydrogen is by far the most desirable fuel because of its high reactivity and simplicity, but can be difficult to handle and store. On the other hand, hydrocarbons are readily available, can be transported and stored easily, but unfortunately are low in reactivity, and are complex. As the term indicates, the compromise fuels are of reasonable reactivity, complexity, and ease of handling and storage. Of the three aforementioned compromise fuels, hydrazine is the closest to hydrogen in reactivity, but is also the most expensive. Ammonia is the next best choice in terms of reactivity and is very inexpensive.

Because of the low reactivity of the hydrocarbons and the complexity of the generated products, it will be very difficult to use them directly. However, we can obtain fuels from them by applying well-known chemical processes. Perhaps the most important such reaction is the steam reformation of carbonaceous fuels to generate hydrogen. The reformation plant may in this case be outside or inside of the fuel cell. Methane is, of course, the prime candidate among the hydrocarbons for reformation, although liquid hydrocarbons and even coal can be used. Methanol is a compormise fuel of lower reactivity that can readily yield hydrogen through steam reformation (in order to convert CO into CO_2). In recent years solid fuels, consisting of reactive metals, have also come into prominence. Of these fuels, aluminum has become the most attractive candidate.

Hence, the prime fuel candidates for fuel cells are pure hydrogen, ammonia, reformed methane and possibly reformed methanol, and aluminum. Some 30 years ago, the emphasis in terms of fuel cell development was placed on pure hydrogen. In the last decade or so, natural gas has become the fuel of choice. Ammonia fuel cells have received intermittent attention all along, but this attitude may be changing at the present time because of the attractiveness of this fuel in automotive applications. Aluminum fuel cells are currently receiving significant attention for automotive applications as well.

We can presently distinguish five different types of fuel cells as follows: alkaline fuels cell (AFC), phosphoric acid fuel cell (PFAC), solid polymer fuel cell (SPFC), molten carbonate fuel cell (MCFC), and solid oxide fuel cell

(SOFC). Within each of these five types there may exist certain variants. The AFC makes use of potassium hydroxide as the electrolyte and operates on pure hydrogen either at atmospheric or elevated pressures. The operating temperature is limited to 80 °C if inexpensive materials such as nickel and silver are used rather than platinum. Practical efficiencies of low-temperature AFCs are on the order of 60% to 70%. A variant of the AFC is the aluminum fuel cell with very high purity (99.999%) aluminum typically constituting the anode of the system, the possible addition of sodium hydroxide to the electrolyte, and an operating temperture in the range of 60% to 80 °C. The anode, being the fuel, is consumed and has to be replaced periodically. The refueling process is simple, safe, and can take place even when the system is in operation. The waste product, aluminum trihydroxide, is collected and sent to a central processing plant, where the aluminum metal is recovered and reused.

The PAFC makes use of phosphoric acid as electrolyte and uses natural gas as fuel. The natural gas is steam reformed outside the fuel cell to avoid carbon monoxide poisoning of the platinum electrodes. PFACs can operate at atmospheric or higher pressures but are temperature limted to about 200 °C. Practical efficiencies of PAFCs are on the order of 45%.

The MCFC makes use of a lithium/potassium carbonate electrolyte which is molten at the operating temperature of 650 °C. This fuel cell can operate on natural gas, methanol, and probably on clean light hydrocarbon fuels all of which may be internally reformed. The electrodes consist primarily of nickel doped with a variety of other metals. Conversion efficiencies of the MCFCs are expected to approach 60%.

The SOFC makes use of an yttria doped-zirconia solid electrolyte that must operate at 1000 °C to attain an appropriate rate of oxygen ion passage through it. The electrodes and interconnections consist of compounds of nickel, zirconium, chromium, manganese, and lanthanum. Expected conversion efficiencies will be in the range 50% to 55%, while the fuel will be natural gas internally reformed.

Finally, the SPFC makes use of a polymer solid-state electrolyte, operates at low temperatures, and employs platinum in the electrodes. Recent development of a solid-state electode which is tolerant to carbon dioxide makes possible the use of externally reformed natural gas or methanol as fuels, in additon to pure hydrogen.

Fuel cells can be used both in power generation and vehicle propulsion. The major advantage of fuel cells for power application will be the significant reduction in emission of pollutants, notably nitrogen oxides, while for automotive applications there will be both significant reducation in pollutants as well as carbon dioxide. Of the five types of fuels cells, PAFCs, MCFCs, and SOFCs are developed for power applications, while AFCs, SPFCs, and PAFCs are better suited for automotive applications.

In terms of stage of development, the types of fuel cells vary. SOFCs require sophisticated ceramic fabrication techniques well beyond the state-of-the-art for production so that commercialization is more than 10 years away. SPFCs need to achieve substantial cost reduction particularly with regard to the presently available membranes. MCFCs still require materials development, in particular cathode integrity improvement so that they are at least 10 years away from commercialization. PAFCs require manufacturing cost reduction, improved engineering, and improved performance. Low-temperature AFCs could be used as soon as competitively priced hydrogen became available. It should be noted that the development of MCFCs, SOFCs, and even PAFCs is driven by the desire to replace present electric power generation methods, such as gas and steam turbines, with a less environmentally damaging techinique that still uses hydrocarbon fuels. By contrast, a hydrogen economy can readily make use of AFCs, which also have the advantage of at least four times higher power densities (for the quoted conversion efficiencies), relatively higher conversion efficiencies at high power densities, and capability of instantaneous start-up because of a lower operating temperature. Of course, in automotive applications AFCs are the fuel cells of choice because of their high power densities (about 1.2 kW/l and 0.7 kW/kg for hydrogen) as well as the much lower operating temperature. The question remains, however, whether to use liquified hydrogen, hydrides, or ammonia as the on-board source of hydrogen.

Several demonstration projects have been or are being developed. Probably the most publicized one has been the 4.5 MW United Technologies PAFC, which operated at the Tokyo Electric Power Company in Tokyo, Japan from April 1983 to December 1985. The success of this project has lead to the commissioning of an 11 MW International Fuel Cells (United Technologies–Toshiba joint venture) PAFC plant in Tokyo to start up in January 1991. In Japan, Sanyo and Fuji Electric are testing several PAFC plants in the 50 to 200 kW range to acquire operational expertise. Plans also exist for testing several PAFCs in the 25 kW to 1 MW range in Italy, using American and Japanese technology. Research on MCFCs and SOFCs continues in the U.S., Japan, and Western Europe, while SPFC development is pursued both in the U.S. and Europe.

In the area of transportation, the U.S. DOE and the U.S. DOT are initiating cooperative programs with industry to explore the use of PAFCs along with batteries in a hybrid mode for urban buses. On the other hand, a consortium has been formed to oversee the introduction of a fuel cell city bus in Amsterdam, Holland using AFCs with a liquid hydrogen fuel storage. A Canadian company (Alupower Ltd.) has modified the standard power unit of an electric car using lead-acid batteries by adding aluminum-air fuel cells so as to make the range of an electric car equal to that of a similar gasoline car. For example, with a typical practical energy density of 3.9 kWh/kg of aluminum, it would take 7 kg of aluminum to increase the range of the GM Impact by 240 miles for a total range of 360 miles (120 miles on batteries and the remainder on aluminum fuel cell). Because of the present high cost of aluminum fuel, however, one may use only the batteries for daily trips under 120 mi and employ the aluminum fuels for daily trips over that range, if there is not sufficient time to recharge the batteries.

In general, engineering and materials development are central to the success of fuel cells. Presently, it is necessary to address costs and reliability matters, while identifying early market niches. Their inherent efficiency leading to reduced carbon dioxide and their low pollutant emissions would be very appealing to the general public. Consequently, politics and legislation could become as important as technological and economic factors to the ultimate success of fuel cells.

10. Conclusions

Energy storage can clearly improve the efficient utilization of available resources and at the same time facilitate the introduction of renewable resources. In particular, the construction of new power plants to supply electricity becomes unnecessary as the existing ones can meet the demand for a long time. In addition, the efficiency of electricity generation can be increased by retiring the older units. Moreover, utilization of the more efficient units not only reduces the energy input for the same output, but it also improves the quality of the environment by reducing the generation of pollution. Lastly, penetration of renewable resources in the energy system results in the reduction of conventional fossil-fuel usage and pollution reduction.

It appears that storage is needed both at the utility level and at the building level. Different systems are more suitable at these two generic levels. At the utility level, pumped hydroelectric, compressed air, and superconducting coils can be used to store energy in the power range of 100 to 1000 MW or more. Smaller loads of the order of 1 to 100 MW, encompassing an aggregation of consumers, can be accommodated by either compressed air or electrochemical batteries. At the individual building level flywheels and possibly batteries may be used. Hence research and development efforts might be directed in the area of 10 to 500 MW compressed air with waste heat recirculation, electrochemical batteries up to 50 MW, and in the more distant future to 1000 MW superconducting coils.

A less expensive electrochemical system for batteries is badly needed. Options include sodium/sulfur or zinc/bromine which use plentiful materials. Lead is too expensive and scarce to be used economically in large scale in lead-acid battery systems. Moreover, research and development is needed for the design of composite flywheels in the range of 10 kWh storage capacity for residential buildings. Despite the satisfactory performance of lead-acid batteries in the 1 kWh storage capacity range, the sodium/sulfur battery must also be developed as a lower cost alternative. Thermal storage either in the form of sensible or latent heat is well applicable at the residential or commercial building level. It is essential therefore that research and development on new sensible storage systems for heat and latent storage systems for both heating and cooling continues. Thermochemical storage, which is appropriate at the utility or industrial customer level, has to also receive more attention.

In the area of transportation the fuel of choice is hydrogen rather than either methanol or ethanol. Even though at the present time hydrogen will be produced from natural gas rather than solar energy, it can be used to eliminate environmental pollution problems in densely populated areas. In the long run hydrogen generation must be based on solar energy and therefore research and development in that area should be pursued vigorously. The major technical factor affecting the introduction of hydrogen, at least in selected locations, is neither generation nor safety, but the complete modification of the present fuel distribution system which would be required. In contrast, it is generally assumed that both methanol and ethanol fuels can be accommodated by the present distribution system with minor changes. This may be, however, totally eroneous. Until the fuel distribution problem gets solved, there is very little chance for hydrogen to be adopted. Because of its extremely low conversion efficiency, the intermal combustion engine must be replaced either by fuel cells operating on hydrogen or batteries charged with solar electricity.

All these activities then, by making use of some form of energy storage or another, result in a more efficient, benign, and self-sustained utilization of energy and material resources. Consequently, significant research, development, and commercialization efforts should be given to a variety of energy storage concepts and systems as explained herein. In this regard then, the concerted actions of academia, industry, and government are necessary for an optimal outcome. It is becoming more apparent that the public is willing to support changes in the present state of affairs in the interest of supporting the environment. Consequently, clear messages must be sent by the policy makers saying that such changes need to be taken.

Appendix: Properties of electrochemical batteries

1. Advanced lead-acid battery

The two electrodes in the form of solid plates consist of lead (Pb) and lead oxide (PbO_2) and the electrolyte between them is sulfuric acid (H_2SO_4). The electrolyte ionizes as follows[16]:

$$H_2SO_4 \rightarrow H^+ + HSO_4^- \ .$$

During discharge the reaction of the cathode (negative electrode) is

$$Pb + HSO_4^- \rightarrow PbSO_4 + H^+ + 2 \, e^- \ .$$

The liberated electrons travel through the external circuit to the anode (positive electrode) where the following reaction takes place:

$$PbO_2 + HSO_4^- + 3H^+ + 2 \, e^- \rightarrow PbSO_4 + 2H_2O \ .$$

Thus, both the cathode and anode during discharge are replaced by $PbSO_4$. The theoretical potential of the standard cell is 1.92 V, but the actual cell potential can be adjusted to 2 V by proper concentration of the reagents. The preceding reaction represent a transfer of 2 moles of electrons per 1 mole of Pb, 1 mole of PbO_2 and 2 moles of H_2SO_4. The 2 moles electrons represent a charge of 1.93×10^5 °C. The work done is equal to 0.386×106 J. Since the reactants have a total molecular weight of 642 g (1 mole Pb, 1 mole of PbO_2, and 2 moles of H_2SO_4), the energy stored per kg of active reactants is 0.386 × 10^6 J/0.642 kg = 0.60 MJ/kg (= 0.170 kWh/kg). In a practical design the energy density realizable is about 15% of the theoretical and can be increased up to 25% in advanced designs. These realizable fractions of the theoretical energy density are typical of all batteries and are due to the presence of materials not participating in the reaction. For instance, in a lead-acid battery such materials include the case, separators of electrodes, water in which the acid is dissolved, and lead alloy grids to support the active electrode and materials.

Advanced lead-acid battery designs have an increased energy density and longer lifetime. The former is achieved by making use of the so-called starved electrolyte concept. A highly porous separator consisting of fiberglass and poly-ethylene immobilizes the electrolyte, while it allows passage of oxygen from the anode to the cathode where it recombines with hydrogen to form water. Hence, the need to add water is eliminated and so is the need to vent hydrogen. Moreover, the electrode grids are made of a lead-calcium alloy rather than lead-antimony and lead-arsenic alloys. The use of lead-calcium not only decreases the weight of the battery, but even more importantly increases the lifetime of it. Notice that antimony slowly promotes the irreversible reaction[16]

$$5PbO_2 + 2Sb + 6H_2SO_4 \rightarrow (SbO_2)_2 SO_4 + 5PbSO_4 + 6H_2O$$

thereby destroying the active material of the anode. However, antimony is usu-ally added to the grid to increase its mechanical strength against vibrations (motive power applications). Advanced batteries are ideal for photovoltaic and other renewable-resources energy storage and are also well suited for residen-tial load leveling, small to medium commercial applications, and electric ve-hicles as they are maintenance free. For example, the GM "Impact" employs 32 advanced lead-acid batteries having a 34.5 Wh/kg energy density for a total storage of 13.6 kWh. Large load-leveling storage schemes at the large commer-cial and utility level make use of the standard open or closed cell design. Table A1 summarizes the characteristics of different lead-acid battery systems.[17]

The larger cost for a utility and a large commercial system are due to the cost of buildings, converters, other equipment, maintenance, labor, etc., even though the cost of the standard battery itself is about \$80 to \$100/kWh. On the other hand, the entire cost of a residential system is the advanced battery itself at about \$150/kWh.

The largest lead-acid storage system in the world is currently operational in the city of Berlin, which until recently had to meet its electricity requirements with its own power stations and could not fall back on the European grid to cope with peak demand. This system, completed in 1987, was designed and built by AEG, has a capacity of 17 MW, and can feed the city power mains in a fraction of a second, if standby power is needed.

2. Sodium-sulfur battery

This cell consists of molten sodium as the anode, molten sulfur as the cathode and has a solid electrolyte.[18-20] This electrolyte is normally made of a beta-alumina or in some designs of fibrous borate glass and has the unique property of allowing the passage of Na ions only. During discharge the reaction at the anode is

$$2\,Na \rightarrow 2\,Na^+ + 2\,e^- \ .$$

Table A1. Performance and cost characteristics of three lead-acid battery systems for stationary applications.

System type	Utility	Large commercial	Residential
Application	Load leveling	Mainly load leveling	Load leveling and renewable energy storage
Battery type	Standard	Standard	Advanced
Nominal energy capacity	100 MWh	6.2 MWh	40 kWh
Maximum energy capacity	100 MWh	7.8 MWh	50 kWh
Power rating	20 MWh	775 MWh	5 kWh
Discharge time	5 hr	8 hr	8 hr
Charge time	9 hr	9 hr	9 hr
Energy/area	29 kWh/m^2	23 kWh/m^2	19 kWh/m^2
Efficiency	85%	82%	80%
Expected lifetime	2000 cycles	2500 cycles	3500 cycles
Installed system cost ($ 1988)	$280/kWh	$200/kWh	$150/kWh

The liberated electrons travel through the external circuit to the cathode where the following reaction takes place:

$$2 \, Na^+ + x \, S + 2 \, e^- \rightarrow Na_2S_x; \quad x = 3, 4, 5 \ .$$

The exact proportion of sodium and sulfur combining to form polysulfides depends on the depth of discharge as well as the operating temperature of the cell. Both Na and S are molten above about 100 °C, but sodium pentasulfide has a boiling point of 275 °C and lower polysulfides have increasing boiling points. Moreover the conductivity of the electrolyte to sodium ions increases with temperatures and at 300 °C is about the same as that of sulfuric acid in a lead-acid battery. Consequently, a sodium-sulfur battery is designed to operate between 300 °C and 350 °C. Higher temperatures become impractical because of the accelerated corrosion of the cell envelope. A sodium-sulfur cell has an open circuit of 2.0 V and a discharge voltage of about 1.7 V. The molten electrodes of the Na-S cell have the advantage over other cells of not being subjected to any permanent damage during charge and discharge. On the other hand, the solid electrolyte may develop cracks and fractures due to thermal stresses that ultimately lead to the failure of the cell as then sodium atoms can also migrate. Improvement of the electrolyte performance is crucial to extending the lifetime of the battery. Another consideration is the thermal management design of a battery so as to maintain the high operating temperature even during extended idle times. A typical Na-S cell is cylindrical, has a 4 cm diameter, a 25 cm length, volume of 300 cm^2, and a 600 g weight. Cell containers can be chromium plated stainless steel (E-Brite) or the less expensive SS410 and low carbon steel.

Table A2. Performance and cost characteristics of two sodium/sulfur systems for stationary applications.

System type	Utility	Residential
Application	Load leveling	Load leveling and renewable energy storage
Theoretical system capacity	131 MWh	62 kWh
Rated system capacity	100 MWh	50 kWh
Power rating	20 MWh	5 kWh
Discharge time	5 hr	10 hr
Charge time	7 hr	7–10 hr
Energy/area	130 kWh/m^2	90 kWh/m^2
Efficiency	75%	85%
Expected lifetime	2500 cycles	5000 cycles
Installed system cost (\$ 1988)	\$200/kWh	\$100/kWh

The development of Na-S batteries is intensively pursued by numerous companies in the U.S., Japan, Germany, and United Kingdom for automotive propulsion, utility load leveling, and renewable energy storage. Significant performance improvements have been attained since 1980, and by the early to mid 1990s, Na-S may become commercially available on a large scale. The characteristics of two systems, a large utility and a residential one are given in Table A2.[17,21] Notice that Na-S batteries are expected to have a much longer lifetime than lead-acid batteries and also be less expensive due to the low cost of sodium and sulfur. In general, it appears that a residential Na-S battery system for both load leveling and renewable energy resource storage is more economical than a Na-S utility system and either a residential or utility lead-acid battery system.

3. Lithium alloy/iron sulfide battery

The typical cell is similar in configuration to a lead-acid battery.[22,23] The cathode consists of a lithium alloy of aluminum or silicon, and the anode consists of iron sulfide (FeS or FeS$_2$). The electrolyte is a mixture of molten LiCl and KCl. The cell operating temperature of the cells is 400 °C to 500 °C. The electrochemical reactions taking place for the two types of anodes are as follows. When the FeS constitutes the anode we have

$$Li\text{-}Al \rightarrow Li^+ + e^- + Al \quad \text{(cathode)},$$

$$2Li^+ + FeS + 2e^- \rightarrow Li_2S + Fe \quad \text{(anode)}.$$

The theoretical specific energy of this reaction is 458 kWh/kg and the open circuit voltage is about 1.3 V. When the anode consists of FeS$_2$ we have instead:

$$Li\text{ -}Al \rightarrow Li^+ + e^- + Al \quad \text{(cathode)},$$

$$4Li^+ + FeS_2 + 4e^- \rightarrow 2Li_2S + Fe \text{ (anode)}.$$

The theoretical specific energy of this reaction is 650 Wh/kg and the open circuit voltage has two values at 1.6 and 1.3 V. Although cells having FeS_2 electrodes have a higher theoretical specific energy than those with FeS electrodes, the high sulfur activity of the former results in high corrosion rates and electrode instability. Thus, the development of cells with FeS has been pursued in the last decade primarily for automotive propulsion systems. Significant energy densities have been attained. Unfortunately, the life cycle of various designs remains extremely limited. Also, the cost of a viable battery remains uncertain because of the limited availability of relatively low cost lithium and the cost of other components of the battery such as the electrode separator, which is made of boron nitride. Consequently, interest in this type of battery has diminished in recent years although research still continues and other anode materials such as titanium sulfide are also investigated.

4. Zinc-bromine battery

The system consists of three main components: The electrochemical module, where the actual electrochemical reactions take place; the circulating electrolyte comprising an aqueous solution of zinc bromide and a bromine complexing agent; and the pumps and reservoirs used to store and circulate the electrolyte.[24-26] During charging, zinc is plated at the negative electrode (cathode) and bromine evolves at the anode according to the reactions

$$ZnBr_2 \rightarrow Zn^+ + 2 Br^- \quad \text{(electrolyte)},$$

$$Zn^+ + e \rightarrow Zn \quad \text{(cathode)},$$

$$2 Br^- \rightarrow Br_2 + 2e^- \text{ (anode)}.$$

The bromine is circulated out of the electrochemical module by a pump and is separated by gravity in a proper reservoir. During discharge, the reservoir valve is opened and Br_2 is fed back to the module. There, zinc and bromine react back to the ZnBr solution, liberating the energy absorbed during charging.

Typical Zn/Br_2 batteries are of the bipolar stack design type with a common circulating electrolyte and make use of plastic conductor electrodes. These features are introduced to reduce the battery cost, although they also pose life cycling problems which need to be solved. The Zn/Br_2 batteries are primarily geared toward renewable energy storage and utility load leveling, even though some consideration has been given to electric vehicle applications. Development of Zn/Br_2 battery systems is currently on-going in the U.S. and Japan. Most recently, a 20 kWh system was designed and built by Exxon Research Corporation and is currently undergoing evaluation at Sandia National Labs in connection with photovoltaic energy storage.

References and notes

1. F.R. Kalhammer, "Energy Storage Systems," Sci. Am. **241**, 56–65 (1979).
2. F.J. Dyson, "Energy in the Universe," Sci. Am. **225**, 50–59 (1971).
3. D.W. Dennis, *Energy: Its Physical Impact on the Environment* (John Wiley and Sons, New York, 1982).
4. R.F. Post and S.F. Post, "Flywheels," Sci. Am. **229**, 17–23 (1973).
5. R.J. Roark and W.C. Young, *Formulas for Stress and Strain*, 5th ed. (McGraw-Hill, New York, 1975).
6. E.C. Kern, Jr. and M.D. Pope, *Development and Evaluation of Solar Photovoltaic Systems*, Report DOE/ET/20279-240 prepared by Lincoln Lab MIT (U.S. DOE, Washington, DC, 1983), pp. 34–38.
7. R.C. Flanagan, "Flywheel Energy Storage Design for Photovoltaic Powered Systems," *Proc., Eighth E.C. Photovoltaic Solar Conf.*, Vol. 1 (Kluwer Academic Publishers, Boston, MA, 1988), pp. 34–38.
8. R.J. Loyd, "Bechtel's Program in Superconductivity Magnetic Energy Storage," Super Currents **2**, 1–6 (1988).
9. *Survey of Thermal Energy Storage Installation in the United States and Canada* (Am. Soc. Heat., Ref. and Air. Cond. Eng, 1984).
10. P.O. Carden, "Energy Correlation Using the Reversible Ammonia Reaction," Solar Energy **19**, 365–378 (1977).
11. E. Bilgen, "Energy Storage and Transportation Based on Sulfuric Acid Decomposition and Synthesis Processes," J. Sol. Eng. En. **109**, 210–214 (1987).
12. J.O.M. Bockris, *Energy Option: Real Economics and the Solar-Hydrogen Option* (John Wiley and Sons, New York, 1980).
13. D. DuBois, SERI Sci. and Tech. Rev. 8 (Sept.–Oct. 1988).
14. A.J. Appleby *et al.*, *Fuel Cell Handbook* (Van Nostrand Reinhold, New York, 1989).
15. *Proceedings, Grove Fuel Cell Conference, 18-21 Sept., 1989, London, UK* (J. Power Sources, London, 1990).
16. J.W. Twidell and D. Weir, *Renewable Energy Resources* (E. & F.N. Spon, London, UK, 1986).
17. D.R. Brown and J.A. Russel, *A Review of Storage Battery System Cost Estimates*, Report PWL-5741 (Battelle Pacific Northwest Laboratory, Richland, WA, 1986).
18. S.A. Weiner, *An Overview of the Sodium-Sulfur Battery*, SAE 750149 (1975).
19. W. Fischer, H. Birnbreirer, and G.N. Benninger, "Performance Characteristics of Sodium/Sulfur Batteries," *21st Inter. En. Conver. Eng. Conf.*, Paper #869229 (Am. Chem. Soc., San Diego, CA, 1986).
20. *Sodium-Sulfur Battery Development*, Publication No. V-6831 (Ford Aerospace, Newport Beach, CA, 1985).
21. H.J. Hashing and C.R. Halbach, "Sodium-Sulfur Load/Leveling Battery System," *15th Inter. Eng. Conver. Eng. Conf.*, Paper No. 809105 (Am. Inst. Aero. & Astron., Seattle, WA, 1980).
22. D.C. Barney, R.K. Steunenberg, and A.A. Chilenskas, "Recent Progress in Lithium/Iron Sulfide Battery Development," *15th Inter. En. Conver. Eng. Conf.*, Paper No. 809041 (Am. Inst. Aero. & Astron., Seattle, WA, 1980).
23. T.D. Kaun and A. Stable, "High Performance Lithium/Iron Disulfide Cell," *21st Inter. En. Conver. Eng. Conf.*, Paper No. 869230 (Am. Chem. Soc., San Diego, CA, 1986).
24. R.J. Bellows, H. Einstein, P. Grimes, E. Kantner, K. Newby, and J.A. Shropshire, "Development of a Bipolar Zn/Br_2 Battery, *Proc. 15th Inter. En. Conver. Eng. Conf.*, Paper No. 809288 (Am. Inst. Aero & Astron., Seattle, WA, 1980).
25. A. Leo, G. Albert, and M. Bilhorn, "Development of Zinc-Bromine Batteries for Stationary Energy Storage Applications," *21st Inter. En. Conver. Eng. Conf.*, Paper No. 869219 (Am. Chem. Soc., San Diego, CA, 1986).
26. T. Fujii, K. Fushimi, T. Hashimoto, Y. Kumai, and A. Hinto, "80 kWh Zinc-Bromide Battery for Electric Power Storage," *21st Inter. En. Conver. Eng. Conf.*, No. 869220 (Am. Chem. Soc., San Diego, CA, 1986).

Energy and transportation:
A technology update

Alan Chachich

1. Introduction

A balanced energy analysis must consider end-use as well as generation. Transportation is a critical end-user. It accounts for 28 percent of the energy consumed in the United States and this fraction is increasing (Figure 1). The nature of this energy is significant: 97% comes from petroleum. Thus transportation consumes almost two-thirds of our yearly requirement. In 1986, the petroleum needed for about half of our transportation energy consumption created one quarter of our trade deficit.[1]

We can reduce consumption by driving more efficiently and by driving less. We can improve efficiency by both advancing vehicular technology and optimizing our transportation system. A holistic restructuring would better integrate the various modes and improve how we select and use them. Alternative technologies (e.g., computers and telecommunications) can make trips more effective, shorter, or unnecessary. Any energy use extracts a financial and environmental cost so conservation is the most beneficial solution where we can apply it.

This chapter reviews the technologies that are increasing vehicle efficiencies and then remarks on the salient policy questions. Automobiles consume about half of our transportation energy as is shown in Figure 2. Trucks are the second largest consumer (with a third of the consumption) followed by aircraft, pipelines, rail, buses, ships, and mass transit with much smaller portions. Clearly technological advances important to cars and trucks will have by far the greatest impact on our total energy picture. Other modes and technologies are barely mentioned or have been eliminated altogether from this discussion due to their smaller impact on total energy consumption. This approach slights the most efficient forms of freight transportation (Figure 3).

2. Propulsion

2a. Internal combustion engines

Engines are distinguished by their method of burning fuel, the type of ignition employed, and the fuel consumed. Diesel engines compress air so that its high temperature ignites the fuel when it is injected. The piston moves during the

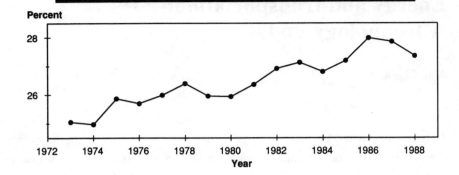

Based on data from the MONTHLY ENERGY REVIEW,
March 1989, Energy Information Administration,
U.S. DOE, Washington DC, DOE/EIA-0035.

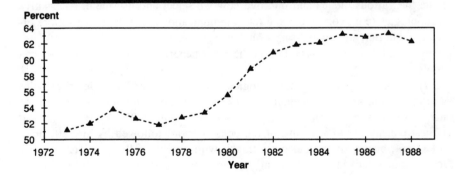

Based on data from the MONTHLY ENERGY REVIEW,
March 1989, Energy Information Administration,
U.S. DOE, Washington DC, DOE/EIA-0035.

Figure 1

burn so the fuel burns at a constant pressure (CD in Figure 4). Gasoline engines
(also Otto cycle or spark ignition) compress an air-fuel mixture to a much
lower pressure than the Diesel (BC in Figure 5). A spark ignites the mixture
which burns so quickly that the cylinder movement during this time is negligible. Thus the mixture burns at nearly constant volume (CD in Figure 5).

1986 TRANSPORTATION STATISTICS

MODE	ENERGY CONSUMPTION Quads	%	VEHICLE-MILES Billion	%	PASSENGER-MILES Billion	%	REVENUE TON-MILES Billion	%
TOTAL DOMESTIC TRANSPORTATION	18.61	100.0	1868	100.0	3488	100.0	3278	100.0
HIGHWAY	15.90	85.4	1838	98.4	3121	89.5	952	29.0
Auto	8.92	47.9	1313	70.3	3008	86.2	–	–
Truck	6.83	36.7	520	27.8	–	–	952	29.0
Bus	0.15	0.8	5	0.3	113	3.2	–	–
AIR	1.65	8.9	3	<.2	320	9.2	6	0.2
Commercial	1.50	8.1	3	<.2	308	8.8	6	0.2
General Aviation	0.15	0.8	n/a		12	0.3	–	–
PIPELINE	0.50	2.7	n/a		n/a		578	17.6
RAIL	0.40	2.1	25	1.3	12	0.3	868	26.5
Passenger	0.01	<.1	<.2	<.1	12	0.3	–	–
Freight	0.39	2.1	24	1.3	–	–	868	26.5
WATER	0.14	0.8	n/a		n/a		873	26.6
MASS TRANSIT	<.02	<.1	2	0.1	35	1.0	–	–

n/a = not available

Compiled from the "National Transportation Statistics Annual Report, 1988," DOT-TSC-RSPA-88-2.

Figure 2

The Otto cycle engine cannot use compression ignition. If the air-fuel charge ignites too early it will fight the compression stroke. The Diesel engine, on the other hand, can be designed to operate with both methods of ignition. Diesels generally burn heavier, higher viscosity, less distilled, and less volatile oils than gasoline. Therefore they have a greater capacity to operate with multiple fuels.[2]

i. Otto cycle engines. The Otto cycle engine consumes the bulk of our transportation energy due to its use in automobiles. Trucks, recreational vehicles, motorcycles, and electrical generators use it as well. Efforts to raise its efficiency focus on improving efficiency when the motor runs at less than full power (part load), combustion efficiency, reduced warm-up time, and higher compression ratios. Compressing more air and fuel in the cylinders makes each stroke of the engine produce more power.

Carburetion: Carburetion is the mixing of air and fuel before combustion in the cylinder. The carburetor adds fuel to the air sucked in by the engine vacuum. Traditionally, a vacuum driven mechanical pump supplied fuel to the carburetor. Electrical fuel pumps came into use in the 1960s and more recently some manufacturers have turned to electronic fuel injection.

The fuel consumption of inexpensive *vacuum carburetors* has improved roughly 11% over the last decade. Modern vacuum carburetors rely on feedback from air pressure changes and electronic controls. Solenoid pumps add fuel when the car accelerates and DC motors supply fuel when the car idles.

Energy Requirements for Passenger Transportation

Mode of Transport	Passengers	Vehicle MPG	Passenger MPG	Fuel energy	Energy Use
		miles/gal	*pass.-mpg*	*BTU/gal*	*BTU/p-mpg*
Bicycle	1	50 *(kcal/mi)*	650 *(equiv)*		200
Walking	1	70 *(kcal/mi)*	450 *(equiv)*		300
Auto - prototype	4	100	400	125,000	300
Bus - intercity	45	5	225	138,700	600
Auto - high econ.	4	50	200	125,000	600
Maglev vehicle	140				800
Comm. Train, 10 cars	800		160	138,700	900
Subway Train, 10 car	1,000		150	138,700	900
Auto	4	30	120	125,000	1,000
Bus - local	35	3	105	138,700	1,300
Intercity Train	200		80	138,700	1,700
(4 coaches)					
A300 jet plane	267		80	135,000	1,700
Motorcycle	1	60	60	125,000	2,100
Auto - Luxury	4	12	48	125,000	2,600
747 jet plane	360		36	135,000	3,800
Light plane (2 seat)	2	15	30	120,200	4,000
Executive jet plane	8	2	16	135,000	8,400
Concorde SST	110		13	135,000	10,400
Snowmobile	1	12	12	125,000	10,400
Ocean liner	2,000		10	150,000	15,000

See sources & note: Ref. 55.

Energy Requirements for Freight Transportation

Mode of Transport	Mileage (ton-miles/gal)	Energy Consumption (BTU/ton-mile)
Oil pipelines	275	450
Railroads	185	670
Waterways	182	680
Truck	44	2,800
Airplane	3	42,000

Courtesy of R. H. Romer, Ref. 55.

Figure 3

Electromagnetic injectors with broad metering ranges are the recent trend in *fuel injection*. They provide greater fuel flow for high power output and more accurate fuel control at low speeds for economy. Increased precision and range in air flow measurement have proved as important as injector design.

Fuel transport after the carburetor affects acceleration, deceleration, and engine speed stability at idle. Operating conditions determine how much gasoline is transported as a vapor, a suspended liquid (atomized), and a thin film on the intake surface. Understanding fuel transport ensures that the air-fuel mixture in the cylinder has the composition which will provide the most efficient burning especially in transient conditions. High voltage *static electricity* can reduce droplet size making a leaner air-fuel mixture. A lean mixture has less fuel per volume of air. The droplet reduction is most effective at low air-flow so carburetor electrification has the greatest effect on fuel efficiency at low engine speeds.[3]

Engines are most efficient at full load, but most transportation use is at part

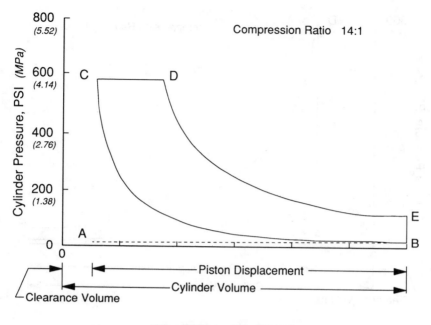

The Diesel Cycle *(Adapted from Ref. 2)*

Figure 4

load. Gasoline engines reduce fuel and air for part load operation by throttling. This makes air enter the cylinders at a lower pressure than it is exhausted, costing efficiency. Diesels are more efficient because they do not throttle at part load.[4]

Combustion: Lean combustion (higher air-fuel ratio) and higher compression generally provide higher thermal efficiency and lower emissions. *Stratified charge combustion* attempts to combine the advantages of Diesels and spark ignition engines by injecting fuel late in the compression stroke. These engines achieve a leaner burn and higher compression ratios without knocking or intake throttling. They can burn many types of fuel including hydrogen, but their combustion is unstable at low load, and they produce hydrocarbon emissions. Advanced sensors and controls may surmount the emission problem but work on most civil applications in the U.S. has ceased.

The *ignition* system starts the burn in the cylinder. It affects cycle-to-cycle combustion stability and the speed and completeness of the burn, particularly in slower burning lean air-fuel mixtures. Stability increases fuel efficiency and complete burning leads to low emissions. The duration, stability and location of the spark control ignition in Otto cycle engines. Multipoint ignition can achieve leaner combustion and reduce hydrocarbon emissions by 35%, but local thermal stresses of the rapid burn can damage the engine.

Ignition design will have to respond to alternative fuels as well as to lean burn strategies. *Plasma Jet Ignition* uses an intense torch that results in very

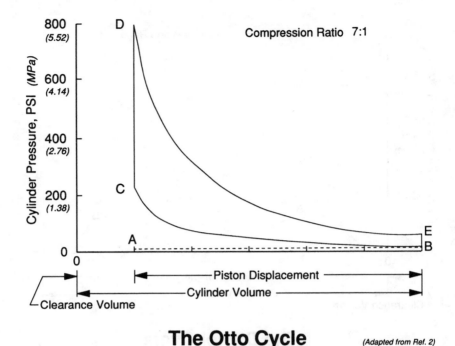

Compression Ratio 7:1

The Otto Cycle

(Adapted from Ref. 2)

Figure 5

clean and complete combustion. *Heated Surface Ignition* does not need a spark generating mechanism. It provides uninterrupted high-temperature ignition that is independent of the compression ratio. At high and medium engine speeds, exhaust gas can heat the ignition surface.

Airflow, turbulence, and combustion chemistry in the cylinder also influence engine efficiency. Major advances in instrumentation and numerical simulation are needed to understand and control them.

Oxygen Separation: A related development by Dow Chemical is a membrane to remove nitrogen from the intake air stream. The oxygen enriched air promises the twin benefits of reduced nitrous oxide emissions and smaller engines.[5]

Engine Design: Designers often gain performance at the expense of fuel economy, but the trend to higher power engines can improve fuel economy by making small lightweight cars desirable.

Power improvements that have benefited fuel economy include:
- lighter pistons, crankshafts, and valve stems,
- low relief piston rings to lower surface tension,
- reduced number of piston rings from 3 to 2,
- less piston ring friction,
- higher surface pressures and speeds on shaft metals,
- roller tappets,

• improved abrasion resistance of sliding materials,
• longer life and lower friction lubricating oils,
• hollow camshafts,
• torsional vibration damper for light crankshafts.

Most of these improvements aid fuel economy by lowering the inertia of moving parts and the weight of the engine.

The trend to higher compression and faster burn for more power increases the significance of heat flux and "knock" caused by premature ignition. Higher pressures increase heat flux out of the cylinder and the likelihood of premature ignition. Advanced knock suppression techniques such as ignition delay, Exhaust Gas Recovery (EGR), or the use of inert gases will determine the success of stratified charge combustion and turbocharging. At present heat transfer is the key factor in piston technology.

Six cylinder engines may be best able to balance demands for smaller size, lighter weight, more power, less noise, and better fuel economy. On the other hand, some believe that it is easier to improve the output of a Wankel rotary engine than a reciprocating engine. Another possibility has automakers racing to incorporate direct fuel injection and charge stratification to two-stroke engines (concurrent intake & compression; power & exhaust strokes). That may reduce emissions enough to put the lighter, more powerful two-stroke engines on the road using up to 30% less fuel.[6]

Other Technologies: Diagnostic intrumentation will prove important to advancing transportation efficiency. Examples include: lasers and high speed photography to study combustion; chromatic laser tracing and resistance probes to study fuel film behavior; laser doppler velocimeter, hot wire flowmeters, smoke wire techniques, and visualization engines to study 3-D gas flow; and an accurate bench test dynamometer to measure fuel efficiency independent of external variables like wind, temperature, and road condition.

Analytical advances are also vital. Examples include: computer hardware and numerical techniques to study combustion and 3-D gas flow; enhanced engine models to study cylinder lubrication, strength, and vibration; new heat transfer models to study lubricant deterioration, thermal stress fractures, thermal expansion, and material heat damage.

System level measurement advances like the bench test dynamometer can guide research by finding how vehicles with different missions need different strategies for optimum fuel efficiency. For example, long-distance trucks benefit most from reductions in rolling resistance and wind resistance, while trucks that stop and start frequently, e.g., buses and delivery trucks, benefit most from weight reduction.[7]

ii. Diesel cycle engines. Diesels power about eighty percent of all ships and trucks, and most railroad locomotives. They are expected to remain the basic powerplant for heavy-duty trucks for at least the next 10 to 20 years. There are three classes of Diesel Engines:

(1) large low speed (< 250 rpm),
(2) medium speed (250–1100 rpm),

(3) high speed (> 1100 rpm).

Large low speed Diesels are primarily found in marine service. They are huge, possessing 5 to 12 cylinders which may have a bore diameter larger than 3 feet, and strokes greater than 6 feet. Such an engine may produce as much as 55,000 hp (41 MW). They burn low-cost residual fuels.

Medium speed engines appear in a wide variety of applications, including marine propulsion, marine auxiliary power, and railroad locomotives as well as stationary power plants, construction machinery, etc. These engines may have stroke and bore dimensions on the order of 10 to 20 inches and produce between 1,000 and 18,000 hp (0.75 and 13.4 MW). They burn a wide variety of fuels including gas-oil mixtures.

The high sulphur content of the fuels used by both medium and low speed diesels poses a substantial acid corrosion threat. To counter this the engines must consume a minimum rate of lubricating oil containing alkaline additives.

High speed Diesel engines are typically used in trucks, automobiles, and on ships to power pumps, generators, winches, lifeboats, bow thrusters, and other machinery. They have stroke and bore dimensions on the order of 5 inches and may produce up to 1500 hp (1.1 MW). They usually burn only the best quality distillate fuels.[8]

Developments in the 1980s raised the thermal efficiencies of diesels to about 50%, near the ideal of 55%. Diesels have been built that operate with one-third the fuel cost of comparable gasoline engines, and prototypes have attained well over 120 miles per gallon. In spite of this, demand for automotive diesels in the U.S. and Japan has declined since 1981 principally due to their poor acceleration response. Other drawbacks include noise, emissions and poor cold starting.[9]

Combustion: Diesel fuel is directly injected into the cylinder (DI) or indirectly into a swirl chamber before the cylinder (IDI). The swirl chamber mixes the air and fuel prior to combustion. The trend has been to DI which may improve fuel efficiency by 15%–20%. A problem unique to diesels is fuel spraying on the cylinder wall which cools the fuel, interfering with its mixing and combustion. Increasing the swirl lessens the problem in an IDI engine. Fuel spraying is more severe for DI engines, especially the automotive diesels, due to their small cylinders. Work to improve the distribution of fuel in the air has focused on various combustion chamber geometries, injection nozzle design, position, and orientation, as well as air heating and glow plugs for faster vaporization, fuel preheating, and static charging to create a finer fuel spray.[10]

Supercharging: Superchargers force air under pressure into gasoline or Diesel engines, raising the compression ratio for increased power and quicker response. Superchargers provide more air for combustion and expansion in the cylinder than in the naturally aspirated engine. Unfortunately, they take mechanical energy from the engine, reducing fuel economy and increasing noise. Turbocharging counters this by using exhaust energy to power the supercharger. Pressure restricts the use of supercharging in diesels while temperature and combustion conditions limit use in gasoline engines.[11]

Pressure wave supercharging is an alternative to turbocharging that performs

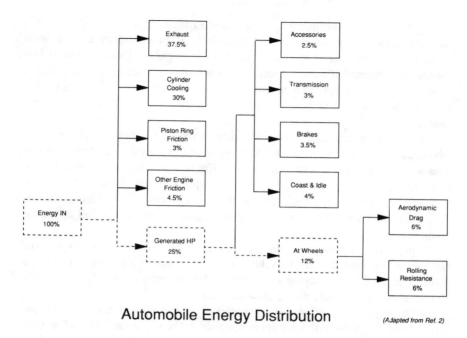

Automobile Energy Distribution

(Adapted from Ref. 2)

Figure 6

better at lower engine speeds. Alternating exhaust gas pressure with intake air pressure generates a pressure wave that compresses the incoming air. It is more commonly used in European trucks but may find more widespread use in the U.S. if the trend to high torque, low rpm engines continues.

Intercoolers further enhance efficiency by cooling the compressed air before it enters the engine. The cooler, denser air carries more oxygen into the cylinder. A recent design that cooled the compressor discharge from 140 °C to 60 °C reduced fuel consumption by 6%. Long-haul truck engines use three types of intercooler: air-to-water, dual radiator air-to-water, and air-to-air. Air-to-water intercoolers which cool air from about 170 °C to 85 °C are the most common. The more complex dual radiators should perform better and will be in production in the next 5 to 10 years. Air-to-air devices achieving temperatures of 43 °C can provide twice the fuel savings of the water-based intercoolers but are more expensive and difficult to design.[12]

iii. Engine thermodynamics. In the conventional diesel engine, 40% of the fuel energy becomes mechanical work, 25% is rejected as cooling, and 35% is wasted in the exhaust stream.[13] Chemical energy in the exhaust is not significant because current emission standards limit incomplete combustion products to minute levels. Figure 6 shows the energy distribution in a typical passenger vehicle. The engine designer strives to shift waste energy from the coolant to the exhaust and to recover energy from the exhaust.

Waste Heat Recovery: A *turbocharger* is a supercharging compressor that is powered by a turbine in the exhaust stream. The major design issues for turbochargers are: (1) broad operating range (low to high engine speeds), (2) rapid load adjustment, (3) endurance and reliability. High efficiency over a wide range of loads and engine speeds is the goal of designers. Exhaust flow at low speeds may not provide enough compression in the supercharger. The reduced volume of air in the cylinders means less power and efficiency. At high speeds pumping loss to back pressure occurs after the compressor achieves its maximum pressure.

Low speed performance has benefited from reduced turbine nozzle area and advanced impellers which increase air volume at low speeds. Another approach uses a free piston between two exhaust manifolds. Reducing turbine nozzle area is extremely effective but it penalizes high speed performance. The variable geometry turbocharger adjusts the swallowing capacity of the turbine to optimize torque, transient response, and fuel consumption over the full operating range of the engine.

To prevent backpressure at high speeds, a waste gate valve will bypass the turbine or the supercharger may be declutched. Smaller engines may be the main benefit of turbocharging even though Volkswagen claims 11% less fuel consumption due to shorter acceleration time.[14]

Turbochargers respond slowly to acceleration especially at low speed or low load. Moreover, if the turbocharger is matched to the engine at full load, engine torque drops at lower speeds. Tolerable in heavy duty applications, these characteristics are a burden to passenger vehicles. Response is improved by reducing rotor inertia, bearing friction, and the turbine nozzle area. Inertia reduction focuses on reducing the diameter and mass of the turbine blades. Recent designs using smaller multiple rotors improve acceleration, torque, and emissions as well as fuel efficiency. Light-weight ceramic rotors under development respond well to rapid speed and load changes. Current problems are insufficient high-temperature strength, vulnerability to hard flying objects, and joining the ceramic to the metal shaft.[15]

Schemes to further use the energy in the exhaust gas, rather than wasting it with a bypass valve, are called *turbocompounding*. The simplest method uses the excess exhaust energy to power a turbine connected to the crankshaft. Unfortunately this system becomes a parasitic load on the engine in typical automotive conditions. A more advanced concept mechanically connects both the compressor and exhaust expander to the crankshaft. Then engine speed controls the compressor, improving torque, acceleration response, and fuel efficiency. A continuously variable drive that maintains a reaction turbine at its most efficient speed throughout the load range would be ideal. Variable geometry turbines, variable speed drives, helical-screw positive displacement compressors and expanders (screw-compounding) are some of the many variations that have been considered. The performance of each varies with load.

Other existing systems including motor/generator systems, steady-flow combustion chambers using bypassed exhaust, and epicyclical gear trains may be too complex for light-duty automotive use. Laminar flow turbos that provide an insulating layer of air to the cylinder walls could reduce heat rejected to the cooling system and possibly control combustion.

Although not always beneficial, turbocompounding raised fuel efficiency 10% for some large diesels. Power and torque can also increase by 20 and 30%, respectively. The costs of the additional hardware have limited commercial development. Accordingly, research to improve the gear system and its cost is a priority. Turbocompound diesels for large trucks will probably enter production in the early 1990s.

Other thermodynamic cycles, called *bottoming cycles*, can also convert exhaust heat into mechanical energy. Vehicle tests of a diesel-organic Rankine ("steam") Cycle compound engine showed a 12.5% fuel savings. Size is the primary impediment to mobile use.[16]

Cooling: Most internal combustion engines are cooled by forcing air on cooling fins or by circulating a liquid coolant through cooling jackets in the engine. Evaporative cooling and cooling without water are now receiving consideration. These advanced cooling systems better fuel efficiency by:

(1) reduced parasitic losses (smaller or no pumps and fans),
(2) less engine friction due to higher temperatures,
(3) better aerodynamics (heat exchanger or smaller radiator),
(4) shorter warm-up time (less coolant),
(5) lower weight.

A conventional cooling system prevents the coolant from boiling. Evaporative cooling boils the coolant to cool the cylinders and dissipates the heat in a condenser. The radiator, when functioning as a condenser, can have 30% more heat transfer capability and so be that much smaller and halve the warm-up time. The temperature gradient along the cylinder is much smaller reducing thermal stress. Thermal stresses are due to temperature differences not absolute temperature.[17]

Cooling without water can use external engine surfaces, intake air, and lubricating oil. *Oil cooling* is the most effective because external surfaces require forced air and intake air cooling is localized. Some claim it will reduce fuel consumption as much as 32%. Oil does not oxidize, freeze, or boil, and is much less likely to cavitate, that is trap air bubbles. Combining the lubricating and cooling systems may also reduce manufacturing costs. Moreover, eliminating water-based systems should enhance reliability as they are responsible for half of all auto breakdowns. *Strategic cooling*, the associated concept of cooling only the hottest components, will require new high-temperature materials. Then cooling the cylinder from the inside with just the lubricant or fresh insulating air might be possible.[18]

Low Heat Rejection: The Low-Heat-Rejection-Engine (LHRE) or *adiabatic engine* restricts the energy flow to the cooling system. An ideal adiabatic engine would have zero heat loss whereas real engines may eventually achieve 40%–50% loss. Most of the energy "saved" must be recovered from the hotter exhaust gas to raise efficiency. The higher operating temperature is the source of the assets and drawbacks of the LHRE. Advantages include 10%–20% better fuel economy, better piston expansion, better exhaust recovery, and greater tolerance for lower cetane fuels. Fuel savings are from the same factors as the advanced cooling systems. Engine volume may be as much as 40% less. Simulations estimate fuel efficiency improvements to be: 2%–4% before cooling system reduction, 6% considering the cooling system, and up to 13% with turbocompounding. Most adiabatic engine work is focused on the diesel but the Wankel rotary engine will benefit greatly from adiabatic improvements due to its large surface area.

Disadvantages of current adiabatic engines include lower reliability, higher cost, uncertain durability, temperatures too high for liquid lubricants, and a volumetric efficiency penalty. The brittle failure mode of monolithic ceramics needed for high-temperature operation and the difficulties in attaching ceramics to metal reduce reliability. Ceramics may eventually improve durability but will raise costs. Volumetric efficiency decreases because hotter cylinders warm the intake air making it less dense. This is a deficit to fuel consumption that the turbocharging must overcome. High-temperature tribology (lubrication) is seen as the greatest single barrier to adiabatic engine development. Expect the internal combustion engine to evolve slowly into an "adiabatic" form by incremental improvements that raise exhaust temperature and the efficiency of waste heat recovery.

The research and commercial development of this technology in Japan and Europe seem better coordinated and more farsighted than in the U.S. Limited commercial production of ceramic engine parts is already underway in Japan (turbocharger) and Europe (exhaust port liners). Ceramic consumer items such as knives and scissors are providing production experience abroad. In fact, U.S. engine companies purchase most of their ceramic parts from Japanese companies.

Ceramic and air gap insulation can reduce heat rejected through the pistons. Ceramic insulation generally takes the form of either a monolithic ceramic or a composite metal-ceramic piston. Monolithic ceramic pistons suffer brittle failure, ring sticking (the high ring groove temperature degrades the oil there), and most seriously, high thermal stress in the crown. The most typical composite piston has a zirconia crown on a cast iron or aluminum body. It suffers from brittleness and difficulties attaching the ceramic to the metal. An air gap piston has an air or vacuum cavity to insulate the crown from the rest of the piston body. At low temperatures the dominant heat transmission across the gap is by conduction and convection. At high temperatures it is by radiation. Attaching the crown to the piston body is a major concern as incomplete sealing allows ingress of dirt which reduces cavity effectiveness.[19]

iv. Transmissions. Gear reduction design determines the operating point of the engine across the load range and so plays a critical role in fuel efficiency. The transmission may have a continuously variable gear ratio (CVT) or use discrete gears. Both manual and automatic transmissions use discrete gears. The driver uses a clutch to shift between gears in a manual transmission. In automatic transmissions, fluid coupling in a torque converter shifts and engages the gears.

Pumping and frictional losses in the fluid make automatic transmissions less efficient. Clutching schemes to bypass the torque converter may lower fuel consumption by 6%–20%. On the other hand, few drivers shift manual transmissions for maximum fuel efficiency. In fact, the driver can perturb fuel economy by as much as 40%. New electronic controls responding to instantaneous engine conditions will provide automatic shifting or signal the driver when to shift and achieve 10% better fuel economy. Future discrete transmissions will have more gears and a greater range so the engine will operate more fully loaded.

The most common CVTs are belt driven. Use in small Japanese cars has improved fuel economy 10%–20%. Major problems for CVTs include narrow gear ratio ranges (< 6:1) and developing belts that can handle the higher power needed for the larger American cars. Hydrostatic CVTs (power transferred by pressurized fluid circuit) can have a gear ratio range of 15:1, but tend to be large, expensive, noisy, and have lower partial load efficiency. Traction drive CVTs (power transferred between rolling elements) present an option that may be less expensive and yield 10%–16% better fuel economy than current transmissions.[20]

v. Electronic controls. Legislation to reduce emissions and fuel consumption in the late 1970s motivated development of electronic engine control (EEC). Microprocessors driving analog control systems replaced intricate, precision manufactured governor timing systems for fuel pumps. At that time, the industry established the fundamental fuel volume and timing algorithms and introduced distributorless ignitions.

More recently digital signals, electromagnetic spill valves, algorithms to optimize injection rate, custom automotive integrated circuits, and integrated direct ignition systems have improved fuel control, response, reliability, emissions, size, weight, and cost. Current advanced electronic systems control air intake, exhaust recirculation, fuel injection, spark timing, and idle speed and can reduce fuel consumption as much as 10%. Future systems will control the injection profile as well. Systems that react to "instantaneous" direct measurements (instantaneous compared to the engine cycle) are replacing slow controls that respond to indirectly measured cycle-averaged engine conditions. In future systems intelligent sensors will process information and intelligent transducer-actuators will adjust for the quality fluctuation and wear of mechanical parts.[21]

Indirect measurements of engine parameters and the proliferation of uncoordinated subsystems limit current EEC. Eventually better coordinated subsystems will lead to totally integrated vehicle control, with fail-safe and fail-soft systems. Advanced sensors that instantaneously measure direct performance are crucial to progress. The use of many types of sensors in relatively small quantities will increase vehicle costs.

Many exciting possibilities are feasible or imminent. A recent adaptive engine controller corrects for both cycle-to-cycle and lifetime variations by perturbing the spark timing to determine the best burn rate and air-fuel mixture. A new cruise control for manual transmission trucks is significant because the great variation in cargo weights results in a much wider range of running resistances than for automobiles. Field-dependent hydraulic fluids that become solid under strong electric fields may replace valves and coils with completely computer controlled hydraulics. A microprocessor-based sound cancellation system detects engine noise with a microphone by the exhaust and drives heat resistant speakers on the exhaust pipe to produce an identical but oppositely phased sound. It would eliminate the traditional muffler and save the 20% of engine power lost to backpressure.[22]

vi. Materials. New materials will lower fuel consumption through reduced weight, advanced sensors, reduced friction, and increased adiabacity. Weight reduction is likely to be the most significant benefit. A 10% weight reduction can improve highway mileage as much as 5% and city mileage by 8%. The typical 4000 lb American car shed about 1000 lbs in the 1980s. Current models could lose another 1000 lbs and achieve a combined city/highway fuel economy of 50 mpg through materials substitution alone. The biggest weight reductions *()* will be from replacing steel and cast-iron with High-Strength Low Alloy (HSLA) steel *(23%)*, aluminum *(55%)*, composite plastic *(65%)*, and magnesium *(75%)*.

Reduced manufacturing and assembly costs may offset the higher costs of these materials. For example, a single plastic piece can replace 40 to 90 steel parts, and aluminum can require only one tenth as many spot welds as steel. Besides use in body parts, plastics can form load bearing parts. Carbon fiber-reinforced plastics and steel fiber-reinforced aluminum reduce the inertia of moving engine parts. Magnesium saves weight but is flammable and subject to corrosion.

Ceramics offer strength and resistance to heat, friction, and corrosion when used in engine parts. As thermal insulators ceramics will increase in-cylinder temperatures. Averaging near 1100 °C at present, the temperature ranges from ambient to 2400 °C with surface gradients of 30 °C. Ceramics, already used in a multitude of engine parts, enhance electronic engine control by their increasing use in many different sensors. Because ceramics are half as dense as cast-iron, most near term applications will be for lowering inertia, friction, or corrosion, rather than for thermal insulation.

Figure 7 lists the pros and cons of the leading ceramics. Silicon nitride shows the best balance of characteristics, at least for the near term. Parts are also made from Alumina ($AlxO_3$) and sialons (composed of silicon, aluminum, oxygen, and nitrogen). Besides the need for high strength, thermal shock resistance, and low thermal conductivity, engine applications will require non-brittle failure modes, good attachment, orthotropic stiffness, high-temperature stability, low thermal expansion (to maintain piston-to-cylinder bore clearances over a wide temperature range), and low thermal stress. Unfortunately, lowering thermal conductivity tends to *increase* thermal stress.

New materials under development include transformation toughened ceramics, custom tailored ceramics, and composites. Composites refer to a fiber or particulate reinforced ceramic or metal matrix, monolithic ceramics attached to metals, or coatings laminated onto ceramics or metals. Several layers of decreasing conductivity can lower the thermal stress on a surface. A compliant

Pros and Cons of Leading Automotive Ceramics

	Benefit	Deficiency
Sintered Silicon Nitride (Si_3N_4)	good strength good thermal resistance	poor insulator
Partially Stabilized Zirconia (ZrO_2)	good insulator	poor high temp. strength poor durability low thermal shock resistance
Silicon Carbide (SiC)	good high temp. strengh	poor thermal shock resistance
Lithium Aluminum Silicate	good insulator good thermal shock property	low flexural strength

Figure 7

fiber layer joining the metal and ceramic can shield the ceramic by absorbing shock.[23]

vii. Alternate Fuels. Petroleum supplies will inevitably end. Thus alternative fuels must extend the oil supply until we transition to other technologies, or replace it altogether. Low oil prices discourage their development for fuel economy though environmental issues motivate some efforts. Fuel options include gas-oil mixtures, low-quality fuel oil, emulsified fuel oil, coal, natural gas, butane, alcohol, hydrogen, and vegetable oil.

Low grade fuel oil requires less refining than gasoline saving energy and money. Obstacles to overcome include: high viscosity, low cetane number, poor atomization, increased wear and corrosion in cylinders and rings (due to higher ash and solid content), and abnormal wear due to ingress of the distillation catalyst. Diesel engines that burn low grade fuels have attained thermal efficiencies equivalent to that of standard diesels. Adding 10% water by volume to emulsify low grade fuel oil can reduce fuel consumption as much as 3%.

Butane (liquified petroleum gas, LPG) is an unavoidable byproduct of oil refining. Adding butane to a gas-oil diesel engine can reduce fuel consumption 5%–6% but raises certain emissions. *Natural gas* has better thermodynamic properties but is not stored as easily. Compressed Natural Gas (CNG) is more popular than the other storage options, Liquified Natural Gas (LNG) and Absorbed Natural Gas (ANG). Natural gas is safer and burns cleaner than methane, ethanol, or propane. Its compression ratio is low for diesel operation which would lower efficiency and range, but it is high for spark ignition which it could enhance. A few cities are testing CNG buses. In greater use in other countries, domestic LPG and CNG vehicles have suffered little market interest. At higher cost, natural gas can be converted into methanol or synthetic gasoline. As with the alcohols and hydrogen, incompatibility with existing fuel distribution systems may be the greatest barrier to its use.[24]

Coal fuels can be liquids, solids, or hybrids (usually coal particles suspended in oil or water). The expertise exists to make high-quality liquid fuels (methanol, substitute natural gas, synthetic gasoline) from indirect liquefaction processes. Diesels and gas turbines can use those with little or no modification. Theoretically more efficient, direct coal liquefaction requires much more research. Micronized dry fuel made from coal particles less than 40 microns in diameter powered a 100 hp automotive gas turbine with excellent results. Handling the dry fuel is the biggest drawback. High quality slurries are being evaluated in gas turbine and diesel engines. The railroads have taken a renewed interest in coal for energy security and cost. Coal may play a more important role as a source for methanol than as a fuel itself. Of all the alternatives, coal would be the most detrimental to the "Greenhouse" effect.[25]

The *alcohols* provide petroleum alternatives that are suitable for compression and spark ignition, have a multi-feedstock base, and produce much less smoke and nitrous oxides than gasoline. Alcohol fuels are ignited by spark, injecting an ignition fuel, or blending with oil. These fuels exhibit poor cycle-to-cycle ignition stability, corrosion, engine wear, lower volume energy content, high ground water contamination potential, invisible flame (engine fire hazard), and an explosive air-fuel mixture in the fuel tank. (The air-fuel mixture in a gasoline tank is too rich to be explosive.) Alcohols also produce 2 to 3 times more aldehydes than gasoline. The corrosion caused by formic acid ($HCOOH$) formed in gas-methanol blends and the swelling of nonmetallic materials require protection for the fuel system. Moreover, accelerated engine wear may result from the hotter combustion temperature.[26]

Methanol comes from natural gas, a fossil fuel, while ethanol derives from vegetable sugar. Only Brazil produces enough ethanol for widespread use. The U.S. currently produces highly subsidized ethanol from corn. A large-scale shift to ethanol would affect feed grain exports, food costs, farm subsidies, and gas tax revenues as well as oil imports. Past poor consumer experiences with ethanol may hamper efforts to introduce any new alternative fuel. South Africa may be developing an additive that would allow ethanol to be used directly in diesel engines without modification.[27]

Using exhaust heat to dissociate (reform) methanol into hydrogen and carbon monoxide prior to combustion can increase the heat of combustion by 20%. Some methanol-fueled prototypes demonstrate thermal efficiencies 20% to 30% better than gasoline engines of comparable compression ratio. Methanol seems to be favored by the engine manufacturers and oil producers for its multi-feedstock base (natural gas, coal, or biomass), hence a larger, more certain, and less expensive supply. In the near term, neither alcohol can be produced in sufficient quantities. Alcohol imports would diversify the imported fuel supply — a marginal benefit.[28]

The Urban Mass Transit Authority and the State of California are exploring alcohol fuel alternatives to solve air pollution problems. ARCO and Chevron have set up a limited distribution network of methanol stations to fuel buses and flexible fuel vehicles (FFV) which can operate on methanol or gasoline.[29]

Engineers at the University of Illinois in Urbana-Champaign have tested oils from corn, peanuts, castor, cottonseed, crambe, linseed, soybeans, rapeseed, sunflowers, sesame, and safflower. Soybean oil returns 4.5 times the energy used to produce it which is over twice the return from most of the other oils. *Vegetable oil fuels* can be a microemulsion or be processed into a diesel-like ester. Neither approach is presently cost effective but they do provide a contingency fuel, especially for farm use.

Hydrogen may provide 6% better efficiency and no emissions but its explosive qualities make storage and handling difficult. Hydrogen can be stored as a cryogenic liquid or at room temperature in a metal hydride. A six passenger van using the latter method tested in 1986 reached speeds of 62 mph with a range of over 120 miles per tank (at 37 mph). Exhaust heat applied to the metal hydride released the hydrogen. Hydrogen is suitable for direct injection, premixed-charge engines, and gas turbines.[30]

2b. Gas turbine (Brayton cycle)

Gas and expansion turbines operate by the Brayton or Joule cycle, a constant-pressure cycle. A fan compresses air which is heated at constant pressure in a combustion chamber. The high-temperature gas imparts mechanical energy to a turbine as it exits the chamber. Excess energy over that required to power the compressor provides the locomotion. Efficiency depends on ambient air temperature, nozzle inlet temperature, and type of cycle as well as the compressor and turbine efficiencies.

Designers can improve turbine engine performance by raising the pressure ratio, turbine temperature, component efficiencies, and by reducing the cooling and leakage flows. Compressor efficiencies have surpassed 91%. New blade designs and attachment methods will reduce compressor weight by 38%. Component efficiencies around 90% at present may ultimately reach 95%. Very sophisticated film cooling schemes force air through holes in the turbine blades allowing operation at temperatures higher than the blade material can withstand (1300–1600 °C). Higher-temperature materials for turbine blades will save 20% of the compressed air used for cooling. Figure 8 suggests future supersonic and hypersonic propulsion system efficiencies over 60%.

Even though it is widely used by industry and utilities, jet aircraft make the most obvious use of the gas turbine engine. Gas turbines suffer lower fuel economy (especially at low speeds) and higher initial cost than gasoline engines. However, their alternative fuel capability, reduced emissions, lower noise, lighter weight, and lower maintenance may eventually put them on the road. Lower emissions, especially of nitrous oxides, result from leaner, hence lower-temperature, combustion and the longer combustion time of the continuous flow operation. The air-fuel ratio of the Brayton cycle is 50 to 60:1 compared to 25:1 for the Diesel and 15:1 for the Otto cycles.[31] Ground-based sys-

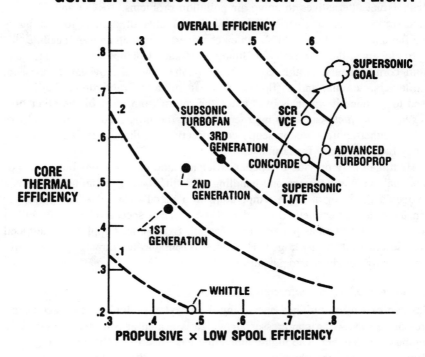

GOAL EFFICIENCIES FOR HIGH-SPEED FLIGHT

Source: NASA Conference Publication 10003, 1987.

Figure 8

tems must convert the exhaust stream kinetic energy to shaft power, further reducing efficiency. Motor vehicle development is expected to be slow over the next 20 years.[32]

Aviation technologies developed in the Aircraft Energy Efficiency (ACEE) program begun by NASA in the mid 1970s will mature in the 1990s. Exciting developments such as the high bypass ratio turbofan and the propfan, based on advanced propeller technology, promise fuel savings of 20% and 40%, respectively. The projected saving of over half a quad of petroleum a year, though significant, is still a small fraction of total energy use. Because fuel cost is a smaller fraction of the direct operating cost than the engine or airframe low oil prices slow application of these technologies. Details of these advances can be found in several references.[33]

2c. External combustion engines

i. Stirling engine. The basic Stirling engine consists of two pistons connected to a shaft so that they may move in different phases. One piston is in contact with a cold reservoir, and the other with a hot reservoir (burning fuel). A regenerator divides the gas-filled space between them into two portions. The regenerator is

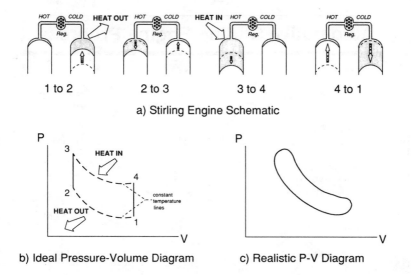

a) Stirling Engine Schematic

b) Ideal Pressure-Volume Diagram   c) Realistic P-V Diagram

Figure 9

typically made of steel wool or a series of metal baffles. It has low enough thermal conductivity to support the temperature difference between the hot and cold sides without appreciable heat conduction. Figure 9 depicts the ideal cycle.

The advantages of a Stirling engine include lower fuel consumption, multi-fuel capability, quiet operation, low emissions, minimal lubrication requirements, potentially high reliability, long life, and reduced maintenance costs. Fuel options for this engine include gasoline, kerosene, alcohol, methanol, diesel fuel, butane, and others.

The Automotive Propulsion Research and Development Act of 1978 mandated the transfer of European Stirling engine technology to the U.S. and then its development for automotive use. The final design of the program, the Mod II, underwent testing in 1986. It is an 83.5 hp engine that burns gasoline and uses hydrogen for the working gas. Figure 10 shows the drastic gains in performance made between 1978 and 1986. The Mod II engine powered a U.S. Postal Service vehicle in 1988. Preliminary results indicate 10% to 14% better fuel economy than the internal combustion engine and much lower emissions. The Post Office ran service tests in 1989 and 1990.

Reliability and sufficiently low manufacturing costs remain to be demonstrated. To lower the financial risk, NASA hopes to guide Stirling engines into starter markets first. Electrical generators, heat pumps, and irrigation pumps can provide the commercial experience to better estimate mass production costs. NASA would also like to introduce Stirling engines to a fleet operator like a courier or trucking service. Many fleet operators have offered to provide the vehicles and drivers for the necessary field test (≈ 50 vehicles). None of the automakers are interested in providing the engines, and the federal government

Automotive Stirling Engine Progress

	1978	1986
Weight/Power	8.52 kg/kW (14 lb/hp)	3.35 kg/kW (5.5 lb/hp)
Manufacturing Cost (300,000 units/yr.)	$5000+	$1200
Acceleration, 0-97 km/h, 0-60 mph; 1417 kg, 3125 lb car	36 sec	12.4 sec
Combined Cycle Fuel Economy	8.1 km/L (19 mgp)	17.5 km/L (41 mpg)
Rare Metals	Cobalt	None
Seal Life	100 hours	2000+ hours

Courtesy of Mechnical Technology Incorporated, Latham, NY.

Figure 10

is satisfied with the demonstration of technical feasibility. As a result Stirling engine development in the U.S. may stall unless another organization steps in to fund the testing. At present, the very low emissions may provide more incentive than energy considerations.[34]

ii. Steam engine (Rankine cycle). The ideal steam engine operates by the Rankine cycle which manipulates the pressure and volume of a constant mass of water as follows: Low-pressure water from the condensor is vaporized, then superheated, at constant pressure in the boiler. It then flows into a cylinder to expand adiabatically against a piston or turbine blades until its temperature and pressure drop to that of the condenser where it began the cycle. Ideally, the work done is equal to the heat input by the boiler minus the heat rejected during condensation. In reality it will be somewhat less due to turbulence, pumping, friction, and thermal conduction through walls. Used in marine engines the Rankine cycle may eventually recover waste heat in turbocompound diesel engines.[35]

2d. Fuel cells for transportation

Fuel cells produce electricity by catalytically reacting hydrogen and oxygen to form water. They are quiet, produce negligible pollution, and can use coal and biomass derived fuels. Most importantly, efficiencies as high as 60% to 80% may make them the most efficient powerplant for transportation (Figure 11). Fuel cell research has concentrated on several cells differing essentially by electrolyte. *Molten carbonate* and *solid oxide* systems are the least developed and not well suited to mobile applications due to their high temperatures. The

Estimated Propulsion System Efficiencies

Power Plant Type [Assumed efficiency, fuel]	Equivalent Coal Economy (km/kg)**
Spark Ignition [7.4 l/100 km, gasoline]	5.0
Compression Ignition [5.7 l/100 km, distillate]	6.4
Stirling/Gas Turbine [5.2 l/100 km, distillate]	7.1
Battery/Electric [0.21 kWh/km]	9.2
Fuel Cell/Electric [0.16 kWh/km]	12.1

Courtesy of American Chemical Society, Rentz, Hagey & Kirk, Ref. 36.

** Includes energy source conversion efficiency
Assumed vehicle: 5 passenger, 150 mile range, 2500 lb curb weight.

Figure 11

alkaline fuel cell has an outstanding record in space applications but its need for extremely pure hydrogen discourages its use for transportation. Efforts to use *reformed fuel* (organic hydrocarbons decomposed into hydrogen and oxygen) have not succeeded.

Phosphoric acid fuel cells are the most developed. They can operate at high efficiency on a variety of processed hydrocarbon fuels including gasoline, diesel, jet fuel, naptha, methane, natural gas, propane, methanol, hydrogen, and ammonia. Advanced cells should be available in the early 1990s. *Ion exchange membrane (solid polymer) technology* is less mature, but promises high power density, low-temperature operation, rigid and contained electrolyte, and cold starting. It is compatible with the same fuels.

Methanol decomposes at only 200 °C using inexpensive catalysts in a simple apparatus, making it an attractive transportation fuel. The heavier hydrocarbons require much higher temperatures and larger equipment for reforming. Figure 12 lists the advantages, disadvantages, and the developmental needs of the various fuel cell technologies. Near term research will emphasize fuel efficient prime movers, fuel flexibility, and environmental acceptability. The long term aim is to use a nonpetroleum fuel.

Hybrid fuel cell/battery power plants are feasible in the near term for buses and passenger cars. The batteries extract a significant weight penalty but are necessary to supply power during the large transients typical of vehicular requirements. The time constant of the reformer, up to 15 minutes, limits present systems. Rapid charging batteries will help performance, but new fast-response reformers being tested should lead to better fuel-cell-only systems.

Fuel Cell Attributes

FUEL CELL

Advantages

Disadvantages

Development Needs

*Courtesy of American Chemical Society,
Rentz, Hagey & Kirk, Ref. 36*

PHOSPHORIC ACID

rejects CO_2 from electrolyte

broad fuel gas tolerance

good thermal integration with fuel processor

potential for low materials costs (except catalysts)

requires precious metal catalysts for electrodes

relatively high operating temperature slows start up

relatively low specific power and power density

improve cathode catalyst & stability for higher efficiency & lower cost

improve anode catalyst resistance to poisons, eg. CO at low temperature

improve electrode-electrolyte interface structure to increase power density

need more durable catalyst support material for longer lifetime

develop high-efficiency pressurization equipment (5 atm)

SOLID POLYMER

relatively high specific power & power density

good fuel conversion density

no electrolyte management problems

potential for long cell life and rugged cell construction

potential for rapid cold-start capability

requires precious metal catalysts

high capital cost of electrolyte & catalysts

requires very low CO content in fuel stream

sensitive to moisture depletion & low temperatures (freeze-thaw cycling)

improve stability & activity of anode catalyst to impurities to lower cost

reduce or eliminate platinum catalyst by substitution or structure modification

develop low cost alternative to Nafion membranes

improve membrane resistane to freeze drying & subsequent electrode shedding

develop high efficiency pressurization equipment

ALKALINE

high fuel conversion efficiency

good performance without precious metal catalysts

relatively long cell life due to electrolyte/materials compatibility

potential for low materials costs

electrolyte does not reject CO2

electrolyte circulates for removal of heat, water &/or CO2

CO2 contamination may require periodic replacement of electrolyte

long life may require noble metals

develop CO2 rejection capability

improve on-board vehicle nitrogen storage

replace platinum with inexpensive, stable electrode catalysts

improve component stability: seals, separator, biopolar plate material

improve system optimization and flowsheet integration

SOLID OXIDE

good fuel conversion efficiency at atmospheric pressure

tolerates high CO levels

no electrolyte management problems

potential for high tolerance to impurities, eg. sulfur compounds

does not require precious metal catalysts

potential for long cell life, low materials cost, simple fuel processing

operates at high temperature

requires cold-start assistance (however, low thermal inertia)

ceramic parts may be sensitive to mechanical shock

improve air electrode: stability & electronic conductivity

improve in situ hydrocarbon fuel processing capability

improve anode performance & impurity tolerance

improve lifetime: adhesion, thermal cycling, aging,interconnects,gas tightness

MOLTEN CARBONATE

good fuel conversion efficiency & power density

tolerates high CO levels

good fuel tolerance

does not require precious metal catalysts

operates at high temperatures

requires cold start assistance

sensitive to sulfur contaminants

requires CO2 recycle from anode exhaust to cathode inlet

dissolution of NiO cathode

current electrolyte migration problems

improve cathode: better conductivity, less costly fabrication

improve cathode: reduce electrolyte loss & raise stability

improve anode: resistance to corrosion&impurities, replace Ni catalyst,strength

improve mech. properties: electrode contacts, seals, thermal cycle, fatigue

improve overall system optimization and flowsheet integration

Figure 12

Mass produced fuel cells will need to generate power at a $500/kW life cycle cost to compete directly with diesel engines. Current technology seems capable of costs in the $300-$600/kW range. Future areas for cost reduction include the use of non-noble metal catalysts and the elimination or simplification of the fuel reformer. Any judgement on market prospects at this time would be premature. Fuel cells are a high risk but very high payoff technology that merit attention as a long range solution to our transportation problem.[36]

2e. Electric propulsion

i. Electric/hybrid vehicles. A hybrid power train incorporates more than one power source. There may be multiple prime power sources but a single prime source and a secondary *mechanical energy storage system* (MESS) is more typical. Vehicles require a small MESS to provide acceleration power rather than drive energy. Possible MESS technologies include flywheels, hydraulic accumulators, and elastomeric storage devices.

A MESS has several benefits. (1) It keeps the engine operating in its most efficient load range by storing energy to prevent the load on the engine from dropping and returning energy when the load increases. (2) It captures energy that wheel-braking would dissipate. That can be 30% of the driving energy for an urban passenger car and as much as 60% for a city bus. (3) It allows the engine to be turned off during idle periods when it is least efficient. (4) It allows use of a smaller engine. Most MESS devices ideally complement the high energy density and low power density of batteries. Furthermore, a MESS can accept braking energy at a more efficient rate than an electric power train.

A heat engine with a battery-electric motor has advantages which include: (1) the use of a smaller engine, (2) the capability for fuel source independence, (3) use of one source when it enhances efficiency, for example, turning the engine off for urban driving, and (4) regeneration of the batteries while driving. The penalty is the additional hardware necessary to integrate two power trains. Energy management control algorithms are crucial to system performance. In fact, as the energy management strategy and propulsion system approach the optimum, urban fuel economy actually surpasses highway fuel economy! The engine is either off or fully loaded, while wind resistance at city speeds is much lower.

Electric vehicles have an enormous potential to reduce both petroleum dependence and pollution. A light-weight electric vehicle could satisfy over 90% of the average family's travel needs. That may be 74% of the annual miles travelled in the U.S. A small number of electric powered trucks and autos are in use, but manufacturers do not see a market warranting large-scale production. Several cooperating organizations plan to put 500 electric powered GM vans on the road in 1990. The vans should have a range of 60 miles, top speed of 52 mph, accelerate to 30 mph in 12 seconds, carry 2000 lbs, and cost $30,000. Operating costs should be similar to those of the gasoline powered van.[37]

ii. Battery storage technology. New algorithms and instruments have advanced battery technology through improved characterization and performance predictions. About 80% of battery problems derive from the conduction path outside of the cells, and in spite of extensive maintenance, many batteries only supply three quarters of their rated output. A permanently installed computer to monitor battery condition and manage maintenence can cost less than the standard labor intensive battery maintenance.

Initial electric vehicle (EV) development focused on passenger cars with lead acid batteries. Interest has since shifted to urban commercial vans with more emphasis on advanced batteries. Market penetration of electric vehicles will depend on advanced batteries having high specific power and energy, low cost, and long life. The initial costs for an EV with Na/S batteries are likely to be 10% more than that of an internal combustion engine powered car but lower energy costs will make the life cycle per mile costs equivalent. The recent "Impact" prototype by GM which combined breakthroughs in battery technology, aerodynamics, weight, and rolling resistance may return the focus to passenger vehicles. The 80 mpg equivalent economy of the 2 passenger Impact results in the same energy use as the commuter train of Figure 3a.[38]

iii. Solar electric vehicles. In spite of the greater emphasis overseas U.S. solar cars are at a high technical level. In November 1987, General Motor's "Sunraycer" won the 1900 mile World Solar Challenge '87 in Australia, crossing that continent on the energy equivalent of 5 gallons of gas. It also broke the

1987 solar speed record of 35 mph with a 5 mile run at 49 mph in June, 1988. Boosted by 5 lightweight 24 volt batteries the Sunraycer set a world record for electric vehicles under 1100 lbs by averaging 75 mph over a 10 km course. All-solar vehicles are unofficially alleged to be capable of 60 mph.

The $8 million car weighs 430 lbs and uses 8800 gallium arsenide and silicon solar cells to power the 11 lb, 2 hp motor. Its development pushed technologies in aerodynamics (0.125 drag coefficient), materials, electric motors, and rolling resistance to the limit. The developers believe their greatest contribution has been to low-speed aerodynamics and high-power efficiency electronics for battery powered EVs. They do not believe solar power is efficient enough for widespread use in transportation although crystalline and thin-film photovoltaics have seen substantial progress recently. Others are attempting to develop practical solar commuter cars to sell for 3–4 thousand dollars by the early 1990s.[39]

iv. Magnetic levitation. Several groups in the U.S. pioneered the work on "electromagnetic flight" or magnetic levitation in the 1960s. Federal support disappeared in 1975 because the Department of Transportation deemed U.S. air and highway transportation systems adequate for anticipated intercity transportation. Some intracity research continued. In 1986, Boeing demonstrated the technical feasibility of two different propulsion systems for 12-passenger, fully automatic, 40 mph intracity vehicles. Despite predicted energy savings (Figure 13) Boeing dropped the work in 1987 for lack of a market. Carnegie-Mellon Institute licensed the technology but lacks a sponsor to continue the work.

Prior to the cutback Kolm and Thornton at MIT demonstrated a $1/25$ scale model of a high speed intercity concept they termed the Magnaplane. Long aerodynamic vehicles resembling aircraft fuselages would "fly" on a traveling wave one foot above the bottom of the troughlike guideways. Overhead masts would support the guideways beside interstate highways. Computer controlled transformers, a mile apart, would supply AC current to the linear induction motor track. The magnaplane would transport passengers at 224 mph, have 6 times the passenger capability of a railroad, and use one-tenth the energy. Concern about airport congestion and air quality has renewed interest in magnetic levitation. Domestic airlines spend a tenth of their energy idling jet engines. Maglev would shift some of our transportation burden from petroleum to electricity.[40]

v. Railway electrification. Railway electrification has been important to transportation on every continent except North America. There are several reasons to consider electrification. (1) Electric locomotives can have three times the tractive horsepower of comparable diesel-electric units and so require fewer locomotive units per train. Better acceleration is a difference that is significant in high speed passenger operations. (2) Maintenance costs are 25 to 56% of diesel-electric costs. (3) It raises locomotive availability from ~ 84% to ~ 96%. (4) Inconsequential shutdown saves energy and wear compared to the start up and idling costs of diesels. (5) Electric trains have lower fuel costs. Proponents

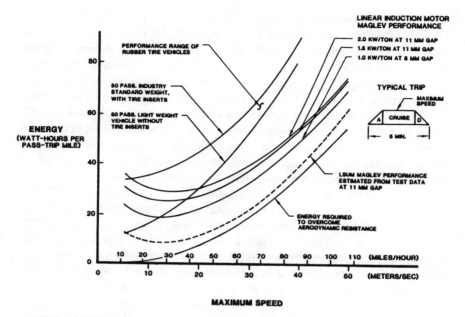

Energy Consumption Comparison: Wheeled Vehicles and Maglev

Source: Gilliland & Pearson, Ref. 40.

Figure 13

argue that electric energy costs will increase at an annual rate 1 to 3% less than that of diesel fuel. (6) Electric locomotives promote petroleum independence because any domestic source can supply electricity. (7) Electric locomotives produce no emissions.

The high capital cost of electrification, $80,000 to $200,000 per mile of track, is the greatest obstacle. To justify the costs a freight railway must have high traffic density and allow high speeds. The American railroads anticipate having to invest $2–4 billion to electrify 10,000 miles of the highest density routes. At present, U.S. railroads do not perceive environmental, social, energy or operating benefits sufficient to justify the massive investment. Electrification offers the biggest advantages to passenger service, so its fate may be determined by the direction taken by passenger rail.[41]

3. Tribology (friction and lubricants)

3a. Lubricants

Engine friction consumes 7.5% of the combustion energy produced by a gasoline engine, almost half at the piston skirts (Figure 6). Diesels are similar but have larger friction losses. The most significant role for high-temperature tribology, though, will be making the adiabatic engine possible.

The highest temperature experienced by the lubricant occurs where the top piston ring reverses direction. The viscosity of a liquid lubricant decreases exponentially with temperature so the film thickness is lowest on the hottest parts. The top ring reversal temperature (TRR) is 200 °C for the conventional water cooled engine but will be 550 °C for the adiabatic engine. A decomposing lubricant leaves deposits and increases wear. All known lubricants decompose by 350 °C but higher TRR is possible due to the short time of exposure at the TRR. Liquid lubricants are now limited to 370 °C but eventually should function with a TRR of 450°C. Additives being studied may raise this to ≈ 500 °C. Future lubricants will require higher-temperature additives. Additives constitute as much as 15% of an engine oil. That is mostly detergents and dispersants with some antiwear agents, antioxidants, corrosion inhibitors, antifoam agents, pour point depressants, friction modifiers, and viscosity improvers.

The advanced synthetics under investigation include: *polyol esters* (TRR = 230 °C or 290 °C with additives) currently used in military jets have good oxidative stability but suffer from viscosity problems; *C-ethers* (TRR > 315 °C) used in supersonic military aircraft have better stability than polyol esters but corrode copper; *polyphenyl ethers* (PPE) have TRR as high as C-ethers, are more stable, and are also used in supersonic aircraft; and *polyperfluoroalkylethers*, used as vacuum pump fluid in corrosive environments, have a TRR from 350 °C to 470 °C, do not leave deposits but react with water to form acids and produce volatile decomposition products.[42] Figure 14 shows lifetime-temperature curves for various liquid lubricants.

Solid lubricants may extend temperatures even further. The solid can be produced by *in situ* decomposition of the primary liquid lubricant or it can be a solid suspension in the oil. Impregnating solid lubricant micropockets in the sliding surface can also form wear resistant materials. Normal wear exposes fresh lubricant. A recently developed ceramic oil-less roller bearing using dry carbon lubrication had half of the friction loss of conventional oil-lubed journal bearings. The wear should be an order of magnitude better as well. Above 540 °C even graphite oxidizes. Some inorganic fluorides and oxides give low friction at higher temperatures, but they have high friction at low temperatures.

Ceramics are highly wear resistant but have high friction when sliding on themselves. Ideally, a ceramic should slide against an alloy or cermet that forms a low friction oxide during wear. The ceramic is the part that requires the greatest wear resistance (e.g., cylinder liner) and the other is the alloy (e.g., piston rings). Reducing carbon monoxide or hydrocarbons on catalytic surfaces can form carbon or graphite coatings.

The primary difficulty with solid lubricants is devising a supply system that replenishes the proper amount across the full range of operating conditions. Suggestions include stick or powder feed, gaseous or liquid suspension feed, incorporated micropockets or retainers, and gaseous surface reactions. Integration into an engine will take at least 5 years. Solid lubricants have friction coefficients of 0.05 to 0.2. Compare that to liquids: 0.04 (when film thickness is greater than surface roughness) and 0.08 (film is same as roughness). Consequently, solids may only function in multilubricant systems (e.g., Figure 15) when liquids cease to function.

Solid lubricants being considered include *phosphate ester decomposition*

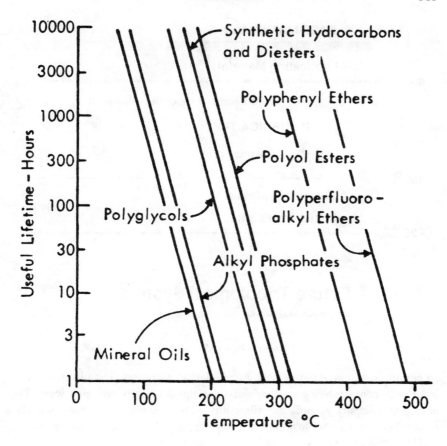

Liquid Lubricant Lifetimes with Temperature

Courtesy of Society of Automotive Engineers, Sutor & Bryzik, Ref. 42.

Figure 14

products (TRR = 317 °C), *PPE vapor film* (TRR = 375 °C), *graphite and graphite fluoride suspensions* (TRR = 480 °C with good life at 540 °C), *metal ceramic paired surfaces* such as those used in Stirling engines (TRR up to 760 °C), and *catalytic surfaces* which offer a coefficient of friction between 0.15 and 0.4.[43]

Another oil-less alternative, *gas lubrication* provides the lowest friction at the highest temperature. Combustion gases provide the film that separates the piston from the cylinder. They decompose at extremely high temperatures and

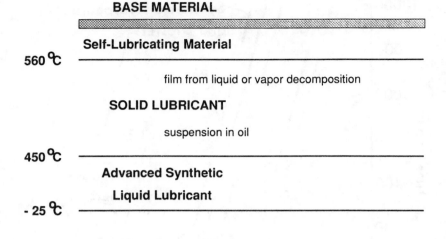

A Future Tribological System

Courtesy of Society of Automotive Engineers, Sutor & Bryzik, Ref. 42.

Figure 15

so permit operation at the highest in-cylinder temperatures. The piston and cylinder only contact during starting and stopping, drastically reducing wear. The ultra low viscosity accords gas films much less friction than liquids but also besets them with a very low load carrying capacity. Challenges to surmount:

(1) Eliminate piston side thrusts (rethink crank shaft).

(2) Ultra high precision manufacture.

(3) Eliminate thermal deviations from concentricity in spite of the temperature gradients.

(4) Maintain extreme clearances, 0.0005 in. (currently 0.005 in.).

(5) Seal upper cylinder from crank case to maintain gas pressure and keep out drivetrain lubricant.

Gas lubrication, a difficult high risk technology that could reduce friction losses by 60%, will probably not be practical in the next 10 years.[44]

3b. Rolling resistance

Rolling resistance can consume as much as 40% of the force generated by a long-haul truck engine. The replacement of over 70% of the bias ply tires on the road with radial ply tires considerably lowered fuel consumption. A radial ply can have over 30% less rolling resistance, hence fuel savings of 15%, highway, and 10% in general. In fact, the 4th generation automobile radials of the 1980s have 45% of the rolling resistance of the 1970 bias ply.

Future replacement of dual radials with new super single radial tires should provide another 8%–10% fuel savings. Other benefits include 10% more tread mileage, lighter weight, fewer flats, better brake cooling (more exposed drum surface), and lower initial cost. Designed for wider axles, the capital cost in wheels and adapters will hinder their adoption. It is difficult for a fleet to use more than one type of tire, so market penetration will be slow (5–10 years).

New low profile tubeless tires were first developed for automobiles. They provide 4% less fuel consumption, 25% more mileage, and lighter weight than conventional tubeless radials. Their lower profile lowers vehicle height, so possibly aerodynamic drag, but it may also present a problem matching trailers to loading docks. They cost 6%–12% more than standard radials, but are expected to replace them over the next 10 years. Molded polyurethane truck tires are being developed in Europe. They offer the advantages of 12% fuel savings, 20% less weight, 50% lower temperatures, and 50% longer tread life compared to the tires in use now. Furthermore, the raw materials can be recycled. They are expected to be available commercially in the next 5 years.

Rolling resistance is also very important to the railroads. Track lubrication has achieved fuel savings of 25%–35% in commercial operation. Several references contain more information on this and related developments.[45]

4. Aerodynamics

Raising efficiency with better aerodynamics is the most desirable option from an environmental perspective. American manufacturers lowered automobile drag coefficients from ≈ 0.55 to ≈ 0.35 mostly by removing sharp edges and making the vehicles smoother with more gradual changes in cross section. Figure 16 translates drag reduction into increased MPG for a 1983 model car. This shows that aerodynamic improvements have a greater effect at higher speeds so the fuel savings is very sensitive to driving pattern. Improvements past 0.3 will require reengineering the vehicle. In fact, solar vehicles, with drag coefficients of 0.125 (better than some jet fighters), are pushing the limits of land vehicle aerodynamics. Production automobiles may always operate in a turbulent flow regime because they have a large rear cross section with respect to body length (unlike aircraft). Ford and General Motors have the lowest drag prototypes (0.137 and 0.166, respectively), yet the drag coefficients of their production vehicles are higher than those of the imports.

By reducing drag the engine will operate at a lower fraction of total power, and so be less efficient. To preserve the aerodynamic savings a designer must modify gear ratios in the transmission, and possibly the axle-ratio. Intelligent matching of the engine and drivetrain is very important to keep the engine at a favorable operating point throughout the full range of vehicle speeds. A mediocre but well matched engine-drivetrain pair may outperform an exceptional engine and exceptional drivetrain if they match poorly.

Performance tradeoffs may also limit aerodynamic advances. Low drag enhances acceleration but sufficiently low drag may compromise handling. For example, a bottom pan may drastically lower the pressure drag of an automobile but make it much more sensitive to wind gusts and truck wakes.[46]

The forward facing surfaces of a *long-haul truck* account for 60% of the total air resistance. As with automobiles, designers have rounded corners, angled windshields and, eliminated or diminished protrusions. Little more can

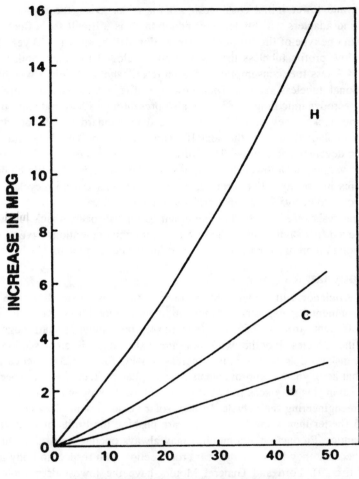

Effect of Driving Cycle on
Aerodynamic Drag Reduction

Representative auto with mpg of:
26 (Urban), 31 (Combined) and 41 (Highway).

Courtesy of Sovran & Society of Automotive Engineers, Ref. 46.

Figure 16

be done to further reduce drag without making major changes to the tractor exterior. Add-on devices are the most significant development in truck aerodynamics. They typically reduce the downflow between the cab and the trailer. The wind deflector is a curved plate on the cab roof that inclines back toward the trailer which reduces drag as much as 15%. The cab-mounted roof fairing is similar but has the sides closed in. Wind tunnel tests show that it can reduce drag by 24% and is also less sensitive to cross winds. Other add-ons include the trailer forebody fairing (streamlines front of the trailer), cab side extenders (plates that extend toward the trailer), and air dam bumpers that direct air around the cab to reduce underbody drag. Tractor manufacturers offer all of these in add-on packages that realize actual fuel savings around 10%. Because of their low cost ($500–$700) they will provide most of the drag reduction for trucks in the near future.

Future add-on devices to expect are tractor and trailer skirts, gap seals, chassis filler panels, and boat tails. The skirts are under test but are not likely to be marketed until maintenance and brake cooling problems are overcome. Furthermore, the wheel assemblies can generate a potentially dangerous spray and they may trap passenger vehicles in an accident. The boat tail extends behind the trailer to reduce turbulence. Recent tests credit them with 10% less drag. These devices will probably reach the market in the next 5 to 10 years and lower fuel consumption by another 17%. Fruehauf manufactured a prototype tractor-trailer unit that combined many techniques to reduce drag 57%, but they do not plan to make it commercially available any time soon. The fuel savings listed in Figure 17 show that in the near term, drag reducing additions will increase efficiency the most for long-haul trucks. In 5–10 years aerodynamic improvements, though still significant, will be overtaken by engine improvements.[47]

Much has been learned about the aerodynamics of railroad trains as well. Space does not permit discussion here other than to note that aerodynamic improvements provide the greatest fuel savings for trains on flat routes when compared with rolling resistance and weight. The greatest fuel savings may not produce the greatest cost savings though, the keyword for implementation. In fact, reducing the tare weight of the train seems to have the greatest cost incentives in more cases.[48]

Turbulence in the boundary layer causes skin friction or viscous drag. It accounts for as much as 50% of the total cruise drag of most subsonic aircraft. Very important developments in Natural Laminar Flow, Hybrid Flow Control, Turbulent Drag Reduction, and Wing Design over the past decade will reduce drag on future aircraft by 50% and result in fuel savings of 25% to 40%.[49]

5. Operational factors

According to a recent study by Argonne National Laboratory, idling trucks may waste up to 400 million gallons of fuel a year. On the average, long distance truckers idle their engines 3 hours a day to keep the engine and fuel warm in the winter and to heat or cool their cabs while they rest. Devices already on the market can perform all the functions of idling at much lower cost. For example, fuel savings can pay for cab and engine-block heaters in less than a year.

Efficiency Improvements for Long-Haul Trucks

IMPROVEMENT: Engine		Aerodynamic	Rolling Resistance	
TIME FRAME:			radials	15%
Past Decade				
Short Term Future (0-5 years)				
turbocharger improvements	1%	add-on devices: 10-20%	low profile	3%
low flow charge cooling	2%	wind deflectors 6-9%		
electronic controls	2%	roof fairings		
reduced parasitics	4%	cab extenders		
design improvements	3%	air dam bumper		
	12%			
Medium Term Future (5-10 years)				
variable geom. turbo.	2%	tractor trailer skirts		
advanced intercoolers	4%	& gap seals 14-17%	super single 5-8%	
electronic controls	3%	boat tail 6%	molded poly. 10%	
reduced parasitics	6%			
design improvements	5%			
turbocompounding/ceramics	10%			
	30%			
Long Term Future (10-20 years)				
		Integrated design 20-27%		
		(incl.: aero. cab, skirts,		
		roof fairing, boat tail,		
Sources: Ref. 47.		*auto. gap seal, front air dam)*		

Figure 17

Traffic management also saves substantial amounts of fuel. A California program that retimed more than 3000 signals saved an estimated $73 million in fuel. The total savings was more than 50 times the initial investment. Developing sensors such as electromagnetic detection loops, doppler radars, pressure sensors, visual/infrared imagers, photoelectric detectors, and microwave radiometers advance traffic management.[50]

Operational factors are vital to air and rail transport as well. Space permits only the mention of a few examples. On-board energy management can reduce aircraft fuel consumption 5% and energy efficient routing another 3% to 8%. Fly-by-wire controls can reduce aircraft weight by 800 lbs. Furthermore, simulation tools now exist to study how fuel consumption is affected by airport design, airspace routing, air carrier scheduling, traffic demands, fleet mix, air traffic control rules, controller decision logic, and aircraft performance. The High Productivity Integral Train (HPIT) concept will reduce fuel consumption as much as 40% for some lines. Further reductions will come from the Advanced Train Control System (ATCS) and energy management programs developed for electric mass transit systems.

A great many other operational technologies fall beyond the scope of this brief overview. The rapidly advancing areas of communications technology and machine intelligence deserve mention. The design of facilities such as highways, airports, ports, pipelines, rail, and bus terminals also affect transportation energy consumption. Environmental considerations such as pollution control, road and track conditions, and weather do so as well. Technologies and policies

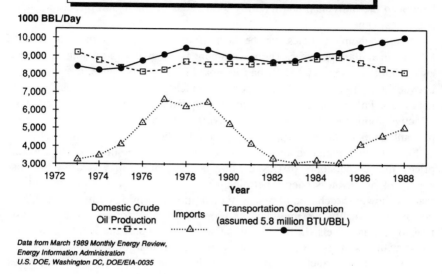

Production / Consumption Gap

1000 BBL/Day

Domestic Crude Oil Production · · · □ · · · | Imports · · · · △ · · · · | Transportation Consumption (assumed 5.8 million BTU/BBL) ———●———

Data from March 1989 Monthly Energy Review,
Energy Information Administration
U.S. DOE, Washington DC, DOE/EIA-0035

Figure 18

concerning these areas will increase in importance as we become more sophisticated in our approach to the energy problem.[51]

6. Summary and conclusions

Transportation efficiency increased substantially in the last decade. Automobile fuel efficiency almost doubled from 1975 to 1986 (16 to 28 mpg) and light truck economy increased as well (14 to 22 mpg), saving over $90 billion. This technology held gasoline use to a 0.3% rise while highway travel increased 31%. The technology for further great improvements exists or is on the horizon. In the 1990s air transport could use 60% less energy, railways 40% less. Automobiles that achieve 60 to 100 mpg exist but await more favorable markets. In spite of our technical advances Figure 1 shows that transportation consumption has increased, both as a portion of our petroleum use and in total.

Technical efforts alleviate, but do not solve, our fundamental transportation energy problem: our exclusive dependence on petroleum. Over 97% of our transportation requires it. Even if the industrial, residential, and commercial sectors became petroleum independent, we would still have to import oil for transportation (Figure 18).

Recognizing the problem is easy; solving it is not. Free market consumer-industry interaction will not protect our long-term interest. The time constant of petroleum price swings is too short. Consumers also discount the value of future energy savings by 20%–30%, weakening the incentive for industry to concentrate on energy efficiency. In addition, as vehicle fuel economy increases, fuel costs become a smaller fraction of the total operating cost. At current gasoline prices a 10 mpg improvement to a 30 mpg auto saves the typical driver less than $100 a year. Policies to resolve the conflict between the short-term benefit to the individual and the long-term needs of the nation present a formidable challenge.

Fuel economy policies, or the lack of them, have an impact of tens of billions of dollars per year. The government is not likely to be able to operate an automobile company better than its board of directors hence, a compromise between the "free market" approach and extensive government regulation is necessary. Policies must be evolved in cooperation with industry to be most effective. Both the industry and the government share responsibility for their past antagonistic relationship, and they should both strive to leave that behind.

Most federal planning and action has been reactive. In times of low oil prices, the "cornerstone" of U.S. contingency response policy is a few months' supply of oil in the strategic petroleum reserve. We need preventive policies that will lower the chance of a supply disruption and mitigate its effects if one occurs. Specifically, we must lay the groundwork now for an optimum transition to alternate fuels. In the meantime we must support the technologies and conservation strategies that will give us the greatest margin.[52]

The executive branch must develop a policy setting forth our goals and ideas for achieving them. Consumer incentives that have been proposed include higher gas taxes, education, and ownership taxes (e.g., gas guzzler tax). Incentives for industry could include research and development support and higher CAFE (Corporate Average Fleet Economy) standards.

Higher gasoline taxes could stabilize energy costs (smoothing the deleterious effects of sudden price swings), encourage conservation, encourage and fund technology development, or fund remedies to social dislocations during the change. Current taxes on gasoline in the U.S. average about 29 cents per gallon, much lower than West European taxes that are measured in dollars. On the other hand, after some point such taxes have little impact on conservation and just become a burden on the lower income population. Ownership taxes discourage use of inefficient vehicles. Combined with an incentive to purchase highly efficient vehicles, they may be more effective at drawing advanced technology into the market.

The CAFE policy succeeded in raising fuel economy because it gave automakers clear long-term targets to achieve and accelerated the introduction of advanced technology into production vehicles. Figure 19 shows that since 1981 improvements in vehicle fuel efficiency have levelled off with the CAFE. Two-thirds of the vehicles in the U.S. are 1980 or older but newer vehicles drive the bulk of the mileage. Thus the replacement of all vehicles older than 1985 with 1985 models, would result in a fuel efficiency improvement of 16.5% for this sector. Surpassing this limit will require a jump in vehicle efficiencies or changes in use patterns. As a result some are calling on the government to raise the CAFE standards to 45 mpg for cars and 35 mpg for light trucks.[53]

The automobile is the key component in our petroleum dilemma. Increasing the mileage of the U.S. automobile fleet just 0.1 mpg would save 20,000 barrels of oil per day. That is equivalent to the estimated production of the contro-

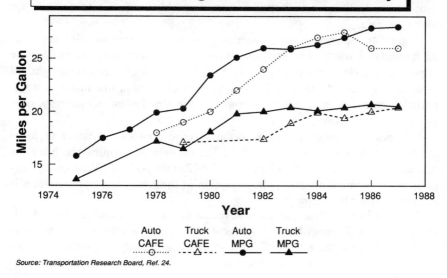

The CAFE and Light Vehicle Efficiency

Source: Transportation Research Board, Ref. 24.

Figure 19

versial Georges Bank fishing grounds off the East Coast. Therefore, the light duty vehicle fuel economy standard is one of the Nation's most important energy policies. We can improve the CAFE standard structure. Applying a fuel economy standard by vehicle class could remove the disadvantage to the domestic automakers and force new technology into all vehicle classes. Increasing the standards in increments that correspond to model development cycles rather than yearly and allowing the industry flexibility in how they meet the standards could also be beneficial. Light trucks must be included as they are the fastest growing segment of the U. S. auto market.

The government has a role identifying and providing long-term support to the most critical areas of research. It could also facilitate the transfer of innovations from research firms to the automakers. Currently, many companies working on advanced automotive technologies are forced overseas to develop and market innovations originally sponsored by the U.S. government. The minimal study of our past efforts will hinder current policy development. We should know what policies failed and succeeded and objectively why.

Another area where the government may have a role is solving the "chicken and egg" dilemma posed by the automotive and petroleum industries. The major automakers have the technology for alternative fueled vehicles but do not want to risk the investment to produce these vehicles without being certain that customers will have sufficient access to the alternative fuel. The petrochemical industry does not want to invest in the distribution network for an alternate fuel without being certain there will be a market for it. The government can assume some of the risk to break this stalemate. One option is to incrementally subsidize the operation of specific vehicle fleets which have central refueling. Another might be fuel price subsidies, but care would have to be taken to prevent narrow political constituencies from squelching superior alternatives for their own gain.

A good transportation policy needs to consider the transportation sector as an interrelated whole. For example, does it make more sense to build more airports in the Northeast or to slow air traffic growth with a magnetic light rail system, or expanded highways? The impact of changes in one mode of transportation on the others needs to be considered in terms of energy security, cost, safety, and other factors.

It is not my intent here to advocate specific policies but rather to focus attention on the transportation problem. The solution will require much serious research and the careful development of deliberate policies. Most energy experts expect some kind of supply disruption in the 1990s and transportation is our most vulnerable sector.[54]

According to the Transportation Research Board we must objectively determine the research needs for an intelligent transition to alternate fuels. We also need to understand the effects of our past efforts, especially conservation strategies. No matter what the source there are environmental and monetary costs for every energy use.

Much of the savings we are realizing now are due to research programs initiated in the 1970s. Since that time the funding for those programs has declined or ended. Regardless, the choices we make now in transportation research and policy will determine our vulnerability to changes in the petroleum supply in the next decade. Too much is at stake to continue in our false sense of security.

Bibliography

Aircraft

Aerospace America printed a very good series of articles on propulsion in September, 1987, and on drag reduction technologies in January, 1988. They provide good technical information but are written for the lay reader.

Bob Jeal, "Moving Towards the Non-Metallic Aero Engine," The Rolls-Royce Magazine **36**, 23–27 (March, 1988). This article is a good comprehensive review of aircraft materials developments.

Auto/truck

Deborah Lynn Bleviss, *The New Oil Crisis and Fuel Economy Technologies, Preparing the Light Transportation Industry for the 1990s* (Quorum Books, Westport CT, 1988). This is a recommended reference on the state of the automobile industry. Bleviss explains the automotive technologies that impact energy consumption, then delineates the technical state of the art. In greater depth, the author explains the social and economic reasons for the status quo and provides a detailed discussion of policy options for corporations as well as governments, both foreign and domestic.

Transportation Research Board, *Research Priorities in Transportation and Energy* (Transportation Research Circular N323, Sept., 1987). Another well researched and detailed discussion of policy considerations.

A. Cubukgil, R. Ridout, and D. Ross, "Recent Developments in Truck Technology: Potential for Improved Energy Efficiency in Long-haul Trucking," Transportation Forum, **2-3**, 53–61 (Dec., 1985). This is a good comprehensive review of energy technology related to long-haul trucks.

Energy Information Administration, *Residential Transportation Energy Consumption Survey: Consumption of Household Vehicles 1985*, DOE/EIA-0464(85) (U.S. DOE, Washington, DC, 1985). This study provides information on auto fleet composition and use.

J. Zucchetto, P. Myers, J. Johnson, and D. Miller, "An Assessment of the Performance and Requirements for 'Adiabatic' Engines," Science **240**, 1157–1162 (1988). A very good assessment of the state of low-heat-rejection engine development.

Magnetic levitation

L.R. Johnson, D.M. Rote, J.R. Hull, H.T. Coffey, J.G. Daley, and R.F. Giese, *Maglev Vehicles and Superconductor Technology: Integration of High-Speed Ground Transportation into the Air Travel System*, U.S. DOE Report ANL/CNSV-67 (Argonne National Laboratory, Argonne, IL, April, 1989). This report explains the various forms of maglev and examines the impact of high temperature superconductors. It is also a well-written source of information on energy use, pollution, costs and market economics for maglev, autos, and aircraft. It includes a comprehensive history and status report of domestic and foreign maglev projects.

Railroad

Railroad Energy Technology Conference II, May 20–22, 1986, Atlanta, Georgia (American Association of Railroads, Chicago, 1987). An excellent collection of papers on the state of energy research for the railroads. Topics include: alternative fuels, engines and components, operating practices, train resistance (track/vehicle interaction and aerodynamics), energy economics, and railroad electrification.

References and notes

Abbreviations used in the references:

AAR:	American Association of Railroads, Chicago, IL.
ACS:	American Chemical Society, Washington, DC.
AIAA:	American Institute of Aeronautics and Astronautics, New York.
AEE:	American Society for Engineering Education.
ASME:	American Society of Mechanical Engineers.
DOE:	U.S. Department of Energy, Washington, DC.
DOT:	U.S. Department of Transportation, Washington, DC.
EIA:	Energy Information Administration, DOE, Washington DC.
FAA:	Federal Aviation Administration, Washington, DC.
IECEC:	Intersociety Energy Conversion Engineering Conference (21st).
IEEE:	Institute of Electrical and Electronics Engineers, Piscataway, NJ.

NASA: National Aeronautics and Space Administration, Washington, DC.
NTIS: National Technical Information Service, Springfield, VA.
SAE: Society of Automotive Engineers, Warrendale, PA.
TRB: Transportation Research Board, Washington, DC.
UMTA: Urban Mass Transit Administration, DOT, Washington, DC.

1. Energy Information Administration, *Monthly Energy Review, March, 1989*, DOE/EIA-0035 (U.S. DOE, Washington, DC, 1989); Robert H. Romer, *Energy Facts and Figures* (Spring Street Press, Amherst, MA, 1985).
2. E. Richard Booser, *CRC Handbook of Lubrication (Theory and Practice of Tribology) V.I. Application and Maintenance* (CRC Press, Inc., Boca Raton, FL, 1983), p. 72.
3. This section based on information in Global Competitiveness Council, *Combustion Research in Japan V. 3: Spark Ignition Engine Research — Part II* (SAE, SP-709, 1987) particularly pages 9–13, 37–39, 51, 75, 82, and 116–117.
4. Deborah Lynn Bleviss, *The New Oil Crisis and Fuel Economy Technologies, Preparing the Light Transportation Industry for the 1990s* (Quorum Books, Westport, CT, 1988), p. 24.
5. This section based on information from Bleviss particularly pages 28, 33, and 40; and the Global Competitiveness Council, V3.
6. This section based on information from the Global Competitiveness Council, V3; and Tom Wilkinson, "Power to the Revolution," Autoweek (Mar. 12, 1990.)
7. For detailed information see Global Competitiveness Council, V3.
8. Information above based on Booser; and A. Cubukgil, R. Ridout, and D. Ross, "Recent Developments in Truck Technology: Potential for Improved Energy Efficiency in Long-haul Trucking," Transportation Forum 2-3, 53–55 (Dec., 1985).
9. See Bleviss, p. 20, and Global Competitiveness Council, *Combustion Research in Japan Volume 4: Diesel Engine Research — Part II* (SAE, SP-711, 1987).
10. See Global Competitiveness Council, V4.
11. Global Competitiveness Council, V3, pp. 41 and 93.
12. See Cubukgil *et al.*, pp. 53–54, Bleviss, p. 26, and Global Competitiveness Council, V4, p. 62.
13. J. Zucchetto, P. Myers, J. Johnson, and D. Miller, "An Assessment of the Performance and Requirements for 'Adiabatic' Engines," Science 240, 1157–1162 (1988).
14. Global Competitiveness Council, V3, pp. 11, 54, and 105, and V4, pp. 23 and 211; Cubukgil *et al.*, p. 53; and Bleviss, p. 24.
15. Charles A. Amann, *Selecting Components for a Compound Low-Heat-Rejection Diesel for Light-Duty Use*, SAE 870025, SP-700 (SAE, 1987), pp. 71–86; Osami Kamigaito, *Ceramic Components for Engines in Japan*, SAE 870019, SP-700 (SAE, 1987), pp. 15–18; Global Competitiveness Council, V3, pp. 53, 54, and 68; and V4, pp. 15, 52, 54, 69, 74, and 100.
16. E.F. Fort, P.N. Blumberg, and J.C. Wood, *Evaluation of Positive-Displacement Supercharging and Waste Heat Recovery for an LHR Diesel*, SAE 870026, SP-700 (SAE, 1987), pp. 91–98; Amann, pp. 71–86; Cubukgil *et al.*, p. 55; Global Competitiveness Council, V4, pp. 16, 37, 39, and 83; Zucchetto *et al.*, p. 1160.
17. Yoshito Watanabe, Hideo Ishikawa, and Makoto Miyahara, *An Application Study of Evaporative Cooling to Heavy Duty Diesel Engines*, SAE 870023, SP-700 (SAE, 1987), pp. 49–56; and Klaus Elsbett, Ludwig Elsbett, Gunter Elsbett, and Michael Behrens, *Elsbett's Reduced Cooling for D. I. Diesel Engines Without Water or Air*, SAE 870027, SP-700 (SAE, 1987), pp. 101–107.
18. Melvin E. Woods, Paul Glance, and Ernest Schwarz, *In-Cylinder Components for High Temperature Diesel*, SAE 870159, SP-700 (SAE, 1987), pp. 185–195; Roy Kamo, Walter Bryzik, and Paul Glance, *Adiabatic Engine Trends — Worldwide*, SAE 870018, SP-700 (SAE, 1987), pp. 1–13; Elsbett *et al.*, pp. 101–107.
19. Alexandros C. Alkidas, *Experiments with an Uncooled Single-Cylinder Open-Chamber Diesel*, SAE 870020, SP-700 (SAE, 1987), pp. 19–29; J.A. Gatowski, J.D. Jones, and D.C. Siegla, *Evaluation of the Fuel Economy Potential of the Low-Heat-Rejection Diesel Engine for Passenger-Car Application*, SAE 870024, SP-700 (SAE, 1987), pp. 57–69; D.A. Parker and G.M. Donnison, *The Development of An Air Gap Insulated Piston*, SAE 870652, SP-700 (SAE, 1987), pp. 265–273; Amann, pp. 71–86; Kamigaito, pp. 15–18; Kamo, pp. 1–13; Woods, pp. 185–195; and Zucchetto *et al.*, pp. 1157–1162.

20. Bleviss, pp. 42–52.

21. Giuseppe Cotignoli and Enrico Ferrati, *A Compact Microcomputer Based Distributorless Ignition System*, SAE 870082, SP-705 (SAE, 1987), p. 35; M.W. Thomson, A.R. Frelund, M. Pallas, and K.D. Miller, *General Motors 2.3L Quad 4 Engine*, SAE 870353 (SAE, 1987); Bleviss, pp. 24 and 31; and Global Competitiveness Council, V3, pp. 51 and 72, and V4, pp. 18, 70, and 107.

22. A.C. Wakeman *et al.*, *Adaptive Engine Controls for Fuel Consumption and Emissions Reduction*, SAE 870083, SP-705 (SAE, 1987), pp. 43–50; Shinichi Inagawa, Junzoh Kuroyanagi, Toshihiro Hattori, and Hitosh Kasai, *Cruise Control with Automatic Gear Selection for Light-Duty Truck*, SAE 870084, SP-705 (SAE, 1987), p. 51; Dina Van Pelt, "Fluids May Improve Automotive Efficiency," Insight, 55 (June 20, 1988); Richard Nass, "Microprocessor-based antinoise system eliminates sound," Electronic Design, 30 (Aug. 25, 1988); Robert Pool, "The Fluids With a Case of Split Personality," Science, 1180–1181 (Mar. 9, 1990); and Global Competitiveness Council, V3, pp. 18, 50, 52, 61, 65, 66, 70, and 120.

23. A.C Chachich, *Energy and Transportation*, 1989, unpublished; Yutaka Ogawa, Takayuki Ogasawara, Minoru Machida, Yooji Tsukawaki, and Kaneyoshi Shimono, *Complete Ceramic Swirl Chamber for Passenger Car Diesel Engine*, SAE 870650, SP-700 (SAE, 1987), pp. 243–250; W.R. Wade, P.H. Havstad, V.D. Rao, M.G. Aimone, and C.M. Jones, *A Structural Ceramic Diesel Engine — The Critical Elements*, SAE 870651, SP-700 (SAE, 1987), pp. 251–264; Bleviss, pp. 52 and 57–60; Global Competitiveness Council, V3, p. 47; Kamigaito, pp. 15–18; Kamo, pp. 1–13; and Zucchetto *et al.*, pp. 1158–1161.

24. Cummins Engine Co., untitled, Richard D. Wilson (EPA), untitled, and Howard B. Padgham (Chrysler), "Statement Regarding the Federal Government's Role in Alternate Fuel Development for Transportation," Testimony before the Energy and Power Subcommittee, Energy and Commerce Committee, U.S. House of Representatives, June 17, 1987; Transportation Research Board, *Research Priorities in Transportation and Energy*, Transportation Research Circular N323, Sept. 1987, p. 20 and 22; Bleviss, p. 204; and Global Competitivenss Council, V4, pp. 49, 53, 80, 224, 232, and 240.

25. A.A. Pitrolo, "Coal-Fueled Locomotives," *Railroad Energy Technology Conference II, May 20–22, 1986, Atlanta, Georgia* (AAR, 1987), pp. 381–404; and Transportation Research Board, p. 22.

26. S.L. McDonald (ARCO), "The Federal Government's Role in Alternative Fuel Development for Transportation," Charles R. Imbrecht (Cal. Energy Comm.), untitled, and Clarence M. Ditlow III, untitled, Testimony before the Energy and Power Subcommittee, Energy and Commerce Committee, U.S. House of Representatives, July 9, 1987; Cummins Engine and Howard B. Padgham, Testimony on June 17, 1987; and Global Competitiveness Council, V3, pp. 32, 58, 61, 84, and 112 and V4, pp. 78, 175, 178, 183, and 184.

27. Philip W. Haseltine, Testimony before the Energy and Power Subcommittee, Energy and Commerce Committee, U.S. House of Representatives, June 17, 1987; A. Barry Carr (Congressional Research Service), "Possible Effects of Legislation Mandating Use of Ethanol in Gasoline," Testimony before the Energy and Power Subcommittee, Energy and Commerce Committee, U.S. House of Representatives, June 24, 1987; J.L. Frank, (Marathon Petroleum), untitled; Vic Rasheed (Service Station Dealers of America), untitled; McDonald; and Ditlow, Testimony before the Energy and Power Subcommittee, Energy and Commerce Committee, U.S. House of Representatives, July 9, 1987; Richard Lipkin, "Diesels Tanked Up on Vegetable Oil," Insight, 50 (July 11, 1988).

28. Global Competitiveness Council, V4, pp. 184 and 194.

29. Testimony before the Energy and Power Subcommittee cited above; James D. Sawyer, "Making a Meth of the Way We Drive," Autoweek, 27–28 (April 23, 1990); Bleviss, p. 197; and Global Competitiveness Council, V3, pp. 131 and 134.

30. "Diesel prescriptions: Eat some veggies...," Science News 134, 123 (Aug. 20, 1988); Lipkin, p. 50; and Global Competitiveness Council, V3, pp. 92, 139, 143, and 166 and V4, p. 223.

31. R.J. Hill (Rolls-Royce), private communication; Bob Jeal, "Moving Towards the Non-Metallic Aero Engine," The Rolls-Royce Magazine, No. 36, 23 (Mar. 1988); Patrick J. Johnston and Gary T. Chapman, "Fitting Aerodynamics and Propulsion into the Puzzle," Aerospace America, 32–37, 42 (September 1987); Griffin Y. Anderson, Daniel P. Bencze, and Bobby W. Sanders, "Ground Tests Confirm the Promise of Hypersonic Propulsion," Aerospace America, 38–42 (September 1987); Richard DeMeis, "An Orient Express to Capture the Market,"

Aerospace America, 44–47 (September 1987); A.K. Gupta, "Propellents and Combustion," Aerospace America, 50 (Dec. 1987); Ciepluch; NASA, 1.18, 23, and 24; Pickerell, p. 154; Reynolds *et al.*; and Van Nostrand's Scientific Encyclopedia, pp. 1145–1146.

32. *Van Nostrand's Scientific Encyclopedia*, 5th ed. (Van Nostrand Reinhold, New York, 1976), p. 1138; Gordon C. Oates, Aerothermodynamics of Gas Turbine and Rocket Propulsion (AIAA, 1988); Cubukgil *et al.*, p. 53; and Pitrolo, pp. 381–404.

33. Murray A. Booth, "Advanced Turboprop Transport Development. A Perspective," 13th Congress of the International Council of Aeronautical Sciences, Aug. 22–27, 1982; Carl C. Ciepluch and David E. Gray "Results of NASA's Energy Efficient Engine Program," Jet Propulsion 3, (Nov–Dec 1987); Gerald L. Brines, "Design Approach for an Optimum Prop-Fan Propulsion System," *1985 Beijing International Gas Turbine Symposium and Exposition, Beijing, People's Republic of China—Sept. 1–7, 1985*, ASME, 85-IGT-57 (ASME, 1985); C.N. Reynolds, A.S. Novick, and R.E. Riffel, "Propfan Propulsion System Development," 8th International Symposium on Air Breathing Engines, June 15–19, 1987, Cincinnati, Ohio; R.E. Owens, W.W. Ferguson, and D.M. Mabee, "Counter-Rotation Prop-Fan Optimization," AIAA/ASME/SAE/ASEE 22nd Joint Propulsion Conference, June 16–18, 1986, Huntsville, Alabama; NASA, *AEROPROPULSION '87 Session 1—Aeropropulsion Materials Research*, NASA Conference Publication 10003, Preprint for a conference at NASA Lewis Research Center, Cleveland, Ohio, Nov. 17–19, 1987; D.J. Pickerell, "Future Trends in Aviation Propulsion," The 1987 Australian Aviation Symposium, Canberra, 18–20 Nov., 1987, pp. 153–157; Douglas L. Bowers, "Aerospace Highlights 1987, Air Breathing Propulsion," Aerospace America, pp. 48–49 (Dec. 1987); James. J. Haggerty, *Spinoff 1987* (NASA Office of Commercial Programs Technology Utilization Division, U.S. Government Printing Office, Washington, DC, 1987); and Chachich.

34. Mark W. Zemansky, *Heat and Thermodynamics*, 5th ed. (McGraw-Hill, New York, 1968), pp. 169, 170; Mechanical Technology Incorporated, Stirling Engine Systems Division, *The Stirling Engine Mod II Design Report*, (NASA-Lewis Research Center & U.S. DOE, Washington, DC, Oct., 1986); Noel P. Nightingale, Congressional testimony before the committee on Veteran's Affairs, HUD, and Independent Agencies, Committee on Appropriations, United States Senate, May 18, 1989 and private communications 6/19 and 6/26/89; Haggerty, pp. 60–61.

35. Zemansky, p. 172; Cubukgil *et al.*, p. 55; Zucchetto *et al.*, p. 1160.

36. H.S. Murray, J.R. Huff, J.F. Roach, and G.R. Thayer, *DOT Fuel-Cell-Powered Bus Feasibility Study Final Report—December 1986*, LA-10933-MS, UC-96 (NTIS, June, 1987); Philip H. Abelson, "Need for Long-Range Energy Policies," Science 240, 1121 (1988); R.L. Rentz, G.L. Hagey, and R.S. Kirk, "Fuel Cells as a Long Range Highway Vehicle Propulsion Alternative," IECEC 869239, *21st Intersociety Energy Conversion Engineering Conference, San Diego, California, Aug. 25–29, 1986*, Vol. 2 (ACS, 1986), pp. 1081–1087; R.E. Parsons, "Alternative Technology for Motive Power," *Railroad Energy Technology Conference II, May 20–22, 1986, Atlanta, Georgia* (AAR, 1987), p. 378; James R. Huff, Nicholas Vanderborgh, J. Fred Roach, and Hugh S. Murray "Fuel Cell Propulsion Systems for Highway Vehicles," submitted to the 14th Energy Technology Conference, Washington, DC, April 14–16, 1987 (proceedings to be published by Government Institutes, Inc., Rockville, MD).

37. S.M. Rohde and T.R. Weber, *On the Control of Vehicles with Multiple Power Sources* (General Motors Research Publication GMR-4749, Warren, MI, Mar. 18, 1985); Howard G. Wilson, Paul B. MacCready, and Chester R. Kyle, "Lessons of Sunraycer," Scientific American, 90–97 (March 1989); Bleviss, pp. 200–203; and Abelson, p. 1121.

38. Robert L. Hodgson and Henry Oman, "Highlights from the First International Workshop on Battery Testing," IECEC 869233, *IECEC, San Diego, Cal., Aug. 25–29, 1986*, Vol. 2 (ACS, 1986), pp. 1062–1065; Wilfried Fischer, Herman Birnbreier, and Gary N. Benninger, "Performance Characteristics of Sodium/Sulphur Batteries," IECEC 869229, *IECEC, San Diego, Cal., Aug. 25–29, 1986*, Vol. 2 (ACS, 1986), pp. 1045–1046; Chachich; Patil and Walsh, "Battery Requirements for Urban Electric Vans, IECEC 869228, *IECEC, San Diego, Cal., Aug. 25–29, 1986*, Vol. 2 (ACS, 1986); and Tom Lankard and J.P. Vettraino, "Even current technology makes significant Impact," Autoweek, 26 (April 23, 1990).

39. "Racing to a place in the sun," Science News **134**, 30 (July 9, 1987); "Sun-Powered Car Sets Speed Record," Insight, 53 (Aug. 15, 1988); P.L. Bustamante, "Solar 'Henry Ford' Wins Cup," The Arlington Advocate, Arlington, MA **116**, 3A (Oct. 20, 1988); Milt Leonard, "Photovoltaic Solar Cell Sets Efficiency Record," Electronic Design, 38 (Sept. 8, 1988); Robert

Pool, "A Bright Spot on the Solar Scene," Science **241**, 901 (1988); Glenn Zorpette, "Photovoltaics: technical gains and an uncertain market," IEEE Spectrum, 42–43 (July 1989); Jayne O'Donnell, "Lawmakers peek at future through Earth Tech fair," Autoweek 22–23 (April 23, 1990); Bleviss, p. 202; and Wilson *et al.*

40. Henry Kolm, Testimony before the U.S. Senate Committee on Environment and Public Works, Subcommittee on Water, Transportation, Infrastructure, Feb. 25, 1988; Richard G. Gilliland, Donald D. Lyttle, and George W. Pearson, *Integrated Magnetic Propulsion and Suspension (IMPS) Final Report* (DOT, UMTA, Office of Technical Assistance, report UMTA-WA-06-0014-86-2, Washington, DC, Dec., 1986); Richard G. Gilliland and George W. Pearson, *Linear Synchronous Unipolar Motor (LSUM) Development Report* (DOT, UMTA, Office of Technical Assistance, report UMTA-WA-06-0014-86-1, Washington, DC, Oct., 1986); H.H. Kolm, R.D. Thornton, Y. Iwasa, and W.S. Brown, "The Magneplane System," Cryogenics (July 1975); Henry H. Kolm and Richard D. Thornton, "Electromagnetic Flight," Scientific American **229**, 17–25 (Oct. 1973); Henry H. Kolm and Richard D. Thornton, "The Magnaplane: Guided Electromagnetic Flight," *Proceedings of the 1972 Applied Superconductivity Conference, Annapolis, Maryland, May 1–3, 1972*, 72CH0682-5-TABSC (1972); L.R. Johnson, D.M. Rote, J.R. Hull, H.T. Coffey, J.G. Daley, and R.F. Giese, *Maglev Vehicles and Superconductor Technology: Integration of High-Speed Ground Transportation into the Air Travel System*, U.S. DOE Report ANL/CNSV-67 (Argonne National Laboratory, Argonne, IL, April, 1989); R. Uher, personal communication; and Gilliland, personal communication.

41. B.A. Ross, "Railroad Electrification—An Opportunity and a Challenge," *Railroad Energy Technology Conference II, May 20–22, 1986, Atlanta, Georgia* (AAR, 1987), pp. 405–412 and Parsons, pp. 368–369.

42. Paul Sutor and Walter Bryzik, *Tribological Systems for High Temperature Diesel Engines*, SAE 870157, SP-700 (SAE, 1987), pp. 165–177; Booser, pp. 33–34 and 563; and Kamo, pp. 1–13.

43. Peter Moorehouse and Michael P. Johnson, *Development of Tribological Surfaces and Insulating Coatings for Diesel Engines*, SAE 870161, SP-700 (SAE, 1987), pp. 211–221; Bruce G. Bunting and James L. Lauer, *The Lubrication of Metals and Ceramics by the Catalytic Formation of Carbon Films*, SAE 870022, SP-705 (SAE, 1987), pp. 41–48; Chachich; Global Competitiveness Council, V3, p. 112; Kamo, pp. 1–13; Sutor, p. 167; and Wade *et al.*, pp. 251–264.

44. Kamo, pp. 1–13; Moorehouse and Johnson, pp. 211–221; Sutor, pp. 165–177; and Zucchetto *et al.*, pp. 1157–1162.

45. J.B. Stauffer, "Reducing Train Rolling Resistance by On-Board Lubrication," pp. 209–217; A.J. Peters and K.C. Rownd, "Wheel-Rail Lubrication Research," pp. 175–198; R. Engdahl, R.L. Gielow and J.C. Paul, "Train Resistance—Aerodynamics—Intermodal Car Application," pp. 225–242; M. E. Smith, "Economics of Reducing Train Resistance," pp. 269–305; R.P. Reiff, "Using Locomotive-Mounted Track Lubricators," pp. 199–208; all articles in *Railroad Energy Technology Conf. II, May 20–22, 1986, Atlanta, Georgia* (AAR, 1987); Bleviss, p. 65; Booser, p. 60; Chachich; and Cubukgil *et al.*, p. 57.

46. Gino Sovran, "Tractive-Energy-Based Formulae for the Impact of Aerodynamics on Fuel Economy Over the EPA Driving Schedules," SAE International Congress & Exposition, Detroit, Michigan, Feb 28–Mar 4, 1983 (SAE Technical Paper 830304, 1983), pp. 6, 8 and personal communication; Bleviss, p. 63; and Wilson *et al.*, pp. 90–97.

47. I. Amato, "Big rigs ease down a long and windy road," Science News **135**, 134 (March 4, 1989); and Cubukgil *et al.*, p. 56.

48. Chachich; Engdahl *et al.*, pp. 225–267; and Smith, pp. 269–305.

49. Bruce J. Holmes, "Progress in Natural Laminar Flow Research," *AIAA/NASA General Aviation Technology Conf. July 10–12, 1984, Hampton, Virginia*, AIAA 84-222 CP (AIAA, 1984); Bruce J. Holmes, "NLF Technology is Ready to Go," Aerospace America, 16–20 (Jan. 1988); Richard D. Wagner, Dennis W. Bartlett, and Dal V. Maddalon, "Laminar Flow Control is Maturing," Aerospace America, 20–24 (Jan. 1988); Jerry N. Hefner and Richard D. Wagner, "Laminar Flow Research Applicable to Subsonic Aircraft," International Symposium on Aeronautical Science & Technology of Indonesia, Jakarta, Indonesia, June 24–27, 1986; Richard D. Wagner, Dal V. Maddalon, and Michael C. Fischer (NASA), "Technology Developments for Laminar Boundary Layer Control on Subsonic Transport Aircraft," Fluid Dynamics Panel

Symposium on Improvement of Aerodynamic Performance Through Boundary Layer Control and High Lift Systems, Brussels, Belgium, May 21–23, 1984; Jerry N. Hefner, "Dragging Down Fuel Costs," Aerospace America, 14–16 (Jan. 1988); Christopher Vaughan, "Saving Fuel in Flight," Science News **134**, 266–269 (October 22, 1988); John B. Anders, Michael J. Walsh, and Dennis M. Bushnell, "The Fix for Tough Spots, "Aerospace America, 24–27 (Jan. 1988); Stephen Budiansky, "How to Fly Like a Fish," U.S. News & World Report (June 1, 1987); C.P. van Dam, "Efficiency Characteristics of Crescent-shaped Wings and Caudal Fins," Nature **325**, 435–437 (Jan. 29, 1987); P.M.H.W. Vijgen, C.P. Van Dam, and B.J. Holmes, "Sheared Wing-Tip Aerodynamics: Wind-Tunnel and Computational Investigations of Induced-Drag Reduction," *5th Applied Aerodynamics Conference, Aug. 17–19, 1987, Monterey, California,* AIAA-87-2481 CP (AIAA, 1987); Art Scheutz, "Aerospace Highlights 1987, Aircraft Design," Aerospace America, 26–27 (Dec. 1987); Chachich; and Haggerty.

50. "...and stop that idling," Science News **134**, 123 (Aug. 20, 1988); Garceau, Le Dinh, Loloyan, and Terreault, "Microwave Vehicle Sensor," *37th IEEE Vehicular Technology Conference, Tampa, Florida, June 1–3, 1987* (IEEE Press, Piscataway, NJ, 1987), pp. 200–205; and Transportation Research Board, p. 19.

51. David B. Middleton, Dennis W. Bartlett, and Ray V. Hood, *Energy Efficient Transport Technology Program Summary and Bibliography,* NASA Reference Publication 1135 (NASA, Sept. 1985); David E. Winer, Manager, Energy Division, Office of Environment & Energy, Federal Aviation Administration, private communication; M.B. Hargrove, "Assessing Railroad Energy Economics," p. 335; S.B. Harvey, "High Productivity Integral Trains," pp. 339–352; and G.B. Mott, "Advanced Train Control Systems," pp. 353–363, all articles in *Railroad Energy Technology Conference II, May 20–22, 1986, Atlanta, Georgia* (AAR, 1987); Richard Uher, "Rail Traction Energy Management Model," *COMPRAIL '87, International Conference, July 7–9, 1987, Frankfurt, Germany* (Computational Mechanics Publications, Springer-Verlag, Berlin, 1987); and Chachich.

52. Data taken from Bleviss and the Transportation Research Board, pp. 4 and 7–11.

53. Janet Raloff, "Energy Efficiency: Less Means More, Fueling a Sustainable Future," Science News **133**, 296–298 (May 7, 1988); Energy Information Administration, *Residential Transportation Energy Consumption Survey: Consumption of Household Vehicles 1985,* DOE/EIA-0464(85) (U.S. DOE, Washington, DC, 1985), p. 16; Susan Dillingham, "Higher Gas Taxes, Fuel Standards Urged," Insight, **26** (Aug. 8, 1988); and Bleviss.

54. Data from Bleviss, Raloff, and Transportation Research Board.

55. Adapted with permission from Romer. The figure compares the relative efficiencies of the various modes. Note the drastic sensitivity of every mode to the number of passengers and to a lesser extent vehicle efficiency. Consequently, the values are approximate. Fuel energy content was taken from the Monthly Energy Review. Capacities and efficiencies from Romer, Johnson, and Chachich.

Agricultural uses of energy

John Dowling

1. Introduction

Although the exact number is difficult to quantify, getting food on the table accounts for roughly 16.5% of the annual use of energy in the United States. McFate's study[1] on energy use in the U.S. food system gives the following figures: the actual production of food, which entails preparing the soil, planting, fertilizing, and harvesting, uses about 3% of the total U.S. energy budget; processing requires 5.4%; transportation expends 0.5%; wholesale/retail employs 2.6%; and preparation utilizes the remaining 5%. Other studies provide similar numbers. Differences among studies arise from variations in the definitions of exactly what constitutes agriculture. For example, is the cultivation of forests properly considered a part of agriculture?[2] Figure 1 presents a schematic of the food production and processing cycle.

The percent of the population involved in agriculture presents a major difference between industrialized nations and developing nations. Industrialized nations have typically replaced agricultural labor with cheap energy for mechanization, thereby freeing workers for industry. Thus the U.S. must maintain the agricultural sector's productivity in spite of other demands on scarce energy supplies, or risk a major shift in its economic priorities. The 16.5% of U.S. energy consumption that goes into food preparation and production represents a relatively small fraction of the national energy budget and the 3% used in crop production is even smaller, but they are critical for social and economic reasons.

This paper looks at the many uses of energy in the agricultural sector. This includes preparing the land, planting, fertilizing, irrigating, weeding or spraying with herbicides and insecticides, harvesting, drying, food processing, transporting to market, storing, packaging, selling to the consumer, and finally cooking.

2. Outlines of energy use in agriculture

Like other end-use sectors, agriculture involves a wide range of technologies used in the production, processing, and preparation of food. Because this study contains chapters which focus on transportation and industrial processes in general, this discussion will concentrate on farming itself and a brief discussion of the food-processing industry.

For purposes of this paper, agriculture is defined as "an activity (of MAN)

The Food Production & Processing Cycle

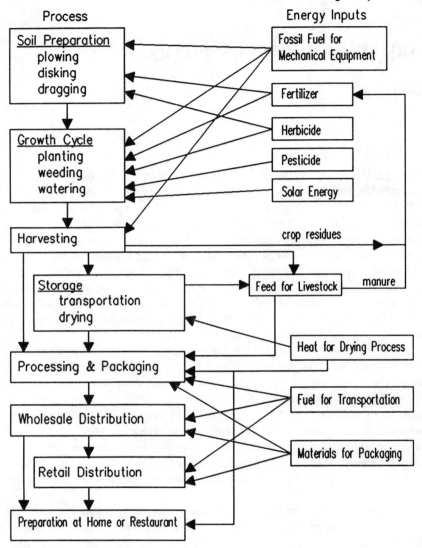

Figure 1

carried out primarily to produce food and fibre (and fuel, as well as many other materials) by the deliberate and controlled use of (mainly terrestrial) plants and animals."[3] This broad definition includes fish farming and forestry as well as the cultivation of silk worms in addition to such more common activities as raising wheat or hogs.

The very diversity of the products produced by agriculture causes a vast array of different technologies to be involved in this sector. Each crop requires individual techniques for its cultivation and hence a new energy analysis. Even a single crop, such as soybeans, requires different energy inputs when it is cultivated on different types of farms or in different states. It costs 2.3 million kcal/ha to grow soybeans in Ohio and 3.7 million kcal/ha to produce them on irrigated land in Nebraska.[4] Differing climates require different cultivation methods. For example, irrigation is needed where rainfall is low. Small farms may use less mechanized equipment than large "agro-businesses."

A simple example will show the complexity of the problem. Consider corn on the cob. Was the cornfield plowed, disked, dragged, and then planted, or was it low-tilled? Was the tractor involved an old gas-guzzler or a more energy-efficient diesel tractor? For either case fertilizer, herbicides, and insecticides were applied. The corn was watered (probably), harvested, transported to market, transported home, cooked, and put on the table and, hopefully, all eaten.

The primary energy activity in agriculture is the rather inefficient conversion of solar energy to organic matter by plants through photosynthesis. The plant material may be consumed directly or used as an input to other agricultural processes. In general, the production of livestock consumes more energy per unit of food energy produced. Most of the corn harvested in the U.S. is fed to animals, which are then slaughtered for their meat. The same planting techniques are used as described for roasting ears (albeit on a much grander scale for field corn than for roasting ears), but the harvesting is different.

Sometimes the ears of corn are machine-harvested, with what remains being plowed under. Sometimes the farmer lets Jack Frost paint the corn a dull-light brown in late autumn. The farmer waits until the moisture content is just right, and then the whole corn stalk, ears and all, is ground up for high grade silage. If the moisture content is not correct the silage may rot in storage. Traditional drying methods force air heated by the combustion of one of the fossil fuels over the silage.

One example of how energy can be saved in drying out a crop is to form a long tunnel with black plastic. The crop to be dried out can be placed in this tunnel and stored in a drying rack connected to the tunnel. A fan is installed in the top of the drying rack. Solar energy heats the interior of the tunnel and this warm air is pulled through by the fan. High winds and bad weather can perturb this system, and lack of space limits this type of process. However, it does work for some farmers. Silage is stored in silos or in the huge plastic sacks (8 feet [2.4 m] in diameter and 150 feet [46 m] long—they look like giant grubworms). There is about a ton of silage per linear foot (0.3 m). Some corn is lost to animals both in the fields and in storage, and some is lost due to rotting or mold.

After consuming the silage, the animals have to be taken to slaughter, processed, and refrigerated. The consumer buys from a store which usually places the meat in an open-air refrigerator (in the local newly remodeled supermarket the dairy products are in a vertical refrigerator with no doors, so that customers can easily reach everything and freeze their socks off while so doing). The customer usually ends up wasting a percentage of what is bought, either in storing, cooking, or in leaving it on the plate. Kling[5] did a study of waste for a variety of foods. His overall conclusion was that 23.7% of the total food produced by a nation was wasted from the time it was gathered until the consumer was done with it.

Animals are above plants on the food chain. They convert food into meat, milk, and eggs. This means they are raised and fed until ready to slaughter, and then less than 20% of the slaughter weight is sold as meat. But this problem is even worse than it seems. In the book *Beyond Oil*[6] the authors put it this way: "Because we feed so much of our crops to animals, crop production has had to increase far faster than population growth to keep pace with the demand for food. In a sense, then, much of the impetus to make heavy use of fertilizers, irrigation, and pesticides comes from the consumer's preference for animal products." Feeding corn to animals, then eating the meat is a very inefficient use of agricultural products.

3. On the farm

Historically, three stages of farming in the U.S. can be distinguished:[7] the *expansionist* period from the pilgrims to pre-World War II where total crop output was increased by expanding the land area cultivated, the *intensification* period starting prior to WW II which used fossil fuels and fertilizers to greatly increase productivity, and the *saturation* period. We have probably reached saturation, a period in which there are effectively limits to what increased fertilizers, mechanization, and pesticides can do to increase production.

In the U.S. farmworkers[8] are about 7% of the population; in Nigeria they are 97%. The average percentages of income spent on food in these two countries are 18 and 70%, respectively. Comparing the total population and the farm population as a function of time for the industrialized world, the total population is seen to grow while the farm population decreases. There are two primary reasons for this. The first is that the fossil fuel energy input enormously increases food energy output per person hour. Commercial corn crops in the U.S. have a food energy output of 3800 Megajoules/person hour, while the corresponding figure for the typical vegetable garden is 4 MJ.

For example, consider what this energy input means in plowing a field. Turning up the soil by hand in a 10 m by 15 m garden would take a very boring day or two. It takes 15 minutes to plow the same plot with a tractor, equipped with a two-bottom plow. This same tractor could easily plow 5 acres in a day, and the big tractors with a six-bottom plow can plow 60 acres/day. Another example of multiplying manual labor many times over is in comparing mechanical versus chemical weeding. Green and McCulloch[9] found that the total energy used for two mechanical weedings was 2.68 GigaJoules/hectare (one hectare = 10000 square meters), while the total energy for one chemical weeding was 0.74 GJ/hectare (including the energy to manufacture the fertilizer).

Although modern mechanized agriculture economizes on labor, it does so at the expense of additional use of energy derived from fossil fuels. In particular, mechanized agriculture depends on motorized farm equipment, which uses gasoline or diesel fuel, and fertilizer, which certainly uses electrical or fossil fuel energy in its production. Natural gas is actually a raw material in the production of nitrogen fertilizer in the United States.[10] Major studies completed during the 1970s attempted to compute the energy needed per unit production of common crops. Most studies ignored the primary solar energy input to photosynthesis, and they varied widely in their treatment of such factors as energy consumed by farm workers and the energy employed in constructing farm buildings. These studies agree that modern, high yield agriculture is far more expensive in energy terms than is its primitive counterpart.

An analysis of agriculture in terms of energy use and labor needed to produce a given product fails to model this end use sector because it neglects the use of land to produce crops. In many areas of the world, expanding populations cultivate nearly all arable land. Therefore the only way to increase food supply is to use existing land more efficiently. Nations with high energy use in agriculture generally obtain much higher yields per hectare than developing nations. In 1983, the average yield of rice grain in developed countries was 5.3 t/ha in contrast to an average yield of 2.8 t/ha for developing countries. Interestingly, the highest yields in the study, 6.1 t/ha, were obtained by the Democratic Peoples Republic of Korea where agriculture is extremely labor intensive.[11] Other oriental nations where there is a centuries long tradition of very intense cultivation of small plots of land use little mechanized equipment but produce very high yields by the use of enormous amounts of human labor.

In many cases, additional inputs of energy, usually from fossil fuels, can actually create a new supply of arable land. There is no better example of this process than the irrigation of the Imperial Valley in California. The Imperial Valley produces rich farm crops but only at the expense of the energy needed for irrigation, not to mention the scarce water that is used.

The second reason for the decline of agricultural workers as a percent of the population of industrialized societies and the increased productivity per hectare is the enlarged use of fertilizers, herbicides, and pesticides to increase yields. The application of the chemicals generally consumes less energy than mechanized or manual weeding even including the energy costs of producing the chemicals. Certainly the increased yields could not have been achieved without chemical fertilizers.

The benefits of modern, mechanized, and chemical intensive agriculture can be summarized in two statements. (1) The cost of manual labor lies near $6000 per Gigajoule, but fossil fuel costs only $15 per GJ[12]; and (2) fossil fuel and fertilizer have increased approximate yields from 1925 to 1980: rice by a factor of 2.7, corn by 4, wheat by 3, and soybeans by 1.7.[13]

But there are limits of growth in productivity achievable using both fossil fuel and fertilizer. Farming can only be mechanized so much. Increased farm production levels off with increased use of fuel and fertilizer. The latter case is obvious to anyone who has inadvertently dumped too much fertilizer on a lawn or a garden plant. Overfertilizing can do more harm than good. Finally the American public is becoming increasingly concerned about the long term effects of agricultural chemicals on public health and the environment. In 1972, the pesticide DDT was banned following the demonstration that it accumulates in the food chain and can be responsible for decimating wildlife such as predatory birds. Last year, Alar, a chemical preservative sprayed on apples to make them easier to ship and more attractive to eat, was banned following a public boycott. Southern Californians split into two violently opposed factions over spraying cities with pesticides to kill the medfly which poses a major threat to the citrus crops in nearby valleys.

Thus while this chapter concerns itself primarily with energy use in agriculture, energy is not the only critical input in this end use sector. It must be traded for extremely scarce and valuable arable land and real risk of damage to the environment.

4. Energy use in agricultural processes

The direct use of fuel on the farm is for ploughing, secondary cultivation, rotary cultivation, light cultivation, drilling/harrowing, spraying, irrigating, combine harvesting, mowing grass, and baling hay. In a report by the Steinharts[14] these activities account for about 44% of the energy for crop production. Two important effects have saved energy: (1) the more energy efficient diesel tractors which appeared in the early 1960s, and (2) the introduction of low-till or no-till cultivation.

Diesel engines typically have an efficiency of 34% versus 17% for a gasoline engine. Keeping the brush cut on about 10 acres of land with a Ford 9N tractor (built circa 1954) used about 16 gallons of regular gasoline. Now using a slightly less horsepower Ford 1510 Diesel (Japanese made, no small tractors are currently built in the U.S.) requires about 9 gallons.

Low or no-till cultivation replaces the process of plowing, disking, dragging, and then planting.[15] In no-till cultivation the tractor passes over the field and plants the seed and disturbs less than 25% of the soil. Low till may work the whole soil area, but soil disturbance is kept to a minimum. In both low and no till, further passes are used to spray pesticides and herbicides, but the whole process is much less energy intensive than standard cultivation. This reduces the compaction of the soil since the number of passes over the ground is decreased. It not only saves fuel, but helps reduce erosion: bare soil is not revealed and the in-place root system holds the soil. The disadvantage of this method is that the earth is not turned over (under low-till the field is plowed at least every four years) and weed seeds have a head start on the new crop. This must be countered by spraying. Labor and fuel costs favor no-till, but are about balanced by the costs of chemical spraying for weeds. Generally the advantages of low till outweigh the disadvantages, but each situation has to be evaluated on its own.

Improved farming techniques such as contour plowing, strip cropping, crop rotation, planting wind breaks, etc., can save energy. These practices are utilized by all good farmers, but there is often a tendency to cut corners: difficult slopes are not contoured properly; crop rotation may be put off because of short term gains; strip cropping requires extra work, and so forth. In general, crop production accounts for about 44% of the energy input to agriculture. The remainder of the energy input on the farm is divided as follows: fertilizing the soil accounts for about 24%, pesticides for 2%, machinery around 20%, and the remaining 10% goes for irrigation, transport, and miscellaneous.[9] Because of the wide variety of farms and crops, these numbers vary widely but they represent a reasonable average. Clearly most farmers could save immediate purchase of fossil fuel energy, but the saving must be balanced against increased consumption of chemicals (whose production costs energy) and the use of more land and water.

Agricultural production uses about 96% of the water that is consumed in the U.S. and 14% of the cultivated land in the world is irrigated.[16] About 10% of U.S. farmland is irrigated.[17] Current problem areas in irrigation concern developing a surface water supply for irrigation, and pumping and delivering water where it is needed. In the southern plains region, irrigation water comes from the rapidly diminishing supply in underground aquifers, while in parts of the desert southwest, water availability has become an issue in development. Certainly lack of water was a major argument against the race-track basing mode for the MX missile.

In addition to scarce water, irrigation uses an appreciable amount of energy. Water must be pumped into the irrigation system and forced through the pipes of the distribution system. If farmers save energy by using open ditches and no pumps, precious water evaporates from the ditches and is lost. Larger pipes save energy because the pump must develop less pressure. Unfortunately, larger pipes cost more money to install, and farmers frequently see the extra capital investment as uneconomic because of the long payback period, particularly at present oil prices. Possibly the best hope for saving energy in irrigation comes from the introduction of more efficient techniques which have been introduced primarily to save water. Center pivot sprinkler irrigation not only loses water to evaporation but also uses large amounts of energy because a pressure head is needed and the system has large amounts of friction. Drip irrigation systems require less head and are more efficient. Energy used for irrigation also depends critically on the depth from which the water must be pumped. For more information on irrigation see the papers by Dvoskin,[18] Smerdon,[19] and Sloggett. [20]

The big area in which to effect energy savings is in fertilizers, but unfortunately this is a tough nut to crack. Fertilizers are very energy intensive. Stout[21] states:

> "a kilogram of nutrient requires about two kilograms of fossil fuel for its manufacture, packaging, transport, distribution and application... In the many areas where the simplest and cheapest source of hydrogen is natural gas, the price of nitrogen fertilizer corresponds to that of natural gas, and much of this reserve is now wasted. It has been estimated that 62% of the natural gas produced by OPEC members in 1972 was flared and that this quantity would be sufficient to produce five times the nitrogen fertilizer consumption of the developing countries in 1978."

The proper farming practices noted above can reduce the need for fertilizer somewhat, but reducing this energy sink requires new breakthroughs in fertilizer technology. This means more research on the efficient production of fertilizer, as well as more studies on what, when, and how much fertilizer should be applied. Crops such as legumes may actually add nitrogen to the soil reducing the need for fertilizer. Fluck[22] and Davis further discuss fertilizer issues.

Beef and dairy farms have a different fertilizer problem. Cows, when grazing, naturally return their droppings to the soil. A visit to a dairy barn during milking time would impress anyone with the aroma and the magnitude of manure handling. A typical cow drinks about 10 gallons of water/day and produces about 80 pounds of manure (both liquid and solid) each day. In snow country cows may only rarely get outside during the winter, so the farmer has to collect the manure. Most distribute it over the fields via a manure wagon — giving rise to the phrase "fresh country air." A few use the manure to generate methane gas and then burn this as a fuel. Doing so reduces the fertilizer returned to the fields, however. There is plenty of manure generated in feed lots which often are not handy to fields that need fertilizing. So energy is lost in collecting the manure and transporting it to where needed.

There are many ways to save energy on the farm. One big impediment is that agriculture is very capital intensive (listen to the evening news when the farm mortgage problem periodically crops up). Two examples are milking cows and making hay. When the milk is taken from the cow (at body temperature) it is pumped into a holding tank. A farmer can use cold spring water to precool the milk. Then a refrigeration tank brings the milk down to storage temperature. An alternative method is to employ a heat exchanger to extract the heat from the milk and then use it to heat water. However, the heat exchanger costs money, and even though it would pay for itself after a few years, many farmers cannot afford the capital outlay.

The other example is making hay. Years ago, the sight of shocks of hay standing in the fields was common. They were prepared by cutting the hay and then raking it into sheaves. The sheaves would be put in the barn until it was filled. The remainder would be stacked outside until needed. Next came bales of hay. To form a bale one needs a baling machine. The baler compacts the hay into a rectangle (about 1 m × 0.3 m × 0.3 m, with a mass of from 15 to 30 kg) and then kicks it out, usually into a wagon which follows behind. Now the move is to round bales. These are actually cylinders with both the length and the diameter being slightly less than 2 m. This requires a different baler and additional attachments to a tractor to handle the bale. Round bales have a mass between 350 and 550 kg. Almost anybody can pick up a rectangular bale; only Superman could pick up a round bale. The advantages of round bales are that you get more bale for the buck (with present fossil fuel costs). This is primarily due to strongly reduced labor costs: rectangular bales have to be picked up, transported, and stored in a barn, but round bales can be left out in the weather (small losses occur due to weathering and to other animals). We are going full circle in making hay: from shocks to small bales to round bales which are really much like shocks. But the mechanization is quite different. They were bringing in the sheaves in biblical times: cutting the hay with scythes and raking it by hand. Rectangular bales require a baler. Round bales require a different baler and special equipment to move the bales. We end up with small shocks, large capital investment, and lower labor costs.

Another aspect of the capital intensiveness of modern farming is the use of mechanical versus manual harvesting. Many complex factors come into play in commercial mechanical harvesting: usually the crop is picked too early and is more likely to be damaged; mechanical picking machines are very capital intensive; and mechanical breakdown at harvest time can be disastrous. One well-known solution to this damage problem has been developed for tomatoes. Researchers have evolved squarish tomatoes with thicker skins. They are easier to pick, they are quite green, and they package well (unfortunately, taste is a secondary consideration here). These mechanical harvesters are often large expensive machines which are used for only a short period each year. Even the giant combines which follow the ripening wheat crop work less than half the year. Another side to the problem is that humans are put out of a difficult, back-breaking job by these mechanical harvesters. While humanitarians may cheer at this solution, the migrant workers often take a different view of things. There are no easy answers.

An important way of saving energy is given by Green:[23] "By far the best hope of achieving significant savings of energy in a practicable and acceptable way is in making the farmer more energy-conscious with advice, help, and instruction on how to save energy. All investigations have shown that there is a great difference between the economy and efficiency with which energy is used on the best managed farms compared with average farms."

Some specific suggested methods for saving energy on the farm can be summarized as follows: The efficiency of agricultural production can be increased by reduction of waste, capital investment in new and efficient technology, and by changes in crop-livestock mix and consumption patterns from high to low energy systems. For the food production areas, better photosynthesis could be facilitated through the use of more energy-efficient varieties or better agricultural practices; reduced harvesting and storage losses produced via better preservation techniques; conversion of grain and forage into animal products could be improved; and more emphasis could be placed on cereals and vegetables instead of less sophisticated food. Regarding fertilizer use there are two approaches to reducing consumption — proper fertilization (where and when the plant needs it) and better utilization of crop residues and manure. Better, more efficient designs, proper tuning, and use of alternative fuels would reduce fuel consumption by machinery. And for irrigation there needs to be more work done on what is needed and when, and better delivery to the plant stock.[24]

The details of energy savings for farms will vary with type of farm, the crops produced, and the geographical location of the farm. Even in a single region where farms produce the same mix of crops, energy use can depend on such variables as soil type and distance to market and storage facilities. Thus although there are many opportunities to save energy on the farm, there are no "magic bullet" solutions which apply to all farms. Energy savings are apt to result in greater use of labor, water, and arable land, all of which are scarce and valuable commodities in their own rights. Growing concern over the environmental side effects of agricultural chemicals will force farmers to limit their use of these aids to increase production while reducing energy use.

An approach to farming which considers all these problems is low input or organic farming. Like agriculture, organic farming is difficult to define precisely. One working definition is:

"Organic farming is a production system which avoids or largely excludes the use of synthetically-compounded fertilizers, pesticides, growth regulators, and livestock feed additives. To the maximum extent feasible, organic farming systems rely upon crop rotations, crop residues, animal manures, legumes, green manures, off-farm organic wastes, mechanical cultivation, mineral-bearing rocks, and aspects of biological pest control to maintain soil productivity and tilth, to supply plant nutrients, and to control insects, weeds and other pests."[25]

The agricultural techniques cited in the above definition of organic farming, along with advances in plant breeding for greater biological efficiency, offer hope for breaking the cycle of saving energy only at the expense of water, arable land, or potential damage to the environment. Many of these measures are already cost effective, and some such as crop rotation are in widespread practice by thrifty farmers. Education of the agricultural community will undoubtedly lead to widespread adoption of more of these measures.

5. The food processing industry

The food processing industry has the problems typical of industries which are discussed elsewhere in this study. The typical energy use distribution for food processing as a whole are as follows: processing the food itself takes about 36.6%, food processing machinery is 0.7%, paper packaging is 4.5%, glass containers are 5.6%, steel and aluminum cans are 14.5%, fuel for transport is 29.3%, and trucks and trailers (manufacture) are 8.8%.[9]

The big place to save here is in the processing of the food itself. Typical techniques for saving energy in food processing are as varied and specific to the product as those for saving energy on the farm. These techniques include use of by-products in agriculture, developing alternatives to current package-sealing techniques (eliminate multiple wrappings of products — local supermarkets frequently put all produce in plastic wrap), heat recovery in processing fruits and vegetables and grains, improved equipment design, use of outdoor winter air for refrigeration in cold climates, etc.[20]

One simple example of a food processing industry is the making of maple syrup, truly a cottage industry in the northeast. In the good old days one drilled into maple trees, hammered a funnel tap into the hole, and hung a two gallon bucket, with protective lid, onto the tap. The next day you gathered the sap and started the boiling process. Today plastic lines run from each tree into feeder lines, and then into a holding tank. The ratio of sap to maple syrup varies from about 20 to 1 for trees with full crown to 40 to 1 for trees in the woods. The water in the sap has to be boiled away until only maple syrup is left. This means raising the temperature of 30 liters of sap from about 0°C to boiling, to get 1 liter of maple syrup. Using a 30/1 ratio requires $30 \times 1000 \times (100 + 540) \times 4.2$ Joules, divided by the efficiency of the stove. For a 20% efficient wood stove this means 4×10^8 Joules/liter. Considering that maple syrup is mostly sugar, the food energy content is about 1.3×10^7 Joules/liter. This gives an output to input ratio of 0.3—without even considering packaging and canning the stuff. And you wondered why maple syrup costs $20 per gallon!

Energy savings in sugaring result from several fairly simple measures. When the sugarer cuts firewood he can store up the branches that aren't worth sawing up for wood stoves. He can leave the sap he collects outside overnight. Since maple syrup has a lower freezing point than water, the layer that freezes (weather cooperating) is mostly water so it can be thrown away. The sap might be boiled outside in a furnace composed of concrete blocks with a piece of bridge grating placed over the blocks. One can surround the boiling pot with bricks to keep the wind off. Processing the sap from a small farm with 13 trees collected during a month requires more wood than it takes to heat a typical farm house all winter. Clearly there are several ways to save energy in this process but they do not appear economically viable. The farmer doesn't process enough syrup to warrant buying a decent stove and boiler outfit, and much of the wood burned is wood that would be wasted otherwise. Many food processing industries have similar excuses of convenience or economics. The net effect is much energy loss. The point of this example is that many food processing procedures need a good comprehensive energy study.

Statistics regarding the energy to produce a 20 ounce can of food illuminate energy use in the food processing industry.[26] The ratio of the food energy output of the contents to the energy to process the food, make the can, and put it into the can is given for several cases: apples in syrup, 0.09; carrots in brine, 0.03; tomatoes, 0.03; baked beans, 0.24; pilchards, 0.44; and beef stew, 0.34. It is clear that the process of canning is very energy intensive. Green[27] states: "Processing of sugar uses as much energy as processing of meat. Production of roasted coffee and cigarettes takes as much energy as production of all frozen fruits and vegetables. Beverages require more energy for production than all agricultural machinery and agricultural chemicals put together." Don't even think about canned gourmet pet food!

6. Conclusions

It is obvious that much needs to be done if energy is to be used efficiently in agriculture. The agricultural industry is driven by two words: cost effectiveness. Energy conservation enters in only if it is really cost effective; and even then it is doubtful due to the already large capital intensiveness of American agriculture. We need comprehensive energy analyses and audits of every phase of energy use in agriculture to find out what can be done. Energy savings can be achieved in all of the following areas: fertilization, field operations, irrigation, crop drying, pesticides, transportation, and frost protection. Careful study is needed to determine the balance between energy savings and use of other scarce resources, notably arable land and fresh water. The battery of agricultural techniques referred to as organic farming offer hope of providing good yields, using less fossil fuel energy and water, as well as avoiding damage to the environment. Because farming varies so widely with different crops, geographic location, and type of farm, energy conservation measures will have to be evaluated for a large number of individual circumstances.

But more importantly, we need to have more balance and more common sense in what we do and how we do it. Lifestyle changes can save energy, but making the changes is not really a technical issue. We have to make better choices: food crops versus tobacco, a home-cooked hamburger versus plastic food encased in plastic at McDonald's, dry versus canned pet food, favor glass bottles over plastic and aluminum cans, waste not want not regarding the food on our plates, cut down on the meat in our diet, refuse plastic bags at the grocery store, etc., etc., and so forth. Bon appetit!

Bibliography

J. Gever, R. Kaufmann, D. Skole, and C. Vorosmarty, *Beyond Oil* (Ballinger, Cambridge, MA, 1987). This is a project of Carrying Capacity, Inc., and is a good introduction to the subject, along the lines of "small is beautiful."

Maurice B. Green, *Eating Oil: Energy Use in Food Production* (Westview Press, Boulder, CO, 1977). This is another good general introduction to the subject along the lines of *Beyond Oil.*

G. Leach, *Energy and Food Production* (IPC Science & Technology Press, Guildford, England, 1976). An interesting overview of the field.

Energy and Agriculture, G. Stanhill, ed. (Springer-Verlag, New York, 1984). This is a very good overview.

Energy in Agriculture: This is an interesting journal published from 1981–1988 which dealt specifically with energy in agriculture. Note the publisher's comment in the final issue: "The energy issue seems to have lost some of the interest it had when we first started the journal."

David and Marcia Pimentel, *Food, Energy and Society* (Edward Arnold, London, 1979). A very comprehensive study of the issues in energy and agriculture.

J.S. and C.E. Steinhart, Science **184**, 307–316 (1974). This is an excellent review article which touches all the main issues.

References and notes

1. Kenneth L. McFate, "Food and Energy: Challenges and Choices," Energy in Agriculture **1**, 91–98 (1981/1982).
2. C.R.W. Spedding, *An Introduction to Agricultural Systems,* second ed. (Elsevier Applied Science, New York, 1988).
3. Spedding, p. 5.
4. W.O. Scott and John Krummel, "Energy Used in Producing Soybeans," in *Handbook of Energy Utilization in Agriculture,* David Pimentel, ed. (CRC Press, Boca Raton, Florida, 1980), pp. 117–121.
5. W. Kling, J. Farm Economics **25**, 848 (1943).
6. J. Gever, R. Kaufmann, D. Skole and C. Vorosmarty, *Beyond Oil* (Ballinger, Cambridge, MA, 1987), p. 171.
7. *Ibid.*, pp. 149–159.
8. Maurice B. Green, *Eating Oil: Energy Use in Food Production* (Westview Press, Boulder, CO, 1977), p. 13.
9. Maurice B. Green and A. McCulloch, J. Sci. Fd. Agric. **27**, 95 (1976).
10. Charles H. Davis and Glen M. Blouin, *Agriculture and Energy,* William Lockeretz, ed. (Academic Press, New York, 1979), pp. 315–332.
11. Spedding, p. 131.
12. Green, p. 35.

13. Gever *et al.*, p. 156.
14. J.S. and C.E. Steinhart, Science **184**, 307–316 (1974).
15. R.C. Fluck and C.D. Baird, *Agricultural Energetics* (AVI Publishing Co., Westport, 1980), pp. 145–146.
16. David and Marcia Pimentel, *Food, Energy and Society* (Edward Arnold, London, 1979), p. 140.
17. Fluck and Baird, pp. 97–98.
18. Dan Dvoskin, Ken Nicol, and Earl O. Heady, *Agriculture and Energy*, William Lockeretz, ed. (Academic Press, New York, 1979), pp. 103–110.
19. Earnest T. Smerdon and Edward A. Hiler, "Energy in Irrigation in Developing Countries," Energy in Agriculture **1**, 99–107 (1981/1982).
20. Gordon Sloggett, "Energy used for pumping irrigation water in the United Sates, 1974," *Agriculture and Energy*, William Lockeretz, ed. (Academic Press, New York, 1979), pp. 113–130.
21. B.A. Stout, *Energy for World Agriculture*, (Food & Agric. Organ. of the UN, Rome, 1979), p. 51.
22. Fluck and Baird, pp. 89–90.
23. Green, p. 113.
24. Stout, pp. 231–248.
25. USDA Report on Organic Farming (1980) as quoted in Spedding, pp. 56–57.
26. J. Hawthorn, Span. **18**, 15 (1975).
27. Green, p. 79.

The potential for reducing the energy intensity and carbon dioxide emissions in U.S. manufacturing

Marc Ross

1. Mechanisms for reducing CO_2 emissions

1a. Overview of energy use and CO_2 emissions

The United States Congress has requested the Department of Energy to analyze two goals for reduction of U.S. emissions of greenhouse gases: 20% and 50% absolute reductions, 5 to 10 and 15 to 20 years in the future, respectively. This article was inspired by that request. A 20% reduction in the carbon dioxide emitted in combustion in the manufacturing sector by the year 2010 is studied. The longer time interval is deemed essential for a realistic attempt to achieve the ambitious goal.

Energy use by manufacturers in the U.S. constitutes 29% of all energy use.[1] It is responsible for a substantial part of the air pollution burden. In particular, the CO_2 emissions from combustion of fossil fuels for heat and power in manufacturing (including the fuel burned to generate the electricity used) was 308 million tonnes of carbon in CO_2 released to the atmosphere in 1985, about 24% of CO_2 releases from all fuel combustion in the U.S.

United States manufacturers accounted for 17.4 quadrillion Btu (or 17.4 quads) of fuel consumption for heat and power purposes in 1985. In addition, they used 3.9 quads of oil and gas for non-energy purposes, that is, as feedstock materials in petrochemicals, solvents, lubes, and waxes. Some of these materials oxidize in the atmosphere creating CO_2, while others do not result in substantial CO_2 emissions (unless the materials are later burned). They will not be considered further in this report.

The major energy form used was electricity (43% of all fuel use for heat and power was for generation of purchased electricity), with direct use of natural gas also heavy (27%). (See Table 1.1.) Direct use of petroleum products and coal for heat and power was 15% each. *The aggregate energy intensity*, that is, the ratio of the energy used by all manufacturers to their production,[2] was steady in the period from 1958 to 1972 and then fell by more than one-third from 1972 to 1985 (Fig. 1.1).

The different sources of energy had quite different histories, however. From

Table 1.1. Fuel used for heat and power and CO_2 released in manufacturing, 1985.

	Fuel use (quadrillion Btu)	CO_2 emitted (million tonnes C)
Coal	2.68	67
Oil	2.80	54
Natural gas	4.66	68
Purchased electricity	7.26[a]	119
Total	17.40	308

*Sources: References 1, 3, and 4.
a. The average input energy to the electrical utility sector is 3.34 × electrical energy delivered (Ref.4). Total input to the electrical utilities is 26.48 quads. CO_2 emissions are allocated to the manufacturing sector in proportion to its purchased electricity (nationally).

1958 to 1985 the aggregate coal and oil intensities fell 70% and 60%, respectively.[3,4] During this period, coal was being eliminated, except at large facilities and in very heavy industries. Petroleum was gradually losing its share to natural gas, and then, with the second oil shock, its use was drastically curtailed. Between 1958 and 1971, the aggregate natural gas intensity rose 30%, and then by 1985 dropped by more than 50% (Fig. 1.2). Natural gas was favored by its low price and convenience until shortages began to be felt in 1971, shortly before the first oil shock.

The combined result of these developments was that the aggregate fossil-fuel intensity fell 15% from 1958 to 1971 because there was fuel-efficiency improvement in energy-intensive sectors, even though real (that is, inflation corrected) fuel prices were low and falling. Beginning in 1971 the decline quickened, and by 1985 the aggregate fossil-fuel intensity declined by another 50%. Part of the reason for this accelerated decline was a relative shift away from energy-intensive production.

The pattern for electricity consumption is similar, except that continuing electrification (new uses of electricity) is combined with the other developments. Thus, the aggregate electricity intensity grew rapidly between 1958 and 1970 (Fig. 1.3), even though the efficiency of electricity-intensive processes, such as electrolysis of brine to produce chlorine and smelting of aluminum, was being substantially improved. This was a period of falling real electricity prices. Since 1970 electricity prices have mostly been rising and the aggregate electricity intensity has gradually declined. The two forces for decline, efficiency improvement and the relative decline in electricity-intensive production, have slightly outweighed ongoing electrification.

In 1958 more than 20% of total electricity used was generated and used on site by manufacturers.[5] On-site generation was falling, however, dropping to less than 8% of the total by 1981. With the Public Utilities Regulatory Policy act of 1978, on-site generation has begun a comeback, rising to 10% of the manufacturing total in 1986. This means that, up to 1981, utility sales of electricity to manufacturers were growing about 0.7% per year faster than total

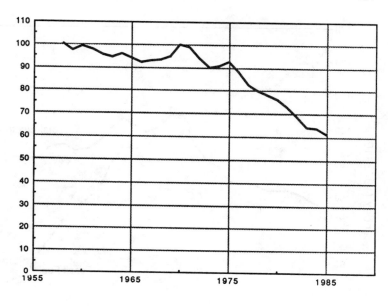

Figure 1.1

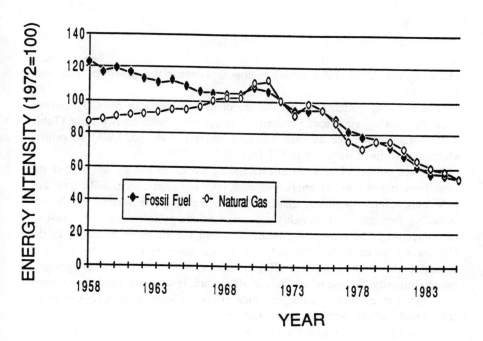

Figure 1.2

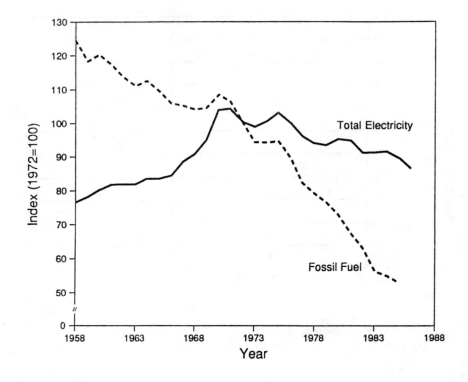

Figure 1.3

electricity use. Utility sales growth is now slower than the growth of total electricity use.

The changing patterns of fuel use are of great importance for CO_2 analysis because the CO_2 emissions per unit of energy vary with energy form (Table 1.2). For example, the average emission characteristic (gC/kWh) for utility electricity declined from 184 to 175 from 1973 to 1985.

Another critical element in understanding trends in energy use is that the manufacturing of bulk materials, including pulp and paper, industrial chemicals, petroleum refinery products, glass, cement and clay products, and metals (not including fabrication), is roughly ten times as energy-intensive as the rest of manufacturing.[6] Moreover, the production and use of these materials in the U.S. have been declining relative to all manufacturing since the early 1970s. It is useful, then, not to analyze CO_2 emissions on the basis of the aggregate energy intensity, the use of all fuels in all sectors. Instead one needs to examine three kinds of underlying changes which affect the intensity of CO_2 emissions from fossil-fuel combustion in manufacturing:

(1) changes in the *real* energy intensity, i.e., in the energy intensities within each sector of manufacturing;

(2) shifts in the composition of production; and

(3) relative changes in the energy forms used.

Table 1.2. Carbon emitted to the atmosphere in the form of CO_2 associated with energy use for heat and power.

Natural gas	14.5 kg/million Btu
Distillate	19.0 kg/million Btu
Residual oil	20.3 kg/million Btu
Coal	25.1 kg/million Btu
Electricity[a]	175 g/kWh
	51.3 kg/million Btu[b]

a. Characteristics of 1985 utility generation in the U.S. The CO2 emissions based on electricity sold are about 6% higher (Ref. 4).
b. @ 3412 Btu/kWh.

1b. Mechanisms for reducing CO_2

The 11 major mechanisms for reducing emissions are indicated at the points of their application in Fig. 1.4. Three of the mechanisms can reduce energy intensity in the separate process steps. Here, *conservation* is defined in the narrow sense of investments for which the major rationale is energy savings. *Housekeeping* refers to no- or low-investment operational changes. *Process change* refers to fundamental change in production processes, change not primarily motivated by energy savings but nevertheless tending to have associated energy savings.

Energy source switching refers to: (a) the potential for substituting natural gas for coal, at least in the short and medium term; (b) the potential for substituting electricity for fossil fuels, if the electricity is non-fossil, or perhaps if it is natural-gas based; (c) the potential for further growth in cogeneration (including substituting high electricity-to-steam technologies for existing modes); and (d) substituting biomass fuels for fossil fuels.

The remaining four mechanisms affect intermediate or final product flows. *Recycling* refers to use of processed bulk materials which have been recovered from later stages of manufacturing or at the post-consumer stage. *Sectoral shift* is the effect of the relative shift of production activity from the energy-intensive (materials manufacturing) sectors to downstream fabrication and assembly sectors. (The export of energy-intensive industry will not be considered.) *Materials substitution* involves substitution among the materials used in the final products, in association with improvements in materials properties and redesign of the final products. *Manufacturing share* refers to the portion of the GNP represented by manufacturing activity.

While the 11 reduction mechanisms are distinct, their effects interfere with one another. For example, the reduction directly associated with process change is less if conservation has been applied to the original process. Because of this complication, a simple model will be used to create a scenario involving the combined effects of the reduction mechanisms.

In this model, each reduction mechanism is specified algebraically in terms of one or two parameters (such as a reduction factor and a penetration factor). The parameters are estimated in part on the basis of historical trends and con-

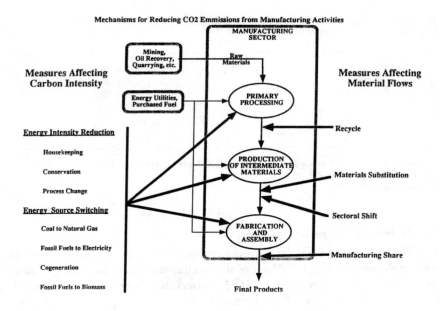

Figure 1.4

servation supply curves. In addition, they depend on physical limitations to implementation and constraints on effective policymaking, as discussed in the next two sections. The model and scenario will be discussed in Section IV and the Appendix.

2. The technical feasibility of the CO_2 reduction mechanisms

The 11 reduction mechanisms are assessed in terms of: (1) their potential to reduce CO_2 emissions, and in terms of (2) technology-related limitations, as suggested in Table 2.1.

2a. Conservation

A conservation supply curve (CSC) is analogous to an oil or gas supply curve: it shows the cumulative capacity to provide energy, e.g., barrels of oil per day, as a function of the price of energy. The capacity to provide energy shown by a CSC is actually the energy supply capacity released by saving, since conservation is not an energy supply. The savings can be expressed in energy capacity units or as a percentage of energy use as in Fig. 2.1.

Conservation supply curves, with today's processes and conservation technologies, show that a 25% reduction in energy intensity is expected to be associated with energy prices two or three times those which apply today.[7] (An example is shown in Fig. 2.1.) Housekeeping and cogeneration are not included in the conservation category; they are discussed below. In addition, fundamental process change and improved efficiency in the use of materials are not included. In all, only about half of the total response to increased energy prices is represented in these narrowly defined conservation supply curves.

Table 2.1. CO_2 reduction mechanisms in manufacturing: Technological opportunity and constraints.

Mechanisms for CO_2 reduction	Overall reduction potential[*]	Physical limitations	Capital limitations	Need for new technology
Conservation	High	Mod	Mod	Mod
Housekeeping	Low	Imp	· · ·	· · ·
Process change	High?	Mod	Imp	Imp
Energy switching				
Other fuels to nat. gas	High	Imp	· · ·	· · ·
Fuels to electricity	High	Imp	Imp	Imp
Cogeneration	High?	Imp	Mod	Mod
Fossil to biomass	Low	Imp	Mod	Mod
Recycle	High	Mod	Mod	Imp
Materials substitution	Low?	Imp	Mod	Mod
Sectoral shift	Low	Imp	Mod	Mod
Manufacturing share	Low	Imp	Imp	Mod

[*]Judged by size of opportunity and potential degree of public policy impact.
Imp: Important. Mod: Moderately important. · · · : Relatively little importance.

The CSC shown here is based on a behavioral or implicit discount rate and corresponds to a capital recovery rate of 33% as shown in Fig. 2.1. Often CSCs are calculated using a socially desirable discount rate corresponding to a capital recovery rate which is, perhaps, only half that used here, implying much more investment at a given energy price.

The gross capital cost of retrofitting today's manufacturing facilities to achieve a 25% reduction (in energy intensity associated with heat and power) would be about $80 billion in 1988 dollars.[8] For comparison, the total *annual* plant and equipment investment in manufacturing is typically also about $80 billion. In actual practice, not all existing production capacity would be retrofitted, but over the time period (to 2010), considerable new capacity would be created with conservation improvements integrated into the design at substantially lower cost. The overall cost would thus be somewhat less than $80 billion. The cost of conservation, to the 25% reduction level, is estimated to be about $60 billion, or an average of about $2.5 billion per 1% reduction.

For comparison purposes, let us ignore CO_2 issues for the moment and consider the capital costs of replacement energy supply systems in the early 21st century. If a new fuel supply system could be brought in (in today's dollars) at $50,000 per barrel per day energy equivalent, then the cost would be about $2 billion for 0.1 quad/year, about a 1% contribution to fuel supply for manufacturing at today's consumption rates. If a new electricity supply system could be brought in at $2000/kw, or at 65% duty factor, $350 per thousand kWh/year, the cost would be about $2.3 billion for 22 trillion Btu electrical energy delivered per year, about a 1% contribution to supply. To these capital costs for supply would have to be added substantial operating costs. Thus, granting the roughness of the estimates, the cost of a 25% reduction through conservation

Electricity Conservation Supply Curve Manufacturing

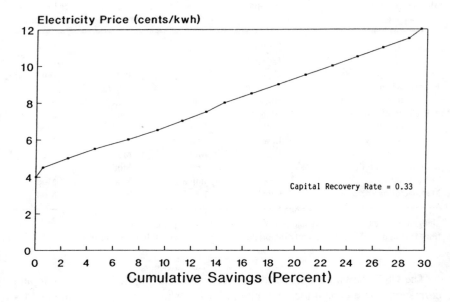

Figure 2.1

investments would be roughly compensated for by reduced needs for new energy supply systems. Thus the net cost of conservation at this level is low, or even negative.

The conservation supply curves discussed here involve current technology. However, if there continues to be rapid development of conservation technology, then in 10–15 years many important new technologies will begin to contribute to conservation. An impression of the importance of new technology can be obtained from examining the technologies which provided the CSC (Fig. 2.1) for electricity in fabrication and assembly industries. (See Table 2.2.[9]) In many cases, decades pass from the time an invention is made to the time a commercial version achieves major market penetration, but development and marketing of energy and electronic technology accelerated in the 1970s. Moreover, the gestation time for small-scale technologies is often much shorter than for the better-known large-scale technologies. In conclusion, there is every reason to expect current CSCs to evolve in the direction of much greater savings potentials (relative to base year energy intensities) in the next decades. Effective technology policies and energy-price signals would accelerate this evolution.

2b. Housekeeping

Housekeeping, while of great importance for secondary forms of energy like steam and compressed air, and for lighting and space conditioning, only has a

Table 2.2. Timing of the introduction of technology for electricity conservation (automotive manufacturing CSC).[a]

New technology (first marketed after 1970)

> Microprocessor-controlled energy management.
> Variable speed drive based on inexpensive power semiconductors.
> Die cushioning using rubber air bags.
> High-efficient lamps and lighting systems.
> Large microprocessor-controlled heating and ventilation systems.
> Aircraft-derivative gas turbine cogeneration.

Old technology

> Isolation valving for segmenting compressed-air systems.
> Small compressors for satellite compressed-air systems.
> High-efficiency motors.
> Cog belts.

a. See Ref. 8.

further overall reduction potential of, perhaps, 5%. The installation of automated energy-management systems, typically requiring expensive rewiring in an existing plant, is included in the conservation category.

2c. Fundamental process change

Fundamental process change has a largely unknown potential. It *is known* that radical new process technology cannot overcome the large minimum thermodynamic requirements for reduction of metal ores or decomposition of brines. (Blast furnace operations, primary aluminum and chlor-alkali manufacture are the most important examples.) Relatively small yet substantial thermodynamic requirements also apply to certain refinery and other organic chemical reformations.[10] But, in principle, most industrial processes have no or negligible thermodynamic requirements. For all these, radical process change could, in principle, greatly reduce energy use. One can, in fact, imagine many processes which require almost no energy use (where now a great deal of energy is used). Four examples of fundamental process change from recent decades are the float-glass process, basic oxygen steelmaking, induction heating of metals, and surface heating with electromagnetic beams. Examples of fundamental process change which may be achieved in the near future are listed in Table 2.3. (Where examples of these technologies are already well established, it is quite new applications which are at issue.) In practice, a great deal of research, invention, development, and major investments would be needed to achieve such changes. Dramatic process changes tend to occur slowly.

Longitudinal analysis of energy intensities in separate manufacturing sectors (summarized as the "real intensity" trend) for the period before the early 1970s shows that fuel intensities were declining about 1.2% per year and electricity intensities increasing about 1.8% per year, even though, for fuels, prices were low and often declining. (Electricity intensities declined, however, in sectors where electricity use is intensive.) Other analysis shows that these developments are part of a long history of technical change. Fundamental process

Table 2.3. Examples of potential fundamental process change with important energy and CO_2 consequences.

Chemicals, paper, and food processes

    Improved separation processes based on membranes, adsorbing surfaces, critical solvents, freeze concentrations, etc.

    Ethylene chemistry based on natural gas feedstocks

    Waste reduction using closed systems

    New and improved catalysts for chemical processing

    Recycling paper and plastics (i.e., into new material and product areas)

    Continued improvements in the processing and forming of materials

Metals processes

    Recycling post-consumer scrap (i.e., into new product areas)

    Near net shape casting

    Surface treatment with electromagnetic beams

    Direct and continuous steelmaking

    Coal-based aluminum smelting

---

changes reduced most factors of production, including fuel, but increased electricity use because of its special role in new technology.[11] In the period after the early 1970s, the real fuel intensity declines quickened and real-electricity-intensity growth ceased. Presumably, in addition to the historical trends just mentioned, price- and policy-induced intensity reductions also occurred in this period. In the analysis below these historical trends are assumed to continue (at moderated rates), underlying the policy-induced energy-intensity reductions which are being modeled.

2d. Switching from coal and oil to natural gas

Several forms of energy source switching can reduce CO_2 emissions. As shown in Table 1.2, natural gas combustion results in less carbon loading of the atmosphere than combustion of other fossil fuels. The potential for natural gas substitution for oil or coal is, however, limited not only because there may be gas supply scarcity. Most of the combustion of oil products in manufacturing is now at remote sites (e.g., in forest products sectors) or is combustion of byproducts (especially still gas and coke) for which there are no other markets at this time. (See Table 2.4.) On this basis the potential for CO_2 reduction from switching from oil to gas in manufacturing is deemed small and will be ignored in the scenario (Section IV). With coal, some uses are also not substitutable at reasonable cost: coking coal for blast furnaces and coal use in remote locations, especially in forest products industries. However, much of the coal use which was built up since 1973 in response to high gas and oil prices and national priorities is substitutable.

2e. Switching from fossil fuels to electricity

The physical CO_2 reduction potential of substitutions of electricity for fossil fuel depends on three coefficients:

Table 2.4. The usage of petroleum in manufacturing.

Application	1985 consumption in quads
Heat and power (except refineries)[a]	
Resid	0.4
Distillate, LPG, gasoline	0.3
Petrochemical feedstocks	3.4
Refinery processes[b]	1.8
Lubes	0.1
Petroleum coke (aluminum smelting)	0.2
Total	6.2

a. Approximately 0.2 of the 0.7 quads in use by forest products industries.
b. Mostly still gas, resid and coke, fuels with low market value.
Source: Ref. 1.

(1) the carbon loading per kWh associated with generating the incremental electricity;

(2) the relative efficiency of electricity, in the substitution application, compared to that of fuel in the original application; and

(3) the carbon loading per unit of fuel value of the fuel replaced.

In the analysis below, it is assumed that these substitutions are attractive only in end uses where the relative efficiency of electricity is three or more times higher than that of the fuel use for which it substitutes, because the current (1985) average price of fuels is $2.50 to $3 per million Btu, while that of electricity is about $14 per million Btu electrical. There are quite a few technologies with such high efficiency.[12] It is also assumed that electricity should be preferentially substituted for fuels with high carbon loading. In the analysis below, it is assumed that the carbon loading per kWh of the *incremental* electricity involved in substitution is likely to be the same as today's average. The average efficiency of use of electricity in the applications in question is assumed to be three times that of the fuel replaced. Under these assumptions, CO_2 reductions occur with substitution of electricity for coal and oil, but not for natural gas. (See Table 1.2.)

A critical issue for this mechanism is capital cost. If, for example, a 20% reduction in coal and oil use were achieved by substituting electricity at $1/3$ of the oil and gas intensity, then, in terms of current energy usage: (1) A reduction of 24 million tonnes of carbon (about 8% of the total from manufacturing activity) would be achieved if generating the incremental electricity involved no CO_2 emissions. (2) The capacity of the incremental utility system (not counting the in-plant equipment) would be about 110 billion kWh/year or about 20 gigawatts (at a 65% utilization rate). The corresponding system capital cost would be, very roughly, $40 billion, if low-cost non-fossil-fuel systems were successfully developed. To this, the cost of the new process equipment needs to be added. Thus, the capital cost of this approach is well over $5 billion per percentage point of CO_2 reduction from manufacturing. (The estimate above for conservation was $2 to $3 billion per percentage point.) The key consideration is that this high capital cost is for substitution where electrical energy can be used 3 times as efficiently as the fuels it replaces. Thus there can be no economic justification for electricity substitution unless the electricity application is highly efficient.

2f. Cogeneration

Cogeneration can enable substantially improved efficiency in the joint production of heat and shaft power, usually steam and electricity. Economical application of cogeneration requires fairly steady heat loads at the site, such as the steam loads in papermaking. The first part of the cogeneration opportunity is simply to exploit these heat loads more systematically in all sectors. The second is to use technology which provides a high ratio of electricity to steam, providing a better match to demands (than old fashioned steam turbine topping-cycle cogeneration) and much larger overall fuel savings.[13] One extraordinary potential, now in the development stage,[14] is replacement of boilers with steam-turbine generation in black-liquor recovery systems at kraft pulp mills with gas turbines which will produce about 3 times more electricity from a given input of black liquor. This black liquor application would enable generation of more electricity without fossil-fuel input.

2g. Switching from fossil fuels to biomass

The final energy switching opportunity considered is to increase the combustion of biomass fuels, usually waste or byproduct fuels in forest-products and food-processing industries, to replace fossil fuels. These opportunities have been heavily exploited in the past 15 years. While further opportunities remain, they are relatively small, unless crops are grown for fuel use or residues now left in the forest or field are aggressively pursued. These latter possibilities are important but are not considered here.

2h. Recycling

Recycling offers an important opportunity for reduction of energy use in manufacturing.[15] The principal sectors affected are primary metals, pulp and paper, organic chemicals/petroleum refining and, to a lesser extent, glass. Where recycled material substitutes for ores in metal manufacture, major energy reductions are achieved. Where they substitute for wood pulp, there can be substantial systems benefits, although the conclusion is complicated by several options in the use of the material. For example, instead of being recycled, waste paper can be used as a fuel, substituting for fossil fuel. Where recycled materials substitute for organic feedstocks, there can be significant CO_2 reductions associated with process energy savings. Glass recycling, unlike glass bottle reuse, has only marginal energy benefits; but has additional CO_2 benefits.

The major issues for the recycling potential are, in order of importance:

(1) the creation of markets for post-consumer recycled material in the manufacture of higher-value products than heretofore; and

(2) the reliable and clean collection of a high fraction of selected post-consumer materials.

Relatively little attention has been paid to point (1). The aluminum beverage can, now made from about 60% recycled aluminum, is the outstanding example of a technologically very demanding product for which a system for the supply and use of recycled material has been developed. Other post-consumer recycled materials are typically used in undemanding products: waste paper in cellulose insulation and pasteboard for items like cereal boxes, and scrapped automobiles in concrete reinforcing bars. The challenge is to convert newspapers into newsprint (as at Garden State Paper Corp.) and autos into sheet for autos, etc., R&D could help achieve such goals.

2i. Materials substitution

A wild card in CO_2 reduction policies would be a systematic effort to substitute materials in final products. In many products: construction components, packaging, white goods, automobiles, etc., different materials are now in head-to-head competition. The CO_2 implications of policy-induced substitutions at this margin are complex and have not been studied. They will not be estimated in this analysis.

2j. Sectoral shift

Sectoral shift has been extensively studied at Argonne National Laboratory.[16,17] The analysis shows that manufacturing activity has been moving relatively downstream, from manufacture of materials as measured in tons to fabrication and assembly, since the early 1970s and before. For some materials (steel and cement) this relative decline has been apparent for decades; now it is clear that bulk materials overall are in long-term relative decline.[18,19] This sectoral shift has been accelerated by the well-known increases in imports of materials like steel and aluminum. But these increased imports have not been the major source of sectoral shift (except in the early 1980s) and will not be considered in the analysis below.

2k. Manufacturing share of GNP

The final mechanism for reduction of CO_2 emissions from the manufacturing sector is a reduced manufacturing share in overall economic activity. The trend in the share of manufacturing in GNP is in dispute because of difficulties in defining price trends for products like computers and price trends for many services, but the impression that the effort to provide services is increasing relative to manufacturing is probably correct. The dispute is not critical for our analysis, however, since the poorly known numerator in manufacturing share appears as the denominator in sectoral shift, and so drops out of the overall analysis. In any case, our focus is policy-dependent CO_2 reduction and we assume for this analysis that sectoral shift and manufacturing share are policy-independent historical trends.

3. The policymaking feasibility of the CO_2 reduction mechanisms

3a. Selection of public policies

Two recent documents describe policies for reducing greenhouse gas emissions from manufacturing: Chapter 8, Volume II of EPA's *Policy Options for Stabilizing Global Climate*,[20] and the Executive Summary of DOE's *Compendium of Options for Government Policy to Encourage Private Sector Responses to Potential Climate Change*.[21] For organizational purposes, we adopt the four broad policy categories from the DOE study: Regulation, Fiscal Incentives, Information Policy, and Technology Development Policy.

In the following, we have selected policy areas on the basis of their effect on CO_2 emissions by the manufacturing sector. We have chosen policy areas, rather than specific policies, for discussion because of the limited scope of this analysis. Most, but not all, the selected policy areas selected were identified and discussed in the reports just cited.

Fiscal incentives form a very broad category of policies affecting prices in the market. Four major policy areas have been selected:

(1) energy taxation of all kinds, including CO_2 emissions fees;

(2) performance-based taxes (a form of excise tax) and subsidies, perhaps using the tax credit mechanism;

(3) disposal fees to influence materials markets and deposit-refund systems and other revenue-neutral schemes to influence materials and energy markets; and

(4) general financial policies to lengthen the time horizon for investors in manufacturing plants and equipment.

Regulation is defined as ". . . rules, supported by sanctions and having the force of law, that are designed to limit the discretion that may be exercised by public and private decision makers." Four policy areas are identified:

(1) utility, or rate of return, regulation, especially of electric utilities;

(2) performance (efficiency) standards for equipment;

(3) emissions controls including tradeable rights; and

(4) materials controls (bans or rationing of specified materials in certain applications, and rules for minimum content of recycled material).

Two policy subcategories in the DOE study are not considered: price controls and licensing.

Information policy is a diverse set of policies lumped into one area for our purposes here. It includes: setting energy-performance goals, collecting performance data, energy-performance labeling of equipment, and exhortation (or moral suasion). More highly technical policies like training, audits, and information about foreign technology are included among technology development policies.

Technology development policy areas selected are: (1) government-funded R&D programs; (2) demonstration programs; (3) education, technical information services and audit programs; and (4) government procurement programs.

The selected policy areas are listed in Table 3.1 and roughly evaluated

Table 3.1. Importance of selected policies to enhance CO_2 reduction mechanisms.

	Conservation	Switch to gas	Switch to elect.	Process change	Recycle
Fiscal incentives					
1. Energy taxes	I[a]	I	I	?	M?
2. Performance based tax/ rebates; investment tax credits	M?	· · ·	M?	· · ·	??
3. Deposit-refund, disposal fees	· · ·	· · ·	· · ·	· · ·	I
4. General finance (reduced discount rate)	I	· · ·	I	I	· · ·
Regulation					
1. Utility regulation (rate of return) energy prices, rebates, bidding	M?	?	I	· · ·	· · ·
2. Performance standards for equipment	M(ltg.)	· · ·	· · ·	?	· · ·
3. Emissions or fuels controls, tradeable rights	M?	I	I	?	?
4. Materials controls (recycle content, rationing)	· · ·	· · ·	· · ·	· · ·	I
Information					
1. Goals, data, labeling, exhortation	M	M	· · ·	· · ·	M
Technology policies					
1. R&D, research centers	I	· · ·	?	I	I
2. Demonstration program	M	· · ·	?	I	I
3. Extension services, audit programs	M	· · ·	· · ·	· · ·	?
4. Procurement by government	· · ·	· · ·	· · ·	· · ·	?

a. I: Important; M: Moderately important.

against five of the most important CO_2 reduction mechanisms. These evaluations are discussed in the next section.

3b. The potential of various policy areas to reduce CO_2 emissions

Nine of the thirteen policy areas will be briefly and qualitatively discussed in terms of their potential impacts on the most important reduction mechanisms.

A critically important issue in examining these policies stems from two characteristics of the greenhouse gas issue: It deals with climate, a phenomenon subject to great fluctuations and uncertainty, and it is a global challenge. No single country is in a position to solve the problem. If the U.S. were to undertake ambitious greenhouse gas reduction policies in the near future, the reductions should be achieved at low net cost. In other words, is it possible that an ambitious CO_2 reduction policy can be carried out in the United States without committing large net resources? While it is clear that strong reductions in certain areas cannot be achieved without substantial sacrifices or costs, the net cost could be very low. In particular, major investments in reducing energy requirements could be balanced by major cost savings in energy supply.

i. Energy taxes including revenue-neutral schemes. Relative to other sectors, industry responds strongly to energy prices. But this response is inhibited by two effects: (1) stiff financial criteria, like a two-year payback for small conservation investments at many firms,[22] and (2) the low priority given energy-cost savings at most firms in industries with low energy intensity, where energy costs are roughly 3% of value added.[6] Nevertheless the response to energy price increases has been substantial in the 1970s and early 1980s, in part because many energy conservation investments have multiple benefits like improving product uniformity or throughput, and reducing maintenance requirements. It appears that the high priority given end-use energy technologies as a result of the strong energy-price signals has been important in motivating the search for projects which both conserve energy and have other benefits.

Analysis of energy intensities at the sectoral level suggests that a doubling in price, results in a 25% to 50% reduction in energy intensity.[23] In the long-term, less energy-intensive industries will respond more than energy-intensive industries, but the latter respond more quickly. A study of electricity use in automotive manufacturing shows that the response is primarily through conservation investments, but also through housekeeping, cogeneration, and fundamental process change.[7,9]

If energy prices are to be raised by taxation, an important question is what would happen to the revenues. There are three principal options: (1) return the revenues to the firms in proportion to their prior energy intensities, so the tax is revenue neutral (discussed below), (2) use part of the revenues to fund incentives and technology policies (discussed below), and (3) use the revenues for general government purposes. Presumably approaches (1) or [(1) with some of (2)] would be easier to enact and would have more predictable consequences; sectoral dislocations and inflationary effects would be relatively small. An analysis of these issues is beyond the scope of this paper.

The political difficulties of substantially raising energy prices, especially their uneven impact, must be carefully considered. There is no escaping the major negative impact on certain energy industries, especially coal mining, and their immediate partners like the railroads. There may be other heavily affected sectors.

In particular, substantial energy price increases could seriously affect energy-intensive manufacturing. Unless the revenues were returned as suggested above, their product prices would increase. In itself, this would probably *not* have an important impact on overall sales,[17] because competition between quite different products is not likely to be affected by modest price increases. For example, plastics based on petroleum did not appear to be adversely affected in their competition with wood-based plastics by the recent oil price shocks. However, the competition between domestic and foreign producers of the same energy-intensive product would be altered, if foreign competitors were not subject to similar energy tax increases. A possible mitigating policy would be refunds such as suggested above or tariffs and refunds for energy-intensive products at the border. If these corrections were only applied, say, to products whose incremental energy costs due to the taxation were over 5% of sale price, relatively few sectors would be affected. Nevertheless, the administration of such tariffs and/or refunds might be complex.

Energy or carbon taxes could be part of a revenue-neutral fee-refund scheme. The tax revenues from each sector could be refunded to firms in that sector on a different basis. For example, the tax might be based on carbon emissions while the refund might be in proportion to production. This approach has two important advantages. While providing the economic inducement to reduce emissions it does so in a way that does not penalize any manufacturing sector and does not create inflationary pressures. There are, however, serious disadvantages in implementation. The outstanding difficulty is caused by the heterogeneity of products and production processes. This means that the administration of refunds would become complex and/or iniquitous as the level and mix of production by any firm changes. Another disadvantage is that little of the carbon tax might be passed on to customers, so they would not get the full price signal.

ii. Energy-performance-based fiscal incentives. Narrowly focused performance-based taxes and rebates to influence equipment purchases by manufacturers are probably not appropriate, but rebates on broadly defined conservation projects, or groups of projects, may be appropriate. (See Utility programs below.) Tax credits for conservation projects were tried (1980–1985), with a 10% credit for qualifying equipment specified on a short list (Energy Tax Act of 1978 and Crude Oil Windfall Profits Tax Act of 1980). A study based on interviews and analysis of detailed data from 15 large energy-intensive firms showed that the 10% credit and short list did not influence firms significantly in their investment behavior.[24] It is plausible that a much larger credit and more generous qualifying criteria would get a strong response. However, this incentive approach would generate a considerable windfall.

iii. General financial policies to reduce discount rates. Conservation investments, process-change investments, and corresponding private R&D and demonstration programs, are widely believed to be strongly inhibited in the U.S. by the short-time horizons of industrial managers and financial markets. Striking differences in plant and equipment investment and improvements in energy and labor productivity between Japanese and U.S. manufacturers can be conveniently described in terms of a difference in time horizons. Much more controversial is any analysis of causes, or of policies to ameliorate the situation. One dramatic analysis, created in large part by a corporate leader in energy-conservation equipment manufacture, associates the short-term horizon with low net national savings due to government policies (income taxation of capital gains, social security as a pay-as-you-go system, and federal deficits).[25] The purpose here is not, however, to propose specific policies, but to point out the potential importance of this policy area. For example, if firms' *de facto* capital recovery rates for conservation projects were reduced from 33% to 20% (payback increased from three to five years), then an energy price increase of 60% could accomplish as much as a doubling of prices under present conditions (assuming the 60% increase would be enough to give energy cost reduction a high priority).

iv. Utility regulation. This issue is too complex to discuss systematically here. Three topics will be mentioned: energy (kWh) versus demand (kW) pricing, and motivating utilities, through variable rate of return policies, to focus on the service costs of consumers. Manufacturers in many industries can do a great deal over a long period of time: either (1) to shift loads away from peak periods (if peak demand charges are high), or (2) to reduce loads in non- or low-production periods (if energy charges are high). If CO_2 emissions reductions would call for one kind of saving or the other, appropriate regulations would produce a significant response.

A specific policy tool for encouraging manufacturers to invest in conservation in spite of their often high discount rates is utility rebates to pay part of the investment cost. The concept is that some conservation investments are more attractive economically than supply investments. With its longer time horizon it may be reasonable for a utility to make part of the investment. The complex issues associated with such proposals have been discussed elsewhere.[26]

More generally, there are major opportunities to remove market imperfections in electricity production and sales. One promising possibility is adjustable returns on the rate base determined by an evaluation of utility performance (relative to other utilities) in minimizing its customers' costs.[27] If most utilities were motivated to do this rather than simply selling kWh as at present, innovative and effective programs might result.

v. Performance standards for equipment. The manufacturing sector is highly heterogeneous and most decisions are made by technically well-informed people. It is therefore not a sector where performance standards are generally appropriate. In the 1978 National Energy Conservation Policy Act energy-efficiency-standards for motors and pumps were proposed and study was mandated, but the concept was dropped when its inappropriateness became clear. In one area, however, performance standards may be effective: lighting equipment. Lighting must be appropriate for production needs, but the energy efficiency of the mass-produced lighting products (lamps, ballasts, fixtures) is of little or no concern to the plant engineer or manufacturer. The incremental first cost would not be substantial if all lighting equipment were required to meet ambitious efficiency standards.[28]

vi. Information. Goal setting (voluntary), data collection, and reporting of energy-intensity improvement, through trade associations in each subsector of manufacturing, were mandated in the 1975 Energy Policy and Conservation Act. They were considered by many energy managers at firms to be remarkably effective, considering the mild nature of the legislation.[29] The requirements expired in 1985. (The compilation and publication of the data by DOE was not considered effective.) The principal benefit from this policy was seen by energy managers as the creation of lines of responsibility within the firm for energy conservation and the creation of energy-consumption reporting systems within the firm, once top management had committed itself to ambitious goals for reduction of energy intensity. One could easily imagine similar requirements in the area of recycled materials.

Statistical information is also needed for public and private planning. For the period 1974–1981 data on energy use by fuel was collected annually for disaggregated sectors of manufacturing (over 400 sectors). Now such data is collected only every 3 years and only for 30 sectors, too few for good analysis.

vii. Research and development programs. While many assume that, since the U.S. R&D effort is so large and generally successful, it is basically sound, there are many who criticize the design and priorities of the entire R&D system as it relates to "civilian" industries.[30] The latter viewpoint can be summarized as follows: The basis of the problem is the underinvestment in research that characterizes the private sector, where a firm cannot expect to receive most of the benefits, i.e., in competitive industries.[31] The federal government's policy of leaving R&D to the private sector except in selected industries has meant massive research support in areas of military interest and in certain other areas with historical precedent. This has left most civilian manufacturing sectors with low research effort. Development of base technologies for most industries, especially energy-intensive industries, has, as a result, been very limited. Manufacturing processes for pulp and paper, inorganic chemicals, construction materials, and metals have not been fundamentally reexamined in light of the revolutions in basic science, modeling technology, and sensing and control technology which have occurred in recent decades. Engineering school education and research in these areas have largely withered.

In light of this analysis a policy option is creation of strong R&D programs, including specialized research centers, in energy-intensive manufacturing processes and in other technology-base areas important to energy-intensity reduction.[32] Examples are shown in Table 2.3. This option differs from DOE conservation programs of the past, which have selectively supported narrow *development* efforts. The policy option here involves creation of coherent programs of general research supported by institutional development.

viii. Demonstration programs. The effectiveness of demonstration programs is controversial. The positive argument is that firms will delay use of a new technology, or new application of an existing technology, until they are assured by at least one demonstration, in the same manufacturing situation that concerns them, that production processes and product will be satisfactory. Public subsidization of demonstrations and dissemination of technical information about them can accelerate the process. There are examples of successful programs.[33] The negative argument is that the public sector tends to do a poor job in selecting technologies for demonstration and a poor job of managing its role in any demonstration. There are important examples of such failures.[34]

Considerable experience has been gained about the design of demonstration programs in the past decade or so in the United States and Europe. Several characteristics of a successful program might be: avoidance of large projects, a relatively small financial role for government, and a strong dissemination activity. (The conventional wisdom that government should reserve its demonstration efforts for projects "too large for the private sector" may be exactly wrong.) Regardless of these specifics, as a result of learning which has occurred, demonstration programs might now be highly effective.

ix. Education, extension programs, and audit services. The DOE has supported an audit program for small-to-medium-sized manufacturing plants for the past dozen years.[35] There are now 13 Energy Analysis and Diagnostic Centers, located at engineering schools, which serve plants within a radius of 150 miles. This is a small but high-quality program. The annual value of energy savings and the energy-conservation investments (over the life of the program) have, perhaps, each been of the order of $100 million. (For comparison, total manufacturing expenditures on energy in 1986 were $43 billion.) The program has several side benefits, the most important of which is the training and commitment of engineers in practical industrial operations. It would be relatively inexpensive to expand such a program to the entire country.

4. A scenario of CO_2 emissions reductions

4a. Results

The scenario results for the year 2010 are developed in a spreadsheet model described in the Appendix. Results are summarized in Table 4.1. If the utility emissions factor (gC/kWh) is unchanged, there is a 20% absolute reduction in CO_2 emissions relative to 1985. If one assumes also that the gC/kWh characteristic is reduced, then one can achieve more than the 20% reduction, as indicated in the table.

It is seen that conservation investment is the largest single reduction mechanism. (Note, however, that the ordering of the mechanisms in the calculation affects their stated contributions.) Moreover, the parameters chosen, although representing strong policy initiatives, do not overstate the potential from this mechanism. Other important mechanisms are seen to be: (1) fundamental process change, (2) reduction in the gC/kWh characteristic of electrical generation, and (3) recycling. Also potentially important are (4) advanced gas-turbine cogeneration, especially if applied with biomass fuels, and (5) a switch from fossil fuels to electricity. Unlike the conservation mechanism, the parameters selected for mechanisms (1)–(5) may well stretch their potential for the 2010 time period.

The switch from fossil fuel to electricity needs to be put in perspective, since it attracts a great deal of attention given the well-established constituencies and federal programs for increased electricity use. The current high gC/kWh characteristic is likely to apply to any *incremental* electricity use in the period 1985–2010. The key policies in the electricity area for CO_2 reductions in the relatively short time scale at issue are: (1) increasing the efficiency of existing electricity uses, and (2) using, if feasible, low emissions generating technology in new and replacement power plants. A third policy of increasing the electricity load on this time scale would be almost certain to mean increas-

Table 4.1. A 20% CO_2 reduction scenario: Summary.

Business-as-usual emissions factor relative to 1985		1.37
Policy-driven reduction mechanism	**Individual reduction (%)**	**Individual reduction factor**
Without CO_2 policies	0	1.00
Conservation and housekeeping	27	0.73
Fundamental process change	8	0.92
Switch from coal to natural gas	2	0.98
Advanced gas-turbine cogeneration	5	0.95
Switch from fossil fuels to electricity	1[a]	0.99
Recycle	5	0.94
Cumulative reduction	42[a,b] (55)[c]	0.58[a] (0.45)[c]
Reduction-scenario emissions factor in 2010 relative to 1985		0.80[a] (0.70)[c]

a. If electrical generation has the same CO_2 implications per kWh as in 1985.
b. The percent reductions cannot be added; the reduction factors must be multiplied.
c. The results in parentheses show the cumulative effect of all mechanisms if the carbon loading per kWh is reduced 23% from 1985 to 2010 for the average electricity used in the manufacturing sector.

ing the continued use of existing coal-fired capacity. In a longer time period, essentially complete substitution of nonfossil-based for fossil-based generation might be achievable. After that, substitution of electricity for fossil-fuel use in manufacturing could yield substantial CO_2 reduction benefits.

Achievement of a 50% reduction in CO_2 would require, in this scenario, roughly a $2/3$ reduction in the gC/kWh characteristic of purchased electricity. To do this by 2010 would involve accelerated replacement of coal-fired generating capacity, with a substantial capital cost. This new capacity would involve technologies or resources not now available (Section II, part E). Each 1% reduction relative to 1985 levels was estimated above (Section II, part A) to cost about $2.3 billion. A 30% reduction in 2010 would involve a capital cost of about 100 billion (1988 $). Some of this investment would be required in any case to replace retired capacity. If one deliberately replaced the coal-fired capacity at retirement then this cost problem could be mitigated. This would, of course, delay achievement of the 50% reduction.

4b. Discussion

The overall lesson from the scenario is that by pushing hard one might reach a 20% reduction in CO_2 emissions from manufacturing by 2010. Of the policies to support conservation investment (in the narrow sense used here), only the massive increase in energy prices would be really fraught with difficulties of implementation. Most of the reduction mechanisms could be partially supported with technology-push policies which might be implemented *without* great difficulties. But the fuel-switching policies appear to have limited rewards *on this time scale* and their implementation *would* raise serious difficulties.

Aside from the problems of adding nonfossil-based generating capacity, the energy-pricing and fuel-switching policy difficulties just cited are of two kinds: (1) Impacted energy industries (especially coal) and their employees would be singled out for sacrifices. (2) Energy-intensive manufacturing would incur cost increases which would damage them competitively, especially with respect to foreign competitors. Difficulty (2) might be mitigated by temporarily rebating the incremental energy costs on the basis of past energy intensities rather than current usage. If such a rebate were restricted to the few energy-intensive industries, it might be administratively manageable.

In conclusion, the scenario suggests that a 20% reduction in absolute emissions of CO_2 associated with manufacturing activity might well be achievable by 2010 with strong and successful policies without large net investments, and without depending on any major breakthrough in energy supply technology. Finally, the scenario suggests that achieving a 50% reduction in the CO_2 emissions associated with manufacturing by 2010 would require massive added investments, using, in many situations, technology not now available. On a longer time scale, the 50% reduction goal might well be achievable at moderate cost.

4c. Uncertainties

The future is of course uncertain and forecasting is inherently inaccurate. However, when thoroughly grounded in a detailed understanding of past trends and current situations, a specialized forecast or scenario can be useful to policy making. For the case in hand some past trends and current situations relating to manufacturing energy use *are* relatively well understood, for example, the opportunities and costs of fuel switching and the downstream shift of manufacturing activity (from materials and energy-intensive sectors to less energy intensive fabrication sectors). Other mechanisms of change are inadequately understood for our purposes. In addition to the paucity of information on conservation, the technical potentials of fundamental process changes (including new recycling technologies) are relatively poorly understood, partly because very little research and development in manufacturing processes of energy-intensive industries is conducted in the United States. One of the benefits of strengthened information and R&D policies (Section III, parts B6 and B7) would be more reliable information on which to base future policy making.

Acknowledgments

I would like to thank my colleagues at Argonne, Cary Bloyd, Gale Boyd, Ron Fisher, Don Hanson, Ed Kokkelenberg, and John Molburg for their help.

Appendix

The model compares product flows, f_i, and energy intensities EI_{ij} in the base year 0 (1985) with those of the target year T (2010). Here $i = 1, 2, 3$ runs over the stage of processing (Fig. 1.4) and $j = 1, 2, 3, 4$ runs over the energy form

Table A.l. Relationships and parameters in the model.

Manufacturing production (manufacturing share effect):
$$f_3(T) = \text{(GNP growth factor)} \times r_9 \times f_3(0)$$

Materials production (sectorial shift effect)
$$f_2(T) = r_8 \times f_3(T)$$

Initial materials production (recycle effect):
$$f_1(T) = f_2(T) \, [1 - f_1^T(T)]$$

Conservation, housekeeping, and historical trend:
$$EI_{ij}(T) = \exp(T \times \ln 0.99) \times r_1 \times EI_{ij}(0)$$
for $j = 1, 2, 3$ (fuels), and for $j = 4$ (electricity):
$$EI_{i4}(T) = \exp(T \times \ln 1.01) \times r_1 \times EI_{i4}(0)$$

Process change:
$$EI_{ij}(T)' = p_3 \times r_3 \, EI_{ij}(T)/r_1 + (1 - p_3)EI_{ij}(T)$$

Switch from coal to natural gas (stages $i = 2, 3$ only):
$$EI_{i1}(T)'' = (1 - p_4)EI_{i1}'(T)$$
$$EI_{i2}(T)'' = EI_{i2}(T)'$$
$$EI_{i3}(T)'' = EI_{i3}(T)' + p_4 \times EI_{ij}(T)'$$

Cogeneration/biomass:
$$EI_{i4}(T)'' = r_6 \ddot{o} \, EI_{i4}(T)'$$

Switch from coal and oil to electricity:
$$EI_{ij}(T)'\,' = (1 - p_5) \, EI_{ij}(T)''$$
for $j = 1, 2$ (coal, & oil), and for $j = 4$ (electricity):
$$EI_{i4}(T)'\,' = EI_{i4}(T)'' + p_5 \times r_5 \, [EI_{i1}(T)'' + EI_{i2}(T)'']$$

Table A.2. Calculation of a 20% reduction scenario for manufacturing (first six columns in million tonnes of carbon emitted).

	Coal	Oil	Natural gas	Fossil total	Purchased electricity[i]	Grand total	Ratio to 1985
1985	67	54	68	189	119	308	
2010							
Historical trends[a]	70	58	75	302	221	424	1.37
Conservation and housekeeping[b]	52	42	54	148	162	310	
Process change[c]	47	39	50	136	149	285	
Coal to gas[d]	33	39	59	131	149	280	
Cogen/biomass[e]	33	39	59	131	134	265	
Fuels to electricity[f]	27	31	59	117	146	263	
Recycle[g]	21	31	58	110	138	248	0.80
Reduced utility emissions[h]	21	31	58	110	106	216	0.70
Reduced utility emissions[k]	21	31	58	110	44	154	0.50

a. Manufacturing growth factor 1.64 (GNP growth factor 1.85), fossil fuel intensity 0.78, electrification 1.29, sectoral shift, $r_8 = 0.80$.
b. Reduction factor, $r_1 = 0.73$.   c. Penetration factor, $p_3 = 0.25$, reduction factor, $r_3 = 0.05$.
d. Penetration factor, $p_4 = 0.50$ process stages 2 and 3 only.   e. Reduction factor $r_6 = 0.90$.
f. Penetration factor, $p_5 = 0.20$, reduction factor, $r_5 = 0.33$, oil and gas only.
g. New recycle fraction 0.50.   h. CO_2 emissions per kWh reduced by factor 0.77 compared to 1985.
i. All except last two rows based on 186 gC/kWh. See Table 1.2.
k. CO_2 emissions per kWh reduced by a factor 0.32 compared to 1985.

(coal, oil, gas, and purchased electricity, respectively). Total use of each energy form at a given process stage is just $f_i \times EI_{ij}$. The product flow unit is arbitrary. Here, in the base year

$$f_3 = f_2 = f_1 / (1-f_1{}^r) = 1.$$

The relationship between the flows and intensities in years T and 0 are defined in terms of parameters as shown in Table A.1.

A particular scenario is developed in Table A.2. Details are described elsewhere.[36]

Bibliography

Policy Options for Stabilizing Global Climate, Executive Summary, Daniel A Lashof and Dennis A. Tirpak, eds., U.S. Environmental Protection Agency (Draft Feb. 1989). This report provides an overview of an ambitious proposal for greenhouse gas reduction in all sectors of activity in the U.S., including discussion of both policies and changes in technology.

Robert Williams, Eric Larson, and Marc Ross, "Materials, Affluence and Industrial Energy Use," Annu. Rev. Energy **12**, 99–144 (1987). The shift by mature industrial societies away from energy-intensive production of materials in bulk toward increased fabrication activities and more knowledge-intensive products is described.

Marc Ross, "Improving the Efficiency of Electricity Use in Manufacturing," Science **244**, 311–317 (April 12, 1989). Trends in electricity use in manufacturing are discussed, with many examples of technological developments.

Robert A. Frosch and Nicholas E. Gallopoulos, "Strategies in Manufacturing," Sci. Am. **261**, 144–152 (Sept. 1989). Creative engineering is described which reduces the requirements for virgin materials and closes manufacturing cycles, reducing pollution.

References and notes

1. *Manufacturing Energy Consumption Survey: Consumption of Energy 1985* (Energy Information Administration, U.S. Department of Energy, Washington, DC, 1988). Energy used for heat and power and as feedstock. The accounting convention adopted here includes the fuel used at power plants to generate electricity consumed; excludes the use of biomass byproducts; includes within heat and power, metallurgical coal used to make coke; and excludes asphalt and road oil. Unless otherwise indicated, energy will refer to heat and power applications.
2. *Productivity Measures for Selected Industrial and Government Sectors*, Bull. 2322, and unpublished supplementary data (Office of Productivity and Technology, Bureau of Labor Statistics, Department of Labor, Washington, DC, 1989).
3. Jack Faucett Associates, *National Energy Accounts 1958–1981* (U.S. Department of Commerce, Washington, DC, 1984); and current data set from the Office of Business Analysis, Dept. of Commerce.
4. Monthly Energy Review, Energy Information Administration, U.S. Department of Energy.
5. Annual Survey of Manufactures and Census of Manufactures, Bureau of the Census, Washington, DC, various years.
6. Marc Ross, "Improving the Efficiency of Electricity Use in Manufacturing," Science **244**, 311–317 (Apr. 21, 1989).
7. "End-Use Technologires in Industrial Applications," Chapter 6, Vol 1 of *CO$_2$ Inventory and Policy Study, a Report to Congress by the Department of Energy* (1989).

8. The total capital cost of conservation is approximately $K=S\,E\,p/\text{CRR}$, where S is the fractional savings (0.25), E is the annual energy consumption, p is the *average* energy price at which conservation takes place, and CRR is the implicit capital recovery rate governing the investments (taken to be one-third). (Note: CSCs are often calculated using low, socially desirable, discount rates, rather than a behavioral discount rate, as here.) Based on 1985 data, for electricity E=640 billion kWh and p=\$0.08/kWh, so K=\$38 billion; and for fuels E=11 quads and p=\$5/(million Btu), so K=\$42 billion.

9. Alan Price and Marc Ross, "Reducing Industrial Electricity Costs—An Automotive Case Study," Electricity Journal **2**, 40–52 (July 1989).

10. Marc Ross, "Industrial Energy Conservation and the Steel Industry of the U.S.," Energy **12**, 1135–1152 (1987).

11. Charles Berg, Science **199**, 608 (Feb. 10, 1978); and "The Use of Electric Power and Growth of Productivity—One Engineer's View," unpublished, Northeastern Univ. (Apr. 1989).

12. Philip Schmidt, *Electricity and Industrial Productivity* (Pergamon, New York, 1984).

13. Robert Williams and Eric Larson, "Expanding Roles for Gas Turbines in Power Generation," see Ref. 17.

14. E.G. Kelleher, "Conceptual Design of a Black Liquor Gasification Pilot Plant," a report to the Office of Industrial Program, U.S. Dept. of Energy (1987).

15. *Energy Savings in Wastes Recycling*, Richard Potter and Tim Roberts, eds. (Elsevier Applied Science, 1985).

16. Gale Boyd, John MacDonald, Marc Ross, and Donald Hanson, "Separating the Changing Composition of U.S. Manufacturing Production from Energy Efficiency Improvements: A Divisia Index Approach," Energy Journal **8**, 77–96 (Apr. 1987).

17. Gale Boyd and Marc Ross, "The Role of Sectoral Shift in Trends in Electricity Use in United States and Swedish Manufacturing and in Comparing Forecasts," in *Electricity*, T.B. Johansson, B. Borlund, and R.H. Williams, eds. (Lund University Press, Lund, Sweden, 1989).

18. Robert Williams, Eric Larson, and Marc Ross, "Materials, Affluence and Industrial Energy Use," Annu. Rev. Energy **12**, 99–144 (1987).

19. Eric Larson, Marc Ross, and Robert Williams, "Beyond the Era of Materials," Sci. Am. **254** 34–41 (June 1986).

20. *Policy Options for Stabilizing Global Climate, Executive Summary*, Daniel A. Lashof and Dennis A. Tirpack, eds. (U.S. Environmental Protection Agency, Washington, DC, Draft Feb. 1989).

21. *A Compendium of Options for Government Policy to Encourage Private Sector Responses to Potential Climate Changes*, Richard A. Bradley and Edward R. Williams, eds., a report prepared by Oak Ridge National Laboratory for the Office of Environmental Analysis, Executive Summary and Chapter 8, Vol. 2 (U.S. Department of Energy, Washington, DC, October 1989).

22. Marc Ross, "Capital Budgeting Practices of Twelve Large Manufacturers," Financial Management, pp. 15–22 (Winter 1986).

23. Gale Boyd, Ed Kokkelenberg, and Marc Ross, *Econometric Analysis, An Overview of the Potential of GHG Reduction Mechanisms in Manufacturing* (Argonne National Laboratory, Argonne, IL, draft 1989).

24. Alliance to Save Energy, *Industrial Investment in Energy Efficiency: Opportunities, Management Practices and Tax Incentives* (Washington, DC, 1983).

25. George Hatsopoulos, Paul Krugman, and Lawrence Summers, "U.S. Competitiveness: Beyond the Trade Deficit," Science **241**, 299–307 (July 15, 1988).

26. Robert H. Williams, "Innovative Approaches to Marketing Electric Efficiency," in *Electricity*, T.B. Johansson, B. Borlund, and R.H. Williams, eds. (Lund University Press, Lund, Sweden, 1989).

27. David H. Moskowitz, "Cutting the Nation's Electric Bill," Issues in Science and Technology **5**, 88–93 (Spring 1989).

28. *Energy Efficient Flourescent Lighting in Canadian Office Buildings: Technologies, Market Factors and Penetration Rates*, a report to the Dept. of Energy, Mines and Resources (Canada) by Marbeck Resource Consultants, Ottawa (1986).

29. *An Assessment of Industrial Energy Technology Development and Response to Federal Fuel-Use Regulations*, a report to the Office of Policy and Evaluation, U.S. Department of Energy by Booz Allen Corp. (1980).

30. Herbert Hollomon, quoted in James L. Penick, Jr. *et al.*, *The Politics of American Science, 1939 to the Present*, revised ed. (MIT Press, Cambridge, MA, 1972), p. 284; Richard R. Nelson, Merton J. Peck, and Edward D. Kalachek, *Technology, Economic Growth and Public Policy* (The Brookings Institute, Washington, DC, 1967); Paul Horwitz, "Direct Government Funding of Research and Development: Intended and Unintended Effects on Industrial Innovation," in *Technological Innovation for a Dynamic Economy*, C.T. Hill and J.M. Utterback, eds. (Pergamon Press, Elmsford, NY, 1979); Deborah Shapley and Rustum Roy, *Lost at the Frontier, U.S. Science and Technology Policy Adrift* (ISI Press, Philadelphia, 1985); Edward E. David, Jr., "An Industry Picture of U.S. Science Policy," Science **232**, 968 (May 23, 1986); Harvey Brooks, "National Science Policy and Technological Innovation," in *The Positive Sum Strategy*, Ralph Landau and Nathan Rosenberg, eds. (National Academy Press, Washington, DC, 1986); Maryellen Kelley and Harvey Brooks, "From Breakthrough to Follow-Through," Issues in Science and Technology **5**, 42–47 (Spring 1989).

31. Richard R. Nelson, "The Simple Economics of Basic Science Research," J. Polit. Econ. **67**, 297 (1959); K.J. Arrow, "Economic Welfare and the Allocation of Resources for Invention," in National Bureau of Economic Research, *The Rate and Direction of Innovation Activity* (Princeton University Press, Princeton, NJ, 1962); Edwin Mansfield *et al.*, "Social and Private Rate of Return from Industrial Innovations," Quarterly Journal of Economics (May 1977); Michael Wolff, "The Why, When and How of Directed Basic Research," Research Management, **29** (May 1981).

32. William Chandler, Howard Geller, and Marc Ledbetter, *Energy Efficiency: A New Agenda*, a report by the American Council for an Energy Efficient Economy, Washington, DC (1988).

33. Howard Geller, Jeffrey P. Harris, Mark D. Levine, and Arthur H. Rosenfeld, "The Role of Federal Research and Development in Advancing Energy Efficiency, a $50 Billion Contribution to the U.S. Economy," Ann. Rev. Energy **12**, 357–396 (1987); Washington, DC (1986); *Community Demonstration Programme in the Sector of Energy Saving in Industry, Commission of the European Communities*, Directorate-General for Energy (1987); *Evaluation of the Energy Demonstration Programme, Energy Efficiency and Renewable Energy Projects*, Commission of the European Communities, Directorate-General for Energy (1988); Department of Energy, United Kingdom, *Energy Efficiency Demonstration Scheme: A Review*, Her Majesty's Stationery Office (1984).

34. John F. Ahearne, *Why Federal Research and Development Fails*, Discussion Paper EM 88-02, Resources for the Future, Washington, DC (1988).

35. F. William Kirsch, *Energy Management Assistance for Small and Medium-Size Manufacturers*, a report to the U.S. Dept. of Energy, University City Science Center, Philadelphia, PA (1987).

36. Marc Ross, "Modeling the Energy Intensity and Carbon Dioxide Emissions in U.S. Manufacturing," in *Energy and the Environment in the 21st Century* (MIT Press, Cambridge, MA, to be published).

Energy conservation in buildings and appliances

David Hafemeister and Leonard Wall

1. Energy usage in buildings and appliances

Since the oil embargo of 1973–1974, the GNP has grown about 35%, while energy consumption[1] has been relatively constant at 75 quads per year. More recently, during the past two years, energy usage has returned to its pre-embargo growth rate of 4%/year, rising from 74 to 80 quads. It remains to be seen if these recent increases will continue or not. Considering the longer period of 1973 to 1986, energy usage actually decreased a very small amount from 74.28 to 74.24 quads.

Since today we are not consuming energy at our former 1973 rate, this can be interpreted as a savings from conservation (Fig. 1) of about $150 billion. If energy were consumed at the same rate, the 35% rise in GNP would have increased the U.S. annual energy bill by $1/3$, or $150 billion per year. *However,* there have also been modal shifts within our society as jobs have shifted from steel mills to the commercial sector. It has been estimated[2] that $2/3$ of the total savings is a result of conservation, and $1/3$ from modal shifts.

There are two kinds of energy conservation, the denial of services and comfort, and the doing of the same task better, with enhanced-end-use efficiency. Conservation from enhanced-end-use efficiency[3,4] has been much more important than conservation by denial. New automobiles, appliances, and buildings (in California and elsewhere) consume about one-half as much energy as their pre-embargo cousins. As before, we have managed to commute to work, take vacations, cool our soft drinks, and stay warm in the winter and cool in the summer. Because cars have lifetimes of 10 years, refrigerators of 20 years, and buildings of 50–100 years, it will take many years to see the full impact of these energy savings. Additional savings of another factor of two are possible with further, up-front investments in energy conservation.

The oil embargo of 1973–1974 stimulated the Congress to legislate energy efficiency standards for automobiles and appliances, and some states, such as California, have created energy standards for buildings. In retrospect, these were wise decisions since conservation has far out-paced the production side of energy in spite of the fact[5] that 10 to 15 times more money has been spent on R&D for energy supply than for energy conservation. By the turn of the century, we will probably be consuming only 50% as much energy as was pre-

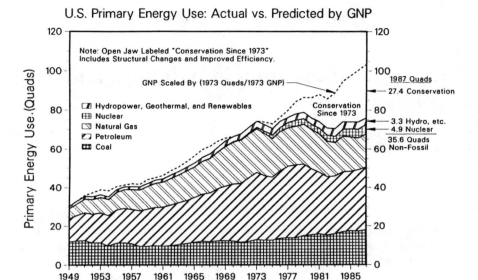

Source: *Economic Report of the President,* Council of Economic Advisors, January 1988, and *Annual Energy Review 1987,* EIA.

Figure 1. U.S. primary energy use: Actual vs predicted by GNP. (Source: Economic Report of the President, Council of Economic Advisors, January, 1988, and Annual Energy Review 1987, EIA. Figure courtesy of Lawrence Berkeley Laboratory, LBL.)

dicted before the oil embargo. These results cause us to question the methods of the energy planners before the oil embargo: *why* were they so wrong? As the outlook for future energy supplies continues to be clouded, it is time to strengthen our efforts for more energy-efficient buildings and appliances.

The total U.S. energy budget is about $440 billion, which amounts to $1800 per person, or $7200 for a family of four. The $440 billion energy budget is about 2–3 times the annual Federal budget deficit, and the U.S. trade deficit, and about 50% more than we spend on defense. The $440 billion is about 11% of the U.S. gross domestic product, as compared to 5% for Japan. About 36% of U.S. energy[1] (27.6 quads/year) is used in residential and commercial buildings. The energy for buildings amounts[4] to $180 billion per year, or about $750 a year per person (Table 1). Because of these large numbers, and because we seem to have been so wasteful in the past, it is prudent to ask the question of "how can we manage our energy future better?"

Table 1 is a disaggregation[5] of the energy costs into fuels and electricity for the U.S. residential, commercial, industrial, and transportation sectors. The data show that $117 billion is used for electricity in buildings, or 78% of the $150B total used for electricity; $60B for residential, $50B for commercial, and $7B for industrial. On the other hand, only $60 billion is spent for fossil fuels used directly for buildings.

Table 1. United States energy expenses (Ref. 5) ($ billions), 1985.

Sector	Fuel	Electricity	Totals
Buildings	60	110	170
Residential	40	60	100
Commercial	20	50	70
Industry	70	40	110
Buildings	3	7	10
Transport	160	0	160
Total	290	150	440
Percent of GNP			11.2%

The 85 million residences in the U.S. have a total floor area of about 110 billion square feet, with another 50 billion square feet in commercial buildings. The average cost of energy per square foot of floor space[5] is about $1 per square foot per year, obtained by dividing the total cost of energy in buildings, $170 billion, by the 160 billion square feet in buildings. The energy cost for residential buildings is about $0.90, and the cost for commercial space is about $1.30. The energy cost for new buildings can be considerably less, about $0.50 to $1.00 for new office space.

The annual energy costs of $1 per square foot over the lifetime of a building (50–100 years) is about the same as the cost of construction of $50–100. By the year 2000, it is projected[5] that the number of residences will grow from 85 to 110 million, and that the commercial floor space will grow from 50 to 70 billion square feet. If the energy efficiency of future buildings and appliances does not improve, energy use in the buildings sector would grow from 27 quads to 32 quads by the year 2000.

As things stand now, the projected expansion of electricity in the U.S. will have to come from nuclear, coal, or natural gas. Otherwise conservation will have to diminish the expansion. The supply side of electricity is strongly driven by the demands of the buildings sector, which consumes 75% of all electricity. One cannot carry out policies for the supply side of electricity, without talking of the demand side. One must examine the uses in detail, determine the effectiveness of additional first-cost investments in terms of the cost of conserved energy (see below), and then determine if investments in end-use efficiency cost more or less than *new, delivered* electricity from coal or nuclear, which is projected[6] to be about 12 cents/kWh at the turn of the century. Policy decisions also involve the broader questions of what is the benefit to the entire society beyond the individual consumer, i.e., factors such as energy security, fiscal responsibility, reliability, and environmental degradation.

2. Technologies for buildings and appliances

Residential energy use per *average* household dropped[7] by 27% between 1972 and 1982. Commercial energy use per square foot rose by 25% between 1960 and 1972, while it dropped by 11% in the next decade. These improvements are the result of many factors. For residences, 25% of the improvement was from improved structures, 10% from improved appliances, 20% from decreased household size, and 30% from changes in household energy use practices (such as thermostat settings). The potential improvements are much more dramatic when one looks at the possibility for *new* buildings and appliances. Improvements in end-use efficiency arise from two causes, (1) better quality with enhanced investments in present technologies, and (2) new technologies. Some of the improvements are as follows.

2a. Buildings

This section is divided into four parts: new residential structures, the retrofit of existing residential buildings, new commercial buildings, and the retrofit of existing commercial and high-rise residential structures. Building envelope technologies are emphasized, but diagnostics, space conditioning equipment, and control systems are also discussed. Actual performance data along with some economic assessments are presented to complement the discussion.

New residential. The steady state heat loss equation for a house can be written as a sum of losses from all the surfaces of the building:

$$dQ/dt = (\Sigma \, UA)\delta T = (\Sigma \, A/R)\delta T,$$

where U is a heat transfer coefficient, R is the thermal resistance value, A is area, δT is the difference between outside and balance point temperatures for the house, and the summation is over the different components of the building envelope including air infiltration. Normal insulation levels for the ceiling and wall have thermal resistance values of R-19 to R-30 and R-11 to R-19 (expressed in the English units of hr ft^2 °F/Btu), respectively, and a typical air infiltration rate is approximately 1 air change per hour (ACH). In California and elsewhere, more stringent insulation standards have cut heat losses by about a factor of 2 since 1975.

Superinsulated homes have become more popular, especially in the northern U.S. and Canada. They typically have from 6 to 12 inches of insulation (R-19 to R-38) in the walls, 12 to 18 inches of insulation (R-38 to R-57) in the ceiling, double or triple glazed windows, and low natural infiltration levels of 0.3 air changes per hour (ACH) or less. Special techniques, such as placing a continuous plastic sheet in the walls and ceiling of a house, can be utilized in new construction to reduce air leakage to a minimum. Since homes should probably have about $2/3$ ACH for good indoor air quality, mechanical ventilation through a heat exchanger with about 75% effectiveness must be used. Such devices cost from $250 to $500 and can save of the order of 100 therms (1 therm = 10^5 Btu) of natural gas in the winter with comparable monetary savings in the summer due to reduced peak power requirements. Because the balance point of the building is considerably lowered, the heating season for superinsulated homes is shortened to a few winter months, and the furnace can be downsized.

Passive solar heating in homes entails solar gain through south-facing windows and thermal storage by some component of the structure itself. In direct gain systems solar radiation strikes interior spaces, often massive floors and/or partition walls, which slowly release heat during the night. An indirect gain system is the thermal storage wall or Trombe wall which generally consists of concrete but may be filled with masonry blocks or water. The Trombe wall is heated up by direct solar radiation and then slowly releases the heat to the interior space. Another indirect gain system is the sunspace (often a greenhouse) which is allowed to rise to quite high temperatures. Thermal storage in the sunspace releases heat energy to the interior space via vents, windows, or a common wall. Roof overhangs on the south side are designed to allow the winter sun to enter but block off the higher summer sun. Most passive solar homes are well insulated but not to the extent of being classified as superinsulated. In a reciprocal fashion most superinsulated homes take advantage of additional windows on the south side and possibly some thermal storage.

As more and more new houses are equipped with air-conditioners, passive cooling systems warrant serious consideration, particularly in the temperate, hot-humid, and hot-arid climatic zones of the U.S. to which more of the country's population is shifting. Four cooling systems — nocturnal convective cooling, nocturnal radiative cooling, direct evaporative cooling, and conductive earth-coupled cooling — utilize various natural cooling sources such as ambient cool air, long-wave radiation to the sky, water evaporation, and soil below the building to reach comfort levels of temperature and humidity within interior spaces.[8] The house must have thermal mass that prevents wide temperature swings, absorbing heat during the warm day, releasing it at night, and cooling to ambient or below temperatures in the process. Preliminary studies[8] suggest that an investment of \$2500 in temperate zones or \$7000 in hot locations would be cost effective in saving from $1/3$ to $4/5$ of the house's cooling load.

Lawrence Berkeley Laboratory (LBL) has compiled monitored performance data for new, low-energy residences in their on-going BECA-A study (*B*uildings *E*nergy-*U*se *C*ompilation and *A*nalysis, part A).[9] The best performers have been found to be superinsulated houses, passive solar residences, and combinations of the two technologies. Their auxiliary heating requirements are typically less than 2 Btu/ft^2 °F day, which is only 10% to 20% that of conventional housing. The superinsulated homes cost approximately \$2000 above conventional practice and lead to extremely low winter heating bills of \$100 to \$200 in extreme climates with a cost of conserved energy well below today's fuel prices. There is one example of a \$20 heating bill in Saskatchewan for a very superinsulated house, undoubtedly with some "extra free heat." It is more difficult to quantify the costs and benefits of passive solar homes, but they tend to be marginally cost-effective.

Existing residential. The most significant technological development in residential retrofits occurred in the mid- to late-1970s with the advent of the blower door and infrared thermography to detect heat losses via air leakage through the building envelope. Princeton and LBL scientists played prominent roles in this development. The discovery of heat bypass losses, including convective loops, explained why insulation often performed far below expectations. Warm air was simply going around it through gaps for pipes or light fixtures, through stairwells and interior partitions, through lowered ceilings, and via other perforations in the envelope. The blower door, a high flow fan that can be attached to an exterior doorway of a house, is used to pressurize or depressurize (10 to 75 Pascals) the interior space so that air leaks through the building envelope are accentuated and easily detected by smoke sticks and infrared cameras.

The leaks are plugged by inserting insulation, caulking, and weatherstripping. Usually the total air leakage is expressed as an area roughly equal to the sum of the areas of the individual leaks. The equivalent leakage area of a typical house is approximately one to one and a half square feet, causing a complete change of air every one to one and a half hours, and necessitating about 250 therms to heat this air during the winter.

A two-person team of "house doctors," armed with a blower door and an infrared camera, can reduce air leakage by 20% to 50% in half a day of work. In addition, they can perform other retrofits on the space and/or water heating systems, including an adjustment of the furnace combustion. The Princeton study[10] demonstrated that an investment of $1300 resulted in an average reduction of space heating energy by 30%, even though the houses already had attic and wall insulation and storm windows.

Besides shell tightening as described above, other retrofits of existing residential structures include the following: adding attic and/or wall insulation, fitting storm windows, wrapping and adjusting the hot water heater, tuning the space heating furnace and distribution system, and installing a clock thermostat. The BECA-B data base[11] at LBL indicates that an investment of $1000 to $1500 in retrofitting a single-family residence would result in an average of 20% to 30% savings of space heating energy, with a cost of conserved energy well below present fuel prices. Approximately $2/3$ to $3/4$ of the predicted savings are actually achieved, as averaged over hundreds of homes.[5] However, for one particularly large project in Hood River, the actual savings were consistently well below (over 50%) the predicted pre-retrofit savings.[12]

New commercial. Principal technological improvements have taken place in the lighting, storage, controls, and the HVAC (heating, ventilation, and air conditioning) systems. Each of these developments will be discussed in modest detail. Since large buildings are dominated by their internal loads, not much emphasis has been placed on envelope strategies except for window glazing.

In the U.S., 30% to 50% of all energy use in non-residential buildings is attributable to lighting. The judicious usage of the natural light of day, "daylighting," accompanied by dimming controls on artificial lighting represents potentially large savings in energy and peak power. The daylighting can take the form of skylights, well-designed fenestration, light shelves, or even light pipes, but the excessive solar gains should be controlled and the light should be

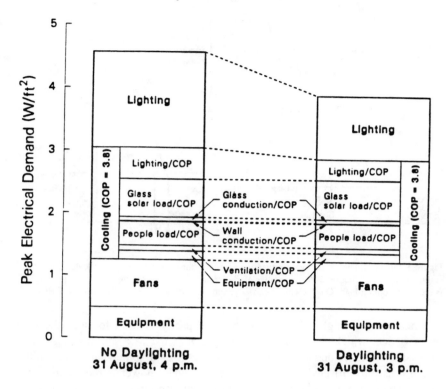

Figure 2. Peak power demand per square foot for a large office building, as simulated for Madison, WI weather by the DOE-2 program. Each floor has 10,000 ft^2 of core (cooled only) plus 6000 ft^2 of perimeter floor area. If built according to ASHRAE Standard 90–75, its peak demand would be 6.7 W/ft^2; if it satisfied Standard 90-E (revised 1985) with daylighting it would use only 4 W/ft^2. Thus its peak demand is down 40%, its yearly energy use is down by 40%, yet its first cost is also slightly down, mainly because of savings by downsizing the air conditioning. Note that about half the peak demand goes to running chillers. With thermal storage this 2 W/ft^2 can be moved entirely off peak for a first cost of about $0.50/W, which is only half of the utility's cost for new peak capacity. The residual peak demand is then down to about 2 W/ft^2. Alternatively, and cheaper, the chiller can be partially (60%) moved off peak for only $(0.00 to 0.25)/W (Ref. 5).

diffused with exterior or interior shading devices. Most of the shading systems are mechanical in nature, but much work is being done on optical switching materials that change transmittance as a function of sunlight intensity or temperature. Daylighting design has become computerized and sophisticated.[13] Due to the application of daylighting, the lowering of illumination levels except at task sites, and the improved efficiency of artificial lighting, connected lighting loads in new commercial buildings have been dramatically lowered from 3–6 W/ft^2 to 1–2 W/ft^2. The rewards are a reduction of HVAC equipment size, decreased energy usage, and lowered peak power charges (Fig. 2).

Energy storage systems[14] have become a cost-effective option for commercial buildings. They provide not only energy savings, but lower peak power demand levels. The cheapest strategy, partial storage (Fig. 3), displaces two-thirds of cooling demand by utilizing a smaller (than conventional) capacity chiller that is sized to run continuously over the 24-hour day.

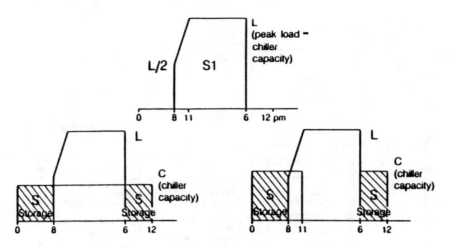

Figure 3. Off-peak thermal storage. Conventional system (top); partial storage (left); demand storage (right). Calculation of the capacities of chiller and storage according to the load profile. S1 is the daily cooling load, C is the chiller capacity, and S is the storage capacity (Ref. 14, LBL).

On a summer night, the excess chiller capacity is used to store the cool water or ice, which then supplements the chiller the next afternoon. A more expensive strategy, demand-limited storage, utilizes a somewhat larger chiller which runs only off-peak, i.e., it is turned off when power is most expensive. This strategy entails not only a larger chiller but also larger storage (both about 40% larger than for partial storage). Water and ice storage are roughly similar in costs and are most popular in the U.S. Success stories, such as Stanford University's cold water storage system at no extra first cost and the resultant $300,000 to $500,000 yearly savings and Alabama Power Company's ice storage system which reduced its building's peak power demand by 50%, are well documented.

Two other storage possibilities are concrete floor/ceiling slabs and phase change materials (PCMs). The Swedes' Thermodeck systems (Figs. 4–6) have popularized hollow-core concrete slabs through which room air is circulated. The concrete mass stores either internally generated heat or nighttime "coolth," depending on the season and has a thermal time constant of approximately 100 hours. Even though Stockholm (almost 6500 heating degree F days) is colder than Chicago (6200 degree days), the Thermodeck office buildings with stored heat from lights and people annually use only about 4 kWh/ft^2 for auxiliary electric resistance heating. PCMs are still in the development stage, but when reliability is established, containers of PCM could be loaded into the hollow cores of concrete floor-ceiling slabs. The material would change phase at 70–74°F, locking the slab at that temperature, and office supply air during summer or winter would be only a few degrees different.

Energy management systems (EMS's) are being installed in an increasing number of commercial buildings. The primary function of an EMS is to save energy and power. The EMS is a computerized system which controls the

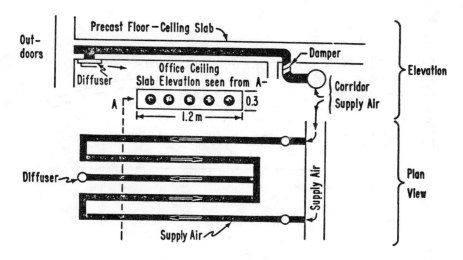

Figure 4. Swedish thermodeck office building. Forced convection and increased thermal surface area enhance the thermal storage. Each 10 m² office module has two slabs of concrete with precast internal air channels (Ref. 5).

major systems within a building, including HVAC, lighting, and miscellaneous operations. The computer and its embedded programs respond to changes in conditions caused by weather, occupants, and equipment. An EMS can turn equipment on and off at predetermined times or according to sensor information, control duty cycles of equipment to minimize energy consumption, adjust position controls for valves and dampers, and a wide variety of other control operations. The second function of an EMS is to monitor building conditions, such as outside air and interior zone temperatures, electrical demand and consumption, boiler operation, water flows and temperatures, and pump and fan status. The data can be logged and then used later for a complete building energy analysis.

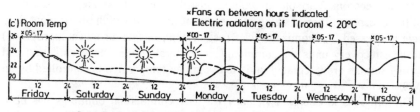

Solid Curve = One winter week, with a cloudy weekend.
Dashed --- = Separate sunny weekend.

Figure 5. Winter. During a winter week, the outdoor temperature varied between -2 and -10 °C. On Friday afternoon, when the internal gains and the fans and radiators are turned off, the indoor temperature starts to fall from 24 °C. By about Monday morning 20 °C is reached. Fans are turned on (the ventilating air system runs with 100% recirculation) and the air is heated one or a few °C depending on the outdoor temperature. At 8:00 Monday morning, the temperature level is still about 20 °C. Each weekday the occupied offices climb 2–3 °C in temperature, and empty rooms remain about 20 °C. Each night the indoor temperature falls 1–2 °C. By Friday afternoon, the cycle is complete (Ref. 5).

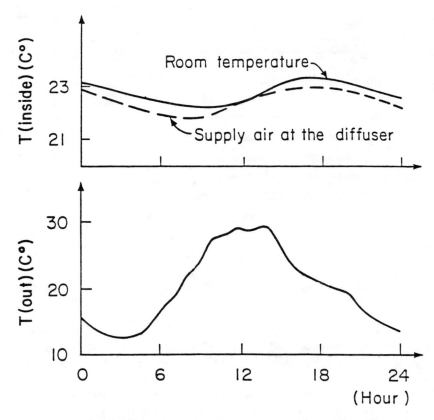

Figure 6. Summer. During the day, the outdoor air temperature varies between 12 and 30 °C and the air going into the system has about the same temperature. The supply air, after passing through the slab cores, has a much smaller variation (22.5 ± 0.5 °C) and the room temperature during office hours increases from 22 to 23 °C (Ref. 5).

The HVAC systems of new buildings are more versatile and efficient than their predecessors. Variable air volume distribution systems are the rule now, not the exception. Proper sizing of equipment to operate with an efficient duty cycle is stressed. Economizer cycles (use of cool outside air when appropriate) and heat recovery units are commonplace. Natural cooling and ventilation and thermal storage are integrated into the overall system, which is controlled by an EMS.

The progress in energy efficiency for new commercial buildings has been quantified by the BECA-CN data base[15] at LBL. New energy-efficient buildings consume about one-half that of buildings in the late 1970s and early 1980s which complied with the ASHRAE 90-75 voluntary standard. These low energy users incorporate most or all of the special features discussed above — efficient lighting, storage, EMSs, and modern HVAC systems — often at little or no additional cost when compared to conventional construction. The efficient performers attain office site intensities of 30 to 70 kBtu/ft^2-yr and source intensities of 100 to 220 kBtu/ft^2 yr, with some below 100 kBtu/ft^2 yr.

Existing commercial and high-rise residential. There is considerable potential for improvement in the energy efficiency of the U.S. stock in the commercial and high-rise residential sectors. Initial retrofit efforts in the commercial sector have been modest.[16] They are mainly low-investment "proven" measures which cost less than $1/ft^2$, save approximately 20%–30% in energy, have relatively fast payback times of 1 to 2 years, and have an associated cost of conserved energy that is much less than present fuel prices. The most popular retrofits have been HVAC operations and maintenance (O & M) and lighting. The O & M is a general category, encompassing a large number of specific measures, ranging from simple actions like changing thermostat setpoints and setbacks to reasonably sophisticated diagnosis and adjustment of boilers, chillers, fans, and distribution systems. Lighting retrofits include delamping, replacing lights with more efficient bulbs or fixtures, and the addition of lighting controls. Installation of an EMS is becoming more common, as is an HVAC modification such as an economizer cycle.

Even though there have been some spectacular successes in the multifamily residential sector (especially adjustments and computerized controls of inefficient HVAC systems), the average retrofit results in moderate savings of 10% to 20%, 6 year payback times for cheaper system measures implemented mainly on fuel-heat buildings, and 20 year payback times for more expensive shell improvements implemented principally on electric-heat buildings.[17] The system measures are mainly heating controls, such as outdoor resets, high-limit outdoor cutouts, and thermostatic radiator vents; and heating system retrofits such as new burners and vent dampers. The shell modifications consist mainly of window retrofits, improving the domestic hot water system, and shell insulation.

The technical potential for saving energy in multifamily residential buildings is quite large, but these savings are not very likely, due to various institutional, financial, and technical barriers. Especially relevant is the reluctance of building owners to invest unless the payback time is of the order of 2 to 3 years. Averaged over groups of buildings, pre-retrofit predictions agreed fairly well with actual savings, although individual buildings might vary up to 20%.[17] At one particular project (Hood River), measured savings were systematically below predicted values by a disappointing 57%. See Figure 7 for the experience with retrofitting schools.

2b. Window technologies

About 5% of the nation's energy (about 4 quads/year, or the equivalent of 2 million barrels of oil per day, 2 Mbd) is lost[13] through the nation's windows. These losses are about evenly split between residential and commercial buildings. Presently available window technologies described below can save at least one-third of this amount, or about 0.7 Mbd. In addition, good window design that enhances natural light, "daylighting," can save about 10% of the energy usage of a commercial building by reducing the need for electrical lighting and the need for air conditioning the heat from the extra lighting. (See Figs. 8 and 9.)

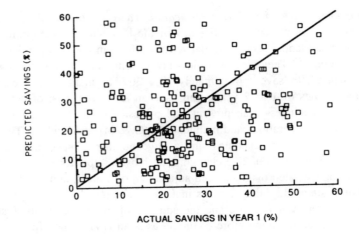

Figure 7. Comparison of predicted and actual energy savings for retrofitting Wisconsin schools (LBL).

The R-value of a typical-single glazed window is about R-1 ($1/R$ in Btu/h ft^2 °F) as compared to a wall of R-11, or R-19. A number of steps can be taken to raise the R values from R-1 to the commercially available superwindows of today of R-3 to R-5. Additional steps can be taken to obtain R-11, but these are not yet cost effective. Improved R values can be obtained by:

(R-1.7) About one-half of the single-glazed loss can be saved by using double glazing, as in "Thermopane" windows, which have a dead air space sealed between the two sheets of glass.

(R-4) Glass strongly absorbs the infrared emissions from the interior of buildings. A thin film of 0.002 in. of low-emissivity material (such as SnO_2:F, In_2O_3:Sn, or Cd_2SnO_4) will reflect the infrared back into the room, acting as a heat mirror. These transparent materials are called "low E" because they reflect about 90% of the infrared, and absorb only about 10%, while glass absorbs 100% of the IR. Since the low-E materials are also poor radiators, they can be used on the outside of the glass to reduce thermal radiation. These low-E materials have a high transmittance in the visible (70%), and a low emissivity (15%) and high reflectance (85%) in the infrared. The R value of a double-glazed window can be doubled by sputtering the low-E material onto the inner surfaces of the double-glazed window at the factory.

It was originally hoped that it would be possible to retrofit all of the nation's windows by spraying or glueing the low-E material onto all existing windows, but unfortunately this is not yet commercially viable. About 10% of the 1985 market of 700 million square feet went into low-E windows at a retail cost of \$2–3 per square foot, corresponding to a cost of conserved energy of \$4/MBtu for natural gas, or 2 cents/kWh for electricity. Some companies anticipate that the cost will drop to \$1 per square foot, and that the market will reach 50% by the early 1990s. When all the windows of the nation have low-E windows, the U.S. will save about 0.7 Mbd.

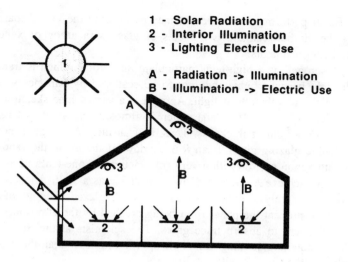

1 - Solar Radiation
2 - Interior Illumination
3 - Lighting Electric Use

A - Radiation -> Illumination
B - Illumination -> Electric Use

Figure 8. Key components and relationships in the daylighting evaluation methodology (LBL).

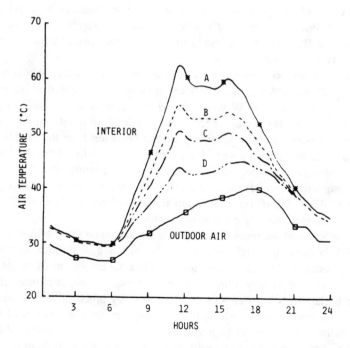

Figure 9. Interior air temperature versus time of day for a sports model automobile under static soak conditions in Phoenix, AZ on a clear day on a typical June 30th. Curves are shown for glazing with varying solar transmittance at normal incidence (%) (LBL).

(*R*-6) Better performance can be obtained by keeping the low-*E* coating and replacing the air between the two sheets of glass with argon or xenon gas, which are better insulators.

(*R*-11) "Aerogel" is a high-tech, particulate insulator[18] which is transparent to visible light because light tends not to be scattered by particles which are smaller than the wavelength of light. Aerogel is a very sparse skeleton of tiny glass particles (5% silica, 95% air) which is almost transparent, yet has a resistance of *R*-5 per cm of thickness. Thus, half an inch of aerogel sandwiched between double glazing gives us an *R*-6 window. If the air in the sandwich is replaced with xenon gas, or with a very soft vacuum of one-tenth of an atmosphere, one can achieve *R*-11, a window as resistive as a wall. Another way to achieve *R*-11 is merely to highly evacuate a double-glazed window, of course, with spacers to prevent the panel from being crushed by the atmospheric pressure and a "getter" to absorb trace gases. The scientists at the Solar Energy Research Institute are looking for a cheap way to laser-seal the panels so thoroughly that the vacuum will last 20 years.

2c. Lighting technologies

About 20% of U.S. electricity is devoted to lighting. The U.S. uses[19] about 500 billion kWh for lighting, which is the equivalent of 100 baseload plants of 1 GWe. (1 GWe plant operating at a 0.6 load factor produces 5 BkWh/year.) About 200 BkWh, or 40 of the 100 GWe plants, are used to power incandescent bulbs, 200 BkWh more are used to power fluorescents, and the rest power high-intensity and outdoor lighting.

Technical advances[19] since the 1973 embargo have considerably increased the efficacy (lumens/Watt, or lm/W) of lighting. The theoretical maximum possible efficacy is 683 lm/W when all the radiation is at the wavelength of greatest response for the eye at 0.555 μ. The maximum possible efficacy is 350 lm/W for the case of uniform spreading all the radiation over the visible spectrum. Incandescent bulbs are very inefficient (about 11 to 18 lm/W) because they are considerably colder than our sun, which predominantly emits in the visible. Thus the thermodynamic efficiency of incandescent bulbs is quite small, about 4% (15/350) for white light lamps, and about 2% (15/683) when compared to maximized single frequency lamps.

On the other hand, mercury fluorescent tubes are much more efficient (about 100 lm/W) because they rely on atomic transition probabilities rather than black-body radiation. A fluorescent tube converts about two-thirds of the input electrical discharge energy into ultraviolet photons. The radiation diffuses through the plasma, being absorbed and reemitted by other mercury atoms, and it is finally converted to visible light by phosphors on the inner wall of the tube. A "ballast" provides the high voltage to initiate the ionization of the mercury vapor, and it also limits the current for safe operation. A number of improvements were incorporated to raise the efficacy of the fluorescent lamp system from its 1973 value of 63 lm/W, to 85 lm/W in 1986. The efficacy of the lamp alone has been improved from 79 lm/W to 95 lm/W, and could, perhaps, reach 200 lm/W by further improvements now being studied. These improvements, listed below, have the long-term potential to reduce the number of GWe power plants needed for fluorescents from 40 to 15. Further savings will accrue because air conditioners will not not have to work as hard to extract the excess heat from the lights. Of course furnaces will have to provide the extra heat during the winter.

High-frequency ballasts (1982). Pre-1980, 60 Hz fluorescent lamps had a 17% energy loss in the ballast, and from end losses because some of the ions hit the cathode and anodes without producing the ultraviolet photons. By operating the lamps at high frequencies (20–40 kHz) with solid-state ballasts (switching power supplies), the ballast efficiency was increased 6%, and the end losses were decreased by about 10%–15%, resulting in an increase in system efficacy of about 20%. In addition, because the power output of the solid-state oscillators are easily adjusted, the light output of the lamp can be varied over a large range. It is then possible to develop lighting control systems that reduce electric lighting when natural light is available or when the office is vacant (occupancy sensors), saving an additional 10%–50%. The Lawrence Berkeley Laboratory initiated the high-frequency ballast program in 1976; commercial production started around 1982; and these ballasts should attain 50% of the market by 1995. The cost of conserved electricity is about 2 cents/kWh, compared with the national average price of 7 cents/kWh.

Narrow-band phosphors (1983). Fluorescent tubes have been designed with a single broad-band phosphor to produce a cool white light. To obtain a warmer, more pleasant color, the efficacy of the lamp was sacrificed by 10% to 20%. There are now available narrow-band rare earth phosphors that emit in narrow regions in the blue-green, yellow, and red regions that can be combined to permit a wide range of warm and cool colors with no sacrifice in efficacy. The manufacturers have now adopted both the high-frequency ballasts and the tuned phosphors, raising the efficacy of the warm color by 10% to about 110 lm/W, ultimately saving about 4 GWe.

Isotopic enhancement (1986). Additional savings can be realized[20] by adding a small amount (3%) of the rare isotope Hg^{196} when filling the lamp. The increase in the amount of Hg^{196} gives an additional spectral pathway for the photons to "diffuse" from the center of the tube to the tube wall, reducing the ultraviolet entrapment by the plentiful mercury isotopes. A gain of 5% has been realized in experiments with enhanced Hg^{196} tubes.

Magnetic enhancement. Experiments have demonstrated that by applying an axial magnetic field to a fluorescent tube, or by using magnetic phosphors, the efficacy has been raised by 10%.

Surface waves. By exciting the plasma with very high frequency surface waves in the near GHz region, it is possible to eliminate the need for electrodes as well as to reduce the optical losses in the volume of the plasma, raising the efficacy by about 40%. Losses in the GHz ballast will be 5% to 15% more than in the 30 kHz supplies, giving a net gain of 25% to 35%.

Two-photon phosphors. Fluorescent tubes are already six times more efficient than incandescent bulbs, but they are still wasteful when one considers that the 5-eV ultraviolet photon from mercury is converted into only one visible photon of 2.5 eV, wasting 50% of the UV energy. A search is currently being carried out to find a phosphor to convert one 5-eV photon into two photons in the visible region. Such a discovery would double the efficiency of conversion of the UV radiation to visible light by the phosphors.

It must be pointed out that these improvements were not developed by the large lighting companies, but rather they were initiated by Federal funds to the Lawrence Berkeley Laboratory (LBL), and by contracts let to smaller firms. It was only after the initial success of these developments that major companies such as Norelco and GE introduced similar products. It is reasonable to assume that the $3 million spent by LBL and its contractors advanced the market penetration by 5 years, saving the U.S. billions of dollars.

Compact fluorescent light bulbs. Small fluorescent lamps yield three-to-four times more lumens/Watt and last 10 times longer than equivalent incandescents. They are selling well in the commercial market, where they save money, both for electricity and for bulb changing. The familiar 100-Watt incandescent bulb costs only about 50 cents, but over its 750-hour life (one month of continuous light) consumes over $5 of electricity. Its efficacy is only 17 lm/W, compared to the 60 lm/W for the new compact fluorescent SL-18. The 18-Watt SL-18 costs $20 and replaces a 75-Watt incandescent (more precisely, a series of 10 of them), and in its whole 7500-hour life (nearly a year of light) uses only $10 of electricity. Thus, the life-cycle cost of the SL-18 is $30 vs $55 for the 10 equivalent incandescent bulbs. In energy terms, the SL-18 avoids burning 400 pounds of coal back at the power plant. The compact fluorescents are best utilized in sockets that have a high annual use; incandescent lamps will still be widely used elsewhere. In addition GE sells a twin-tube lamp and ballast together for a total cost of about $18, but replacement lamps can be purchased separately for $6, which is just the price of the 10 incandescents that it replaces. The advances in the larger fluorescent and compact fluorescent lamps should save the U.S. a minimum 20 GW, or 20% of the 100 GW that the U.S. devotes to lighting.

2d. Refrigerators

Refrigerator energy consumption[21] has varied widely over the years as shown in Fig. 10. Energy consumption increased by about a factor of 3 from 600 to 1800 kWh/year between 1950 and 1972, due to a combination of increased size from 7 ft^3 to 17 ft^3, and the addition of extra features such as automatic defrost, and a reduction of 40% in actual efficiency. After the energy embargo of 1973 and the resulting California standards, the energy consumption of refriger-

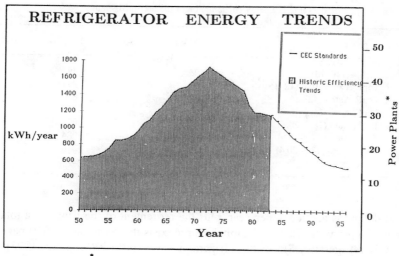

*Plants needed to run 125 million U.S. refrigerators and freezers.

Figure 10. Refrigerator energy usage in kWh/year (1950–1983) and California Energy Commission Standards (CEC, 1983–1995). The number of GWe electrical power plants needed to power the 125 million U.S. refrigerators and freezers is given in the right-hand margin (LBL).

ators reversed direction, dropping to 1200 kWh by 1983, and it will further drop to 600–700 kWh/year by 1993.

The refrigerators are improved by using:·
- polyurethane foam insulation which is less conductive than fiberglass,
- thicker foam in the doors and walls,
- more efficient motors and compressors,
- increased heat exchanger surface area,
- double gaskets,
- evacuated panels, bottom mounted condensors,
- mounting fan motors outside the freezer compartment,
- elimination of case heaters or provision of a switch.

It is instructive to estimate the number of electrical power plants that will ultimately be saved from the conservation measures adopted between 1975 and 1993 on the nation's 125 million refrigerators and freezers (r&f):

kWh/year saved=(1800–700 kWh/r&f)(125 M r&f)=1.4 × 10^{11} kWh/year,

1 GWe power plant produces = $(10^6$ kW)(8766 h/year)(0.6 duty factor)
= 5.3 × 10^9 kWh/year,

Number of base-loaded plants saved = (1.4 × 10^{11} kWh/year)/(5.3 × 10^9 kWh/GW-year) = 26 GWe.

How does one evaluate the additional cost of $150 to improve the refrigerator that saves 1100 kWh per year? The cost of "conserved energy" is a useful way to evaluate the return on the additional costs. The cost of money in con-

Table 2. Energy consumption of the best refrigerators.

Type (ft^3)	Best model (kWh/year)	Average of better (kWh/year)
Manual (10–13)	498	497
TopFreeze, Partial (12–15)	661	709
TopF, Auto (14.0–16.4)	767	800
TopF, Auto (16.5–18.4)	767	839
TopF, Auto (18.5–20.4)	840	893
TopF, Auto (20.5–24.0)	945	984
SideF, Auto (18.5–20.4)	1156	1218
SideF, Auto (20.5–25.0)	1158	1210

stant dollars is about 5% per year, plus the retirement of the debt, for a total of about 7% per year. This total capital charge costs the owner about $10 per year for this refrigerator. Since the refrigerator saves 1100 kWh (1975–1993), the cost of conserved energy is about (1000 cents/1000 kWh), or 1 cent per kWh, far below the U.S. average price of electricity of 7 cents per kWh.

Many models of refrigerators consume considerably less than the side-by-side freezer, automatic defrost model discussed above. The energy consumption of the best refrigerators is listed in Table 2.[22]

2e. Least cost analysis

The Lawrence Berkeley Laboratory has the lead within DOE to calculate the economic impacts of the Federal Standards for Residential Appliances (Fig. 11).

By using both the costs of the various improvements for the refrigerators and the concomitant energy savings, Levine[23] has determined the life-cycle costs for each situation, and then minimized the life-cycle cost to obtain the optimal result. LBL obtained the life-cycle cost savings for the optimized refrigerator and freezer designs shown in Table 3.

3. The conservation market place

A variety of factors impede conservation in the market place. Here are a few of the problems.

3a. Fuel prices and energy taxes

This chapter does not deal directly with oil, but it is useful to point out some of the effects of oil pricing. To this day, the U.S. has considerably cheaper gasoline prices as compared to Western Europe and Japan as shown in Table 4.[24] Amendments as small as one cent/gallon to raise money for energy R&D have been defeated by the U.S. Congress. The political process correctly realizes that a man's automobile is his moving castle, and that only in extreme circumstances can the gasoline tax be increased in order to save energy. In the near term, the demand for energy is relatively inelastic (unresponsive) to small increases in energy prices, but in the longer term, the elasticities rise, as the

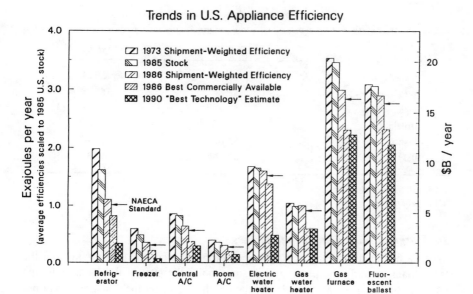

Figure 11. Trends in U.S. appliance efficiency (LBL).

society responds more favorably to conservation through enhanced end-use efficiency. In view of the higher prices, it is no wonder that Europe and Japan use much less energy per unit GNP as compared to the U.S. In Japan's case, it uses about 50% as much energy per unit GNP as does the U.S., but the *comparison is difficult* because the Japanese use both more efficient technologies and have a more modest life style. We conclude that energy conservation through higher taxes will not be achieved because (1) it is unlikely politically, and (2) the effects will be small because the energy market place is relatively inelastic to price changes. Thus, energy standards seem to be the more practical avenue to increase end-use efficiency.

Table 3. Life-cycle cost savings for optimized refrigerator and freezer designs.

Appliance	LCC savings discount at 5%	Payback period	Cost of conserved energy (c/kWh)
Refrigerator/freezer			
Top freezer, automatic	$249	2.12 years	1.22
Side freezer, automatics	$192	1.02 years	0.59
Freezer			
Chest, manual	$96	1.75 years	0.95
Upright, manual	$241	0.87 years	0.47

Energy conservation

Table 4. Comparative gasoline taxes (Ref. 24).

Country	Consumer cost[a]	Tax component[b]
United States	$0.98	$0.29
West Germany	$2.28	$1.48
Japan	$3.60	$1.61
Britain	$2.60	$1.68
France	$3.10	$2.44
Italy	$4.12	$3.31

a. Per gallon, including tax.
b. Per gallon.

3b. Marginal pricing and capital access

When a new coal or nuclear plant is being constructed, the utility takes out the large loans to build the plant, and at some point the cost of the loans is folded into the rate base. If the plant is expensive, the effect on the consumer may not be too large since the customer pays the average cost of electricity. For example, if electricity from a new facility costs twice as much as the "old electricity," it will only raise the price of electricity by 10% if the new facility comprises 10% of the grid. The average pricing of energy, and the easy access to capital markets can create an environment for non-optimal economic solutions. On the other hand, the individual consumer directly pays the additional costs for enhanced-end-use efficiency (R-19 vs R-30 walls), and in principle should be able to react to good economics by minimizing life-cycle costs by proper purchases at the economic margin.

Good economics would have the society compare the marginal cost of the production of a kWh versus the marginal cost of conservation of the same kWh. Some British and American utilities[25] have used a continuously varying "spot price" of electricity with a heavy reliance on simple computers to match the marginal values of supply and demand more equitably. Because the capital cost for new electrical generation is in the region of $1–2 (or more) per Watt, an additional 75-Watt light bulb that is operated continuously will cost the utility about $100 in up-front capital. Various idealistic schemes have been discussed which would pass these up-front costs onto the consumers during construction, forcing them to use more energy efficient technologies. If the consumers had to pay part of the extra capital for the inefficient bulb as compared to the 18-Watt compact fluorescent bulb, it would amount to about $1.5/W \times 57$ W = $85, which might affect their decision.

3c. Diverse ownership

Because Americans move on the average about every seven years, the typical homeowner does not strongly consider those conservation measures that have payback periods of more than five years. In addition, many Americans do not own their own homes, and, thus they are not encouraged to make energy improvements in their rental homes and apartments. Since landlords do not pay the utility bills, they also have little incentive to make investments in energy conservation for their rental units. Because of the problems of transitory and diverse ownership, investments with a payback period of 3 years for a return of 30%/year are neglected while the consumer typically invests his money elsewhere in 10% bonds. It is for these reasons that the market forces are quite inelastic for energy in the residential sector. Governments have had only modest success in encouraging landlords to improve the energy efficiency of rental units. A couple of examples are: (1) The weatherization program has donated free insulation for lower-income dwellings. (2) The city of Portland, Oregon has required existing buildings to be reasonably "energy efficient" before they can be sold. (3) Some states have used an energy labeling system which requires the labeling of the annual energy consumption by buildings. These are obviously only partial solutions.

4. Energy economics

The energy-conservation laws often require that energy standards must be carried out on a cost-effective basis. If the payback period is too long for a particular conservation measure, then the law usually does not require its implementation.

4a. Life-cycle costs

The life-cycle cost (LCC) is the sum of the initial purchase cost and the future fuel, operation, and maintenance costs. The usual approach is to calculate the LCC for an appliance without improvements and with a variety of energy saving measures. The future costs and benefits are discounted[26] to their present value according to the standard formulas. For example, if money is worth 5%/year in constant dollars, $100 today has the same value as $105 one year from now. Thus a $105 future profit (or loss) must be "discounted" to its present value of $100 if we are to compare the future profits (discounted benefits) to capital costs at the time of the purchase.

The choice of the discount rate must be done very judiciously since the choice of this parameter can have a very large effect on our conclusions and the resultant energy standards. A large discount rate dramatically reduces the value of future benefits, while a zero value for the discount rate neglects the return on invested money. The discount rate is necessary when comparing the value of money in hand that one spends for an improved appliance versus the avoided costs of longer term reduced fuel costs that result from buying the improved appliance. A choice of 5% (in constant dollars) is usually considered about the right value for the discount rate since the prime rate of interest is about 5% plus the rate of inflation. Some have used a lower value of 3%, based on long-term dividends from bonds, and others have used higher values if risks are involved.

4b. Supply curves

When mining a resource, supply curves (Fig. 12) that plot the available resource (such as barrels of oil) as a function of the cost of extraction are used to

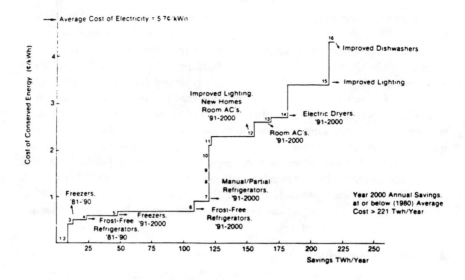

Figure 12. Supply curve for electrical appliances in the year 2000. The unit of conserved energy (in constant 1980$) assumes that all increased costs are amortized over the useful life of the measure, using a 3% (real-dollar) interest rate (Ref. 5).

determine the resource that is economically recoverable. In a similar fashion, conservation "supply curves" can be used[27] to establish the amount of energy which can be saved through conservation as a function of the cost per unit energy expended on conservation measures. For example, one could obtain one point on the conservation supply curve from refrigerators as follows: Improvements in refrigerators that cost 1 cent/kWh can yield an energy saving of 125 BkWh/year, generated by a total of 25-GW power stations. Other conservation measures would fill in the rest of the curve.

5. Environmental effects

When chosing energy technologies, one should compare the environmental impacts of each choice. The National Environmental Protection Act of 1970 established the Environmental Impact Statement (EIS) procedures which allow for public comment and judicial review. In general, a production option impacts adversely on the environment, much more than the conservation option that replaces production, but the details must be analyzed on a case-by-case basis. The substantial amounts of pollutants from large power plants must be compared to the pollution of the comparable conservation option. For example, a 1-GW coal power plant will consume about 2,500,000 tons of coal a year, releasing about 50,000 tons of SO_2 a year and 10,000,000 tons of CO_2; a 1-GW nuclear plant will produce about 30 tons per year of spent fuel, and have a finite probability of an accident. Clearly the pollutants from a *safely operating* nuclear plant are more manageable than those from the coal plant. Even the apparently pleasing hydro-electric plants can flood valleys, and distort rivers, while wood burning has been known to pollute a valley.

The main detrimental environmental effect of conservation in buildings is the degradation in indoor air quality from increased pollutants and from increased levels of radon[28] in houses. As it turns out, the source strength for radon is much more important than the degree of tightness (the number of air exchanges per hour, or ACH) of the house. Houses vary greatly, by factors as large as 10–20, in the number of ACHs, while the radon levels vary by a factor of 1000 or more. There is little correlation between the tightness of houses and their radon levels (pCi/liter) when one makes a scatter plot of all houses. In practice, the houses with the highest rates of radon naturally contain high amounts of radon, primarily because they are in high source regions, rather than because they are naturally tight, energy conserving houses. However, if a house has a problem, and one wishes to maintain very low air infiltration rates, but yet also flush out the interior air, this can be accomplished with air-to-air heat exchangers at a cost of about $400/house.

It is clear that there are tradeoffs. Energy losses from infiltration are proportional to the number of air exchanges per hour, while adverse health effects are inversely proportional to the number of ACH (or proportional to the residence time). Thus, there is a point of diminishing return as to the air tightness of housing, and one can choose a standard based on various judgment calls. Recently the Environmental Protection Agency has estimated that about 5000 to 20,000 deaths from lung cancer are caused annually by radon. This figure is considerably lower than the 100,000 lung cancer deaths from smoking, but it may well be the largest cause of lung cancer for nonsmokers. (See Fig. 13 for radon distributions in houses.)

The EPA has recommended reducing the radon level in houses with radon levels exceeding 4 pCi/liter, which is about three times the average value of 1.5 pCi/l. The EPA data has been criticized for using data obtained in basements, without reducing these values for the living areas above the basement. Perhaps about 6% of houses,[29] and not the 20%–30% as stated by EPA, exceed the 4 pCi/l level, and remedial methods should be applied to these houses.

6. Mandatory standards

6a. Automobile fuel economy standards

In 1975, the Congress passed the Corporate Average Fuel Economy (CAFE) standards that require a fleet averaged energy economy of 27.5 miles per gallon. Our present auto fleet averages about 19 to 20 mpg. If one considers that the U.S. will have 160 million autos and light trucks by the year 2000, each traveling 10,000 miles per year, the U.S. would consume in the *steady state*:

- 8 Mbd at the 1973 rate of 13.5 mpg,
- 4 Mbd at the CAFE rate of 27.5 mpg,
- 2 Mbd at a doubled CAFE rate of 55 mpg.

Congress is now considering raising the standard to 35 mpg, and many are calling for a standard of about 50 mpg, a regime where most of the manufacturers have developed new prototypes. At the time of passage of CAFE in 1975, the automobile industry fought very hard to prevent these standards from

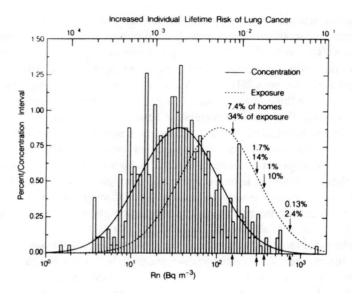

Figure 13. Histogram of the aggregate radon distribution data from samples obtained in 552 houses. The solid curve is the log-normal function with the geometrical mean and geometrical standard deviation, calculated directly from the radon concentration data. The top axis, lifetime risk for individuals, also refers to this solid curve. The dashed curve is the distribution of total population exposures to radon progeny, as a function of indoor radon concentration. The labeled arrows refer to the fraction of homes and the fraction of the total population exposure associated with radon concentrations equal to or greater than the indicated values (LBL).

being passed. One of the arguments that was widely used at the time was: "The new energy efficient cars will be more expensive, this will prolong the life of the gas guzzlers and therefore we should not have standards." Of course, this effect is both small and transitory, and therefore irrelevant. Now 15 years later, the U.S. auto industry is having a difficult time competing against the supposedly more reliable energy efficient imports which are limited by quotas in order to salvage the U.S. industry. Most observers agree that our competitive situation would be much worse without the standards.

6b. Building standards

This section is divided into three parts. First, the different types of standards are defined and briefly described. Next, the principal participants in the development of standards are introduced and their roles are discussed. Finally, comparisons between the various regulations are made as to emphasis, stringency, predicted effects, and estimated results.

Building standards fall into three different categories: prescriptive, performance, and some combination of the two approaches. Prescriptive standards involve the specification of each individual building component that influences energy consumption (levels of insulation, type of glazing, efficiency of conditioning equipment, etc.), usually in the form of various component "packages." Performance standards require that the design energy consumption for a building fall below a numerical energy budget established by building type, occupancy level, and climate zone. Computer simulation is required to demonstrate compliance. Combinations of prescriptive and performance standards can be handled by a "points" system whereby the effects of component trade-offs have been calculated by computer modeling. The prescriptive approach has the advantage of being easy both to implement and to demonstrate compliance, whereas the performance approach allows much more design flexibility in reaching a required efficiency level. In recent years computer methods and performance standards appear to be growing in popularity.

Mandatory energy efficiency standards applicable to all new buildings throughout the country have never been established by the U.S. government. Instead, building codes incorporating standards developed by the American Society of Heating, Refrigerating, and Air Conditioning Engineers (ASHRAE) have been adopted and are enforced by state and local governments. In addition to ASHRAE, the state of California and the Northwest Power Planning Council have shown leadership in the development of building standards. Energy regulations for all new federal buildings are in the process of being updated and implemented. The organizations involved in standards development (ASHRAE, California Energy Commission, N.W. Power Planning Council, and DOE) have interacted with each other, with private industry, and with the public to arrive at continually evolving building efficiency rules and guidelines.

In 1974, at the urging of NCSBCS (National Conference of States on Building Codes), ASHRAE accepted the challenge of writing a consensus energy standard for new buildings based on a criteria document drafted earlier by the National Bureau of Standards.[30] This model standard was published in 1975 and became known as ASHRAE Standard 90-1975, "Energy Conservation in New Building Design." Most of the original standard was revised and released in 1980 as ASHRAE Standard 90A-1980. Both 90-1975 and 90A-1980 were primarily prescriptive standards, but allowed compliance via the performance approach.

The next revision of the standard began in 1982, with the support of DOE. There are separate documents for the energy efficient design of new commercial and new high-rise residential buildings (Standard 90.1P) and for the design of low-rise residential buildings (Standard 90.2P). Differences accounting for 15%–30% energy savings exist between the proposed revision and the 1980 revision.[31] These new standards allow for compliance by any of the three basic approaches. As of early 1989, Standard 90.1P has undergone a third public review and is nearing the completion of the very slow ASHRAE consensus process, whereas 90.2P is about to start public review. Compliance with ASHRAE Standard 90 is voluntary, but building energy codes and standards throughout the country at the local and state government level are based on Standard 90. It is clearly the dominant influence on efficiency standards in the U.S.[32] In addition to Standard 90 for new buildings, ASHRAE has established energy conservation standards for existing buildings, ASHRAE Standard 100.

Even though there are no mandatory national building energy standards, the federal government has been active for many decades in promoting energy efficiency in buildings. The HUD (Housing and Urban Development) minimum property standards have existed since the 1950s and have been revised several times so that its residential building standards are reasonably energy efficient.[33] In the early to mid-1970s, the GSA (Government Services Administration) designated several federal office buildings as energy conservation demonstration projects. New technologies were tested, and the design resource energy budget was set at 55 kBtu/ft^2 yr, an ambitiously low usage rate for that era of mainly 250–300 kBtu/ft^2 yr buildings. Even though the demonstration buildings enjoyed only moderate success, the GSA adopted energy efficiency guidelines for all new federal office buildings. Consumption data in subsequent years have shown some low consumers but no definite trend towards efficient new buildings in the federal sector. In 1976 Congress passed the Energy Conservation Standards for New Buildings Act and required DOE to develop mandatory federal standards for new buildings. This was to be applicable to all buildings constructed with any federal assistance (which would be essentially *all* new buildings) and was to take the form of building energy performance standards (BEPS). DOE released the draft standard in late 1979, but the objections from private industry were so strong that Congress in 1980 backed down and designated the standards as models for voluntary state adoption.[32] More recently in May of 1987, DOE published a notice that more stringent energy-conserving design requirements for new commercial buildings were being developed in parallel to ASHRAE Standard 90.1P and would be mandatory for all new federal buildings (approximately 10% of the total commercial construction). These new standards are essentially completed and should be in force sometime in 1989. They will allow for compliance by all three basic approaches.

The state of California has been at the forefront of the development of building energy standards and codes since 1975. In that year the California Energy Commission released Title 24, which required certain energy efficiency measures in all buildings, both residential and non-residential. There have been numerous revisions of the standards, many of them innovative and also based on minimum life-cycle cost. From an early prescriptive standard, the regulations evolved into a flexible set, allowing compliance via prescriptive, performance, or "points" approaches. In many cases the California standards are more stringent than ASHRAE Standard 90.

Also providing leadership in the development of standards and in serving as a model for other regions has been the Northwest Power Planning Council (Washington, Oregon, Idaho, Montana). Although the Council's model conservation standards (MCS) are based on ASHRAE Standard 90, its cost-effective regulations go beyond ASHRAE's standards, particularly in the case of lighting efficiency. The implementation of conservation programs by the Bonneville Power Administration (BPA) in response to the Council's policy decisions has clearly demonstrated the potential savings from improved building energy efficiency.

Primary emphasis for residential buildings has been for increased levels of insulation in the thermal envelope, restrictions on the amount and type of glazing, decreased rates of air infiltration, and better appliance efficiencies, including space conditioning equipment. For non-residential buildings the measures have included improved efficiencies for HVAC equipment and service hot water, decreased artificial lighting levels and power densities, and more efficient energy distribution systems, in addition to the usual thermal tightening of the building envelope. Ideally the standards should be set at the economic optimum for the building, which would vary depending on local construction costs, fuel choices, and local fuel costs.

ASHRAE Standard 90-1975 was not a very restrictive nor economically optimal standard. It was the initial effort, and evolved through a consensus procedure. The present California Title 24 standards are the most stringent in the country for residential buildings.[32] In addition, the Northwest Power Planning Council's standards and the new ASHRAE standards are also close to the optimum conservation combination of residential measures. Allowances are given in California, but not by ASHRAE, for passive solar features such as thermal mass. In the non-residential sector the most discussed feature in the standards is that of lighting, with the difficult factors of illumination levels and mixing between fluorescent and incandescent sources added to equipment efficiency requirements. Lighting standards are typically expressed in terms of power density — connected watts per ft^2 of illuminated area — and also are dependent on the particular space usage or task. Whole-building power density limits or room-by-room calculations are allowed. The ASHRAE lighting standards are not as stringent as the California standards.

ASHRAE Standard 90-1975 was estimated to lower energy usage in new residential buildings by an average of 11% and in new non-residential buildings by 40%-50%, when compared to conventional practice in the mid-1970s.[33] The application of ASHRAE 90-1975 did not appear to increase initial construction costs, but did increase design costs. Overall, the annual energy savings clearly offset any increase in design costs during the first year of building operation. The proposed (but never implemented) BEPS by DOE was estimated to result in energy savings of 22% to 51% in single family residences and 17% to 52% in commercial and multiresidential buildings when compared to buildings designed and constructed in the mid-1970s.[34] The first ASHRAE Standard 90 update (90A-1980) was calculated to provide small (4%–5%) energy and cost savings over Standard 90-1975, but the second revision (90.1P) has been estimated to result in an average of 16% savings over 90A-1980 for new commercial buildings.[35] Standard 90.1P has been shown to be economically feasible via payback and life-cycle cost analyses. The new proposed DOE standards for federal buildings have been developed in parallel to 90.1P and should result in comparable energy savings.

Actual measured energy performance data for buildings verify that energy is being saved due to the promulgation of standards over the past 14 years. In the Northwest, BPA's new MCS houses have been shown to use 45% less space heat than newly constructed conventional "control" houses.[36] Lawrence Berkeley Laboratory's compilation of energy data[15] indicates that new office buildings which comply with the ASHRAE 90.1P standards use about 50–80 $kBtu/ft^2$ yr of site energy for all end uses, or about one-half the average energy use for existing U.S. office buildings. The "best" new buildings, which have energy features beyond the standards, are considerably below this level, using as little as 20–40 kBtu/ft^2 yr in site energy (or less than 100 kBtu/ft^2 yr in resource energy).

6c. Appliance standards

In 1976, the State of California adopted[7] energy standards on refrigerators, freezers, and air conditioners, and in 1983 this was extended to water heaters, furnaces, heat pumps, and other gas-fired products. The California Energy Commission has estimated that these standards will reduce California's peak power needs by 1750 MW, and save consumers more than $600 million by 1987, with the savings to rise beyond this as the older appliances are replaced, and as California experiences new growth. New York and Kansas have also adopted standards for minimum efficiency standards for air conditioners. In 1987 the U.S. Congress amended the Energy Policy and Conservation Act to establish energy-efficient appliance standards[37] for the entire nation. The Reagan Administration had originally attempted to avoid these Congressionally mandated standards, but in 1985, the U.S. District Court (including Justice Bork) ruled in favor of the plaintiffs that the "no-standard standard" violated the requirement that the nation establish energy-efficient standards for appliances when they were cost effective. Since the appliance industry realized that one Federal standard was preferable to 50 state standards, they negotiated with the environmentalists to establish the details of such standards. The choice of the discount rate used for these calculations was part of the ruling of the Appellate court. Various estimates have been given for the net present benefit from national standards. The Lawrence Berkeley Laboratory has the lead within the Department of Energy on these calculations, and they estimated that[23] the present net benefit from the standards would be about $20 billion from refrigerators and freezers purchased between 1990–2015. The LBL group estimates that the total savings from all the home appliances will be between $35 and $48 billion for the same period, and that 16 GW of power can be avoided. Others[37] have considered further improvements, such as reducing the consumption of refrigerators from today's 1000 kWh/year down to about 220 kWh/year.

6d. Lighting standards

The starting voltages and operating currents for fluorescent lamps are controlled by the ballasts. The older magnetic ballasts use a coil of wire surrounding an iron core, dissipating about 20% of the energy. The electronic ballasts, introduced around 1980, dissipate about 10% of the light's energy. This improvement reduces the power of a typical two bulb, 48-in. fluorescent system from 96 watts to 86 watts. The magnetic ballasts cost about $14 as compared to $30 for the electronic ballasts, but the electronic ballasts consume about $56 less electricity (at 7 cents/kWh), or about four times the additional cost of $16. In 1982, the State of California adopted a standard[38] to require electronic ballasts, and in 1988 Congress adopted, without objection, these standards for the nation.

7. Findings on energy usage in buildings and appliances

In this chapter we have described the improved technologies for buildings and appliances. Some of the results are:

• Insulation values in residences have been increased by about a factor of 2, reducing the balance point of buildings. The total energy usage for all households dropped about 27% between 1972 and 1982.[7]

• Superinsulated homes which use $200 or less of fuel per year have become commercially viable.

• Passive solar design features (glass plus mass) have been widely adopted by the architects. In gentle climates like California and many other parts of the country, it is relatively easy to use passive solar for most of the heating energy.

• Large office building resource energy intensiveness has dropped from a high of about 500 kBtu/ft^2 yr for some buildings and an average of 280 in 1979, to 100, or less, in new buildings.

• Computer simulation programs, such as DOE-2 and PEAR, can estimate energy usage to within about 15%, helping architects to reduce energy costs of buildings.

• Thermal storage, as in the Thermodeck system described earlier, allows the heat of the day to be used to combat the coolth of the night, negating the need to hook these buildings to the Stockholm district heating system.

• For large buildings, off-peak cooling and storage of water and ice in the evenings for use in daytime air conditioning is being widely adopted to save peak power.

• Time of day rates for commercial and industrial electrical power is saving considerable peak power.

• Research on indoor air quality has revealed the radon problem, and methods to mitigate it.

• Refrigerators and freezers will be consuming only one-third as much energy by 1993 as their 1973 predecessors, ultimately saving about 25 large power plants. Furnace efficiencies have risen from 65% to 90%, and central air-conditioners use one-half as much energy as the 1973 models.

• Low emissivity double-glazed windows with R values of 3 to 4 have strongly penetrated the market. Values as high as R-10 have been developed, but are not yet commercially viable.

• High-frequency ballasts and improved phosphors have improved the efficiency of fluorescent lights by about 25%. Other techniques such as isotopic enhancement and surface waves will make further improvements.

• National energy standards have been adopted for major appliances, lighting ballasts, and automobiles. The national appliance standards will save about 25 GWe, and the automobile standards will save about 4 Mbd in the steady state compared to 1973 values. There are no national building standards, but a number of states have adopted building codes which are more stringent, or comparable, to the standards adopted by the American Society of Heating, Refrigerating, and Air Conditioning Engineers (ASHRAE). Because the lifetimes of buildings, appliances, and automobiles are 50, 20, and 10 years, respectively, it will take many years to see the complete benefits of these standards.

• Life-cycle costing which takes into account the properly discounted costs and benefits has been accepted in judicial proceedings. Comparisons between production and conservation technologies are now evaluated in terms of the cost of conserved energy. Market forces for these economic advantages have been relatively weak, as society as a whole has invested in large-scale plants through rate bases.

• About 10 to 15 times as much money has been spent on developing the production technologies as compared to the conservation technologies, with the latter producing much more tangible results. Conservation has done much better than the "renewable" energy sources.

8. Conclusions

8a. Progress since the oil embargo

Energy consumption has been approximately constant since the oil embargo of 1973–1974 at about 75 quads per year. If energy consumption and GNP were still locked in-step, one would have expected an increase of about 35% in energy usage. Perhaps $2/3$ of these savings, or $100 billion per year are due to improved energy efficiency, with the remainder coming from modal shifts in our society. The large overestimates of 1972 for very large future energy usage were very wrong. Our society is now more aware of finite resources, energy planning, and life-cycle costing. Our predecessors should be given very low marks for concentrating only on what it takes to maintain exponential growth into the future without examining the integrals of the exponential growth.

8b. Enhanced production versus improvements in end-use efficiency

The savings from enhanced end-use efficiency have far outpaced improvements in the production of energy. The production technologies all have some distinct drawbacks. Clearly, the U.S. will need further energy supplies, even with im-proved end-use efficiency. However, the easiest, and least-expensive gains can come from further improvements in end-use efficiency. When the cost of con-serving energy is less than the cost of additional supplies, taking into account all the costs, then end-use efficiency should be favored. The Department of Energy has spent[39] about 10 to 15 times more money on the production of energy as compared to monies spent for improvements of end-use efficiency, only $166 million in 1988. See Fig. 14 for DOE's funding[40] for conservation.

It is surprising that conservation has made the much larger progress[41] than the production industries, in spite of its being funded so poorly. It should also be pointed out that conservation (end-use efficiency) has far outperformed the renewable energy sources. As a nation we have been short-sighted to favor to such a great extent the production technologies over enhanced energy effi-ciency. The ethic that "America won't conserve itself to greatness" must be diminished. *It is our opinion that the Department of Energy should, over the next five years, move to greatly expand R&D funding for the enhanced end-use efficiency technologies.*

DOE Energy Conservation R&D
Requests and Appropriations (1982 $)

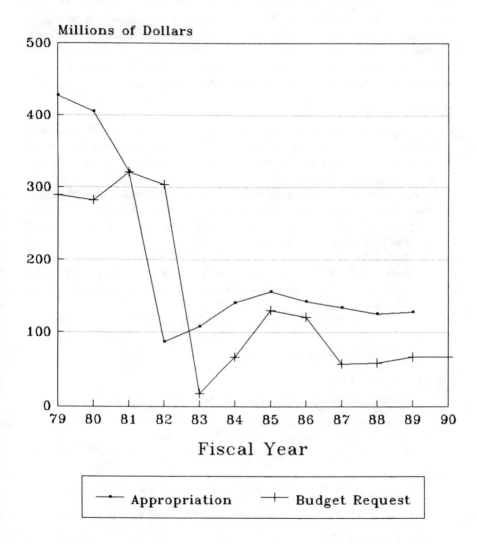

Figure 14. DOE budget requests and appropriations for energy conservation R&D (Ref. 40, 1979–1990).

8c. Energy standards

The energy standards for automobiles, appliances, and lighting are now in place and operating at a cost advantage to the American public. The automobile standards will ultimately save us 50% of the auto fuels, or about 4 Mbd, twice the Alaskan pipeline. The refrigerator standards will save $2/3$ of its former energy usage, amounting to the power from about 25 GW of power plants. As improvements in technologies enter the market place, the standards should be strengthened on a cost-effective basis. Several states now have energy efficiency standards for buildings, and they seem to be very successful. The Administration and the Congress should determine if such standards would be advantageous on a national basis. To the extent that the Building Efficiency Performance Standards (BEPS) can be shown to have net benefits to the society, taking into account cost, environmental gains, and enhanced energy security, these standards should be adopted on a national basis.

Bibliography

Energy Sources: Conservation and Renewables, D. Hafemeister, B. Levi, and H. Kelly, eds., Amer. Instit. Phys. Conf. Proc. **135** (American Institute of Physics, New York, 1985). A 700 page collection of technical articles on the conservation technologies.

E. Hirst, J. Clinton, H. Geller, and W. Kroner, *Energy Efficiency in Buildings: Progress and Promise* [Amer. Council Energy-Efficient Economy (ACEEE), Washington, DC, 1986]. A review of the data on energy in buildings and appliances, with a discussion on policy.

A. Meier, J. Wright, and A. Rosenfeld, *Supplying Energy Through Greater Efficiency* (Univ. CA Press, Berkeley, CA, 1983). Supply curves are developed to determine the potential for conservation as a function of the price of conserved energy.

H. Geller, J. Harris, M. Levine, and A. Rosenfeld, Anun. Rev. Energy **12**, 357 (1987). A review of the progress on enhanced-end-use efficiency of energy usage in buildings and appliances.

References and notes

1. *Annual Energy Review*, and *Energy Conservation Indicators* (Energy Information Administration, Washington, DC, 1988).
2. R. Mowris, private communication, Jan. 1989.
3. A. Rosenfeld and D. Hafemeister, Sci. Am. **258**, 78 (April 1988).
4. *Energy Sources: Conservation and Renewables*, D. Hafemeister, B. Levi, and H. Kelly, eds., Amer. Instit. Phys. Conf. Proc. **135** (American Institute of Physics, NY, 1985).
5. A. Rosenfeld; A. Rosenfeld and D. Hafemeister, in Ref. 4.
6. California Energy Commission, *Relative Cost of Electricity Production* (Calif. Energy Commission, July 1984), Ref. 4, p. 597.
7. E. Hirst, J. Clinton, H. Geller, and W. Kroner, *Energy Efficiency in Buildings: Progress and Promise* [Amer. Council Energy-Efficient Economy (ACEEE), Washington, DC, 1986], p. 7.
8. J.G. Ingersoll and B. Givoni, Ref. 4.
9. J.F. Busch and A.K. Meier, *Monitored Performance of New, Low-Energy Homes: Updated Results from the BECA-A Data Base*, LBL-18306 (Lawrence Berkeley Laboratory, Berkeley, CA, 1985).
10. G.S. Dutt, M.L. Levine, B.G. Levi, and R.H. Socolow, *The Modular Retrofit Experiment — Exploring the House Doctor Concept*, PU/CEES Rept. 130 (Princeton University, CEES, Princeton, NJ, 1982).
11. L.W. Wall, C.A. Goldman, A.H. Rosenfeld, and G.S. Dutt, Energy and Buildings **5**, 151 (1983).
12. E. Hirst, *Cooperation and Community Conservation: Final Report, Hood River Conservation Project*, DOE/BP-11287-18 (U.S. Department of Energy, Washington, DC, 1987).

13. S. Selkowitz, Ref. 4.
14. A. Rosenfeld and O. de la Moriniere, ASHRAE Transactions **HI-85-15**, No. 4 (1985).
15. M.A. Piette, L.W. Wall, and B.L. Gardiner, ASHRAE Journal **28**, 1 (1986).
16. B.L. Gardiner, M.A. Piette, and J.P. Harris, *Measured Results of Energy Conservation Retrofits in Non-Residential Buildings: An Update of the BECA-CR Data Base* (ACEEE 1984 Summer Study on Energy Efficiency in Buildings, Santa Cruz, CA, 1984).
17. C.A. Goldman, K.M. Greely, and J.P Harris, *Retrofit Experience in U.S. Multifamily Buildings: Energy Savings, Costs, and Economics*, LBL-25248 (Lawrence Berkeley Laboratory, Berkeley, CA, 1988).
18. J. Fricke, Sci. Am. **258**, 92 (May 1988).
19. S. Berman, p. 247, Ref. 4.
20. J. Maya *et al.*, Science **226**, 435 (1984).
21. H. Geller, Ref. 4.
22. American Council on an Energy-Efficient Economy, *The Most Energy-Efficient Appliances* [Amer. Council Energy-Efficient Economy (ACEEE), Washington, DC, 1988].
23. M. Levine, Ref. 4. M. Levine *et al.*, *Federal Efficiency Standards for Residential Appliances* (Lawrence Berkeley Laboratory, Berkeley, CA, Sept. 1987).
24. D. Yergin, Foreign Affairs **67**, 110 (Fall 1988).
25. R. Peddie and D. Bulleit, Ref. 4.
26. D. Hafemeister, Am. J. Phys. **50**, 29 (1982).
27. A. Meier, J. Wright, and A. Rosenfeld, *Supplying Energy Through Greater Efficiency* (Univ. CA Press, Berkeley, CA, 1983).
28. A. Nero, Sci. Am. **258**, 42 (May 1988). *Radon and Its Decay Products in Indoor Air*, W. Nazaroff and A. Nero, eds. (Wiley, New York, 1988).
29. A. Nero, Phys. Today **42**, 32 (April, 1989).
30. D.L. Grumman, "Standard 90: Review, Preview, and Update," ASHRAE Journal **25**, 10 (1983).
31. J.W. Jones, "Special Project 41: Development of Recommendations to Upgrade ASHRAE Standard 90A-1980, 'Energy Conservation in New Building Design,' " ASHRAE Journal **25**, 10 (1983).
32. D.B. Goldstein, "The American Experience With Establishing Energy Efficiency Standards for New Buildings: Case Studies of California and National Energy Standards," Third Soviet-American Symposium on Energy Conservation, Yalta, Crimea, U.S.S.R. (1988).
33. C. Lentz, "ASHRAE Standard 90-75: Impact on Building Energy Usage and Economics," ASHRAE Journal **18**, 4 (1976).
34. W.S. Fleming and H. Misuriello, "BEPS and ASHRAE 90: Evaluation and Impact," ASHRAE Journal **22**, 6 (1980).
35. D.B. Crawley and R.S. Briggs, "Standard 90: The Value," ASHRAE Journal **27**, 11 (1985).
36. A. Meier and B. Nordman, *Incremental Value of Measured Data from the Residential Standards Demonstration Program*, LBL-24213 (Lawrence Berkeley Laboratory, Berkeley, CA, 1988), pp. 3–19.
37. D. Goldstein *et al.*, *Developing Cost Curves for Conserved Energy in New Refrigerators and Freezers* (Natural Resources Defense Council, San Francisco, Jan. 1987).
38. H. Geller and P. Miller, *1988 Lighting Ballast Efficiency Standards* (ACEEE, Washington, DC, Aug. 1988).
39. H.R. Heddi, Subcomm. on Energy, Conservation, and Power, House of Representatives, U.S. Congress, June 1985.
40. H. Geller, *Federal Energy R&D* (ACEEE, Washington, DC, March 1989).
41. H. Geller, J. Harris, M. Levine, and A. Rosenfeld, Annu. Rev. Energy **12**, 357 (1987).

Efficient use of electricity:
The impact of materials technologies
on motor-drive systems[1]

Samuel F. Baldwin[2]

1. Introduction

Within just one generation after its introduction in the 1880s, electricity replaced steam as the preferred means of providing motive power to factories (Figure 1). The initial motivations to electrify manufacturing included the freedom to site a plant anywhere — not just by a stream or a coal mine — and the savings in energy that were possible. It quickly became clear, however, that the production advantages were even more significant. Steam and water power required complicated shaft, pulley, and belt systems to distribute mechanical drive from the central power plant throughout the entire factory. Equipment had to be positioned where it could be reliably coupled to the central drive, not according to the inherent logic of the manufacturing process. And a difficulty at any station often meant shutting down the entire factory.[3]

Electric motor drive changed all that. It allowed each piece of equipment to be driven independently and to be positioned so as to best meet manufacturing process needs. This revolutionized factory design and layout, provided much better process control and equipment utilization, and improved working conditions — all of which led to dramatic increases in productivity.[3]

Today, electric motors are the workhorses of modern industrial society. They run home refrigerators; drive office heating and ventilation systems; power industries' pumps, fans, and compressors; and keep cities' water supplies flowing. In 1986, the fraction of U.S. primary energy (fossil heat equivalent basis) devoted to the electric sector was 36%, and of the electricity generated, nearly two-thirds (Table 1) was used in electric motors. The total annual cost of electricity to U.S. consumers is currently about $150 billion, and for electric motors alone therefore approaches $100 billion.[4–6]

Despite their important role in the economy, however, electric motors have been technologically quiescent during the past several decades,[7] in sharp contrast to the revolutionary changes that have occurred in communication and information technologies. Now these same technologies are being combined with new developments in power electronics, magnetic materials, and system design to dramatically transform electric motor drive. These changes promise to

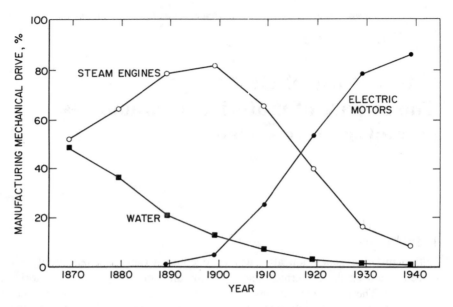

Figure 1. Percentage of manufacturing mechanical drive from water power, steam engines, and electric motors by year. Adapted from Barkenbus (Ref. 13).

improve significantly the manner and the efficiency with which we use electricity.

1a. Electric motor fundamentals

Electric motors have traditionally been classified as direct current (dc), synchronous, or induction (Figure 2). Direct current motors have been used where the value of their easy and precise control was greater than their expensive maintenance.[8] Synchronous motors have been reserved mainly for large installations where their higher efficiencies compensated for their higher installed costs. In contrast, induction motors have been used where the need for reliability and low cost was most important, and have consequently been the predominant choice in residential (one-phase), commercial, and industrial (three-phase) applications for decades.

Electric motors work much like a pair of bar magnets in which the attraction or repulsion between poles causes the magnets to move. In motors, these magnetic fields are usually generated by electric currents in stationary (stator) and rotating (rotor) coils, although permanent magnets are sometimes used to generate one set of these fields. The torque of a motor is established by the strength and spacing of these stator and rotor fields and the angle between them.[9]

In dc motors, constant magnetic fields are generated in the stator by dc or by permanent magnets. Fixed carbon brushes slide over contact points on the rotor windings as the rotor turns in order to mechanically switch the dc current between different rotor coils. This keeps the dc-generated magnetic field stationary in space and with respect to the stator. The carbon brushes and rotor coils

Table 1. Estimated motor drive as a fraction of total U.S. electricity use.

	Little, 1976 (%)	DOE, 1980 (%)	Masher, 1980 (%)
Residential	7.1	2.8	19.6
Refrigeration*	. . .	. . .	8.2
Space heating*	. . .	. . .	4.1
Air conditioning*	. . .	. . .	4.5
Other	. . .	. . .	2.8
Commercial	11.9	14.5	14.7
Air conditioning*	6.5	. . .	8.0
Refrigeration*	4.2	. . .	5.2
Other	1.2	. . .	1.5
Industrial	27.9	26.1	30.0
Pumps*	8.5	. . .	9.6
Compressors*	4.9	. . .	5.3
Blowers and fans*	4.3	. . .	4.9
Machine tools	2.4	. . .	2.5
Other integral hp	3.1	. . .	3.5
dc drives	2.1	. . .	3.1
Fractional hp	1.2	. . .	1.1
HVAC	0.7	. . .	. . .
Public and miscellaneous	17.3	14.4	3.5
Municipal water works*	6.9	0.7	. . .
Electric utilities*	8.7	6.5	. . .
Mining and construction*	1.7	2.0	. . .
Other	. . .	5.2	. . .
TOTAL	64.2	57.8	67.8
Reference (13)	5	7	33

*Primarily consists of pumps, fans, and/or compressors.

are oriented so that the interaction between the stator and rotor generates the maximum possible torque. The speed and torque of dc motors are easily controlled[10] by varying the stator (field) and/or rotor (armature) current and/or voltage.

Unlike the stationary magnetic fields of the dc motor stator, ac motors establish magnetic fields in the (stationary) stator coils that are constant in magnitude but effectively rotate around the stator. Three-phase motors accomplish this by applying one phase of alternating current to each of three different coils wound on the stator. Each of these coils is wrapped around the stator at a different point: in the most simple case, the coils are set 120 degrees apart on the stator coil. By offsetting the alternating current in time (phase) and space (on the stator winding) in this manner, a single magnetic flux wave rotating around the stator at the line frequency (60 Hz or submultiples) is generated.

This can be visualized most easily by imagining the case where the coil at

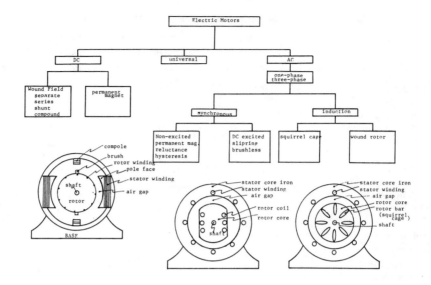

Figure 2. Classification and characteristics of electric motors.

the top of the stator has maximum current through it; the two coils 120 degrees on either side also have current through them that is ±120 degrees out of phase with the top coil. The magnetic fields generated by these other coils have components in the same direction as the top coil with half the magnitude each; and components perpendicular to that of the top coil which are opposite each other and cancel. Summing the contributions from all three coils leaves a single component in the direction of the top coil with twice the amplitude of the top coil by itself. The same analysis a fraction of a second later when the current through one of these side coils has reached its peak shows a net field in the direction of this coil. In this way, summing the sinusoidally varying fields through each individual coil leaves a magnetic field effectively rotating in space.[11]

In contrast, single-phase ac cannot produce an effectively rotating magnetic flux by itself. Applying single phase power to a single stator coil simply creates a magnetic field that is fixed in space and increases and decreases sinusoidally. A single phase motor therefore uses auxiliary starting windings (split phase, capacitor start) or windings (shaded pole) to produce an asymmetric torque to start the motor turning. Once rotation begins, the rotor is subject to a net positive rotating stator flux in its reference frame.

Synchronous motors typically use three-phase ac to generate a magnetic field of constant magnitude rotating in the stator as described above and have a constant magnetic field fixed and rotating with the rotor. The magnetic field of the rotor is typically generated either by applying a dc current through slip rings to the rotor coil or by making the rotor out of a permanent magnet material. The rotor of a synchronous motor turns in lock-step with the rotating magnetic field of the stator.[12]

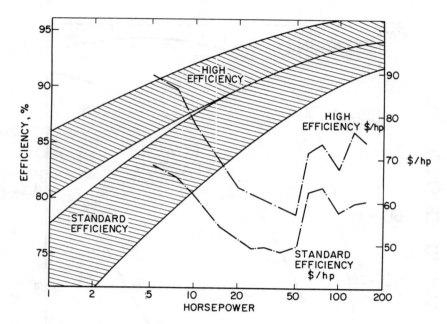

Figure 3. Approximate ranges for full-load efficiency vs size, and costs (in 1986$/hp) for one manufacturer vs size for National Electrical Manufacturers Association (NEMA) design B standard and energy-efficient 1800-rpm, three-phase induction motors. Adapted from Andreas (Ref. 13) and motor catalogue data.

Induction motors also use three-phase ac to generate a magnetic field constant in magnitude and rotating around the stator as described above. The rotor, however, turns at a slightly slower speed than the magnetic field generated by the stator. By Faraday's law of induction, this relative motion between the stator field and rotor induces a current, and thus a magnetic field, in the rotor as well. Torque is generated by the interaction between the stator field and this magnetic field induced in the rotor coil.

Universal motors have brushes and can be operated on either single-phase ac or on dc.

As seen in Figure 3, standard three-phase ac induction motors have efficiencies ranging from about 73% for 1 hp motors to 93% at 150 hp. The losses that occur can be categorized as conduction, magnetic core, friction and windage, and stray losses; they are depicted in Figure 4. Conduction losses are due to resistive (I^2R) heating in the stator and rotor. Magnetic core losses are due to a combination of eddy current and hysteresis losses induced by the 60-Hz field in the stator and rotor iron. Friction losses occur in the motor bearings, and the windage losses are caused by the ventilation fan and other rotating parts. The remaining losses are usually lumped together in the catchall term "stray load" losses.[13]

With the increasing cost of electricity in recent years, many manufacturers have begun to produce lines of energy-efficient motors with typical improvements in efficiency as depicted in Figure 3. Techniques used to achieve these

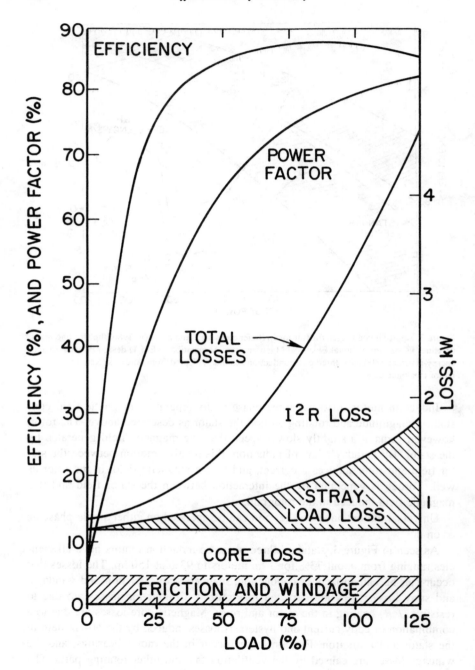

Figure 4. Typical components of loss, and approximate efficiency and power factor vs load for a three-phase, design B, 1800-rpm induction motor. From A.D. Little, Inc. (Ref. 13).

savings include: using higher conductivity and larger cross section conductors in the stator and rotor to reduce resistive heating losses; both increasing the core length and using thinner laminations of better magnetic materials to reduce eddy current and hysteresis losses in the core; and reducing the air gap between rotor and stator to reduce the required magnetomotive force and its generating currents.[14] Optimization studies examining more than 100 induction motor design parameters have detailed the lifetime savings possible under various conditions. In the case of single-phase motors, simply optimizing motor dimensions could raise efficiencies an average of 4.5% over present-day standard motors at no additional cost.[15]

Although there is a price premium for energy-efficient motors (Figure 3), they often pay for themselves quickly by lowering the high cost of operation. A standard 50-hp motor, for example, might cost $2500 to purchase and $2.05 per hour to operate (91% efficient and $0.05/kWhr). In contrast, a comparable energy-efficient motor might cost $2875 to purchase and $1.96 per hour to operate (95% efficient). Thus, in a single 6000 hour (full-load) operating year, the difference in operating costs is $540, for a net saving of $165. Even scrapping a working motor and replacing it with an efficient model would have a simple payback of just over five years. Of course, shorter operating hours would stretch out these payback times.

It is important to note, however, that decreasing resistance in the rotor circuit to reduce resistive heating losses also reduces the starting torque and increases the rotor speed at its load point (Figure 5). High-efficiency motors can have higher transient starting currents — increasing the load on the power supply.[16]

Low starting torques and thermal capacities may require choosing oversized motors to start high-inertia loads such as fans.[17] This can be especially important where there is significant variation in the line voltage and the combination of low voltage and large load inertia can cause the motor to stall (Figure 5) and even burn out. Oversizing itself may have little or no direct efficiency penalty. For example, a standard 5-hp motor at full load may have an efficiency of 87.5% compared to 87% for a 10-hp motor at half-load. Similarly, a 75-hp motor may have an efficiency of 94% compared to 94.5% for a 125-hp motor at 60% load (Figure 3). At very low load, of course, the efficiency drops rapidly (Figure 4), but the amount of power that is wasted is relatively small. However, an oversized motor leads to a lower power factor.

The power factor of an electric motor is the cosine of the phase angle between the voltage and the current in the rotor. A load with a low power factor[18] requires larger generation, transmission, and distribution equipment to handle the increase in kVA (kilovoltamperes). It increases resistive heating losses in this equipment, and can lower system voltage regulation.[19] Up to one-fifth of the total losses in power distribution in the United States, or as much as 1.5% of total U.S. electrical generation,[20] can probably be attributed to a lagging power factor,[21] and utilities often penalize users who have excessively low ones.[22] Correction for a lagging power factor is possible by inserting shunt

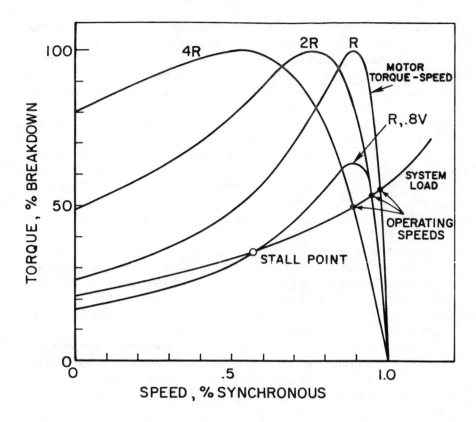

Figure 5. Motor speed vs torque curves for various values of rotor resistance and applied voltages. Also shown is a system load curve and the corresponding operating points for each motor torque-speed curve.

capacitors or an overexcited synchronous motor on the line to change the relative phase of voltage and current. However, so doing always entails a capital expense, the cost-effectiveness of which varies from case to case.

Finally, a higher rotor speed at the load point will increase the power demand of speed-dependent loads such as fans and pumps unless compensated for by appropriate adjustment of pulleys, gears, etc., and may cancel a significant fraction of the savings of using an energy-efficient motor. For example, the standard 10-hp motor from a major manufacturer has an efficiency of 87.5% and a full-load speed of 1740 rpm, while the high-efficiency version has an efficiency of 90.2% at a speed of 1750 rpm. For a load, such as a pump or fan, whose power demand scales as the cube of rotational speed,[3] this additional speed will increase power demand by $2/3$ of the power that might have been saved by using an efficient motor. Improving motor efficiency can thus have undesirable side effects requiring special attention.

1b. Conventional system design

Electric motors are just one element of very large and complex systems, each

Table 2. Typical pumping system losses.

Location	Losses, % of input at that device	Remainder, % of starting total
Coal	. . .	100.0
Electric power plant	70	30.0
Power distribution lines	8	27.6
Power transformers	4	26.5
Electric motors	12	23.3
Shaft couplings	2	22.5
Pump	28	16.4
Pump throttling valve	14	14.2
Piping system	20	11.3
Reference 13		

component of which must be taken into account to optimize overall energy efficiency. Table 2 lists estimated national average losses and net available power at different points in U.S. pumping systems. As seen in this example, each kWhr saved by reducing piping system losses saves roughly 9 kWhr of power plant fuel. These average values, however, include many systems which are not throttled. As shown below, the losses in specific cases can be much higher. Careful design and operation of end-use equipment thus has enormous leverage for saving energy upstream.[23] Similarly, reducing the load can save considerable capital investment at each point of the system upstream.

Pumps, fans, and compressors are the most common motor driven equipment (Table 1), and each specific design has a unique performance map determined by the detailed fluid mechanics of its operation.[24] The system operating point is then the intersection of this performance map with the system load curve — the sum of kinetic energy, static pressure, and gravitational potential that the equipment must overcome (Figures 6 and 7).

In principle, equipment is chosen so that the optimal efficiency is matched to the desired operating point. In practice, there are numerous complications when systems are designed.

1c. System sizing

Systems are usually deliberately designed to be oversized, that is, to have excess capacity. Pipes, ducts, pumps, fans, and other driven equipment and components are oversized, partly because it is extremely difficult to predict system flow rates and friction factors accurately in advance, and to allow for the effect of the buildup of deposits on duct and pipe walls over time.[25] Motors are oversized to handle starting electrical, mechanical, and thermal stresses, particularly with high-inertia loads such as fans;[26] to provide a safety margin for the worst-case load over their lifetimes; and occasionally to handle plant

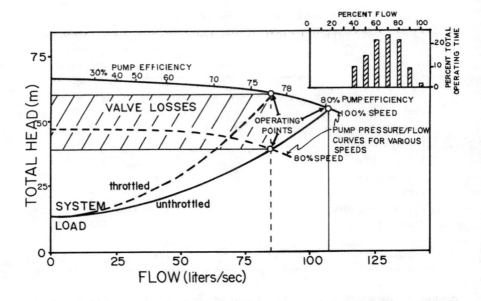

Figure 6. Throttling losses in a variable-flow, variable-pressure pumping system at 80% flow as compared to reduced speed operation. The energy losses due to throttling are roughly proportional to the shaded area. Inset is a typical pump duty cycle. Note that throttling also causes the pump to operate at a lower efficiency. Adapted from Armintor and Connors (Ref. 32).

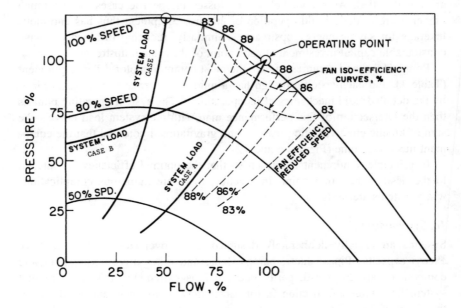

Figure 7. Characteristic pressure and flow curves for a fan at different speeds; fan isoefficiency curves; and system loads for unthrottled, throttled, and high-static-pressure cases. Note that for a high static-pressure drop, changing fan speeds does not allow the highest fan efficiency point to track the load curve. Adapted from Reason (Ref. 25).

expansion or perhaps to be widely interchangeable within the plant.[27] In developing countries, oversizing motors is important for preventing motor stall and possible burnout when the line voltage drops.[28] More generally, systems are oversized because the increased energy and capital costs are usually less than the risk of equipment failure. In manufacturing, for example, average electricity costs in the U.S. are equivalent to just 2.8% of the value added;[29] the cost of motor failure and unplanned shutdown of the entire process line can be a much more severe penalty.[30]

The intent of this design process is understandable and reasonable. Manufacturers, design engineers, and users all want to ensure that the system can meet the demands placed on it at every stage of the process. Although oversizing can provide direct safety margins for equipment, it can have unintended negative side effects. For example, when each successive element of a drive system is sized to handle the load presented by the previous component plus a safety margin, oversizing can quickly become excessive and require throttling valves or vanes to limit flow. Throttling not only wastes power itself, but may also move the fan or pump off its optimal operating efficiency (design) point (Figures 6 and 7), making its operation inefficient as well. Standard engineering design rules, together with manufacturing safety margins, sometimes automatically lead to significant inefficiencies of this type which the user may be unaware of.[31] Oversizing motors also leads to a low power factor (Figure 4). Re-evaluating traditional safety margins may therefore offer significant savings of both capital and operating costs.

1d. Part- and variable-load operation

Although system sizing is determined by the load that must be driven, conventional system operation is determined by the torque-speed characteristics of the induction motor. These usually give essentially fixed-speed operation (Figure 5) as determined by the winding (number of poles), the line frequency, and, to a lesser extent, the rotor resistance and other factors. Oversizing, as explained earlier, leads to part-load operation at all times. In addition, many systems need variable outputs. Space heating and cooling, manufacturing, municipal water pumping, and most other motor drive loads vary with the time of day, the season, and even the health of the economy. Such variations can be quite large. A centrifugal pump duty cycle might vary as shown inset in Figure 6.[32] A boiler might vary from 10% load 10% of the time, to 40% load 20% of the time, to 80% load 5% of the time, with its fan loads varying an appropriate fraction of this.[33] Similar load variations can be found for compressors[34] and machine tools.[35] Table 3 lists estimated average loads and other characteristics of motors in the United States. Driving such part- or variable-load systems with fixed speed motors has often led to extremely inefficient control strategies.

In some systems, output is controlled by turning the motor on and off as needed. In an air-handling system with turbulent flow, for example, the power loss due to pressure drop in the ducts varies nearly as the cube of the flow rate. A system that has a duty cycle of 50% could instead provide the same total flow by running continuously at half the flow rate. This would reduce the

Efficient use of electricity

Table 3. Characteristics of U.S. motor population, 1977.

| | Electric motor size, horsepower | | | | | |
	<1	1–5	5.1–20	21–50	51–125	>126
Total number, millions	660.0	55.0	10.5	3.3	1.7	1.0
Average life, years	12.9	17.1	19.4	21.8	28.5	29.3
Weighted average size, hp	0.26	2.1	11.9	32.5	86.7	212.0
Average efficiency, %	65.0	77.0	82.5	87.5	91.0	94.0
Average load, % full load	70.0	50.0	60.0	70.0	85.0	90.0
Average capacity factor, %	3.5	7.0	17.4	27.7	37.5	43.2
Total annual use, 10^9 kWhr	30.5	33.9	103.4	155.2	337.7	573.3

Reference 13

power loss due to duct friction by 87.5%, and the corresponding total energy consumption by 75%.

Other systems run the motor continuously at part-load. Both motor efficiency and power factor, however, decrease for part-loads as shown in Figure 4.[36] This is particularly important for loads less than 50%.

Finally, in many systems control is exercised by letting the motor run at its normal speed and then throttling the input or output with a control valve or vane. Throttling valves (on pumps) or outlet dampers (on fans) reduce system flow by raising the static pressure the pump or fan is working against (Figures 6 and 7). Inlet vanes (fans) impart a spin to air entering a fan and thereby effectively change the pitch of the fan vane and thus its characteristic curves. These strategies allow flow reductions typically of 50%–60% and sometimes more.[37]

Industrial and commercial pumps, fans, and compressors have estimated average losses of 20%–25% or more when throttled or otherwise inefficiently controlled.[38] In reviews of entire plants, losses in motors and drives have been estimated to be as large as 40%–80% of input power.[39] To reduce these enormous energy losses, a method of varying induction motor speed without sacrificing the reliability of brushless operation is needed.

The control of motor speed has been a perennial problem for engineers. In November 1896, Harry Ward Leonard said to the American Institute of Electrical Engineers, "The control of the speed of an electric motor from a state of rest to a state of full speed is a problem of rapidly growing importance to the electrical engineer."[40] In the nearly 100 years since then, a variety of techniques have been used to control induction motor speed; some of these are listed in Figure 8.[41]

Unfortunately, until recently no drive or coupling technique proved sufficiently reliable, efficient, flexible, and widely applicable or low enough in cost to be generally accepted; recent advances in power electronics are now changing this situation.

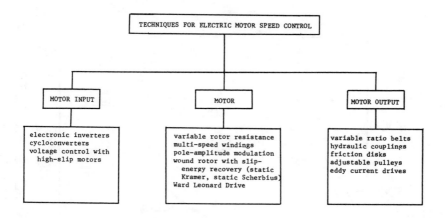

Figure 8. Classification of various techniques for electric motor speed control.

2. The materials revolution

Materials technologies have improved rapidly in recent years as understanding of the physics of condensed matter has advanced. Improvement has been most dramatic in solid state electronics devices — mainly computer and communications technologies. But similar advances are being made in technologies of use in motor-driven systems. Among these technologies are power electronics that allow precise control and matching of motor speed and torque to the load — thereby eliminating, for example, throttling losses; soft magnetic materials to reduce core losses in transformers and motors; improved permanent magnets to eliminate rotor resistive heating losses; and many others. The recent advances in high-temperature superconductors may one day lead to the elimination of resistive heating losses throughout electric power systems as well as providing numerous other advantages.

2a. Power electronics, sensors, and controllers

As discussed above, a crucial factor in achieving higher efficiencies in motor drive systems is the ability to precisely control motor speed. One technique is to supply the power going into an induction motor at an adjustable frequency. *Adjustable-speed electronic drives (ASD)* driven by mercury-arc rectifiers such as thyratrons and ignitrons were designed and tested in the 1930s,[42] but found little acceptance because of their high cost and inconvenience. The development of solid state devices with large power-handling capabilities, reductions in device costs, and rising electricity prices have made solid state electronic motor drives both possible and increasingly desirable. By 1986, over 200,000 electronic ASDs ranging in size from approximately 0.25 to 67,000 hp had been sold in the United States.[43]

During the 40 years since the invention of the point-contact transistor in 1947 and the 30 years since the invention of the thyristor in 1957, the capabilities of power electronic devices have improved dramatically (Figure 9).[44] Today, thyristors 10 cm in diameter, weighing 2.3 kg packaged, and capable of

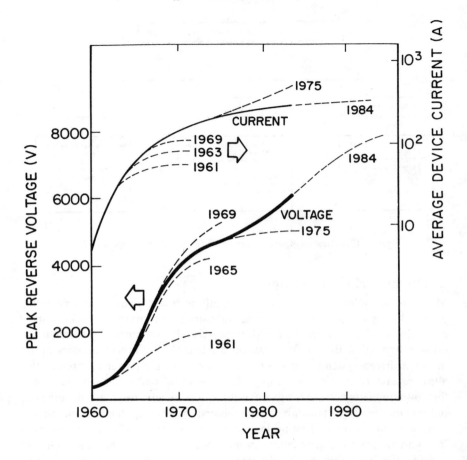

Figure 9. Maximum voltage and average current ratings of thyristors, and performance predictions by year. Adapted from Adler *et al.* (Ref. 44).

handling 3000 A at 4000 V have been commercialized, and 3000-A, 8000-V devices are under development.[45]

Recently, bipolar junction transistors and Gate Turn-Off thyristors have similarly undergone significant improvements in their high-power capability. These devices greatly simplify the circuitry of ASDs and correspondingly, can also reduce the cost and increase the reliability of ASDs. In the near future, new power devices such as the insulated gate bipolar transistor and the MOS controlled thyristor may provide further improvements in ASD performance.[46–48]

2b. Adjustable speed electronic drives

Numerous types of ASDs have been developed which can be classified in a variety of ways including: by the type of semiconductor device, circuit, or motor used; by the control strategy; or by the voltage or power level.[49] Figure 10 illustrates the classification used here and shows block diagrams and example waveforms for several of the systems.

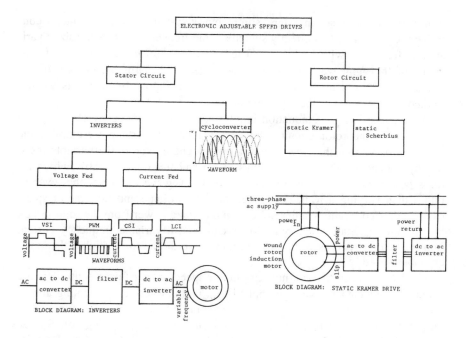

Figure 10. Classification of electronic adjustable speed drives, and block diagrams and waveforms for selected cases. Descriptions are in the text.

In the voltage-source inverter (VSI) drive, line power is converted to dc, smoothed by a capacitive filter, and then inverted to variable-frequency and variable-voltage ac power. The frequency is adjusted as desired to control the motor speed; the voltage is adjusted to keep the ratio of voltage to frequency constant (except at low speed, when the voltage is raised to compensate for the rotor resistance) and thus maintain constant motor torque over the desired speed range. The VSI produces a stepped-voltage waveform.[50]

The current-source inverter (CSI) also converts line power to dc, then smooths it with an inductive filter to give a constant current to the inverter, which in turn produces a stepped-current waveform for the motor. VSIs and CSIs are the most common drives for medium (approximately 50–200 hp) and high (approximately 200–1000 hp and above) powers.[51]

The load-commutated inverter (LCI) is similar to the CSI but is used with a modified synchronous motor. LCI drives are typically used in systems with very large power ratings (1000 hp).[52]

Although pulse-width modulated (PWM) drives similarly adjust the frequency to control motor speed, in contrast to these other inverters the signal is a chopped full voltage of varying width rather than a continuous reduced voltage. Motor inductance provides filtering to leave the average voltage-to-frequency ratio constant. The current drawn by the motor is nearly sinusoidal. PWM drives are the most common type now in use at low powers.[53]

There are a variety of other important types of electronic motor controllers in use, including cycloconverters, slip recovery systems, and power factor controllers. Cycloconverters directly switch ac line frequency to a synthesized variable ac frequency with no dc link. The output waveform thus consists of segments of the input waveforms. Because of the high harmonic content of the output voltage and input currents, the maximum output frequency is usually limited to less than half the input frequency.

As discussed earlier, induction motor speed can be varied by changing the resistance of the rotor circuit. Slip energy recovery systems accomplish the same thing, without the corresponding loss of energy across the variable resistor, by connecting the rotor windings to a fixed frequency inverter which returns power back to the motor supply. By varying the output of the inverter, rather than varying a resistor's value, the rotor voltage and thus motor speed can be adjusted. Static Kramer drives can only take power from the rotor circuit by this means. Static Scherbius drives can feed a variable frequency current into the rotor circuit as well, and can thus drive the motor at higher than synchronous speed.[54]

Finally, although they do not vary motor speed, power factor controllers are useful for reducing the voltage applied to a motor when it is operating at part load. These operate by varying the voltage applied to the motor as a function of its power factor, and thus its load. This results in energy savings of, for example, 25% at 40% load for single-phase motors and 10% at 40% load for three-phase motors. Savings increase rapidly for small loads.[55]

2c. Power integrated circuits

By precisely matching motor speed to its load, ASDs offer significant improvements in the performance of motor-drive systems, but their cost is still sometimes too high for general use — particularly in low-power devices. Costs can be lowered significantly by combining the power devices and their controller in a single package, known as a "power hybrid" or "power module." More significant still are new "power integrated circuit" (PIC) technologies which put the power devices and their controllers on the same chip.[56]

These PIC technologies, just beginning to emerge from the laboratory, combine power devices running at hundreds of volts and tens of amperes on the same chip as their microprocessor controller running at low voltages and currents. In addition to reduced costs, these technologies promise to greatly increase ASD reliability and performance, among other things. When combined with sensors for pressure or force, temperature, acceleration, sound or vibration, light, magnetic field, or even gas analysis on the same chip, a complete control and drive system is possible. Some chips even monitor their own temperatures and regulate them by reducing operating currents or even shutting down if temperatures get too high. Adding such intelligent functions can be done at little or no additional cost.[57]

The technology of power integrated circuits is progressing rapidly. Devices capable of handling 1.5 kVA are available; by 1995 devices handling 100 kVA are expected. By 1992, this is expected to be a $1.2 billion-per-year industry.[58] As volume increases and prices drop, smart power chip technology can be expected to have a significant impact on low- to medium-power motor-drive systems.

2d. Soft magnetic materials

Significant efficiency improvements in motors and transformers are possible through the use of improved soft magnetic materials to reduce hysteresis and eddy current losses in their cores. Core losses in generators, transformers, and motors have been estimated to be about 3, 48, and 30 billion kWhr, respectively, during 1977. Combined, these losses were 3.8% of total net U.S. electricity generation. Core losses in lamp ballasts cost a further 16 billion kWhr annually or 0.75% of net U.S. electricity generation.[60]

Soft magnetic materials have been steadily and dramatically improved during the past 100 years as indicated in Figure 11.[61] Today, non-oriented electrical steels, the predominant type used in standard motors and transformers, have losses of 2–4 W/kg at 60-Hz and 1.5-T maximum flux density. Grain-oriented electrical steels have typical losses of 1.3 W/kg at 1.7 T.[62]

Recent innovations in Japan have reduced losses in production units to 1.1 W/kg at 1.7 T and in the laboratory to as low as 0.57 W/kg at 1.7 T and approximately 0.34 W/kg at 1.4 T.[63]

The recent development of amorphous alloys of FeNiBSi offers further performance gains. The lack of structure in amorphous materials minimizes hysteresis losses. Further, these alloys have a moderately high electrical resistivity, which minimizes eddy current losses. Finally, the very rapid cooling required to freeze the material in an amorphous state, 10^5–10^6 °C per minute, requires that they be produced in very thin laminations, leading, in principle, to inherently simple, fast, and less capital-intensive production systems: one simply squirts the material out to freeze on a rotating drum, with few special precautions, and all in one step. This is in marked contrast to today's oriented silicon steels, which require as many as 6–10 steps of careful melting, ingot casting, and precise schedules of hot and cold rolling and high-temperature annealing.[64]

Amorphous alloys of FeNiBSi have now achieved losses as low as 0.25 W/kg (1.4 T) in preproduction units of 25-kVA distribution transformers and 0.02–0.09 W/kg (1.4 T) for the lowest laboratory measurements. Extensive work has been done on the use of amorphous materials to reduce core losses in transformers and motors. And with the reduction in core losses there is a reduction in operating temperature, leading to longer insulation life and further reductions in the losses of the copper windings. Amorphous alloys could all but eliminate the annual $7 billion in magnetic core losses in the United States and could potentially dramatically restructure the $1 billion-per-year U.S. soft magnetic materials industry.[65]

2e. Permanent magnets

Permanent magnets have seen similarly dramatic improvements in recent years (Figure 12). In the 1930s, Japanese discovered alnicos. In the 1940s ferrites were discovered and developed in the Netherlands. Rare-earth permanent magnets (REPMs) were first demonstrated by Strnat (U.S.) in the 1960s, and the

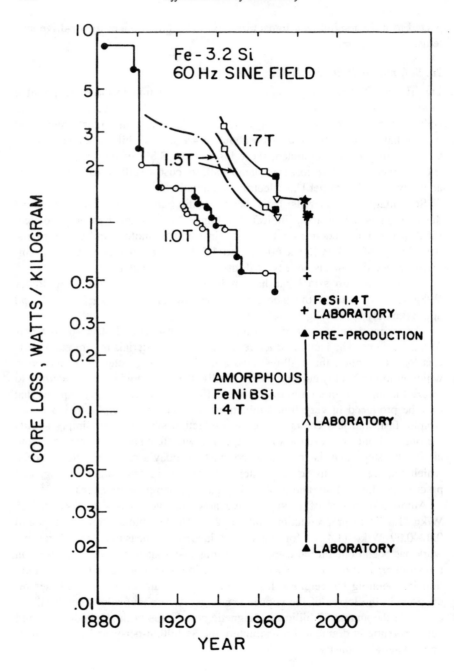

Figure 11. Power loss for magnetic cores vs year. Values marked * are from National Materials Advisory Board (Ref. 62), #, Fukada and Hoshi (Ref. 63), +, Takahashi *et al.* (Ref. 63), ^ Tan (Ref. 65). Other values are from Luborsky, Frischmann, and Johnson; and Jacobs (Ref. 61); and National Materials Advisory Board (Ref. 62).

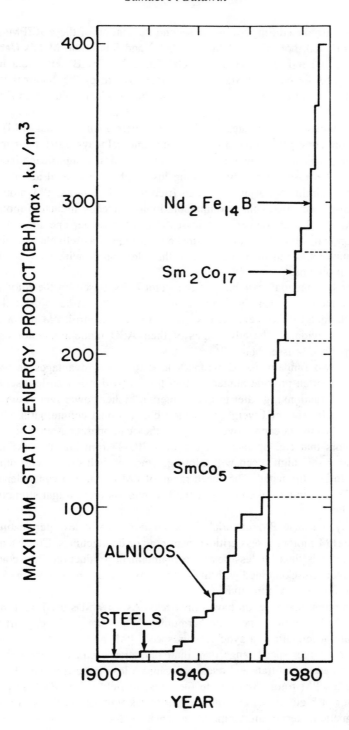

Figure 12. Energy products for selected permanent magnets since 1900. From Strnat (Ref. 70) and M. Sagawa, S. Hirosawa, H. Yamamoto, S. Fujimura, and Y. Matsuura, "Nd–Fe–B Permanent Magnet Materials," Jpn. J. Appl. Phys. **26**, 785–800 (1987).

Japanese introduced Sm_2Co_{17} — the second generation of these REPMs — in the 1970s. In 1983, both General Motors (U.S.) and Sumitomo Metals (Japan) announced $Nd_2Fe_{14}B$ permanent magnets. These $Nd_2Fe_{14}B$ permanent magnets, according to the National Materials Advisory Board of the National Research Council, have an energy product "so large that it will revolutionize the design of motors."[66]

Electronically commutated, or brushless, permanent magnet (PM) motors differ from conventional motors in two fundamental ways. First, their rotors are made of permanent magnet material. This provides a significant increase in efficiency because the resistive heating losses of conventional ac or dc motor rotors are eliminated, as is the brush friction of dc motors. PM motors have 18%–35% lower losses than comparable high-efficiency induction motors, and possibly even lower. Lower rotor losses also mean cooler operation. Brushless operation and solid PM rotors may make such motors as reliable as their induction counterparts and far more reliable than dc motors, with their brush maintenance problems.

Second, sequential switching of the stator coils, based on the rotor position as determined by optical encoders on the rotor shaft, is used to drive the rotor. This technique provides excellent speed and torque control. PM motor controllers also provide all the advantages of their ASD induction motor controller counterparts, discussed below.[67]

These two fundamental differences lead to other advantages as well. PM motors can often provide higher starting torque (which is also less sensitive to voltage fluctuations), greater torque to inertia, higher power factor, and significantly smaller size and weight than their conventional counterparts. Unlike dc motors, PM motors cannot overspeed, yet they can operate at very high speeds. PM motors that can operate at more than 20,000 rpm are now available. In some cases this high speed will allow gearmotors, belt drives, and other such devices to be eliminated. The speed range of PM motor and controller systems is also large, limited by cogging at the low end and centrifugal strength at the high end.[68]

Finally, because PM materials have an intrinsically low permeability, the airgap in PM motors is less critical than in ac or dc machines. Consistent motor performance is thus far less costly to maintain in production. Part counts can also be reduced. Combined with their other advantages, these have made brushless PM motors very attractive.[69]

PM motors, however, do have drawbacks. As for ASDs used with induction motors, their controllers generate harmonics. When sta⸱ ⸱ directly off the line and while accelerating to synchronous speed, PM mo⸱ suffer torque pulsation over a larger speed range than their inductio⸱ ⸱⸱⸱ counterparts. Excessive overloads, even transient ones, can cause a loss in performance by demagnetizing the PM rotor. And due to the constant field of the PM itself, driving the motor at high speeds can induce large back voltages in the stator. Finally, the magnetic materials are temperature-sensitive; they can suffer long-term irreversible demagnetization if operated at moderately high temperatures, above 150–200 °C. This is particularly true for the new $Nd_2Fe_{14}B$ magnets, although the addition of small amounts of cobalt may raise the operating temperatures considerably.[70]

Even before the development of PM materials (such as $Nd_2Fe_{14}B$ in 1983) with the potential for relatively low cost and high performance, manufacturers of appliances, computer peripherals and office equipment, machine tools, robotics, and other equipment, had responded dramatically to the potential presented by PM motors. The world market in 1985 for PMs was $1 billion of which Japan produced 40%, the United States about 32%, and Europe 20%. As of 1984, sales of PM motors and controllers were expected to grow more than 21% annually during the next five years, as the market climbs from $740 million in 1982 to $1.96 billion in 1987. Due to their many attractive features, including plentiful raw materials and easy machinability, $Nd_2Fe_{14}B$ magnets are expected to rapidly become a major force in the PM market.[71]

2f. Superconductors

The 1986 announcement by Bednorz and Muller of superconductivity at temperatures as high as 35 K in La_2CuO_{4-y} ceramics has led to explosive growth of theoretical and experimental research in this field around the world. Since then, superconducting transition temperatures as high as 125 K in $Tl_2Ba_2Ca_2Cu_3O_{10}$ have been observed.[72]

Much work has already been done on applications of low-temperature superconductors for electric power generation, transmission, distribution, and storage, and with good results. A superconducting 1750 MVA distribution transformer with an amorphous core, for example, had losses of just 200 W compared to 1200–1500 kW for a conventional transformer.[73] Cooling the transformer, however, required 150 kW. Such applications are generally severely constrained by the economics of liquid helium production and use necessary to reach the required low temperatures, and the high capital cost of vacuum vessels to thermally isolate the equipment from the environment. The new high-temperature superconductors instead use low-cost liquid nitrogen for cooling, broadening their potential applicability.

In addition to reducing the estimated 4% of total U.S. electricity lost in the generation, transmission, and distribution of electric power, superconducting generators will offer reduced size and weight and will have lower synchronous reactance – increasing system voltage stability. Superconducting transmission and storage systems might also allow greater use of alternative energy resources now constrained by their intermittant nature, including photovoltaics and wind power. Finally, if room-temperature superconductors were to eventually become available and broadly applicable in motors, an additional 3% of total U.S. electricity use could be saved and perhaps provide other significant advantages as well. A recent review of the potential of the new superconductors to reduce electricity use is given by Bergsjo and Lars.[74]

3. Modern system design

ASD technology allows a significant change in system operation. In contrast to the conventional practice of oversizing systems, using constant speed motors to drive them, and throttling excess flow with vanes or valves, ASDs and their associated sensors and controllers allow precise matching of motor speed to load. This can significantly improve overall system efficiencies and process control, and provide many other advantages as discussed below. Furthermore, this flexibility may allow design rules that lead to extreme oversizing to be relaxed, potentially reducing capital investments.

In the simplest terms, ASDs provide significant energy savings by taking advantage of the affinity laws for fans, pumps, and some other equipment. That is, changing the rotational speed of a fan or a pump changes the flow rate proportionally to the fan speed, the pressure drop or head by the square of the speed, and the power by the cube of the speed. Thus, to meet the demands of a system at half its maximum flow rate, an ASD could be used to reduce the motor speed to half its line-synchronous speed, with a corresponding drop in power demand to $(0.5)^3 = 0.125$ of its maximum (Figures 6 and 7). For systems with highly variable loads and/or duty cycles; steep system and/or equipment pressure-flow curves; and large powers and annual operating hours, the resulting savings with ASDs can be quite large.

In practice, the application of ASDs requires attention to a variety of factors. The system load may demand constant pressure and variable flow; constant flow and variable pressure; or both variable flow and variable pressure. Different analyses are required for each case. For systems with large static pressure drops, the system load will not reduce to zero at zero flow and speed. The result can be a large overestimate of the potential energy savings of an ASD. More important, this characteristic makes it very difficult to match the maximum efficiency point of the fan or pump to the system load across the entire range of operation (Figures 6 and 7). Such difficulties can arise even in heating, ventilation, and air-conditioning (HVAC) systems, for example, where a static pressure must be maintained at the terminal boxes to allow proper flow control to each space.[75]

System efficiency calculations must now also include losses within the ASDs themselves as well as additional losses in the motors. Losses of a standard induction motor, for example, can be expected to increase by 20% (or 2% of total energy consumption for a 90% efficient motor) when the motor is used with a six-step VSI because of harmonics; most manufacturers recommend that standard ac motors be torque derated 10%–15% when used with VSI or CSI. When operating near full-load, by-pass switches can sometimes help avoid the excess losses in both ASDs and motors.[76]

Adjustable-speed drives can also be used with intelligent controls, and offer the potential to greatly change the way motive power is used. For example, an intelligent air-handling system could measure in real time the motor and fan efficiencies as well as the pressure drops in the duct as a function of volume blown, then choose how fast to run to meet ventilation needs and minimize energy use. At the expense of lower furnace or heat-exchanger efficiency, air of a larger temperature difference might allow slower flow rates using less energy overall. Even the quality of the energy could be taken into account, sacrificing more low-grade thermal heat for smaller gains in high-grade electrical energy.

3a. Case studies

Adjustable-speed drives are being successfully used in numerous applications. In industry they are used in boiler fans in chemical plants, utility plants, packaging equipment, glass-blowing machines, cement factories, and many others. In commerce they are used in commercial refrigeration systems, HVAC systems, and other applications. In residences, they are used in air-conditioning and heat pump systems.[77]

Numerous case studies in the literature document typical energy savings of 30%–50% in fan and boiler feedpumps, 20%–25% in compressors, 30%–35% in blowers and fans, 20%–25% in pumps, 25%–35% in central refrigeration systems, and 20% in air-conditioning and heat pumps. A residential heat pump system with ASD/permanent magnet motor-driven compressor and fans, due on the U.S. market in 1989, is expected to offer heating and cooling efficiencies 30%–40% better than today's average model. Of course, ASDs will not provide savings in constant speed full-load applications, so the national energy savings will be less than might be calculated by assuming universal adoption at the above efficiencies.[78]

Payback times of 2–3 years or less are common for ASDs. For example, a 1000-hp ASD might have capital and installation costs of $100/hp each, or $200,000 total.[79] Retrofit to a 1000-hp motor running full load (but throttled) at 95% efficiency on an 80% duty cycle, and assuming 30% energy savings and an electricity cost of $0.05/kWhr, this ASD would save $82,000 per year — a simple payback of just 2.5 years.

At the other end of the scale, the cost of mass produced ASDs used with residential heat pumps and other appliances is reportedly down to $25/hp.[80] For a standard induction motor with an efficiency of 75% (Figure 2) running on a 30% annual duty cycle, and an electricity cost of $0.05/kWhr; and arbitrarily assuming 20% savings, such an ASD would have a payback time of just one year. Of course, the actual savings may well be higher as permanent magnet motors might be used, and the design of the compressor or other equipment might be improved, among other changes. Further, the cost of electricity is usually closer to $0.10/kWhr.

In addition to higher efficiency and frequently lower lifetime costs, electronic ASDs offer a number of advantages over operating induction motors directly off the line. Intelligent controls can be incorporated to improve process quality. By permitting slow, controlled starts, ASDs reduce electrical stresses on motors, transformers, and switchgear, and reduce mechanical stresses on motors, gears, and associated equipment. In contrast, starting an induction motor directly off the line typically leads to current in-rushes six times the normal full-load operating current, and for larger motors overheating can limit line starts to twice per hour. Operation at reduced speeds can further extend motor and equipment lifetimes directly, but supplemental cooling may be necessary at low speeds, since direct cooling by the motor's fan is then less effective. Indirectly, ASDs increase lifetimes of motor-driven equipment by avoiding the back-pressures generated by conventional throttle valves or dampers used to limit the output of pumps or fans, and by permitting constant bearing lubrication—in contrast to equipment operated in an on–off mode. Users who have converted to ASDs have often found savings in maintenance of 10%–20%.

Other ASD advantages include isolation of the motor from the power line, which can reduce problems caused by varying or unbalanced line voltage. In some cases power failure for a few cycles allows fixed-speed motors to slow to a point at which they cannot be reaccelerated without decreasing their loads. In contrast, ASDs can often compensate for such interruptions and thus provide a "ride-through" capability. ASDs can also allow operation at higher speeds than the 60-Hz line frequency allows. Finally, ASDs are both relatively compact and can be located conveniently—unlike many mechanical speed controls—so they are easy to retrofit.[81]

3b. Utility operation

ASDs can improve power plant operation and may someday assist load management efforts. ASDs, however, may also increase uncertainty about future load growth and may cause problems by their generation of power line harmonics.

ASDs have been used with a variety of power plant motors—such as forced-draft and induced-draft fans, boiler-feed and condensate pumps, and many others—to provide significant energy savings as well as other operational advantages discussed above. Use of a 2000-hp ASD on the #2 boiler-feed pump at the Ft. Churchill, Nevada, power plant (a 110-MW gas fired unit used for spinning reserve), for example, allowed the minimum load to be reduced from 16 to 12 MW and the auxiliary power demands to be reduced by 0.9MW—saving $1.6 million per year at an installed cost of just $375,000.[82]

ASDs will be used increasingly in household and commercial refrigeration and air-conditioning systems due to their efficiency and other advantages. By appropriately modifying the design of these ASDs and by establishing standard protocols, direct implementation of load management techniques via power-line carriers or other means may then be possible at a very low marginal cost. Load management techniques are already being used in large-scale systems in the U.S. and Europe. Southern California Edison, for example, currently has 100,000 air conditioners in a load management program networked via a VHF-FM radio system. Despite the high cost of establishing the system—approximately $120 per participating household just for the control devices, their installation, and marketing—benefits to the utility in terms of peak-load reduction and other factors have been almost 4 times greater. Further, the program has been very well received; just 1.8% of the participants have withdrawn voluntarily.[83]

The total electricity savings possible by the use of ASDs is not yet well understood and will increase uncertainty in load growth forecasts in the near-term. Total savings will depend on a variety of factors including the rate of cost reduction of ASDs and corresponding market penetration, and the types of part or variable-loads driven. Further, the individual savings achieved may depend on the development of new engineering design rules that fully exploit the opportunities presented by ASDs. As noted above, in reviewing entire plants, the losses in motors and drives have been found to be as much as 40%–80%.[84]

The harmonics generated by electronic devices can also cause problems for utilities; problems that have long been with us. One of the first applications of rectifiers was for electrochemical processing at a copper refinery in Utah during the 1920s. When energized, the rectifier's harmonics induced voltages in telephone lines paralleling the power lines sufficient to disrupt communications.[85]

Harmonics are in fact generated by many sources. The non-linear characteristics of magnetic cores, for example, introduce harmonics. In recent times, competitive pressures have forced manufacturers to design equipment more and more critically, pushing the operating points further into the non-linear characteristics of the magnetic materials and thereby greatly increasing harmonic generation.[86]

Similarly, the non-linear characteristics of the electronic devices in ASDs generate harmonics. Converters that use SCRs cause line notching—a brief short circuit that sends the line voltage abruptly to zero at that point. Furthermore, the current drawn is closer to a square wave than a sine wave. Converters that use diode bridges draw current in a periodic pulse and also cause distortion of the line voltage. A six-step inverter will generate characteristic harmonics at $nK \pm 1 = 5, 7, 11, 13, 17, \ldots$ times the line frequency, each with maximum theoretical amplitudes equal to (1/harmonic) of the fundamental. Twelve-step inverters can, in principle, significantly reduce the generation of harmonics, but in practice system reactance and converter phase shifts somewhat reduce their advantage.[87]

The harmonics generated can cause a variety of problems. Within the motor, they can increase total motor losses by several percent; increase voltage stress on windings and their insulation; sometimes cause torque pulsation at low speeds, localized heating, and acoustic noise; excite resonant torsional vibrations (shortening equipment shaft lifetimes); and cause other problems. On the line, they can cause localized heating within the distribution transformer, and increase losses throughout the generation and distribution system. Harmonics can also excite series or (more often) parallel resonance in the power circuits, particularly in the presence of distributed power factor correction capacitors. Even the induction watthour meters used to measure household electricity consumption are affected, consistently recording values as much as 1% too high when harmonics are present. Finally, harmonics on the line can interfere with computers, communications, and other equipment. The harmonics generated by ASDs may be particularly dangerous because the amounts of power being handled are so large.[88]

Numerous technologies are being developed to reduce these harmonics. For example, harmonics can be minimized by appropriately adjusting switching patterns according to precomputed values in look-up tables stored in read-only memories (ROM), or even by doing real-time sampling and appropriate adjustment of the synthesized waveform. Of course, standard harmonic filters can also be used.[89]

Many countries have established legislation limiting the line harmonics allowable from converters. Total harmonic distortion is typically limited to 5% with sublimits on individual harmonics. In the United States, FCC rules regulate only those devices that generate or use timing signals greater than 10 kHz. Devices that run at a slower clock rate are exempt even if they generate unacceptable line interference. In contrast, IEEE standard 519 specifically addresses line noise generated by converters on the basis of line notch size and voltage distortion, and has been used with considerable success. A standard specifically for ASDs, P958, will soon be proposed by an IEEE working group now being formed.[90]

4. Conclusion

These various opportunities for reducing electricity use can now be compared as shown in the hypothetical estimates of Table 4, which in turn are based on the pumping system of Figure 6.

As seen in Table 4 for 80% flow, just 5% of the input coal is converted to useful energy (to overcome static pressure drop) at the end of the piping, and losses are spread throughout the system. ASDs achieve the greatest individual savings, and do so by eliminating shaft coupling and throttle valve losses, and by raising pump efficiencies by operating closer to the design point. Surprisingly, simply increasing the pipe diameter by 25% also offers large savings due to the excessive piping losses of this particular system (Figure 6) and due to the great leverage such reductions have upstream in the system.[91]

Repeating the calculations for 65% flow (see the inset load profile of Figure 6) gives similar savings: the increased throttling losses and reduced pump efficiency (72% due to being further from the design point) are roughly counterbalanced by decreased piping losses. For this particular system, combining all of the various improvements listed leads to electricity savings of some 70% (Table 4); large capital savings would obviously follow if in a new installation. Achieving such large savings generally, however, is highly unlikely: many systems operate at or near full-load most of the time; many will have appropriately sized piping — balancing the capital and energy costs. But by thinking of these as systems rather than as collections of individual components, numerous opportunities can be found for reducing electricity use. This is an area worthy of much more research.

The technologies described here are either available commercially, with sales growing rapidly, or in field testing. In Japan, for example, sales of ASD driven heat pumps are reportedly greater than those of constant speed systems.[92] Energy savings are, however, but one factor driving the dissemination of these technologies, ASDs in particular. As important are their numerous other advantages noted above, including better control of process, often reduced maintenance, and relatively easy retrofits. Intelligent controls will be increasingly prominent and important in the future.

In spite of a variety of potential market barriers, sales of energy-efficient induction and permanent magnet motors and their ASD controllers are growing fairly rapidly in the United States and other developed countries (DCs). PM motor technologies, for example, are expected by some to dominate certain

Table 4. Hypothetical savings for various pumping system configurations (Figure 5).

Location	Base line effr	Base line mn	EN EF[f] motor eff	Amor[g] core eff	Power[h] factor eff	PM[i] motor eff	Pump[j] speed eff	Pipe[k] size eff	ASD[l] eff	All[m] eff
Coal	...	100	...	...	...	...	...	...	...	...
Generation	33[a]	33	...	*	33.8	...	...	...	...	33.8
Transmission & distribution	90[b]	30	...	91.5	*	...	...	...	...	91.5
ASD	...	...	...	...	...	...	...	...	95	95
Electric motor	91[c]	27	94	91.5	*	95.5	−	−	90	94.5
Shaft coupling	98[d]	26	...	...	...	...	...	...	100	100
Pump	77[e]	20	...	...	...	...	80	...	79	83
Throttle valve	66[e]	13	...	...	...	...	...	...	100	100
Piping system	35[e]	5	...	...	...	...	...	56	35	56
Coal reqd.		100	96.8	97.8	97.6	95.3	96.2	63	67	37
Savings		0	3.2	2.2	2.4	4.7	3.8	37	33	63

"eff" is the device efficiency in %. "rmn" is the remaining energy as a % of the starting total. "Cold Reqd." is the input coal energy required and "Savings" are the percentage point reductions in coal use made possible for the different component and system improvements while still providing the same useable energy at the end of the piping system.

References: Base line:: a−(Hingorani in note 74); b−(Werner in note 60); c−(32) and Figure 3; d−(US DOE in note 13); e−(32) for the case of 80% flow under variable pressure and flow, see Figure 6. The piping system efficiency is given here by the ratio of system static pressure at zero flow to total head less valve losses. f−Energy-efficient motor (32) and Figure 3. g−Amorphous core transformers and motor (Werner in note 60 and Figures 4, 11), *generator savings are additional but are not detailed here. h−Power factor improvements all *power factor losses (A.D. Little Inc. in note 13) have been lumped together and assumed cut in half. i−Permanent magnet motor losses assumed to be 25% less than an energy efficient motor (Richter and Cornell in note 67) and Figure 3. This is for a PM motor run directly off the line so that there are no controller or harmonic losses. With a controller, losses will be slightly less than the ASD case. j−Pump speed twice the base line specific speed (Karassik, I.J., Krutzsch, W.C., Fraser, W.H., and Messina, J.P., *Pump Handbook* (McGraw-Hill, New York, 1986). k−Pipe size 25% greater than base line reduces pipe friction by 59%, assuming laminar flow (Krassik, *et al.*). l−Adjustable speed drive (32), note that the expected decrease in electric motor efficiency due to ASD harmonics (Mohan in note 38, Rice in note 76, Kumagi *et al.* note 74) is in part compensated for by operating the motor at a higher effective load (Figure 4) and that the pump efficiency is increased by running closer to the design point than is possible by throttling (Figure 6), but is still less than the design point efficiency of 80% due to the pump affinity laws not tracking the load curve with static pressure drop; shaft couplings and throttling valves are assumed to be eliminated. m−All, combining the PM motor (with efficiency reduced slightly for controller harmonics), amorphous core transformers, ASD, 25% larger pipe, operating at the higher power factor, and running the pump at twice the specific speed. Note that other potential synergistic savings or losses (such as lower efficiency due to reduced size) between components are not considered. Actual savings may differ significantly from these estimates.

markets, such as appliances, within a very few years. In addition, more than 200,000 ASDs have been sold in the United States in recent years.[93]

In less-developed countries, market barriers such as shortages of capital and high effective interest rates, high import duties and taxes, and the lack of knowledge of and access to these new technologies, among other factors, may more severely constrain their use than in DCs. Yet their use is more important for these nations. An average annual investment of $130 billion ($1982) for all energy equipment (but most of it in the electric sector) has been projected by the World Bank as needed in less-developed countries over the next decade. Improving the efficiency of motor-drive systems could significantly reduce this expenditure.[94]

In Brazil, for example, it has been estimated that more efficient motors could reduce electrical demand in new installations at a cost of $150/kW, and in replacing less efficient motors at a cost of $525/kW. Adjustable-speed drives can show similarly large returns, depending on the load and annual usage. In contrast, to supply 1 kW of "firm baseload" hydroelectric power now costs about $3375. Ways to overcome market barriers inhibiting use of these efficient motor-drive technologies are desperately needed. Strategies that might be tried include minimum efficiency standards, efficiency labeling, and utility rebates, among many others.[95]

Modern solid state electronics, magnetic materials, and other technologies promise to revolutionize motor-drive system design and operation worldwide. Although these technologies are already proving irresistible in many applications, much work remains to be done: to further improve equipment performance and reduce costs; to develop standards for the use of ASDs;[96] and to develop new rules for system design and operation so that the full potential of these technologies can be realized.

Acknowledgments

The author wishes to acknowledge the Hewlett Foundation, the seven New Jersey gas and electric utilities, the New Jersey Department of Energy, and especially the Center for Energy and Environmental Studies, Princeton University for their support while much of this work was being done.

Bibliography

J.C. Andreas, *Energy-Efficient Electric Motors* (Marcel Dekker, Inc., New York, 1982). This book provides useful empirical information for the lay reader interested in electric motor-drive systems. Although it is now somewhat dated, given the rapid advances in the field, it provides a useful overview of electric motors, mechanical and electronic speed changing equipment, and methods of applying these in efficient drive systems.

J.M.D. Murphy and F.G. Turnball, *Power Electronic Control of AC Motors* (Pergamon Press, Elmsford, NY, 1988). This text provides the interested technical reader an extensive description of how motor drives work while minimizing the use of equations.

Steven Nadel, Michael Shepard, Gail Katz, Steven Greenberg, Anibal de Almeida, "Energy-Efficient Motor Systems: A Handbook on Technology, Program, and Policy Opportunities," American Council for an Energy-Efficient Economy, Washington, D.C., forthcoming.

References and notes

1. Portions of this paper have been reproduced, with permission, from the Annual Review of Energy, V 13 – Copyright 1988 by Annual Reviews Inc.; and from "Electricity: Efficient End-Use and New Generation Technologies, and Their Planning Implications," Proceedings of an International Symposium, Thomas B. Johansson, Birgit Bondlund, and Robert H. Williams, eds. (Lund University Press, 1989).

2. The author is currently with the Office of Technology Assessment, United States Congress, Washington, DC 20510-8025. The views expressed are those of the author alone and not necessarily those of the Office of Technology Assessment.

3. W.D. Devine, Jr., "Historical Perspective on Electrification in Manufacturing," in *Electricity Use, Productive Efficiency and Economic Growth*, S. Schurr and S. Sonenblum, eds. (EPRI, Palo Alto, CA, 1986); J.N. Barkenbus, *Roles of Electricity: A Brief History of Electricity and the Geographic Distribution of Manufacturing* (EPRI, #EM-5298-SR, Palo Alto, CA, 1987); S.H. Schurr, "Energy Use, Technological Change, and Productive Efficiency: An Historical Interpretation," Annu. Rev. Energy **9**, 409–425 (1984).

4. The $ value for electricity is less than two-thirds the total due to the higher than average percentage of motor drive in industry and the corresponding lower average cost of electricity.

5. S.F. Baldwin, *Energy-Efficient Electric Motor-Drive Systems: I. High Performance Materials, Adjustable Speed Drives, and System Design* (Center for Energy and Environmental Studies, Princeton University, Princeton, 1988).

6. USDOE Energy Info. Admin., *Annual Energy Review 1986* (U.S. Dept. of Energy, Washington, DC, 1987).

7. This is not to say that there have been no advances at all. Major advances include vacuum pressure impregnation with synthetic resins, more rugged stator and rotor construction, and the extensive use of computer design techniques. See Alfred O. Staub and Edward L. Owen, "Solid State Motor Controller," IEEE Trans. Ind. Applic. **IA-22**, N.6, 1113–1120 (1986) and E.L. Owen, "Power Electronics and Rotating Machines – Past, Present, and Future," *15th Annual Power Electronics Specialist Conference* (1984), pp. 3–11.

8. This includes the difficulty of developing the high level of experience needed to work with brushes and slip rings.

9. A.E. Fitzgerald, C. Kingsley, Jr., and S.D. Umans, *Electric Machinery* (McGraw-Hill, New York, 1983).

10. S.A. Nasar, *Electromagnetic Energy Conversion Devices and Systems* (Prentice-Hall, Inc., Englewood Cliffs, NJ, 1970).

11. The above discussion is mathematically expressed as follows. In a standard two-pole machine, the three-phase voltages (currents) applied to the stator windings, because they are applied to coils $2/3$ radians apart, sum to generate a single flux wave rotating around the stator at the synchronous speed, w_s. Thus,

    $I_A = I_M \cos w_s t,$

    $I_B = I_M \cos (w_s t - 2/3),$

    $I_C = I_M \cos (w_s t + 2/3),$

    $Glux = I_A + I_B e^{-j2/3} + I_C e^{+j2}/3 = -\frac{3}{2} I_M e^{-jwt},$

    and the voltages driving these currents are a phase angle, 0, ahead (behind) of these currents.

12. Note that the power factor of a synchronous motor can be controlled by adjusting the level of dc rotor excitation.

13. A.D. Little, Inc., *Energy Efficiency and Electric Motors* (U.S. Dept. of Commerce, NTIS PB-259, Washington, DC, 1976), p. 129; USDOE, A.D. Little, Inc., *Classification and Evaluation of Electric Motors and Pumps*, DOE/CS-0147 (U.S. Department of Energy, Washington, DC, 1980); W.J. McDonald and H.N. Hickok, "Energy Losses in Electrical Power Systems," IEEE Trans. Ind. Applic. **IA-21**, 124–136 (1985); J.C. Andreas, *Energy-Efficient Electric Motors* (Marcel Dekker, Inc., New York, 1982); A.A. Jimoh, R.D. Findlay, and M. Poloujadoff, "Stray Losses in Induction Machines: Part I. Definition, Origin and Measurement; Part II. Calculation and Reduction," IEEE Trans. Power App. Sys. **PAS-104**, 1500–1512 (1985).

14. J. Reason, "Powerplant Motors," Power, S1-S16 (1986) and Andreas as cited in note 13.

15. J. Appelbaum, E.F. Fuchs, and J.C. White, "Optimisation of Three-Phase Induction Motor Design. Part I: Formulation of the Optimization Technique" and with I.A. Khan, "Part II: The Efficiency and Cost of an Optimal Design," IEEE Trans. En. Conv. **EC-2**, 407–422 (1987); E.F. Fuchs, J. Appelbau, I.A. Khan, J. Hall, and U.V. Frank, 1987. "Optimisation of Induction Motor Efficiency," two volumes (EPRI EL-4152, Palo Alto, CA, July 1985 and May 1987); N.H. Fetih and H.M. El-Shewy, "Induction Motor Optimum Design, Including Active Power Loss Effect," IEEE Trans. En. Conv. **EC-1**, 155–160 (1986).

16. F.A. Scheda, "Transient Inrush Currents in High-Efficiency and Standard Motors," IEEE Trans. Ind. Applic. **IA-22**, 145–147 (1986).

17. D.M. Mullen, "Specify Motors By Starting Duty, Not by Running Duty," Power 55–57 (February 1985) and J.W. Griffith and R.M. Mckay, *Improved Motors For Utility Applications: V. 6: Squirrel-Cage Rotor Analysis* (EPRI EL-4286, Palo Alto, CA, 1986).

18. Note that the power factor of energy-efficient motors is lowered generally due to the higher flux levels (although at a lower flux density due to the increase in core dimensions). Personal communication, Ed Owen, General Electric, March 7, 1988.

19. U.S. DOE, Arthur D. Little, Inc., *Classification and Evaluation of Electric Motors and Pumps* as cited in note 13.

20. See note 6.

21. A. D. Little, Inc., *Energy Efficiency and Electric Motors* as cited in note 13.

22. "Computing Power Factor Penalties," Energy User News **3** (August 19, 1985).

23. H.N. Hickok, "Adjustable Speed—A Tool for Saving Energy Losses in Pumps, Fans, Blowers, and Compressors," IEEE Trans. Ind. Applic. **IA-21**, 124–136 (1985) and U.S. DOE and A.D. Little Inc., *Classification and Evaluation of Electric Motors and Pumps* as cited in note 13.

24. A.J. Stepanoff, *Centrifugal and Axial Flow Pumps*, 2nd ed. (John Wiley and Sons, New York, 1957).

25. G.S. Settles, J.T. Hamrick, M. Summerfield, and M. Gunn, "Energy-Efficient Pump Utilization," J. Energy **1**, 65–72 (1977); J. Reason, "Fans," Power S1–S24 (September 1983) and U.S. DOE and A.D. Little, Inc., Classification and Evaluation of Electric Motors and Pumps, as cited in note 13.

26. See Mullen as cited in note 18.

27. See A.D. Little Inc., *Energy Efficiency and Electric Motors* as cited in note 13.

28. H.S. Geller, *End-Use Electricity Conservation: Options for Developing Countries* (World Bank En. Dept. Paper 32, Washington, DC, 1986).

29. M. Ross, "Trends in the Use of Electricity in Manufacturing," IEEE Technology and Society, 18–22 (March 1986).

30. See A.D. Little Inc., *Energy Efficiency and Electric Motors* as cited in note 13.

31. S.F. Baldwin and E. Finlay, "Analysis of Electric Motor Driven Systems in the Sugar Industry: Electricity Conservation in Boiler Fans and Irrigation Pumps," Presented at Meeting of the Jamaican Assoc. of Sugar Technologists, Ocho Rios, Jamaica, Nov. 5–6, 1987.

32. J.K. Armintor and D.P. Connors, "Pumping Applications in the Petroleum and Chemical Industries," IEEE Trans. Ind. Applic. **IA-23**, 37–48 (1987).

33. J.D. Rozzner, P.W. Myers, and C.J. Robb, "The Application of Adjustable Frequency Controllers to Forced Draft Fans for Improved Reliability and Energy Savings," IEEE Trans. Ind. Applic. **IA-21**, 1482–1489 (1985).

34. A.C. Price and M.H. Ross, *Reducing Electricity Costs in Automotive Assembly and Stamping Plants* (University of Michigan, unpublished, 1988).

35. N. Ladomatos, N.J.D. Lucas, W. Murgatroyd, and B.C. Wilkins, "Industrial Energy Use—III. The Prospects For Providing Motive Power In A Machine Tool Shop From A Centralized Hydraulic System," Energy Research **3**, 19–28 (1979) and Ross as cited in note 34.

36. A.D. Little, Inc. as cited in note 13.

37. P.W. Brothers and M.L. Warren, "Fan Energy Use In Variable Air Volume Systems," ASHRAE Transactions **92**, Pt. 2 (1986).

38. N. Mohan, *Techniques for Energy Conservation in ac Motor-Driven Systems* (EPRI EM-2037, Palo Alto, CA, 1981); D.P. Masher and J.S. Smith, *Impact of Advanced Power Semiconductor Systems on Utilities and Industry* (EPRI, EM-2112, Palo Alto, CA, 1981); "Pacing Plant Motors For Energy Savings," EPRI Journal 22–28 (March 1984); Hickock as cited in note 23; and Settles *et al.* as cited in note 25.

39. W. Murgatroyd and B.C. Wilkins, "The Efficiency of Electric Motive Power in Industry," Energy **1**, 337–345 (1976).

40. B.L. Jones and J.E. Brown, "Electrical Variable-Speed Drives," IEE Proceedings **131A**, 516–558 (1984).

41. Jones and Brown as cited in note 40 and Andreas as cited in note 13.

42. Jones and Brown as cited in note 40 and E.L. Owen, M.M. Morack, C.C. Herskind, and A.S. Grimes, "ac Adjustable-Speed Drives With Electronic Power Converters—The Early Days," IEEE Trans. Ind. App. **20**, 298–307 (1984).

43. *Adjustable Speed Drives: Directory, Manufacturers and Applications*, Second ed. (EPRI, Palo Alto, CA, 1987) and EPRI Journal as cited in note 38.

44. D.Y. Chen, "Power Semiconductors: Fast, Tough, and Compact," IEEE Spectrum 30–35 (Sept. 1987) and M.S. Adler, K.W. Owyang, B.J. Baliga, and R. Kokosa, "The Evolution of Power Device Technology," IEEE Trans. Elec. Dev. **ED-31**, 1570–1591 (1984).

45. Y. Ikeda and T. Yatsuo, "Recent Progress in Power Electronic Devices," Hitachi Review **36**, 5–12 (1987) and J. Douglas, "Sealed in Silicon: The Power Electronics Revolution," EPRI Journal 5–15 (Dec. 1986).

46. Thomas A. Lipo, "Recent Progress in the Development of Solid-State ac Motor Drives," IEEE Trans. Power Electronics **3**, 105–117 (1988).

47. B.J. Baliga and D.Y. Chen, *Power Transistors: Device Design and Applications* (IEEE Press, Piscataway, NJ, 1984).

48. G. Kaplan, "Thyristors: Future Workhorses in Power Transmission," IEEE Spectrum 40–52 (Dec. 1982).

49. D.A. Jarc and D.W. Novotny, "A Graphical Approach to ac Drive Classification," IEEE Trans. Ind. Applic. **IA-23**, 1029–1035 (1987); S.B. Dewan, G.R. Slemon, and A. Sraughen, *Power Semiconductor Drives* (John Wiley and Sons, New York, 1984); J. Reason, "Large Multi-Speed Motor Drives— The Alternatives," Power 68–71 (January 1985); P. Cleaveland, "Motor Controls: A Technology Update," Chiltons I&CS 50–55 (March 1984); A.O. Staub and E.L. Owen, "Solid-State Motor Controllers," IEEE Trans. Ind. Applications **IA-22**, 1113–1120 (1986); *Adjustable Speed ac Drive Systems*, Bimal K. Bose, ed. (IEEE Press, Piscataway, NJ, 1981), pp. 1–21; Steve Greenberg, Jeffrey P. Harris, Hashem Akbari, and Anibal de Almeida, *Technology Assessment: Adjustable-Speed Motors and Motor Drives (Residential and Commercial Sectors)* (Lawrence Berkeley Laboratory, Berkeley, CA, March 1988); Mohan as cited in note 38; Jones and Brown, Reference 40.

50. References cited in note 49.

51. Mohan, Ref. 38; Adjustable Speed Drives, Ref. 43; Jarc *et al.*, Dewan *et al.* and Reason in references cited in note 49.

52. Reason as cited in note 49.

53. Reason and Cleaveland as cited in note 49; References 23 and 43.

54. Dewan *et al.* as cited in note 49.

55. Andreas as cited in note 13 and Mohan, and Masher and Smith as cited in note 38.

56. E. Duane Wolley, "Power Semiconductors: a Perspective of the Power Semiconductor Committee," IEEE Trans. Ind. Applic. **22**, 970–971 (1986).

57. H. Brody, "Smart Power," High Technology 18–24 (Dec. 1985); S.P. Robb and J.L. Sutor, "Comparing Power ICs," Machine Design 119–122 (Feb. 21, 1985); V. Rumennik, "Power Devices Are In The Chips," IEEE Spectrum 38–42 (July 1985); and R. Allan, "Sensors in Silicon," High Techn. 43–50 (Sept. 1984).

58. B. Marshall, "The Future of Smart PICS," in *Recent Advances in Discrete MOS-Gated Power Structures* (Electro/86, Boston, May 13–15, 1986).

59. Robert H. Williams and Eric D. Larson, "Steam-Injected Gas Turbines and Electric Utility Planning," IEEE Technology and Society, pp. 29–38, 1986.

60. F.E. Werner, "Electrical Steels: 1970–1990" in *Energy Efficient Electrical Steels*, A.R. Marder and E.T. Stephenson, eds. (Proc. AIME Metal. Soc. Pittsburgh, Oct. 5–9, 1980); USDOE End. Info. Admin., *Monthly Energy Review, July 1987* (U.S. Dept. of Energy, Washington, DC, 1987); and F.E. Luborsky, P.G. Frischmann, and L.A. Johnson, "The Role of Amorphous Materials In The Magnetics Industry," J. Magn. Magn. Mater. **8**, 318–329 (1978).

61. L.R. Bickford, "Magnetism During the IEEE's First One Hundred Years (1884–1984)," IEEE Trans. Mag. **MAG-21**, 2–9 (1985); M.F. Littman, "Iron and Silicon-Iron Alloys," IEEE Trans. Mag. **MAG-7**, 48–60 (1971); I.S. Jacobs, "Magnetic Materials and Applications—A Quarter-

Century Overview," J. App. Phys. **50**, 7294–7306 (1970); and Luborsky *et al.* as cited in note 60.

62. National Materials Advisory Board, *Magnetic Materials*, NMAB-426 (National Academy Press, Washington, DC, 1985).

63. T. Fukuda and M. Hoshi, "New Technology for Large Power Transformers," Hitachi Review **33**, 137–140 (1984); N. Takahashi, Y. Ushigami, M. Yabumoto, Y. Suga, H. Kobayashi *et al.*, "Production of Very Low Core Loss Grain-Oriented Silicon Steel," IEEE Trans. Mag. MAG-**22**, 490–495 (1986); and Ref. 62.

64. J.J. Gilman, "Magnetic Glasses," Science **208**, 856–861 (1980); H.S. Chen, "Glassy Metals," Rep. Prog. Phys. **43**, 353–432 (1980); K. Moorjani and J.M.D Coey, *Magnetic Glasses* (Elsevier, New York, 1984); Ref. 5 and Luborksy as cited in note 56 and Jacobs as cited in note 61.

65. K.S. Tan, "Amorphous Ribbon Domain Refinement by Scribing," IEEE Trans. Mag. MAG-**22**, 188–191 (1986); D.S. Takach and R.L. Boggaverapu, "Distribution Transformer No-Load Losses," IEEE Trans. Power App. Sys. PAS-**104**, 181–193 (1985); D.W. Whitley, "Type Testing of Amorphous Metal Distribution Transformers," IEEE Trans. Power Deliv. PWRD-**2**, 827–830 (1987); M.M. Saied, "Feasibility of Using Magnetic Amorphous Metals As Core Materials in Power Transformers," Elec. Mach. Power Sys. **12**, 325–341 (1987); L.A. Johnson, E.P. Cornell, D.J. Bailey, and S.M. Hegyi, "Application of Low Loss Amorphous Metals in Motors and Transformers," IEEE Trans. Power App. Sys. PAS-**101**, 2109–2114 (1982); Ref. 62; and Moorjani as cited in note 34.

66. S. Chikazumi, "Evolution of Research in Magnetism in Japan," J. Appl. Phys. **53**, 7631–7636 (1982); Bickford as cited in note 61; and Ref. 62.

67. E. Richter, T.J.E. Miller, T. Neumann, and T.L. Hudson, "The Ferrite Permanent Magnet ac Motor—A Technical and Economical Assessment," IEEE Trans. Ind. Appl. **21**, 644–650 (1985); E.P. Cornell, "Permanent Magnet ac Motors," *Proc. Conf. Drives/Motors/Controls 83*, P. Lawrenson and A. Hughes, eds. (1983); F.N. Klein and M.E. Kenyon, "Permanent Magnet dc Motors Design Criteria and Operation Advantages," IEEE Trans. Ind. Appl. **20**, 1525–1531 (1984); Appliance, Special Issue **41**, Pt. 2 (1984); J. Barber, "New dc Brushless Entries Touted for Low-Speed Uses," Energy User News (March 5, 1984); T.J. Miller "Rare Earth Magnets Contribute to Small Size, High Torque of New Motor Designs," Control Engineering 90–91 (May 1983).

68. K. Strnat, F.M. Tait Professor of Electrical Engineering, University of Dayton, Dayton, Ohio. personal communication, November 19, 1985 and Refs. cited in note 67.

69. Special issue of Appliances cited in note 67.

70. N.A. Demerdash, T.A. Nyamusa, and T.W. Nehl, "Comparison of Effects of Overload On Parameters and Performance of Samarium-Cobalt and Strontium-Ferrite Radially Oriented Permanent Magnet Brushless dc Motors," IEEE Trans. Power App. Systems PAS-**104**, 2223–2231 (1985); A.L. Robinson, "Powerful New Magnetic Material Found," Science **223**, 920–922 (1984); K.J. Strnat, "Rare-Earth Permanent Magnets: Two Decades of New Magnet Materials," Symp. Soft and Hard Mag. Mater. with Appl., Amer. Soc. for Metals, Orlando, Fla., Oct. 4–9, 1986; Richter *et al.* and Miller as cited in note 67; and Ref. 46.

71. G. Graff, "Remaking the Magnet," *New York Times*, F23 (Sunday, March 30, 1986); K. Strnat, *Study and Review of Permanent Magnets for Electric Vehicle Propulsion Motors*, DOE/NASA/0189-83/2 (Department of Energy, Washington, DC, Sept. 1983); M.A. Rahman and G.R. Slemon, "Promising Applications of Neodymium Boron Iron Magnets in Electrical Machines," IEEE Trans. Mag MAG-**21**, 1712–1716 (1985); *New York Times*, "Supermagnets: More Pull, Less Weight," F23 (March 30, 1986); References 62 and 67.

72. C.C. Toradi, M.A. Subramanian, J.C. Calabrese, J. Gopalakrishnan *et al.*, "Crystal Structure of $Tl_2Ba_2Ca_2Cu_3O_{10}$, a 125 K Superconductor," Science **240**, 631–633 (1988).

73. M. Yamamoto, N. Mizukami, T. Ishigohka, and K. Oshima, "A Feasibility Study on a Superconducting Power Transformer," IEEE Trans. Mag. V MAG-**22**, 418–420 (1986).

74. Narian G. Hingorani, "Superconductivity: The State of the Art," IEEE Power Engineering Review 3 (May 1988); M. Kumagi, T. Tanaka, K. Ito, Y. Watanabe, and Y. Gocho, "An Approach to Optimal Thermal Design of Superconducting Generator Rotor," IEEE Trans. En. Conv. V. EC-**1**, 217–224 (1986); Nils-Johan Bergsjo and Lars Gertmar, "Superconductivity and the Efficient Use of Electricity," in *Electricity: Efficient End-Use and New Generation*

Technologies, and Their Planning Implications, Thomas G. Johansson, Birgit Bodlund, and Robert H. Williams, eds. (Lund University Press, Lund, 1989). See also Ref. 5.

75. J.R Pottebaum, "Optimal Characteristics of a Variable-Frequency Centrifugal Pump Motor Drive," IEEE Trans. Ind. Applic. **IA-20**, 23–31 (1984); W.R. Wood, "Beware of Pitfalls when Applying Variable-Frequency Drives," Power 47–79 (February 1987); L. Norford and S. Englander, personal communication, CEES Princeton; Refs. 23, 32, and 37.

76. D. Scholey, "Induction Motors for Variable Frequency Power Supplies," IEEE Trans. Ind. Applic. **IA-18**, 368–372 (1982); D.E Rice, "Adjustable Speed Drive and Power Rectifier Harmonics—Their Effect on Power System Components," IEEE Trans. Ind. Applic. **IA-22**, 161–177 (1986); C. James Erickson, "Motor Design Features for Adjustable-Frequency Drives," IEEE Trans. Ind. Applic. **24**, 192–198 (1988); Mohan as cited in note 38; and Ref. 40.

77. N. Mohan, K.F. Brooks, and W. Montgomery, *Energy Conservation in ac Motor Driven Systems By Means of Solid State Adjustable Frequency Controllers* (Dept. Elec. Engn., U. of Minn., Minneapolis, 1985); J.E. Wasilewski, "Solid-State ac Drives: New Reliability for Packaging Equipment," IEEE Trans. Ind. Applic. **IA-21**, 107–111 (1985); S.S. Cuffe and P.W. Hammond, "A Variable Frequency ac Blower Drive Installation for Efficient and Accurate Control of Glass Tempering," IEEE Trans. Ind. Applic. **IA-21**, 1047–1052 (1985); T.R. Nilsson, B. Sinner, and O.V. Volden, "Optimized Production and Energy Conservation," IEEE Trans. Ind. Applic. **IA-22**, 442–445 (1986); G. Ponczak, "Energy Management in Supermarkets: Variable Speed Compressor System to Save $3000/year," *Energy User News* (May 4, 1987); P. Raffaele, "New Controls; HVAC Retrofit to Save Firm $93K per Year," *Energy User News* **12** (April 27, 1987); H.S. Geller, "Energy-Efficient Residential Appliances: Performance Issues and Policy Options," IEEE Tech. and Soc. **5**, 4–10 (1986); G. Raymond, "EPRI/Carrier Advanced Heat Pump," ACEEE First National Conference on Appliance Efficiency, Oct. 27, 1987, Arlington, VA; N. Mohan and J.W. Ramsey, *Comparative Study of Adjustable-Speed Drives for Heat Pumps*, EPRI EM-4704 (EPRI, Palo Alto, CA, 1986); and Ref. 33.

78. F. Watson, "ac Speed Drives Catching On In Industrial Process Use," *Energy User News* **9** (Aug. 20, 1984); B.J. Harrer, M.J. Kellog, A.J. Lykes, K.L. Imhoff, and Z.J. Fisher, *Electric Energy Savings From New Technologies*, PNL 5665-rev 1 (1986); J. Oliver and R. Ferraro, "Reduce Plant Auxiliary Load, Save Fuel with ASDs," Electrical World 42–44 (June 1986); "The Advanced Heat Pump: All the Comforts of Home . . . and Then Some," EPRI Journal (March 1988); Raymond as cited in note 76; USDOE as cited in note 13; Reason as cited in note 49; Refs. 33 and 38.

79. See Ref. 43.

80. See Greenberg *et al.* as cited in note 49.

81. Reference 23; EPRI Journal as cited in note 38; Reason and Cleaveland as cited in note 49; and Watson as cited in note 78.

82. Jim Oliver and Ralph Ferraro, "Reduce Plant Auxiliary Load, Save Fuel With ASDs," Electrical World 42–45 (June 1986); Wilbur Montgomery, George Tator, James A. Oliver, and Ralph J. Ferraro, "Ft. Churchill Station Boiler Feed Pump Conversion to Adjustable Speed Operation," 1985 PCEA Engineering and Operating Conference, March 19–20, 1985, Los Angeles; and Mohan *et al.* as cited in note 77.

83. Dennis J. Gaushell, "Automating the Power Grid," IEEE Spectrum 39–45 (October 1985); Gary F. Strickler and Sharon Kau Noell, "Residential Air Conditioner Cycling: A Case Study," IEEE Trans. Power System **3**, 207–212 (1988); and Per Bohlin, Mats Edvinsson, Gustav Lindbergh, and Carl-Gustav Lundqvist, "Successful Implementation of A Nation-Wide Load Management System," IEEE Trans. Power Systems **V.PWRS-1**, 90–95 (1986).

84. Reference 39.

85. R.P. Stratford, "Analysis and Control of Harmonic Current in Systems with Static Power Converters," IEEE Trans. Ind. Applic. **IA-17** (1981) and J. Douglas, "Quality of Power in the Electronics Age," EPRI Journal 6–13 (Nov. 1985).

86. J. Arrillaga, P.A. Bradley, and P.S. Bodger, *Power System Harmonics* (John Wiley and Sons, New York, 1985); and A.A. Mahmoud, "Power System Harmonics: An Assessment," in *Power System Harmonics*, A.A. Mahmoud, W.M. Grady, and M.F. McGranaghan, eds. (IEEE Press, Piscataway, NJ, 1984).

87. D.A. Jarc and D.P. Connors, "Variable Frequency Drives and Power Factor," IEEE Trans. Ind. Appl. **IA-21**, 771–777 (1985); D.A. Bradley, P.J. Morfee, and L.A. Wilson, "The New

Zealand Harmonic Legislation," IEE Proceedings **132A**, 177–184 (1985); D.A. Jarc and R.G. Schieman, "Power Line Considerations for Variable Frequency Drives," IEEE Trans. Ind. Appl. **IA-21**, 1099–1105 (1985); R.G. Schieman, "Power System Requirements for Thyristor Drives," IEEE Trans. Ind. Applic. **IA-21**, 849–852 (1985); Mohan and EPRI Journal as cited in note 38; and Reason and Cleaveland as cited in note 49.

88. D.A. Gonzales and J.C. McCall, "Design of Filters to Reduce Harmonic Distribution In Industrial Power Systems," IEEE Trans. Ind. Applic. **IA-23**, 504–511 (1987); E.F. Fuchs, D.J. Roesler, and F.S. Alashhab, "Sensitivity of Electrical Appliances to Harmonics and Fractional Harmonics of the Power System's Voltage. Part I: Transformers and Induction Machines; Part II: Television Sets, Induction Watthour Meters and Universal Machines," IEEE Trans. Power Deliv. **PWRD-3**, 437–453 (1987); J.C. Appiarius, D.W. Houghtaling, D.F. Lackey, R.M. McCoy, and N.W. Miller, *Improved Motors for Utility Applications, V.4: Impact of Harmonics*, EPRI EL-4286 (EPRI, Palo Alto, CA, 1986); I.D. Hassan, "Specifying Adjustable-Speed ac Drive Systems and Currently Available Industry Standards," IEEE Trans. Ind. Applic. **IA-23**, 581–585 (1987); H.O. Nash and F.M. Wells, "Power Systems Disturbances and Considerations for Power Conditioning," IEEE Trans. Ind. Applic. **IA-21**, 1472–1481 (1985); C.L. Philibert, "Watch Out for Hamonics When Specifying SCR Motor Drives," Power 49–50 (August 1986); F.H. Wolff and A.J. Molnar, "Variable-Frequency Drives Multiply Torsional Vibration Problems," Power 83–86 (June 1985); Rice as cited in note 76; and Refs. cited in note 87.

89. M. Hombu, S. Ueda, A. Ueda, and Y. Matsuda, "A New Current Source GTO Inverter with Sinusoidal Output Voltage and Current," IEEE Trans. Ind. Applic. **IA-21**, 1192–1198 (1985); D.M. Divan, "Optimum PWM Waveform Synthesis – A Filtering Approach," IEEE Trans. Ind. Applic. **IA-21**, 1199–1205 (1985); J.T. Boys and S.J. Walton, "A Loss Minimized Sinusoidal PWM Inverter," IEE Proc. **132B**, 260–268 (1985); S.R. Bowes and A. Midoun, "Sub-Optimal Switching Strategies for Micro-processor Contolled PWM Inverter Drives," IEE Proc. **132B**, 133–145 (1985); J. Hamman and L.P. DuToit, "A New Microcomputer Contolled Modulator for PWM Inverters," IEEE Trans. Ind. Applic. **IA-22**, 281–285 (1986); and Arrillaga *et al.* as cited in note 86.

90. D.J. Pileggi and A.E. Emanuel, "An Examination of Existing Harmonic Recommended Limits, Guides and Standards," in *Power System Harmonics*, A.A. Mahmoud, W.M Grady, and M.F. McGranaghan, eds. (IEEE Press, Piscataway, NJ, 1984); Mahmoud as cited in note 86; Bradley *et al.* and Jarc and Schieman as cited in note 87; and Hassan as cited in note 88.

91. This does not, however, take into account changes in pump or motor efficiency, etc. It also does not provide for the complex tradeoffs in capital costs for larger pipe (less the smaller motor and pump) versus the energy savings.

92. S.M. Jeter, W.J. Wepfer, G.M. Fadel, N.E. Cowden, and A.A. Dymek, "Variable Speed Drive Heat Pump Performance," Energy **12**, 1289–98 (1987).

93. References 5 and 42; Erickson as cited in note 76; U.S. DOE and A.D. Little Inc. as cited in note 13; and special issue of Appliance as cited in note 67.

94. World Bank, *The Energy Transition in Developing Countries* (World Bank, Washington, DC, 1983); and Ref. 28.

95. H.S. Geller, *The Potential For Electricity Conservation In Brazil* (American Council For An Energy Efficient Economy, Washington, DC, 1984); H. Geller, J. Goldemberg, R. Hukai, J.R. Moreira, C. Scarpinella, and M. Yoshizawa, *Electricity Conservation in Brazil: Potential and Progress* [American Council for En. Eff. Econ., Washington, DC and Companhia Energetica de Sao Paulo (CESP), Sao Paulo, Brazil, 1987]; J. Goldemberg and R.H. Williams, *The Economics Of Energy Conservation In Developing Countries* (Princeton University, Center for Energy and Environmental Studies, Report No. 189, Princeton, NJ, 1985); and Ref. 28.

96. Hassan as cited in note 88.

97. I.J. Karassik, W.C. Krutzsch, W.H. Fraser, J.P. Messina, "Pump Handbook," New York, McGraw Hill (1986).

Appendix: Energy units

David Bodansky

1. Introduction

Many sorts of units are used in energy discussions. They fall into two broad categories: (a) those whose definition is not related to a particular fuel, which we here term "basic" units; and (b) those whose definition is related to idealized properties of a specific fuel, which we here term "source-based" units. These units, along with special issues related to electricity, are discussed in succeeding sections. Table 1 gives conversion factors between units, as well as the energy content of specific fuels.

2. Basic units

Joule (J). The unit of energy in the International System, here equivalent to the MKS system, is the joule. It is ultimately defined in terms of the meter, kilogram, and second.

Kilocalorie (kcal). The kilocalorie is defined in terms of the joule. The definition is related to the historical definition of the mechanical equivalent of heat, but standard practice is to adopt an exact conversion factor. Two definitions are in widespread use, with no clear consensus as to the preferred definition. These are:

thermochemical calorie: 1 kcal = 4184 J (exact),

International Table calorie: 1 kcal = 4186.8 J (exact).

The former is used, for example, by the U.S. National Bureau of Standards and in the major energy study by the International Institute for Applied Systems Analysis (IIASA),[4] while the latter is used for United Nations and OECD publications.[5] The difference between these two values is rarely of more than aesthetic significance, given the imprecision of determinations of energy use and supply.

There is a potential for confusion between the kilocalorie (4184 joules) and the gram-calorie (4.184 joules), because each sometimes is referred to as the "calorie." The food "calorie" always corresponds to the kilocalorie or Calorie.

Table 1. Conversion of units.[a,b]

General

1 short ton (ton) = 2000 lb = 0.907185 tonne
1 metric ton (tonne) = 1000 kg
1 barrel = 42 U.S. gallons = 159.0 litres

1 Btu (British thermal unit) = 1055 J (Joules)
1 kWh (kilowatt hour) = 3.6 MJ = 3412 Btu
1 kWh of electricity requires on average
  10,253 Btu to produce, corresponding to a
  mean thermal efficiency of 33% (1988 U.S.
  fossil-fuel average)

Large units

1 quadrillion Btu = 10^9 MBtu = 10^{15} Btu
1 exajoule (EJ) = 10^3 PJ = 10^{12} MJ = 10^{18} J
1 terawatt-yr (TWyr) = 10^9 kWyr
  = 8.76×10^{12} kWh

	Quad	EJ
1 Quad	1.000	1.055
1 EJ	0.948	1.000
1 TWyr (100% conversion)	29.89	31.54
1 TWyr (33% efficiency)	90.6	95.6
10^9 tonne coal equiv (Gtce)	27.76	29.29
10^9 barrel oil equiv (bboe)	5.80	6.12
10^9 tonne oil equiv (Gtoe)	42.43	44.76
10^9 tonne oil equiv (Gtoe)[c]	39.69	41.87

Fuel values

1 barrel of crude oil = 0.137 metric ton
1 million barrels per day of crude oil
  = 2.12 quad/yr = 2.23 EJ/yr

	MBtu	GJ
Nominal or standard equivalents:		
1 barrel of crude oil (boe)	5.8	6.12
1000 cu. ft. of natural gas	1.000	1.055
1 short ton of coal	25.18	26.57
Average heat content (U.S. 1988):		
1 barrel of petroleum products	5.408	5.705
1000 cu. ft. of natural gas	1.029	1.086
1 short ton of coal	21.53	22.72
1 cord of dry wood (1.25 ton)	21.5	22.7
1 barrel of natural gas liquids	3.812	4.022
1 barrel of aviation gasoline	5.048	5.326
1 barrel of motor gasoline	5.253	5.542
1 barrel of distillate fuel oil	5.825	6.145
1 barrel of residual fuel oil	6.287	6.633

a. Adopted from Ref. 1.
b. Based on *Annual Energy Review* 1988 (Ref. 2), *Monthly Energy Review* (Ref. 3), and IIASA report (Ref. 4).
c. Alternate equivalent, used by OECD (Ref. 5).

British thermal unit (Btu). This is the English system analog of the calorie. Again, several definitions are used, but a large majority of sources give the same approximate equivalence:

$$1 \text{ British Thermal Unit} = 1055 \text{ J.}$$

This is not an exact definition, but agrees to within 0.01% with the "International Table Btu," which is defined exactly as 251.996 IT calories (equal to 1055.06 J).[6] The kilowatt-hour (kWh) is then related to the Btu by:

$$1 \text{ kWh} = 3412 \text{ Btu.}$$

Quadrillion Btu (quad). This is a more convenient unit than the Btu for discussing amounts of energy on the scale of, say, annual energy consumption. It is widely used in U.S. publications. By definition:

$$1 \text{ quad} = 10^{15} \text{ Btu.}$$

Exajoule (EJ). This is another unit widely used for describing large amounts of energy. By definition, 1 EJ = 10^{18} J, and thus the exajoule and quad are roughly equal in magnitude:

$$1 \text{ quad} = 1.055 \text{ EJ.}$$

Terawatt-year (TWyr). Many of the discussions of world energy budgets have adopted the terawatt-year as a standard, perhaps because it is a somewhat larger unit than either the exajoule or quad. By definition:

$$1 \text{ TWyr} = 10^9 \text{ kWyr} = 31.54 \text{ EJ} = 29.89 \text{ quads.}$$

3. Source-based conversion factors

Actual energy content. In discussing energy production and use it is often convenient to speak in terms of the bulk amount of fuel, e.g., a barrel of oil or a ton of coal. These terms are then sometimes used as more general energy units. While useful in putting a primary focus on the fuel of interest, there is an inherent imprecision in such an approach because "oil" and "coal" embrace a variety of products, whose energy content per unit mass differs. The differences often exceed 10%. For example, the energy content of 1 barrel of motor gasoline averaged 5.25 million Btu (MBtu) for the U.S. in 1988, while the energy content of 1 barrel of residual fuel oil averaged 6.29 MBtu (see Table 1).

For "petroleum products" as a whole, embracing a range of products which includes not only gasoline and fuel oil but also asphalt and petrochemical feedstock and sometimes natural gas liquids (NGL), the EIA reports an overall average for 1988 of 5.408 MBtu per barrel. Average values for coal and natural gas are listed in Table 1.

Equivalent fuel values. Given the wide variations in actual energy content for coal and oil, it is common to introduce nominal equivalents, which are intended to provide a uniform measure. Thus, a nominal conversion factor is used for crude oil, which is close to its actual energy content:

$$1 \text{ barrel of crude oil (equiv)} = 5.8 \text{ MBtu.}$$

With this definition, a correspondence can be established between millions of barrels of oil per day (mbd) and quads per year (actually units of power rather than energy, but the distinction need not cause confusion):

$$1 \text{ mbd} = 0.0058 \times 365 = 2.12 \text{ quad/year,}$$

which is sometimes rounded off to: 1 mbd = 2 quad/year.

An energy equivalence for oil is often specified in terms of energy per metric ton (tonne). Here, however, there is no standardization and prestigious sources variously define a tonne of oil equivalent at values ranging from 10.0 to 10.7 million kcal (with either definition of the kcal.). Thus, there is no standard meaning for the term "tonne of oil equivalent." Two examples, which span the range in common use, are given in Table 1. They are there expressed in terms of the gigatonne because the gigatonne of oil equivalent is a common unit for describing large energy budgets. Roughly:

$$1 \text{ gigatonne oil (equiv)} \approx 40 \text{ quad.}$$

With the large variations which exist in the literature, it is necessary in each case to ascertain the particular definition being used by the authors. No standard meaning can be safely assumed.

The situation is somewhat better for the definition of the tonne of coal equivalent. Here it is quite standard to set this equal to 7 million kcal, although there remains the minor ambiguity in the definition of the kilocalorie as well as a potential confusion between specification in terms of the short ton (ton) and metric ton (tonne). At a level of precision which makes the kilocalorie ambiguity irrelevant:

$$1 \text{ gigatonne of coal (equiv)} = 27.8 \text{ quad,}$$

$$1 \text{ gigaton of coal (equiv)} = 25.2 \text{ quad.}$$

4. Special issues relating to electricity units

Gross and net electricity production. The gross electricity generation (in kWh) is the amount of energy measured at the terminals of the generating unit; the net generation is measured at the output transformer of the power plant. They differ by the amount of electricity consumed within the plant itself, roughly 4%.

Electricity Sales. Electricity sales differ from net generation due to transmission losses, net imports, and (in a given time period) time lags between generation and billing. In standard DOE publications, the listed "net generation" is the generation by utilities. Thus, the electricity purchased by utilities from small producers for resale appears as a difference between sales and generation. In recent years, the ratio of sales to net generation has been about 0.95.[3] Transmission and distribution losses amount to about 8%, but are partly balanced by imports and the purchases from small producers (in roughly equal amounts).[2]

Electricity output vs fuel consumption. At 100% efficiency, the conversion from heat to electricity is at a rate of 3412 Btu per kWh. Actual generation efficiencies, limited by considerations of the Second Law of Thermodynamics and design practicalities, are in the neighborhood of 30%. More specifically, for U.S. power plants during 1987 to 1989, the average heat input per kWh of net generation was approximately 10,253 Btu/kWh for fossil-fuel plants and 10,776 Btu/kWh for nuclear power plants.[3] These heat rates correspond to efficiencies of 33% and 32%, respectively.

In considering energy consumption for electricity, a separation is sometimes made between the electrical energy consumed at the point of use (converted from electricity sales at the rate of 3412 Btu/kWh) and the energy losses incurred in electrical generation and transmission. The end-use energy and the losses are then individually tabulated, with the losses roughly twice as great as the end-use energy. However, this separation can be misleading, especially in comparisons of fuels, because losses in the end-use of the fuel can also be substantial. These end-use losses (e.g., the inefficiency of gas furnace systems or gasoline engines) are often substantially greater for fossil fuels than for electricity. Thus there is a significant asymmetry in the treatment of electricity and fossil fuels when losses prior to delivery are included and losses after delivery are ignored.

Further, from the standpoint of energy resources the interesting number is the total energy consumed at the generating plant, whether used efficiently or wasted. For these reasons, it is usually desirable to consider energy budgets in terms of the energy content of the resource originally consumed, i.e., the primary energy and not the end-use energy.

Energy equivalent for hydroelectricity, etc. To facilitate comparisons between hydroelectricity and other electricity sources, a conversion factor is assigned to hydroelectric power which translates kWh generated to a nominal primary energy, based on the fossil-fuel energy displaced (presently 10,253 Btu/kWh). The same is customarily also done for the small amounts of electricity generated by biomass, wind, solar photovoltaic, and solar thermal power.[3]

Gigawatt-year (GWyr). The total electrical generating capacity of the U.S. is several hundred million kW, or several hundred gigawatts (GW), and large individual plants have capacities close to one GW. This makes the gigawatt-year (GWyr) a natural unit to use in discussions of total electricity production. By definition:

$$1 \text{ GWyr} = 8.76 \times 10^9 \text{ kWh.}$$

Assuming an average conversion factor of about 10,300 Btu of primary energy per kWh generated, it follows that:

$$1 \text{ quad} = 11.1 \text{ GWyr.}$$

It is to be noted that a 1000-MW plant does not normally generate 1 GWyr of electricity per year. The ratio of the actual electricity generated to the amount which would be generated were the plant to operate at full capacity for the full year is the *capacity factor*. Typical coal and nuclear plants operate at capacity factors of about 60%. Thus, 1-GW capacity in such plants normally gives 0.6 GWyr per year, equivalent to the consumption of about 0.054 quad of fossil fuels.

References

1. H. A. Bethe and D. Bodansky, "Energy Supply," in *A Physicist's Desk Reference*, 2nd Edition of *Physics Vade Mecum*, H. L. Anderson, ed. (American Institute of Physics, New York, 1989).
2. *Annual Energy Review 1988*, Report DOE/EIA-0384(88) (U.S. DOE, Washington, DC, 1989).
3. *Monthly Energy Review, September 1989*, Report DOE/EIA-0035(89/09) (U.S. DOE, Washington, DC, 1989).
4. *Energy in a Finite World, A Global Systems Analysis*, W. Häfele, Program Leader, Report by the Energy Systems Program Group of the International Institute for Applied Systems Analysis (IIASA) (Ballinger, Cambridge, 1981).
5. *Energy Balances of OECD Countries 1985/86*, International Energy Agency, Organization for Economic Cooperation and Development (OECD, Paris, 1988).
6. *Handbook of Chemistry and Physics*, 66th Ed. (CRC Press, Boca Raton, 1985), p. F-299.

Glossary

Adjustable speed drive (ASD). A technology for varying the output speed of an induction motor. Electronic ASDs work by converting the constant frequency and voltage of incoming power to a variable voltage and frequency.

AEC. Atomic Energy Commission.

Average annual energy flux. A term in wave energy conversion technology that quantifies the power available from the ocean waves in a particular region. It is usually expressed in units like kW/m, i.e., the average power available per unit length of wave crest. It is proportional to the square of the average wave height in a random sea.

Band-to-band Auger recombination. Recombination of an electron and a hole occurring between bands of the same energy in which no electromagnetic radiation is emitted.

Base power. Power generated by a utility unit that operates at a very high capacity factor.

Battery. A device which directly converts the energy stored in chemical compounds into electricity either in an irreversible (primary battery) or reversible (secondary battery) manner.

Biochemical conversion. The transformation of biomass to fuels and chemicals using chemicals, enzymes, and/or microorganisms. Examples include producing ethanol by hydrolysis and fermentation and producing methane by anaerobic digestion.

Biocrude. Plant-derived oil consisting of a heterogeneous mixture of hydrocarbons generated from biomass using thermochemical processes.

Biofuel. A solid, liquid, or gaseous fuel generated from biomass.

Biomass. Organic material of a non-fossil origin (living or recently dead plant and animal tissue) including aquatic, herbaceous, and woody plants, animal wastes, and portions of municipal wastes.

Biomass fuel pathway. The integrated sequence of processes through which biomass is grown, moved, stored, processed, refined, and distributed as a biofuel.

Biotechnology. The use and manipulation of life forms for man's benefit, often at the cellular or subcellular level.

Bitumen. A very viscous hydrocarbon occurring in tar sands. It is soluble in organic solvents, which are sometimes used to extract it from the rock matrix.

Breakeven. The point at which power generated by fusion reactions in a plasma equals the heating power that must be added to sustain the plasma's temperature. Breakeven measures only power actually injected into the plasma, and does not include other power consumed in fusion experiments such as that needed to create the confining magnetic fields or that lost in generating the plasma heating power.

Capacity factor. The ratio of the average power load of an electric power plant to its rated capacity.

Capital charge rate. The factor which annualizes or amortizes the up-front, or capital, cost of a project, i.e., the fraction which must be paid each year, assuming equal installments over the economic life of the project.

Carcinogen. An agent with the ability to cause cancer.

CC OTEC. Closed-cycle OTEC. The working fluid in the OTEC thermodynamic cycle is in thermal contact, but not actual contact with the ocean water.

Central power tower. A configuration of independently tracking solar collectors focusing all the reflected energy onto a receiver placed atop a tower.

Chemical vapor deposition (CVD). A method of depositing thin semiconductor films. With this method, a substrate is exposed to one or more vaporized compounds, one or more of which contain desirable constituents. A chemical reaction is initiated, at or near the substrate surface, to produce the desired material that will condense on the substrate.

Cogeneration. Joint production of work and heat, most often electricity and steam.

Cold fusion. Nuclear fusion reactions that do not require generation of extreme temperatures on the order of millions of degrees Kelvin. Cold fusion refers to an approach called muon catalysis as well as to the reported, but not yet convincingly reproduced, claims of fusion reactions in electrolytic cells at room temperature.

Combined cycle gas turbine system. A system utilizing a gas turbine to produce both heat and electricity.

Compressed air storage. A method of storing reserve energy by compressing air in underground caverns.

Cost of conserved energy. The cost of energy saved, e.g., about 1 cent/kWh for new refrigerators as compared to older models.

Cost of avoided power. The price that utilities must pay, depending on circumstances, for purchasing electrical power from small producers of power.

Cultural techniques. Activities related to crop production such as fertilization, weed control, pest control, spacing of plants, and timing of harvests.

Curie (Ci). A measure of the radioactivity of a substance; one curie equals the disintegration of 3.7×10^{10} nuclei per second and is equal to the radioactivity of one gram of radium-226.

Dangling bonds. A chemical bond associated with an atom on the surface layer of a crystal. The bond does not join with another atom of the crystal, but extends in the direction of exterior of the surface.

Direct current motor. A motor which has a constant magnetic field in the stator generated by either direct current or by permanent magnets.

Direct solar gain. A type of passive solar heating design consisting of large south-facing windows and appropriately sized and located internal thermal storage.

Distributed collector system. A configuration of modular parabolic trough concentrating collectors with an interconnected absorber pipe network to carry the solar heating working fluid to a heat exchanger.

Doe-2. The computer model developed by the Lawrence Berkeley Laboratory to calculate the energy intensiveness of buildings.

Dopant. A chemical element (impurity) added in small amounts to an otherwise pure semiconductor material to modify the electrical properties of the material. An n-dopant introduces more electrons. A p-dopant creates electron vacancies (holes).

Electrification. Increased intensity of electricity use associated with new applications of electricity, an historical process which is still continuing, although, perhaps, at a reduced rate.

Electrodeposition. Electrolytic process in which a metal is deposited at the cathode from a solution of its ions.

Electron volt. An energy unit equal to the energy an electron acquires when it passes through a potential difference of one volt; it is equal to 1.602×10^{-19} joule.

Energy crop. Plants selected, developed, and cultivated primarily for energy uses.

Energy intensity. The ratio of energy use in a sector to activity in that sector, for example, the ratio of energy use to constant-dollar production in manufacturing.

Enhanced-end-use efficiency. The conservation of energy by improved technology, as compared to conservation by denial.

Enhanced oil recovery. Several techniques for assisting oil extraction including injection of steam, polymers, surfactants, carbon dioxide, and other agents into the oil bearing formation.

Enrichment. The process by which the percentage of the fissile isotope uranium-235 is increased above that contained in natural uranium.

Enthalpy. Internal energy of a vapor plus the product of pressure and volume.

EPA. Environmental Protection Agency.

Feedstock. A fuel which, instead of being burned, is used as a material in products such as petrochemicals, asphalt, and lubricants.

Fission. The splitting of an atomic nucleus with the release of energy.

Flashing. The process where a superheated liquid suddenly expands into gas.

Flat plate collector. A solar collection device for gathering the sun's heat, consisting of a shallow metal container covered with one or more layers of transparent glass or plastic; either air or a liquid is circulated through the cavity of the container, whose interior is painted "black" and exterior is well insulated.

Fluidized-bed combustion. A technique used in coal furnaces that uses air streams to make pulverized fuel act as a fluid; a chemically active substance such as limestone is added to the process, reducing the sulfur production during the process. Unburnt particles are constantly recycled. The result is increased efficiency of combustion, and reduced sulfate emissions.

Flywheel. A device made of either simple or composite materials used to store energy in the form of rotational kinetic energy.

Forebay. Water storage area created by a dam.

Fuel cell. A device combining a fuel with oxygen in an electrochemical reaction to generate electricity directly without combustion.

Fusion. The combining of two atomic nuclei with the release of energy.

Geothermal reservoir. Large volume of hot rocks lying above a heat source in which heat is trapped by insulating, impermeable cap rocks and which contains fluid.

Gibrat site ratio. This is the ratio of the length of the required dam in a tidal power installation to the average annual expected energy output. It is a rough measure which varies inversely as the economic viability of the project.

Greenhouse effect. The global warming resulting from the absorption of infrared solar radiation by carbon dioxide and other trace gases present in the atmosphere. (The term is a misnomer in that in actual greenhouses the warming comes primarily from restrictions on air flow.)

Greenhouse gases. Gases which contribute to the *greenhouse effect* by absorbing infrared radiation in the atmosphere. These gases include carbon dioxide, nitrous oxide, methane, water vapor, and a variety of chlorofluorocarbons (CFCs).

Gross hydraulic head. Approximate difference in elevation between the water from the forebay and the turbine in a hydroelectric plant.

Half-life. The time required for the disintegration of half of the atoms in a given amount of a specific radioactive substance.

Heat engine. A device converting heat into mechanical energy with the aid of a fluid transferred between two reservoirs at different temperatures according to one of three thermodynamic cycles (Brayton, Rankine, Stirling).

Herbaceous. Non-woody plants, both annual and perennial species.

High-frequency ballasts. By operating fluorescent bulbs at higher frequencies, about 10% of the electrical energy can be saved.

Hot dry rock geothermal reservoirs. Artificial geothermal reservoirs created by drilling separate injection and production wells into hot dry rocks, fracturing the rocks, and injecting fluid which extracts heat from the rocks and is recovered to produce electricity.

Hybrid (or fission/fusion hybrid). A reactor using a fusion core to produce neutrons that in turn either induce fission reactions or breed fissionable fuel. Most of the energy ultimately converted to electricity—either in the hybrid reactor itself or in fission reactors fed by fuel bred in the hybrid—would come from fission reactions, with fusion neutrons serving to improve the efficiency of the fission process.

Hydraulic head. Height through which water falls as at a hydroelectric dam.

Hydrogen economy. Use of hydrogen gas as a universal, virtually non-polluting, fuel which is generated by renewable energy resources, is transported via pipelines in a gaseous form or in a liquid hydrogen-rich compound (ammonia, methylcyclohexane) form, and stored accordingly.

Hydrolysis. The conversion, by reaction with water, of a complex substance into two or more smaller units, such as the conversion of cellulose into smaller sugar units.

Ignition. The point at which a fusion reaction becomes self-sustaining. At ignition, fusion self-heating is sufficient to compensate for all energy losses; external sources of heating power are no longer necessary to sustain the reaction.

Induction motor. An alternating current motor in which currents and corresponding magnetic fields are induced in the rotor in order to generate a torque.

Inertial confinement. An approach to fusion in which intense beams of light or particles are used to compress and heat tiny pellets of fusion fuel so rapidly that fusion reactions occur before the pellet blows itself up. The pellet's inertia confines it long enough for fusion energy to be produced.

Injection well. A well through which a material such as steam is introduced into an underground reservoir or ore deposit as an aid in extracting a desired product.

Isotopes. Forms of the same chemical element that have the same numbers of protons in their nuclei but different numbers of neutrons. Isotopes have near-identical chemical behavior but can have very different nuclear properties.

***In situ* processing.** The extraction of a product such as shale oil or bitumen from the ore while it is in its original location underground.

Kerogen. The organic material in oil shale. In its original form it is insoluble in common organic solvents, but it can be processed to produce a good quality refinery feedstock.

Knock. A premature detonation that occurs when the fuel-air mixture in an internal combustion engine ignites before compression is complete.

Landscaping design. A type of passive solar cooling design resulting in a lower building cooling load both by direct shading and by reduction of ambient air temperature through plant evapotranspiration.

Life-cycle costs. The entire cost of an energy device, including the capital cost in present dollars, and the costs and benefits, discounted to the present.

Linear induction motor. A motor in which electric current in one winding creates a traveling magnetic wave which induces currents in a second winding. The magnetic field from the secondary currents interacts with the primary field resulting in a force between the two windings. In the linear motor, one winding is the track against which the other winding moves.

Low-*E* windows. Windows that reflect the infrared back into the room, instead of absorbing and transmitting the IR heat to the outside.

Low or no-till cultivation. Method of planting and alteration which disturbs less than 25% of the soil but uses large quantities of pesticides and herbicides.

Luz solar thermal power. A practical and already commercial solar thermal electric power generating concept consisting essentially of a distributed collector system with significant engineering improvements so as to be cost competitive to fossil-fuel power generation.

Magnetic confinement. Any means of containing and isolating hot plasma from its surroundings by using magnetic fields.

Methane. CH_4, the simplest hydrocarbon compound, and major constituent of natural gas.

Micropropagation. The use of plant tissues or cells to reproduce new plantlets.

Moderator. A material, such as ordinary water, heavy water, or graphite, used in a reactor to slow down high-velocity neutrons, thus increasing the likelihood of further fissions.

Modified *in situ* processing. The extraction of a product such as shale oil or bitumen from the ore without mining but after rubbling the ore to improve access to the product material.

Muon catalysis of fusion. The process by which a negatively charged muon binds to an atomic nucleus, screening its electric charge and allowing it to approach another nucleus closely enough for a fusion reaction to take place. The muon is usually ejected following the reaction, permitting it to catalyze further reactions.

Mutagen. An agent that causes mutation in offspring of exposed individuals.

Natural gas. Gaseous fuel found in nature and believed to result from the decay of organic matter. The primary constituent of natural gas is methane (CH_4).

Natural gas liquids. Hydrocarbons associated with natural gas, which are commonly found or handled in liquid form. Some components are gases at standard pressure and temperature, and are liquids at elevated pressures. Examples include ethane (C_2H_6), propane (C_3H_8), and pentane. (Natural gas liquids are not to be confused with *liquid natural gas*, namely natural gas transported and stored in liquid form at low temperature.)

Negative feedback. A mechanism whereby a cause produces an effect that reduces the original cause — results in the damping of the change in conditions; e.g., warming of the Earth increases cloud production, which may (not yet established) have a net effect of reducing the amount of heat reaching the Earth, thus tending to reduce the original warming.

NRC. Nuclear Regulatory Commission.

OC OTEC. Open-cycle OTEC. The working fluid in the OTEC thermodynamic cycle is the ocean water itself. This innovation makes the OTEC cycle more efficient and allows the production of desalinized water as a by-product.

Off-peak storage. By cooling water, or freezing ice at night, the cooling ability of air conditioning equipment can be expanded by a factor of about 2, and the less valuable evening electricity is used.

Oil shale. A rock containing an organic material called kerogen, which can be extracted from the rock by heating.

Organic farming. Food production system which avoids the use of synthetic fertilizers, pesticides, growth regulators, or livestock feed additives.

Oscillating water column. A particular generic type of wave energy conversion device in which the ebb and flow of a water column in and out of the device acts like a piston to compress air and force it through a turbine.

OTEC. Ocean Thermal Energy Conversion. A process which exploits the natural temperature gradient between shallow and deep ocean waters as the driving potential for a simple thermodynamic cycle that can extract work out of the temperature difference.

Part load. Operation of an engine to provide less than 100% of its rated power output. It will be less efficient than when fully loaded.

Passive solar design. A building design that makes use of structural elements, using no moving parts except possibly fans, to heat or cool spaces in the building.

Peak power. Power generated by a utility unit that operates at a very low capacity factor; generally used to meet short-lived and variable high demand periods.

Penstock. Large pipe conveying water from the forebay to the turbines in a hydroelectric plant.

Petroleum. A generic term for liquid and, in some usages, gaseous hydrocarbon fuels. The liquid fuels include crude oil and its products, natural gas liquids, and heavy oils such as bitumen. The terms *petroleum* and *oil* are often used interchangeably. In such usage, *oil* includes both crude oil and its products and *petroleum* does not include natural gas. (The term literally denotes, from its roots, "rock-oil.")

Polycyclic hydrocarbons. Hydrocarbon compounds containing more than one benzene-like ring; some are carcinogenic in minute quantities.

Positive feedback. A mechanism whereby a cause produces an effect that increases the original cause — can result in "runaway" where a totally new equilibrium occurs; e.g., warming of the Earth will reduce the reflectance of solar energy due to snow cover, resulting in more heat absorbed and a still warmer Earth.

Power factor. The cosine of the phase angle between the voltage and current of a circuit.

Primary production. The process of extracting oil from the ground without the introduction of any material from the surface.

Production well. A well through which a desired product is extracted from an underground reservoir or ore deposit.

Pumped hydrostorage. A technique to store electric energy in the form of potential energy by utilizing a dam and pumping water into a reservoir behind it.

Pumped-storage facility. Hydroelectric plant that uses water that has been pumped to an elevation higher than the powerhouse. Pumped is usually done during times of low power demand and electricity produced at hours of high demand.

PURPA. Public Utilities Regulatory Policy Act.

Pyrolysis. The process of heating an organic material in the absence of oxygen to produce a mixture of organic gases, liquids, and solids.

Recoverable conventional resources. Fossil-fuel resources that are extractable through conventional drilling and mining techniques.

Reinjection. Process by which cooled geothermal fluid is pumped back into the geothermal field in order to maintain pressure.

Rem (Roentgen equivalent man). The unit of biological radiation dose given by the product of the absorbed dose in rads and the relative biological efficiency of the radiation. A rad is the unit of absorbed dose of radiation and corresponds to an energy absorption of 100 ergs per gram.

Reserves. The *resources* of a given fuel whose location has been identified but which have not yet been extracted.

Residual fuel oil. Highly viscous oil that is a residue from the distillation process; generally used only in large boilers.

Residues. Biomass which remains unused or uncollected in the harvest or processing of plant material for a particular end use.

Resistive voltage drop. The voltage developed across a cell by the current flow through the resistance of the cell.

Resources. The total available amount of a fuel, past and future, which can be recovered under a given set of costs and technologies. These are often tactily or explicitly taken to be the affordable costs and technologies of the present or near future. The total amount of fuel, independent of feasibility of extraction, is sometimes termed the *resource base*.

Reverse osmosis. An advanced method of water desalination relying on semipermeable membrane to separate undesirable components and pollutants in water.

Rotor. The rotating portion of a motor.

Rotary engine. An internal combustion engine that uses "rotating pistons" or rotors to contain and compress the fuel-air charge in lieu of the conventional linear motion pistons. The combustion spins the rotor providing rotational energy directly to an output shaft. Its claimed advantages stem from having many fewer moving parts and not having to convert linear piston motion to rotational motion. The Wankel is a specific type of rotary engine that has been more successful than others.

Saccharification. The conversion of long chain carbohydrates into their component fermentable sugars. See HYDROLYSIS.

Scrubber. A device for removing noxious sulfate and nitrate emissions from fossil-fuel plants or other combustion sources.

Sea clam. A novel wave energy device that is made of a series of air bags affixed to a floating rigid spine. The surging action of the waves on the air bags causes the air to be forced through turbines.

Secondary recovery. The process of assisting oil extraction by injecting water or natural gas into a formation in order to increase driving forces.

Sectoral shift. The ongoing shift, in mature societies, from production of bulk materials to fabrication of products, in other words, the trend toward reduction in material content (by weight) in final products.

Series resistance. Parasitic resistance to current flow in a cell due to mechanisms such as resistance from the bulk of the semiconductor material, metallic contacts, and interconnections.

Short-rotation silviculture. Production of trees under agricultural-like conditions using close spacings, weed control, and 3- to 10-year intervals. Short-rotation systems frequently use genetically improved hardwood trees.

Solar concentrator. A device which uses reflective surfaces in a planar, parabolic trough, or parabolic bowl configuration to concentrate solar radiation onto a smaller surface.

Solar still. A device consisting of one or several stages in which brackish water is converted to potable water by successive evaporation and condensation with the aid of solar heat.

Solar subdivision. Building site layout resulting in an optimal availability of solar radiation for building heating and cooling.

Solar thermal electric power. The indirect conversion of solar radiation into electricity by solar collectors, a heat engine, and electrical generators.

Solar toxic waste destruction. Effective technique to destroy hazardous wastes with the aid of concentrated solar radiation.

Source term. The inventory of radionuclides released to the environment, especially in a nuclear-power plant accident.

Spray pyrolysis. A deposition process whereby heat is used to break molecules into elemental sources that are then spray deposited on a substrate.

Stator. The non-rotating portion of the motor which generates a magnetic field.

Steam injected gas turbine. A turbine that recycles some produced steam into the combustion chamber to gain extra efficiency.

Superconducting coil storage. A method of storing electric energy in the form of a magnetic field generated by an electric current circulating in a large coil maintained at liquid helium temperatures in order to be superconducting.

Synchronous motor. An alternating current motor in which the rotor has a constant magnetic field rotating with the rotor.

Synthesis gas. A mixture of hydrogen, carbon monoxide, and sometimes other gases that is manufactured by reacting coal with steam. It is an intermediate product in the preparation of synthetic liquids and gases.

Synthetic fuel. Fuel product, usually a liquid or gas, manufactured to have the same properties as a naturally occurring fuel.

Tailwater. Water that exits the turbines of a hydroelectric plant.

Tar sand. A rock containing an organic material called bitumen, which can be extracted by means of heat or a solvent and then used directly or processed to produce a refinery feedstock.

Temperature coefficient of reactivity. A measure of the change in power generation per unit change in temperature of a reactor. A negative temperature coefficient means that a reactor has less ability to sustain a chain reaction as the temperature is increased.

Teratogen. An agent that causes deformities in fetuses while *in utero*.

Thermal neutrons. After a number of scattering collisions, the velocity of a neutron is reduced to such an extent that it has approximately the same average kinetic energy as the atoms, or molecules, of the medium in which it is undergoing elastic scattering. Neutrons whose energies have been reduced to values in this region are designated thermal neutrons, and the process of reducing the energy of a neutron to this region is known as thermalization.

Thermal storage. Any of several techniques to store heat energy by utilizing either the heat capacity of materials, the latent heat of phase change, or the heat of chemical dissociation.

Thermochemical conversion. The use of various combinations of heat, pressure, and catalysts to transform biomass into fuels and chemicals.

Thermocline. The water temperature below a point on the ocean surface typically decreases as a function of depth until it reaches a minimum value which remains constant as greater depths are attained. The depth at which the temperature levels off varies from point to point over the ocean with typical values from 500–1000 m. The depths at which the temperature levels off form a hypothetical surface or contour called the thermocline.

Tidal power. The use of the hydraulic head created by the difference between high and low tides to generate hydroelectric power.

Tidal resonance. Certain coastal estuaries, by virtue of their topography, amplify the local effects of the tides that flow in and out of them. This phenomenon is known as tidal resonance.

Tribology. The science of sliding surfaces, includes friction, wear, and lubrication.

Unconventional natural gas resources. Resources of natural gas that require special and new technology for economical extraction. Examples are: gas trapped in low permeability rock, such as coal; gas trapped in permafrost; and gas trapped in Devonian shale layers.

Undiscovered resources. Fossil-fuel resources that are estimated on the basis of geological consideration, but not known to be there.

Utility function. A hypothetical function of many economic variables, maximized when the rationally best choice among alternatives is made.

Ventilation. The purposeful utilization of atmospheric air to cool the interior of a building predominantly at, but not necessarily limited to, the night.

Water heating. The process of generating domestic hot water by employing a flat plate collector and utilizing solar radiation.

Wave energy conversion. Any one of a number of schemes which convert the rocking, pitching, or heaving motion of ocean waves into electrical energy.

Wind power plants. Arrays of interconnected wind turbines which generate large amounts of standardized electricity for delivery to a utility transmission-distribution system.

Wind turbines. Electro-mechanical machines which utilize rotating airfoils to convert power in a wind flow field to electrical power.

Index